油气光学系列丛书

油气光学的研究与进展
Research and Progress of Oil and Gas Optics

赵 昆 詹洪磊 祝 静 陈 儒 编

科学出版社

北 京

内 容 简 介

本书提出油气光学概念，开拓了油气光学工程的研究方向，将光学前沿技术与油气资源的勘探开发相结合，开展了油气物质光学性质的应用基础研究及光学方法在油气领域应用的重大关键技术、前瞻性技术研究，涵盖了太赫兹光谱、斜入射光反射差、激光感生电压、激光超声等光学新技术在石油勘探、油气储运、石油化工及环境污染和安全领域的研究与应用进展。

本书可作为油气光学工程及相关专业的高年级本科生、研究生的参考用书，也可作为油气光学及相关专业科研人员及管理人员的参考资料。

图书在版编目(CIP)数据

油气光学的研究与进展 = Research and Progress of Oil and Gas Optics/赵昆等编. —北京：科学出版社，2019.5
（油气光学系列丛书）
ISBN 978-7-03-061071-3

Ⅰ.①油… Ⅱ.①赵… Ⅲ.①油气资源评价–光学分析法
Ⅳ.①TE155

中国版本图书馆 CIP 数据核字(2019)第 074442 号

责任编辑：万群霞 陈 琼 / 责任校对：王 瑞
责任印制：吴兆东 / 封面设计：无极书装

科学出版社 出版
北京东黄城根北街 16 号
邮政编码：100717
http://www.sciencep.com

北京建宏印刷有限公司 印刷
科学出版社发行 各地新华书店经销

*

2019 年 5 月第 一 版　开本：787×1092　1/16
2019 年 6 月第二次印刷　印张：30 3/4
字数：716 000

定价：298.00 元
（如有印装质量问题，我社负责调换）

前　言

　　以石油、天然气为典型代表的油气资源是世界各国现阶段最重要的能源，是各国制造业、交通运输业、国防工业的基础，是维持经济发展和社会秩序的物质前提，在国民经济中占据着非常重要的地位。全球油气资源种类丰富，储量巨大，但勘探、开发和加工自然界存在的各类油气资源都依赖先进的科学技术，如首先对自然资源的分布、品位等储存状况进行探测，其次对开发过程中的资源状况进行有效监控，保证能源生产安全和环境友好，最后对能源产品的品质进行检测。我国油气资源具有地质条件复杂、地表条件多样、富集程度低、油气埋藏深和原油含水率高等特点，勘探开发与高效利用面临许多困难和挑战。如何解决油气资源勘探开发过程中的技术难题，降低开采成本，提高经济效益，已经成为业内研究者关注的焦点，研究并推动油气资源勘探的新理论、新方法的发展完善势在必行。

　　近年来，光学技术以其非接触、高灵敏、不受电磁干扰等优势，在石油领域已经展现重要的科研及应用价值，其中，以遥感、电磁波等方法为代表的勘探测井技术是勘探油气资源的基础，以各类质谱、色谱及紫外、红外、可见光、荧光光谱方法为代表的光谱技术是液态、气态能源物质的分析基础。随着光学技术的不断发展，光学方法可以成为油气资源现有表征和评价方法的补充手段，也有可能为油气资源的评价提供新的数值指标，在未来有着光明的应用前景。

　　目前，针对国家油气能源战略的重大需求，油气光学科研创新团队已在油气资源光学探针表征与评价、油气资源光探测物理与器件、流体光学、纳米岩石物理、大气环境光学等方面取得了重要进展。

　　油气光学领域的发展尚处于初始阶段，要实现跨越式发展和产业化应用还需假以时日。只要坚持科学的世界观和方法论，油气光学探测技术一定可以为能源产业提供重要的支撑和保障，油气光学这一新兴交叉学科一定能够大放异彩。

　　由于作者水平有限，本书难免存在不妥之处，敬请读者批评指正。

赵　昆

油气光学探测技术北京市重点实验室

石油和化工行业油气太赫兹波谱与光电检测重点实验室

中国石油大学（北京）

2018 年 10 月

目 录

前言

第一篇　油气光学探测技术

Application of THz technology in oil and gas optics ··3
Less than 6 GHz resolution THz spectroscopy of water vapor ···································8
Determining the humidity-dependent *ortho*-to-*para* ratio of water vapor at room temperature by terahertz spectroscopy ··19
Size effect on microparticle detection ··30
Surface phase-transition dynamics of ice probed by terahertz time-domain spectroscopy ······39

第二篇　油气储层潜能的光学技术表征评价

Characterizing the rock geological time by terahertz spectrum ··51
Characterization of inclusions in evolution of sodium sulfate using terahertz time-domain spectroscopy ··57
$CaCO_3$, its reaction and carbonate rocks: Terahertz spectroscopy investigation ··················66
Spectral characterization of the key parameters and elements in coal using terahertz spectroscopy ··78
Applying terahertz time-domain spectroscopy to probe the evolution of kerogen in close pyrolysis systems ··90
Discriminating the mineralogical composition in drill cuttings based on absorption spectra in terahertz range ··95
Oil yield characterization by anisotropy in optical parameters of the oil shale ···············108
Optimization of pyrolysis efficiency based on optical property of semicoke in terahertz region ··119
The mechanism of the terahertz spectroscopy for oil shale detection ··130
Layer caused an anisotropic terahertz response of a 3D-printed simulative shale core ········141
Characterizing the oil and water distribution in low permeability core by reconstruction of terahertz images ··151
Laser-induced voltage of oil shale for characterizing the oil yield ··155
Ultraviolet laser-induced lateral photovoltaic response in anisotropic black shale ···············166
Ultraviolet laser-induced voltage in anisotropic shale ··174

Transient laser-induced voltaic response in a partially illuminated dielectric core ········· 184
Characterizing the rock perforation process by laser-induced voltage response ············ 193
页岩各向异性的全光学检测 ··· 198
Characterizing of oil shale pyrolysis process with laser ultrasonic detection ············ 209
Evaluation of simulated reservoirs by using the oblique- incidence reflectivity
　　difference technique ··· 218
Real-time detection of dielectric anisotropy or isotropy in unconventional oil-gas reservoir
　　rocks supported by the oblique-incidence reflectivity difference technique ············ 222
Supporting information ··· 234
Oblique-incidence reflectivity difference application for morphology detection ············ 241
In situ monitoring of water adsorption in active carbon using an oblique-incidence
　　optical reflectance difference method ··· 250
Real-time monitoring the formation and decomposition processes of methane hydrate
　　with THz spectroscopy ··· 257

第三篇　光学技术在油气储运中的应用

Simultaneous characterization of water content and distribution in
　　high-water-cut crude oil ··· 263
Non-contacting characterization of oil-gas interface with terahertz wave ·············· 274
Terahertz dependent evaluation of water content in high-water-cut crude oil using additive-
　　manufactured samplers ··· 277
Non-contact measurement of the water content in crude oil with all-optical detection ········· 288
The detection of water flow in rectangular microchannels by terahertz time-
　　domain spectroscopy ··· 297
Pattern transitions of oil-water two-phase flow with low water content in rectangular
　　horizontal pipes probed by terahertz spectrum ······································· 309
Reliable evaluation of oil-water two-phase flow using a method based on terahertz
　　time-domain spectroscopy ··· 318
Probing disaggregation of crude oil in a magnetic field with terahertz time-domain
　　Spectroscopy ··· 331
Characterization of morphology and structure of wax crystals in waxy crude oils by
　　terahertz time-domain spectroscopy ··· 341
Optical characterization of the principal hydrocarbon components in natural gas using
　　terahertz spectroscopy ··· 353
Water adsorption dynamics in active carbon probed by terahertz spectroscopy ············ 365

Terahertz double-exponential model for adsorption of volatile organic compounds
　　in active carbon ·······373
Adsorption dynamics and rate assessment of volatile organic compounds in
　　active carbon ·······383

第四篇　油气产品及污染物的光学技术表征评价

A spectral-mathematical strategy for the identification of edible and swill-cooked dirty oils
　　using terahertz spectroscopy ·······393
Qualitative identification of crude oils from different oil fields using terahertz time-domain
　　spectroscopy ·······403
The spectral analysis of fuel oils using terahertz radiation and chemometric methods ·······414
Evaluating $PM_{2.5}$ at a construction site using terahertz radiation ·······432
Non-contacting characterization of $PM_{2.5}$ in dusty environment with THz-TDS ·······447
Terahertz assessment of the atmospheric pollution during the first-ever red alert period in
　　Beijing ·······451
Terahertz-dependent $PM_{2.5}$ monitoring and grading in the atmosphere ·······458
Fuel properties determination of biodiesel-diesel blends by terahertz spectrum ·······476

第一篇 油气光学探测技术

Application of THz technology in oil and gas optics

Xinyang Miao   Honglei Zhan   Kun Zhao

(Beijing Key Laboratory of Optical Detection Technology for Oil and Gas, China University of Petroleum, Beijing 102249, China)

Fossil fuels currently satisfy ~88% of China's energy demands. Oil-gas, known as the hydrocarbon occurred from various types of underground strata, is the most significant part of primary energy sources. Problems in oil-gas exploration, transportation and petrochemical affect the economic viability of the petroleum industry. Recently, the increasing complexity of stratigraphic condition and the fall in oil prices has made the situation more critical. Thus, questing for new technologies is never ending. Optics, a general method of passing information to and from materials, can be served as an appropriate method to promote the development of oil-gas industry. Terahertz (THz) radiation, which is located between far-IR (infrared) and millimeter-wave bands of the spectrum, is willing to be a "silver bullet" in various aspects of oil-gas resources owing to the unique advantages. By using a THz system with appropriate scan ranges, the mixture of complicated components can be precisely characterized[1]. In this paper, some recent progresses of applied THz technique in various research areas of oil-gas optics detection are presented, including exploration and development of oil-gas reservoirs, transportation of oil-gas as well as evaluation of petrochemicals and pollutants (Figure 1).

Characterization of oil-gas reservoirs underground includes the properties of organic matters and rocks. As a kind of hydrocarbon-generating parent material, kerogen is a complicated mixture of organic components, with the composition and pyrolysis process obscure to researchers. THz measurements have built a relationship between the absorption coefficient (α) and hydrocarbon evolutionary stage (R_o). Simultaneous characterization of the kerogen's oil- and gas-generating peaks was achieved by plotting THz-α with R_o (Figure 1(b)), indicating great promise of THz technology to improve kerogen analysis[2]. Oil shale, defined as a finely grained sedimentary rock with kerogen contained, yields substantial amounts of oil and combustible gas by destructive distillation. As an important indicator to optimize the comprehensive utilization, oil content of the pyrolytic oil shale (semicoke) was observed to be associated with THz-α, indicating THz technology could be an effective selection for evaluating the oil content of oil shale[3]. In addition to the organic matters, characterizing the mineral's evolution of a reservoir is also essential to the development of the resource.

Diagenetic stages (e.g. eogenetic stage, telogenetic stage and metamorphic stage) of halite rock were stated by THz-α with the temperature[4], proving that THz wave was effective in geological research.

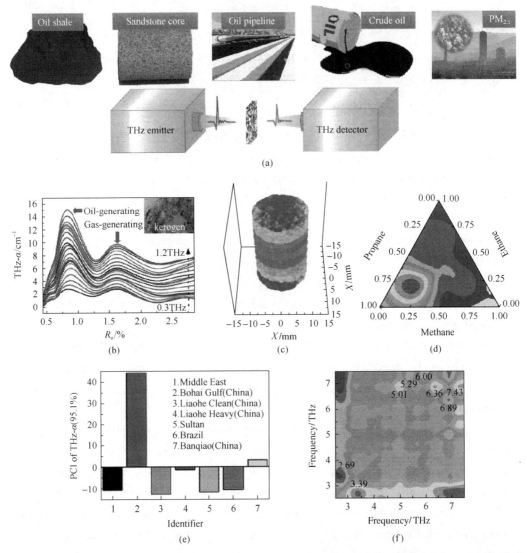

Figure 1　(a) Schematic diagram of THz spectroscopy and various test results; (b) maturity dependent THz-α of the kerogen; (c) THz tomography characterized the oil and water distribution in a core; (d) phase projection picture of the $CH_4 + C_2H_6 + C_3H_8$ system; (e) crude oil classification by the first principle component (PC1) of THz-α; (f) 2D correlation plot of $PM_{2.5}$ over the frequency range 2.5~7.5THz

Better understanding of the reservoir's seepage law during the mining process is essential to improve oil recovery and predict future performance of the oil field, especially for the low permeability reservoirs. An argillaceous sandstone reservoir core with the permeability 1.0 millidarcy was measured with a reflective THz-TDS. Slices of the core were numbered according to their original positions and scanned by THz-TDS via mechanical motion to obtain the distribution of oil and water in the entire core. As shown in Figure 1(c), the existence of kerosene, water and remaining oil was clearly displayed by THz tomography, with the gradually varied color related to the changing of water content in the core[5].

After exploited from wells and pretreated preliminarily, crude oil and natural gas will be transported with pipelines and other tools. Identifying the sources of crude oil and components of natural gas in the pipeline is of great importance for subsequent processing. Crude oils from different oil fields and principal components of natural gas were investigated with the combination of THz technique as well as multivariate statistical methods including cluster analysis and principle component analysis. The combination was able to classify the crude oils with different geographical locations[6] and quantify the hydrocarbon components of natural gas[7], which were exhibited in Figures 1(d) and (e). Besides, oil and water two-phase flow in horizontal pipes is a common occurrence in the petroleum industry for the transportation of crude oil. The investigation of pattern transitions is very important in horizontal oil-water two-phase flow system, especially for the measurement of flow parameters and optimization of industrial production process. THz-TDS was employed to characterize the water content and the pattern transitions in the oil-water two-phase flow simultaneously. The amplitude of the THz pulse was correlated to the water content, and the critical flow rates of the flow pattern transitions point show a non-linear trend with the water content. This work provided an in *situ* means to investigate the flow system in the process of crude oil transportation.

Additionally, non-destructive testing of the pipeline as well as the oil tank is a preparatory work to prevent oil leaks, efficient and safe methods are also required for gas detection, storage and transportation. A simulated sample with crude oil filled was illuminated with THz-TDS, and the oil-gas interface was clearly shown by THz images and spectra[8]. We also monitored the formation and decomposition processes of methane hydrates by THz amplitude changed over time. In our experiment, the cage-structure of hydrates formed gradually with the decreasing temperature of 272.2~266.2K, and remained stable when the temperature goes up to 272.2K owing to the van der Waals forces between CH_4 and H_2O molecules. Accordingly, THz amplitude rose during the initial freezing stage, and maintained a nearly constant value with the decreasing pressure and rising temperature[9]. The results demonstrated the unique capability of THz-TDS to monitor the synthesis of gas hydrate, which were helpful to the gas storage and transportation researches.

As one of the most important sources of fuels and raw materials, petrochemicals have made remarkable contributions to the development of human civilization. Owing to the sensitivity of low-frequency vibration and rotation modes, THz technology is considered to be a reliable analytic method in liquid petrochemicals detection. Qualitative analyses of lubricant, gasoline and diesel were performed. Quantitative detection of their additives was also carried out by analysis with BP artificial neural networks (BPANNs) and support vector machine. Typical concentrations of sulfur in gasoline (ppm-grade) could be detected, together with methyl methacrylate in diesel below 0.5%[10, 11].

$PM_{2.5}$, defined as the particulate matters in an atmosphere with diameters of no more than 2.5 microns, has attracted extensive attention recently owing to the health impacts to human beings. Therefore, techniques for monitoring $PM_{2.5}$ are of great significance. In our studies, THz spectra of $PM_{2.5}$ with various concentrations were measured by Fourier transform infrared spectrometer[12] and THz-TDS[13, 14], with larger signal to noise ratio in relatively high and low frequencies, respectively. Partial least squares and BPANN were employed to quantitatively characterize the $PM_{2.5}$ content with the input of absorbance over 0.1~2THz and 2.5~7.5THz, the elemental compositions were also studied by monitoring $PM_{2.5}$ masses in conjunction with two-dimensional correlation spectroscopy [Figure 1(f)]. The results could be used by environmental policymakers and experts to develop appropriate plans and techniques for PM control.

In summary, THz spectroscopy has been proved to be a promising means for determining oil-gas resources and their products. Therefore, as an emerging technology, THz technique is willing to promote the development of oil and gas optics in the future.

Acknowledgements

This work was supported by the National Natural Science Foundation of China (Grant No. 11574401), the National Basic Research Program of China (Grant No. 2014CB744302), the Specially Founded Program on National Key Scientific Instruments and Equipment Development, China (Grant No. 2012YQ140005), and the China Petroleum and Chemical Industry Association Science and Technology Guidance Program (Grant No. 2016-01-07).

References

[1] Zhan H L, Sun S N, Zhao K, et al. Less than 6 GHz resolution THz spectroscopy of water vapor. Science China: Technological Sciences, 2015, 58: 2104-2109.

[2] Bao R M, Wu S X, Zhao K. Applying terahertz time-domain spectroscopy to probe the evolution of kerogen in close pyrolysis systems. Science China: Physics Mechanics & Astronomy, 2013, 56: 1603-1605.

[3] Bao R M, Li Y Z, Zhan H L, et al. Probing the oil content in oil shale with terahertz spectroscopy. Science China: Physics Mechanics & Astronomy, 2015, 58: 114211.

[4] Bao R M, Meng Q, Wang C L, et al. Terahertz spectroscopic characteristics of the geological diagenetic and metallogenic. Scientia Sinica Physica, Mechanica & Astronomica, 2015, 45: 084203.

[5] Bao R M, Miao X Y, Feng C J, et al. Characterizing the oil and water distribution in low permeability core by reconstruction of terahertz image. Science China: Technological Sciences, 2016, 59: 664201.

[6] Zhan H L, Wu S X, Bao R M, et al. Qualitative identification of crude oils from different oil fields using terahertz time-domain spectroscopy. Fuel, 2015, 143: 189-193.

[7] GeL N, Zhan H L, LengW X, et al. Optical characterization of the principal hydrocarbon components in natural gas using terahertz spectroscopy. Energy & Fuels, 2015, 29: 1622-1627.

[8] Wang D D, MiaoX Y, Zhan H L, et al. Non-contacting characterization of oil-gas interface with terahertz wave. Science China: Physics Mechanics & Astronomy, 2016, 59: 674221.

[9] Miao X Y, Sun S N, LiY Z, et al. Real-time monitoring the formation and decomposition processes of methane hydrate with THz spectroscopy. Science China: Physics Mechanics & Astronomy, 2017, 60: 014221.

[10] Zhan H L, Zhao K, Zhao H, et al. The spectral analysis of fuel oils using terahertz radiation and chemometric methods. Journal of Physics D: Applied Physics, 2016, 49: 395101.

[11] Qin F L, Li Q, Zhan H L, et al. Probing the sulfur content in gasoline quantitatively with terahertz time-domain spectroscopy. Science China: Physics Mechanics & Astronomy, 2014, 57: 1404-1406.

[12] Zhan H L, Zhao K, Bao R M, et al. Monitoring $PM_{2.5}$ in the atmosphere by using terahertz time-domain spectroscopy. Journal of Infrared, Millimeter and Terahertz Waves, 2016, 37: 929-938.

[13] Zhan H L, Li Q, Zhao K, et al. Valuating $PM_{2.5}$ at a construction site using terahertz radiation. IEEE Transactions on Terahertz Science and Technology, 2015, 5: 1028-1034.

[14] Zhan H L, Zhao K, Xiao L Z. Characterizing the oil and water distribution in low permeability core by reconstruction of terahertz images. Science China: Physics Mechanics & Astronomy, 2016: 59: 644201.

Less than 6 GHz resolution THz spectroscopy of water vapor

Honglei Zhan[1,2]   Shining Sun[2]   Kun Zhao[2]   Wenxiu Leng[2]   Rima Bao[2]
Lizhi Xiao[1]   Zhenwei Zhang[3]

(1. State Key Laboratory of Petroleum Resources and Prospecting, China University of Petroleum, Beijing 102249, China; 2. Beijing Key Laboratory of Optical Detection Technology for Oil and Gas, China University of Petroleum, Beijing 102249, China; 3. Department of Physics, Capital Normal University, Beijing 100048, China)

Abstract: The application of terahertz (THz) technique in many fields like petroleum industry requires the high resolution and multi-function to realize the precise detection of different materials. In this study water vapor in atmosphere was scanned with short-and long-path THz spectrometers with different scan ranges. The full width at half maximum of the THz characteristic lines in long-path system reached less than 7GHz and that in short achieved ~5.5GHz, indicating that the THz spectrum of water vapor could be used to calibrate THz devices. THz systems with right scan ranges could be a promising selection for the identification of solid, liquid and gaseous materials with very similar structures due to the high resolution. The research demonstrated that the multifunctional THz test could be realized through the balance of resolution and sampling space in many fields, such as oil-gas industry.

Keywords: terahertz; water vapor; resolution; full width at half maximum

1 Introduction

Most THz devices are used in an atmospheric environment when applied in different fields, such as biology, food safety, chemical engineering and petroleum industry. The propagation of THz pulse in atmosphere is mainly influenced by the absorption of THz electromagnetic radiation in water vapor. Especially for the gas detection which often needs a long optical path to obtain stronger signals, the remote sensing of molecules in the THz region is a highly challenging problem[1-3]. To reduce the absorption effect of water vapor, THz system is often filled by nitrogen or dry air before the materials are measured. However, the absorption caused by vapor can still not be eliminated absolutely.

At the same time, due to the evident absorption features of water vapor in THz range, THz technique is a promising selection to be used in industry for the detection of gas mixtures including water vapor. Bidgoli et al.[4] employed THz transmission spectroscopy to monitor the steam and CO contents of the raw gas from industrial biomass gasifiers. Based on the combination of THz equipment and a designed gas spectrometer, gaseous mixtures can be precisely detected and the THz system yielded high-quality data with an acceptable temporal resolution. In our opinion, water vapor can act as a standard substance to calibrate the THz system and finally validate the stability and resolution, not only because of the ubiquity of vapor in atmosphere, also because of the evident absorption features with small bandwidth in THz range.

There was no strong incentive to develop new technologies without first identifying a "killer application". To date THz technique achieved a rapid development in last decades owing to a series of THz devices and one or more fields will be necessary to act as the "killer application" and realize the revolutionary development of THz technology[5]. At present, THz spectroscopy has been applied in many fields, such as biology, material, chemistry, medication, petroleum, and so on[6-27]. Taking petroleum for example, high feasibility and practicality have been proved about THz technique applied in this industry. Rapid characterization of oil-gas resources and potential evaluation of oil-gas reservoir can be realized through THz technique with a simple way according to the previous work[28-36]. The crude oils from different oil fields can be directly identified by the combination of THz and principal component analysis. It is of great importance to rapidly determinate oils in pipelines. In addition, THz was also used to quantitatively characterize the principal components of natural gas, the oil content of shale, the water content of crude oil, adsorption dynamics and so on. However, multifunctional THz equipment is necessary for the promotion of this technique in oil-gas industry. Oil detection needs stable output of signals and gas determination needs long path and high resolution. Also, the evaluation of oil-gas reservoir requires large space of sampling area and high resolution. In this research, the THz spectra of air with similar humidity were measured using two transmission THz setups, whose optical paths were relatively short and long, respectively. Different scan ranges were set in the process of measurement. It was proved that the resolution, defined by full width at half maximum (FWHM), reached below 6GHz when the scan range was appropriate. Therefore, the results indicated that THz equipment could be designed for the detection of oil, gas and other materials with high-resolution.

2 Experimental methods

The experimental setups are comprised of two transmission terahertz time-domain spectroscopy (THz-TDS) systems, Z1 and Z2, and a femtosecond Ti-sapphire laser (MaiTai) from Spectral-Physics(Figure 1). The Laser is diode-pump mode-locked with a repetition of 80MHz and generates femtosecond laser pulses whose duration is 100fs and center wavelength equals 800nm. As shown in Figure 1, the femtosecond laser beam is initially split into two beams by polarization beam splitter (PSB1) to act as the optical resources of Z1 and Z2 with different light paths. Taking Z1 for example, the laser power is attenuated so that the average power of input laser is less than 150mW. After split by PBS2, two beams are obtained as the pump pulse and probe beam, respectively. The pump beam is used to generate THz radiation through a p-type GaAs with ＜100＞ orientation. The diffused THz pulse is initially focalized by a hyper-hemispherical len (HHSL) and then reflected by an indium titanian oxide (ITO1) to obtain collimated THz radiation. After focalized by lens (L2 and L3) and reflected by mirrors (M9 and M10), the collimated THz pulse transmits into the detection system. The probe beam from PSB2 initially reaches and transmits through the automatic delay stage. After reflected by M5~M8, the probe beam reaches the detection system. In the system, the probe laser is focalized by L4 and reaches ITO2. Then, probe beam and THz pulse are collinear and pass through hyper spherical lens (HSL); the hybrid beams are focalized onto a 2.8mm thick ＜110＞ZnTe, whose index ellipsoid will be changed by THz electric field; therefore, the polarization state of probe beam with linear polarization is altered due to the electro-optic

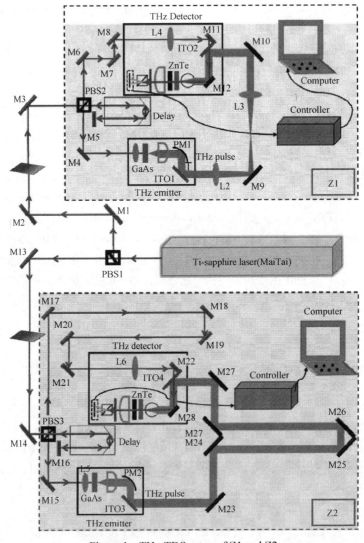

Figure 1　THz-TDS setup of Z1 and Z2

ZnTe crystal. After passing through a quarter wave plate and a Wollaston prism, the pulse beam is divided into two orthogonal beams including s- and p-polarization. The beams reach a differential detector silicon. Deviation of optical powers will be converted into a differential current intensity. Current difference is proportional to the THz pulse electric field. In addition, the delay stage can change time delay between THz pulse and probe laser so that time domain spectroscopy, THz electric field as a function of time, can be detected. In this setup, THz signal is detected and amplified by a lock-in amplifier which can greatly improve the signal-to-noise ratio. The amplifier and other relative devices are integrated in a controller, which is connected with a computer. Therefore, computer can control the controller and measurement process, such as paramenter setting, experimental operation and signal acquisition.

As shown in Figure 1 the laser source, THz emitter and detector of Z2 are as same as those of Z1. Z1 is a typical TDS system, while Z2 is rebuilt as a long-path system on the bases of Z1 and has a longer path of THz pulse. The distance equals 60cm from M24 to M25. Besides, the THz beam between PM2 and M28 keeps collimated light and its direction is fixed. Therefore, Z2 can act as a multipurpose equipment used for gas and gas hydrate detection due to its long optical path[22,23]. The scan ranges of Z1 and Z2 are alterable according to the actual need. In this study, the water vapor corresponding to different scan ranges were tested by setting the beginning and ending points of scan. Detailed values were listed in Table 1. Six groups of THz spectra of air were finally obtained through both Z1 and Z2 at room temperature with a relative humidity of ~8.6%. The measurement condition kept unchanged, but the scan ranges were altered according to Table 1. What we did was to optimize the optical paths and to augment the scan ranges.

Table 1 Details about THz measurement in Z1 and Z2

Z1			Z2		
Beginning/mm	Ending/mm	Range/mm	Beginning/mm	Ending/mm	Range/mm
9.5	13.5	4	20	25	5
8.5	17.5	9	20	29	9
7.5	21.5	14	16	30	14
6.5	25.5	19	12	30	18
5.5	29.5	24	8	32	24
4.5	32.5	28	4	32	28

3 Results and discussion

The THz field amplitude of air as a function of time was shown in Figure 2 with special scan lengths. Due to the different beginning points of scan, THz peaks were located at distinct time and the difference of time delay was proportional to that of scan range. The THz amplitudes of all the measurements were highly closed to each other with the values of

~180mV and ~48mV for Z1 and Z2, respectively, indicating the stability of the THz systems performance. Here, the higher THz peak in Z1 than Z2 was due to the longer THz path in Z2 than Z1. The larger optical path was in THz system, the stronger absorption air had. Thus Z2 can be widely used for the detection of many materials, especially the sample which had little absorption in THz range. For example, the principal components of natural gas, including methane and ethane, can be qualitatively and quantitatively determined by THz spectroscopy based on the combination of Z2 and a special long-path cell[22].

The frequency dependent THz power spectra were calculated using fast Fourier transform (FFT) as shown in Figure 3. According to the THz signal amplitude in Figure 2, the THz power of THz frequency-domain spectroscopy (THz-FDS) in Z1 was larger than that in Z2 over the frequencies. The sharp decrease of THz power was observed at many frequencies. The decline scope varied with the different frequency positions; yet, Z1 and Z2 had the sharp change at same THz frequencies. It was revealed that water vapor in air had many significant absorption features in THz range as reported[1,2].

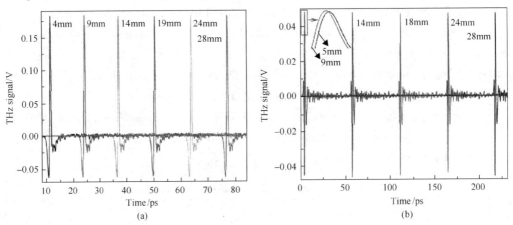

Figure 2 THz-TDS of air for different scan range of Z1(a) and Z2(b)

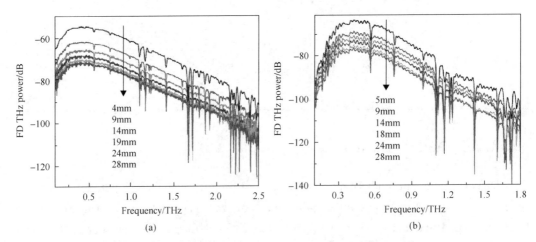

Figure 3 THz-FDS for different scan range of Z1(a) and Z2(b)

FWHM refers to the band width of half maximum height relative to the baseline. Generally, the narrower FWHM is, the higher resolution rate obtained. is Some similar materials had the absorption features in THz range and their feature locations were close to each other. The lower FWHM was significant for the qualitative detection of these materials and their mixtures. To confirm the optimal condition of measurement, the FWHM of characteristic lines at randomly selected frequencies were extracted and related to the scan range. The detailed relations of FWHM and scan range in both Z1 and Z2 were shown in Figure 4, where the frequency dependent THz-FDS around the selected frequency were depicted in the illustration pictures. With the increasing of scan range, FWHM decreased gradually when the range was narrow, and then remained unchanged with the scan range. When range was 5mm, the FWHM equaled ~15GHz and ~18GHz, respectively. When the scan range exceeded 10mm, FWHM reached below 7GHz. In addition, FWHM in Z1 even achieve ~5.5GHz in Figure 4(a). Increasing the scan range can greatly improve the resolution of equipment and the detection accuracy of samples.

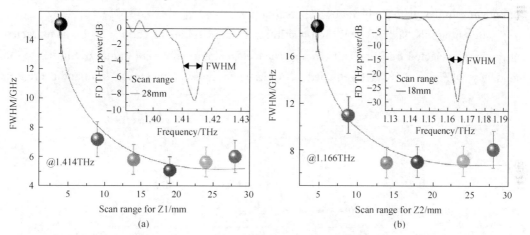

Figure 4 Scan range dependent FWHM for Z1 (a) and Z2(b) based on THz-FDS

Calibration is necessary for the THz devices to validate the preciseness and stability of the equipment. Carbon oxide (CO) is one of the typical substances. As previous articles reported, water vapor in air showed maltiple absorption features at special frequencies[37]. —OH⋯O of vapor molecules formed the hydrogen bond network, which was a special intermolecular or intramolecular interaction and a type of strong molecular link. Due to the ceaselessly forming and breaking of hydrogen bonds on the ps timescale, THz spectroscopy was much sensitive to the vibration mode of hydrogen bond network. In this research, we also scanned the THz-TDS in the condition without humidity by constantly introducing dry nitrogen gas to the system. Both the systems were covered by transparent plexiglass so that the medium of measurement remained nitrogen. THz-FDS can be then calculated using FFT based on THz-TDS. Herein nitrogen was considered as reference and air as sample, so the absorbance spectra can be obtained by $-\ln(A_{Sam.}/A_{Ref.})$, where $A_{Sam.}$ and $A_{Ref.}$ were the THz frequency-domain amplitudes of sample and reference, respectively. The absorbance spectra of water vapor were measured by Z1 and Z2 and shown in Figure 5[38]. Z2 had a narrower range of effective frequency due to its lower THz signal, which is different from Z1. The absorbance spectra displayed sharp absorption peaks at 0.56THz, 0.75THz, 0.99THz, 1.10THz, 1.17THz, 1.21THz, 1.41THz, 1.60THz, 1.68THz and 1.72THz in both Z1 and Z2. These absorption features were closely agreement with the previous reports[39-41]. Compared to Z1, Z2 showed more evident absorption features of vapor due to its longer path, especially at the frequencies where the absorption was relatively weak, such as 0.56THz, 0.75THz, 0.99THz and 1.23THz. Consequently, long-path became often necessary for gas detection to precisely characterize principal components especially when the total pressure was small. The results not only proved the accuracy and stability of equipment, but also demonstrated that water vapor was a suitable selection for the calibration of THz equipment. Also, Figure 5 could help judge whether absorption peak attributed to the tested materials, which was an important issue when unknown samples were analyzed by THz technique.

To validate the stability and high-resolution of the equipment, the FWHM was extracted at randomly selected frequencies from absorbance spectra as shown in Figure 6. Similar to the tendency in Figure 4, FWHM equaled ~20GHz when the scan range was 4mm and decreased

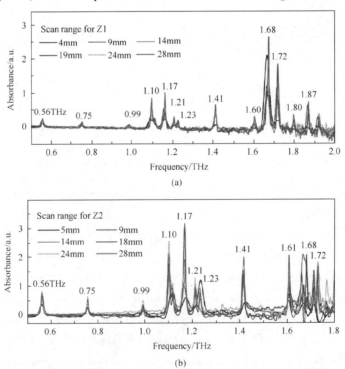

Figure 5 Frequency dependent absorbance for different scan range of Z1 (a) and Z2 (b)

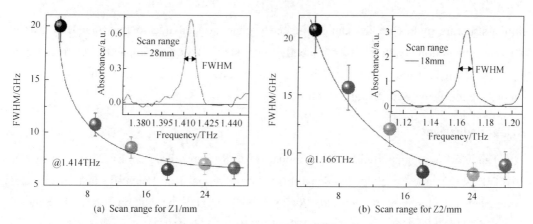

Figure 6 Scan range dependent FWHM for Z1 (a) and Z2 (b) based on absorbance spectra

gradually to 9GHz, 6.3GHz and 6.5GHz at, 19mm and 29mm in Z1, and 12GHz, 8.2GHz as well as 7.9GHz at 14mm, 18mm as well as 24mm in Z2. If the scan range were suitable, the minimum FWHM should reach in absorption spectra. A small FWHM in absorption spectra reflected a high resolution of equipment and a strong ability to distinguish the similar subjects whose absorption features were located at frequencies very close to each other. A conclusion can be drawn that water vapor, a normal material or gas existed in atmosphere environment, was a suitable selection to calibrate and validate the THz equipment.

It is evident that scan range has an influence on the resolution of transmission THz systems with both short and long path. Both systems can reach the FWHM below 7GHz. Relative to long-path setup, short-path system has a higher resolution according to the minimum FWHM (~5.5GHz) shown in Figures 4 and Figures 6. Such accuracy, sensitivity and optical path would show promise for THz monitoring of different kinds of materials, including solid, liquid and gas in some fields. THz systems with the optical paths in Figure 1 and suitable scan range can be considered for the establishment of standard database to identify the similar-structure materials in the future.

4 Conclusions

In summary, this research shows a high-resolution THz spectrum of water vapor measured by short- and long-path THz transmission setups. The minimum FWHM can reach ~5.5GHz. The high-resolution measurement can not only provide an accurate method for the calibration and validation of new THz setups, but also prove a promising selection for the rapidly qualitative identification of solid, liquid and gaseous materials whose structures are very similar to each other. Using such THz systems with appropriate scan ranges, the mixture with complicated components, such as oil-gas resources which are composited by many similar-structure hydrocarbons, can be precisely characterized due to the small FWHM of adjacent absorption features. Consequently, this study is quite impactful to prepare a high-resolution THz setup for the multifunctional detection of materials in biology, food safety, chemistry, petroleum industry, and so on.

Acknowledgements

This work was supported by the National Basic Research Program of China (Grant No. 2014CB744302), the Specially Funded Program on National Key Scientific Instruments and Equipment Development (Grant No. 2012YQ140005), and the National Natural Science Foundation of China (Grant Nos. 61405259, 11204191 & 11574401).

References

[1] Yang Y, Shutler A, Grischkowsky D. Measurement of the transmission of the atmosphere from 0.2 to 2THz. Optics Express, 2011,19: 8830-8838.

[2] Melinger J S, Yang Y, Mandehgar M, et al. THz detection of small molecule vapors in the atmospheric transmission windows.Optics Express, 2012, 20: 6788-6807.

[3] Yang Y, Mandehgar M, Grischkowsky D. Time domain measurement of the THz refractivity of water vapor. Optics Express, 2012, 20: 26208-26218.

[4] Bidgoli H, Cherednichenko S, Nordmark J, et al. Terahertz spectroscopy for real-time monitoring of water vapor and CO levelsin the producer gas from an industrial biomass gasifier. IEEE Transactions on Terahertz Science and Technology, 2014, 4: 722-733.

[5] Mittleman D M. Frontiers in terahertz sources and plasmonics. Nature Photonics, 2013, 7: 666-669.

[6] Liu J F, Zhou Q L, Shi Y L, et al. The rotation of polarization of a terahertz wave through subwavelength metallic structures. Science China: Physics, Mechanics & Astronomy, 2013, 56: 514-518.

[7] Yang Y P, Lei X Y, Yue A, et al. Temperature-dependent THz vibrational spectra of clenbuterol hydrochloride. Science China: Physics, Mechanics & Astronomy, 2013, 56: 713-717.

[8] Ala'a A A A, Sergei N Y, Jonathan T, et al. Terahertz spectroscopy of hydrogen sulfide. Journal of Quantitative Spectroscopy and Radiative Transfer, 2013, 130: 341-351.

[9] Cai H, Wang D, Shen J L. Study on terahertz spectra of SO_2 and H2S. Science China: Physics, Mechanics & Astronomy, 2013, 56: 685-690.

[10] Ma Y H, Wang Q, Li L Y. PLS model investigation of thiabendazole based on THz spectrum. Journal of Quantitative Spectroscopy and Radiative Transfer, 2013, 117: 7-14.

[11] Horiuchi N, Zhang X C. Searching for terahertz waves. Nature Photonics, 2010, 4: 662-662.

[12] Dragoman D, Dragoman M. Terahertz fields and applications. Progress in Quantum Electronics, 2004, 28: 1-66.

[13] Zhang L J, Zhang Y, Gao Y, et al. Design of terahertz band-stop filter based on a metallic resonator on high-resistivity siliconwafer. Science China: Technological Sciences, 2013, 56: 2238-2242.

[14] Takenori T, Takahiro O, Ikumi K, et al. Estimation of water content in coal using terahertz spectroscopy. Fuel, 2013, 105:769-770.

[15] Xiong W, Yao J, Li W, et al. Hybrid terahertz metamaterial structure formed by assembling a split ring resonator with a metalmesh. Science China: Physics, Mechanics & Astronomy, 2013, 56: 882-887.

[16] Gómez R J, Schotsch C, Haring B P, et al. Enhanced transmission of THz radiation through subwavelength holes. Physical Review B, 2003, 68: 201306.

[17] Bao R M, Wu S X, Zhao K, et al. Applying terahertz time-domain spectroscopy to probe the evolution of kerogen in closepyrolysis systems. Science China: Physics, Mechanics & Astronomy, 2013, 56: 1603-1605.

[18] Chun Y, Jeanna B, Iouli E G, et al. Oxygen, nitrogen and air broadening of HCN spectral lines at terahertz frequencies. Journal of Quantitative Spectroscopy and Radiative Transfer, 2008, 109: 2857-2868.

[19] Li J, Shi X F, Gao F, et al. Filtration of fine particles in atmospheric aerosol with electrospinning nanofibers and its sizedistribution. Science China: Technological Sciences, 2014, 57: 239-243.

[20] Siegel P H. Terahertz technology in biology and medicine. IEEE Microwave Theory and Techniques Society, 2004, 52: 2438-2447.

[21] Fang H N, Zhang R, Liu B, et al. Temperature dependence of the point defect properties of GaN thin films studied by terahertz time-domain spectroscopy. Science China: Physics, Mechanics & Astronomy, 2013, 56: 2059-2064.

[22] Al-Douseri F M, Chen Y Q, Zhang X C. THz wave sensing for petroleum industrial application. International Journal of Infrared and Millimeter Waves, 2006, 27: 481-503.

[23] Li N, Huang P, Zhao H J, et al. The quantitative evaluation of application of hyperspectral data based on multiparameters jointoptimization. Science China: Technological Sciences, 2014, 57: 2249-2255.

[24] Zhao H, Zhao K, Bao R M. Fuel property determination of biodiesel-diesel blends by terahertz spectrum. Journal of Infrared Millimeter and Terahertz Waves, 2012, 33: 522-528.

[25] Qin F L, Li Q, Zhan H L, et al. probing the sulfur content in gasoline quantitatively with terahertz time-domain spectroscopy.Science China: Physics, Mechanics & Astronomy, 2014, 57: 1404-1406.

[26] Ikeda T, Matsushita A, Tatsuno M, et al. Investigation of inflammable liquids by terahertz spectroscopy. Applied Physics Letters, 2005, 87: 034105.

[27] Li Q, Zhao K, Zhang L W, et al. Probing PM2.5 with terahertz wave. Science China: Physics, Mechanics & Astronomy, 2014, 57: 2354-2356.

[28] Zhan H L, Wu S X, Bao R M, et al. Qualitative identification of crude oils from different oil fields using terahertz time-domain spectroscopy. Fuel, 2015, 143: 189-193.

[29] Jin Y S, Kim G J, Shon C H, et al. Analysis of petroleum products and their mixtures by using terahertz time domain spectroscopy. Journal of the Korean Physical Society, 2008, 53: 1879-1885.

[30] Jiang C, Zhao L J, Zhao K, et al. Probing disaggregation of crude oil in magnetic field with terahertz time-domain spectroscopy.Energy & Fuels, 2014, 28: 483-487.

[31] Ge L N, Zhan H L, Leng W X, et al. Optical characterization of the principal hydrocarbon components in natural gas using terahertz spectroscopy. Energy & Fuels, 2015, 29: 1622-1627.

[32] Jin W J, Zhao K, Yang C, et al. Experimental measurements of water content in crude oil emulsions by terahertz time-domainspectroscopy. Applied Geophysics, 2013, 10: 506-509.

[33] Leng W X, Zhan H L, Ge L N, et al. Rapidly determinating the principal components of natural gas distilled from shale withterahertz spectroscopy. Fuel, 2015, 159: 84-88.

[34] Zhan H L, Wu S X, Bao R M, et al. Water adsorption dynamics in active carbon probed by terahertz spectroscopy. RSC Advance, 2015, 5: 14389-14392.

[35] Greeney N S, Scales J A. Non-contacting characterization of the electrical and mechanical properties of rocks at submillimeter scales. Applied Physics Letters, 2012, 100: 124105.

[36] Dorney T D, Rossow M J, Symes W W, et al. Single-cycle terahertz electromagnetic pulses: A new test bed for physical seismicmodeling. Geophysics, 2003, 68: 308-313.

[37] Castro-Camus E, Palomar M, Covarrubias A A. Leaf water dynamics of Arabidopsis thaliana monitored in-vivo using terahertz time-domain spectroscopy. Scientific Reports, 2013, 3: 2910.

[38] Hiromichi H, Takamasa S, Toshiyuki I, et al. Precise measurement of pressure broadening parameters for water vapor with a terahertz time-domain spectrometer.Journal of Quantitative Spectroscopy and Radiative Transfer, 2008, 109: 2303-2314.

[39] Yang Y, Mandehgar M, Grischkowsky D. Determination of the water vapor continuum absorption by THz-TDS and molecular response theory. Optics Express, 2014, 22: 4388-4403.

[40] Slocum D M, Slingerland E J, Giles R H, et al. Atmospheric absorption of terahertz radiation and water vapor continuum effects. Journal of Quantitative Spectroscopy and Radiative Transfer, 2013, 127: 49-63.

[41] van Exter M, Fattinger C, Grischkowsky D. Terahertz time-domain spectroscopy of water vapor. Optics Letters, 1989, 14: 1128-1130.

Determining the humidity-dependent *ortho*-to-*para* ratio of water vapor at room temperature by terahertz spectroscopy

Xinyang Miao[1,2]　Jing Zhu[2]　Kun Zhao[1,2]　Honglei Zhan[2]　Wenzheng Yue[1]

(1. State Key Laboratory of Petroleum Resources and Prospecting, China University of Petroleum, Beijing 102249, China; 2. Beijing Key Laboratory of Optical Detection Technology for Oil and Gas, China University of Petroleum, Beijing 102249, China)

Abstract: The origin of the water spin isomers observed under various physic-chemical conditions is of great interest, including that of H_2O molecules in the gas phase. Here, Terahertz time-domain spectroscopy (THz-TDS) was used to study the humidity-dependent *ortho*-to-*para* (O/P) ratio of water vapor at room temperature. The relative contents of *para* and *ortho* molecules were obtained by fitting the absorption lines of water vapor showing the relationship between the spin isomer contents and humidity. Larger O/P ratios with values of ~3.2 were observed at lower humidity (<20%) due to the stronger attractive forces of *para* molecules. The concentration of the ortho isomers then began to decrease at higher humidity (>20%) due to the preferential formation of dimers and clusters at increasing concentrations. Thus, the ratio gradually decreased with increasing humidity.

Keywords: water vapor; terahertz; THz spectroscopy; spin isomers; humidity; *ortho*-to-*para* ratio

1 Introduction

Water is a mixture of two nuclear spin isomers with different spin state orientations. In terms of quantum mechanics, two H_2O molecules (*ortho* and *para*) are distinguished by the magnitude I of the total nuclear spin: $I = 1$ and $I = 0$, respectively[1]. Studying the mechanisms of spin-selective processes in H_2O molecules is important to many basic and applied areas of science[2-4]. The thermodynamic probability of the *ortho* state is three times larger than that of *para* state; thus, it is commonly accepted that the H_2O spin isomer molecules are in a statistical quantitative ratio of 3∶1 (named the *ortho*-to-*para* [O/P] ratio) under ambient conditions[1]. Varying conditions can lead to a shift in the *ortho*-to-*para* ratio, which makes it a valuable tracer for physiochemical characteristics and physical conditions of the molecules[5,6]. The differences in the H_2O spin isomers can be quantified via the O/P ratio, including by kinetic studies, adsorption and desorption in porous materials[7], evaporation and crystallization[8], and cluster and smog formation[9]. Information about interstellar clouds can also be obtained by this ratio and is helpful in astronomical observations[10]. However, until now, the origin of the *ortho* and *para* states observed in various physicochemical conditions remains an open question—especially for spin-selective processes of H_2O molecules in the gas phase[5,6,10].

Due to the different symmetries of the spin components in water, the wave functions of the isomers in vibrational ground states differ from each other[11]. Therefore, the spin isomer ratio of gaseous H_2O can be determined by featured absorption lines[12]. Various absorption features of water vapor have been observed in the terahertz (THz) range, which makes THz spectroscopy a promising tool for detecting spin isomers[13-15].

Studies of H_2O isomer enrichment and conversion have been performed based on the selective spectral features in THz frequencies. For example, backward wave oscillator (BWO) spectroscopy can determine the *ortho*-to-*para* ratio in water vapor via the two rotational lines located near $37cm^{-1}$. Deviations from the normal 3∶1 *ortho*-to-*para* ratio were observed for water vapor in dynamic sorption processes[16].

Four photon laser spectroscopy (FPLS) can also provide the temperature-dependent intensity ratio of the rotational lines assigned to the *ortho* and *para* isomers from $36\sim41cm^{-1}$ and $78\sim90cm^{-1}$[17]. Compared to the above methods, THz time-domain spectroscopy (THz-TDS) can provide abundant information on the intermolecular and intramolecular vibration modes. It quantitates absorption features across a much wider frequency range from $0.1\sim2THz$ with spectral resolution <6GHz[18].

A previous study measured the absorption profile of *ortho* and *para* isomers in water vapor using THz-TDS with increasing humidity. The results showed the separate variation tendencies in *ortho*- and *para*-H_2O molecules as a function of water vapor concentrations[19]. However, the relative content of the two spin isomers was not determined in their study and quantitative characterization of the H_2O spin isomers was necessary for THz measurements. Here, we studied the dependence of humidity on the O/P ratio in water vapor at room temperature. The relative contents of *para* and *ortho* molecules in water vapor were obtained by measuring and fitting the absorption lines of water vapor; the results showed that the ratio was related to the humidity. Thus, the THz technique is a promising tool for detecting spin isomers in H_2O molecules.

2 Materials and methods

The experimental setup was comprised of a transmission THz-TDS system and a LEO-50 femtosecond laser from Daheng Optics. Figure 1 showed that the femtosecond laser beam (800nm) was initially split into two beams. The pump beam was focused onto the surface of a biased GaAs photo-conductive antenna for generation of THz waves. Collimated THz pulses were focused with a lens, reflected by mirrors, and then finally transmitted onto the detection system with a THz path length of 80cm. The hybrid beams composed of the THz pulse and the probe laser beam were focused onto a 2.8mm thick <110> ZnTe for electro-optic detection[20].

The set-up was placed in an enclosure and filled with laboratory air with a relative humidity of ~55% at 25.2℃±0.5℃. Various quantities of nitrogen were purged into the box to reach different humidity levels (55.0% to 0%±1.0%). The measurements were performed in replicates of four under ambient atmospheric pressure.

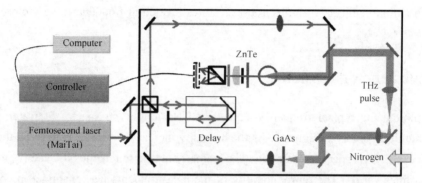

Figure 1　Top view of THz-TDS setup for the detection of water vapor

The THz measurements allow us to resolve the strong rotational absorption characteristics of the H_2O molecules, but the device needs higher frequency resolution. Previous studies have shown that the frequency resolution of a THz measurement is mainly determined by the scanning duration (T_s) of the spectrometer[18]. According to the Fourier transform theory, T_s in the time domain is equivalent to the convolution of the spectrum in the frequency domain; higher spectral resolution is obtained by extending the temporal measurement window.

The spectral resolution of the THz-TDS can be improved by increasing the duration of the temporal measurement, but is limited by the noise in the measurement and the dynamic range of the system in the time-domain[21]. Here, the effect of T_s was presented in Figure 2. The THz frequency-domain spectra of the atmosphere at 25.4℃ and ~31.1% relative humidity were shown as T_s varing from 25.4ps to 132.3ps.

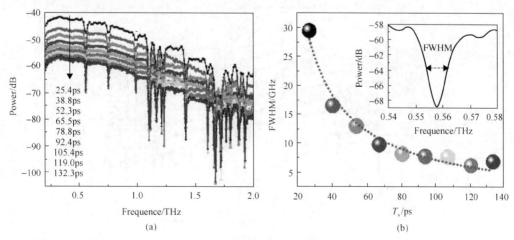

Figure 2　(a) Frequency-domain spectroscopy for water vapor at different scan ranges; (b) Scan range-dependent FWHM based on frequency-domain spectroscopy

An inversely proportional relationship between the full width at half maximum (Δf) and T_s was observed and was fitted by the equation $\Delta f=0.709/T_s$. Here, Δf decreases as T_s increases; it has a minimum of ~6GHz indicating that the increasing T_s can greatly improve the resolution of equipment. Smaller Δf reflect the higher resolution of the equipment and a better ability to distinguish similar subjects with absorption features located at frequencies very close to each other.

3 Results and discussion

We plotted the typical frequency-dependent THz amplitude spectra of water vapor at different humidity levels in Figure 3. As the humidity increases from 4.5% to 55%, the overall shape of each THz spectrum remains the same, but the intensities of the water vapor absorption lines varied. The reproducibility of the measurements was examined by obtaining the spectra of H_2O vapor at various humidity values four times. The results from different experiments were similar. In addition, the THz pulse excites a coherent ensemble of molecules in rotational and vibrational states when it propagates through water vapor. This results in the spectrum's absorption lines. The 16 modes of absorption lines are depicted in Figure 3 from 0.1THz to 2.0THz, and the modes and types at some typical frequencies are presented in Table 1[19,22].

To express these values more clearly, the peak values of the measured absorption lines were extracted from the spectra. Due to the intense absorption of the water vapor at higher frequencies, the spectra were not only composed of absorption features, but also many types of noise. Therefore, the effective frequency range was reduced to 0.1~1.5THz for acceptable signal-to-noise ratios. Variations in the peak intensity with the humidity at the selected frequencies are depicted in Figure 4 and correspond to the transition of *ortho*- and *para*-H_2O vapor.

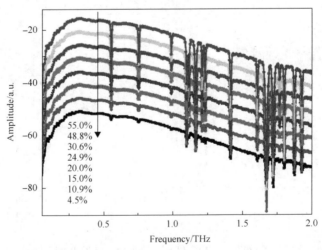

Figure 3　Frequency-domain spectroscopy of water vapor at different humidity values

Table 1 Rotational transitions in water vapor from 0.2~2.0THz at room temperature

Number	Frequency /THz	Upper			Lower			Type
		J	K_{-1}	K_{+1}	J	K_{-1}	K_{+1}	
1	0.558	1	1	0	1	0	1	Ortho
2	0.754	2	1	1	2	0	2	Para
3	0.985	2	0	2	1	1	1	Para
4	1.096	3	1	2	3	0	3	Ortho
5	1.111	1	1	1	0	0	0	Para
6	1.161	3	2	1	3	1	2	Ortho
7	1.226	2	2	0	2	1	1	Para
8	1.412	5	2	3	5	1	4	Ortho
9	1.603	4	1	3	4	0	4	Para
10	1.669	2	1	2	1	0	1	Ortho
11	1.714	3	0	3	2	1	2	Ortho
12	1.794	6	2	4	6	1	5	Ortho
13	1.865	3	1	0	4	0	4	Ortho
14	1.920	3	2	2	3	1	3	Para

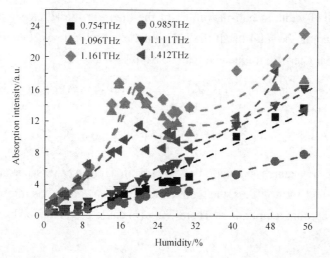

Figure 4 The variation of *para* and *ortho* rotational transitions intensity with humidity values of 0%~55%

Usually, the absorption of material follows the Beer-Lambert law, which states that there is a linear relationship between the absorbance and the concentration of an absorbing species. However, the absorption will deviate from linearity at higher concentrations as the molecules become closer and begin to interact with each other. Such concentration effects were observed here. The *ortho*- and *para*-H_2O molecules had different absorption values as a function of water vapor concentration (Figure 4). Three *para* transitions (0.75THz, 0.98THz and 1.11THz) increased linearly with the humidity; meanwhile, the *ortho* absorption intensities at 1.09THz and 1.16THz increased rapidly from 0 for the nitrogen to ~16 at 20.0% humidity. It then decreased to ~12 at approximately 32% and finally increases again with increasing humidity. Similar trends were also observed at 1.412THz with the same turning points at 20.0% humidity. The increase in humidity changed the amplitude ratio of the *ortho* and *para* absorption lines indicating marked changes in the relative contents between them.

We then modeled the rotational absorption spectra to measure the O/P ratio of H_2O molecules in water vapor. The absorption of the resonant lines are calculated as the sum of all resonant lines with the corresponding line intensity function ($S_j(v)$) taken by launch simulation results from HITRAN[22]. Therefore, the absorbance line of water vapor resonant line $\alpha(v)$ can be written as the product of the number density of water vapor in molecules (N_{H_2O}), the THz path length (l), $S_j(v)$, and the van Vleck-Weisskopf (VVW) line shape function $f(v,v_j)$ in which v_j is the resonant line center wavenumber[23].

$$\alpha(v) = N_{H_2O} S_j(v) f(v, v_j) l \tag{1}$$

The VVW line shape is given as

$$f(v, v_j) = \frac{1}{\pi}\left(\frac{v}{v_j}\right)\left(\frac{\Delta v_j}{(v-v_j)^2 + \Delta v_j^2} + \frac{\Delta v_j}{(v+v_j)^2 + \Delta v_j^2}\right) \tag{2}$$

where Δv_j is the half-width at half-maximum of the line. For the HITRAN database of water vapor, the total $\alpha(v)$ is the total of all the individual absorption lines; therefore, the sum line shape of all the characteristic frequencies can be expressed as

$$f(v,v_j) = \sum_{i \in j} f(v, v_i) = \sum_{i' \in para} f(v, v_{i'}) + \sum_{i'' \in ortho} f(v, v_{i''}) = F(v, v_{para}) + F(v, v_{ortho}) \tag{3}$$

Here, we define $F(v, v_{para}) = \sum_{i' \in para} f(v, v_{i'})$ and $F(v, v_{ortho}) = \sum_{i'' \in ortho} f(v, v_{i''})$.

Additionally, the launch simulation of $S_j(v)$ from HITRAN was performed under ideal conditions (296K and 1atm①); thus, the number density of *ortho*-H_2O is three times larger than that of *para*-H_2O at room temperature. Therefore, $\alpha(v)$ could be calculated by

$$\begin{aligned}\alpha(v) &= N_{H_2O} S_j(v)\left[F(v, v_{para}) + F(v, v_{ortho})\right] l \\ &= (N_P + N_O) S_j(v) F(v, v_{para}) l + (N_P + N_O) S_j(v) F(v, v_{ortho}) l \\ &= S_j(v) l \left[4 N_P F(v, v_{para}) + \frac{4}{3} N_O F(v, v_{ortho})\right] \\ &= N_P S_j(v_{para}) F(v, v_{para}) l + N_O S_j(v_{ortho}) F(v, v_{ortho}) l \end{aligned} \tag{4}$$

Where, $S_j(v_{para})$ and $S_j(v_{ortho})$ are defined as the line intensity functions for *para*- and *ortho*-H_2O vapor, respectively:

---

① 1atm=101.325kPa.

$$S_j(v_{para}) = 4S_j(v)$$
$$S_j(v_{ortho}) = \frac{4}{3}S_j(v) \quad (5)$$

According to the intensity of the *ortho*- and *para*-H_2O at each absorption line from 0.7THz to 1.5THz, the N_O and N_P values at each measured line was first fitted by Equation (4). The average fitted values of N_O at 1.096THz, 1.161THz, and 1.412THz as well as N_P at 0.754THz, 0.985THz, and 1.111THz were then calculated. Next, we obtained the N_O/N_P values from the THz absorption spectra measured at different humidity values. Examples of the simulated curves plotted with the experimental results are shown in Figure 5 at 3.8%, 20.0%, and 48.8% humidity, respectively. Slightly larger differences existed between the calculated and experimental absorption intensities for *ortho*-H_2O compared to *para*-H_2O, but there were differences for the peak position for all lines. The N_O/N_P values of all measured spectra determine the humidity-dependent *ortho*-to-*para* ratio of water vapor at room temperature (Figure 6). The error bars come from testing errors (e.g. instrumental errors and humidity measurement error) as well as fitting errors (mismatch of the fitting lines and the measured lines).

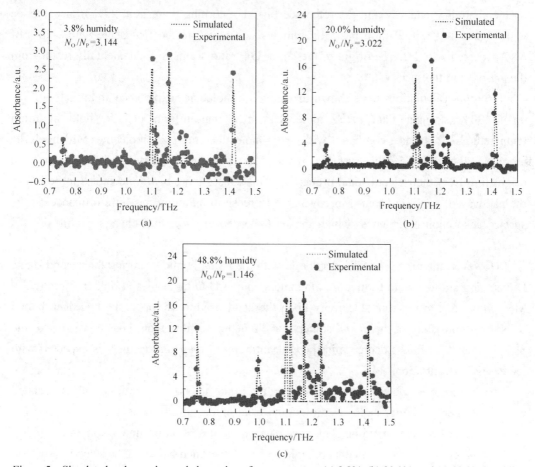

Figure 5   Simulated and experimental absorption of water vapor at (a) 3.8%, (b) 20.0%, and (c) 48.8% humidity

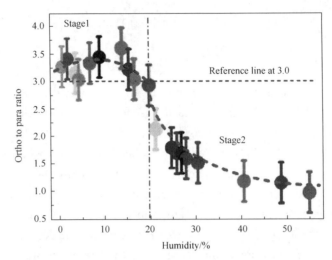

Figure 6 Variation of the O/P ratio in water vapor with the humidity at room temperature

As depicted in the figure, the ratio of N_O and N_P varies as a function of the H_2O concentration. Generally, the variation can be divided into two stages (0.8%~20.0% and 20.0%~55.0%) based on the changing tendency and the reference line. The turning point is at 20% in accordance with Figure 4. The O/P ratio basically remains unchanged in the first stage with values of N_O/N_P located near 3.2. The origin of the larger O/P ratio seems to fundamentally result from the presence of the surfaces.

Ab initio calculations have shown that the *para* molecules experience an attractive force up to 30% stronger than *ortho* molecules in a non-uniform electric field at room temperature[11,16]. Hence, compared with *ortho* molecules, the *para* molecules will likely be more adsorbed by the solid surfaces in the THz setup (e.g. mirrors and metal surfaces) originating from their electrostatic properties. This causes more *ortho* molecules to be left in the pathway of the THz beam. Moreover, the increase in humidity has little influence on the surface adsorption divergence, which makes the *ortho*-to-*para* ratio change slightly in this stage.

However, the ratio gradually decreased with the rising humidity during the second stage. Dimers and clusters were formed by monomers among H_2O molecules at higher humidity.[19] Meanwhile, due to the spin selectivity upon the condensation of vapor, the hydrogen-bound complexes consisting of the *ortho* isomers have a better stability than those consisting of the *para* isomers[17]. Thus, a decrease in the concentration of the *ortho* isomers is expected with increasing concentration.

The response of the electromagnetic radiation in the THz range is intrinsically associated with low energy events such as molecular torsion or vibration as well as inter- and intramolecular hydrogen bonding. This creates a series of absorption lines. Numerous studies of H_2O isomer enrichment and conversion have been performed based on the selective spectral features in THz frequencies. Backward wave oscillator spectroscopy has been used to determine the O/P ratio based on two rotational lines located at 36.5cm^{-1}, 37.2cm^{-1}, and 40.3cm^{-1} (1.096THz, 1.111THz, and 1.226THz in this study)[11,12]; there was a deviation from the typical 3∶1 O/P ratio in favor of *ortho* molecules in saturated water vapor above the glycerol film. The O/P ratio in the saturation vapor of the treated solution was shifted from ~3.2 to ~2.2 due to the enhanced fugacity of *ortho* molecules and enriched *para* molecules due to pumping[12]. Four photon laser spectrometry could measure from 36~40cm^{-1} and 78~90cm^{-1} (1.08~1.2THz and 2.34~2.7THz). An increase in the temperature causes a variation in the FPLS amplitude ratio of the *ortho* and *para* lines from about 1∶4 to about 1.05∶1[17]. Though diverse means are taken to study O/P ratios of water vapor, THz methodology is preferred because THz wave is sensitive to the hydrogen bond. Thus it is hard for us to have a benchmark by other means. Previous studies mainly focus on other varieties that may have impacts on this ratio and in this study we concerned about concentration, and we failed to find the humidity dependent O/P ratios of water vapor at room temperature in previous works. We believe that the THz technology is a valuable means for the spin isomers evaluation in water vapor, and more data can be obtained by further researches.

Importantly, compared with the above measurements performed inside the designated gas cells, our tests were performed entirely inside the THz-TDS chamber. The *para* molecules had more of a chance to be trapped by the solid surfaces inside the setup. Nevertheless, we aim to perform the THz technique to measure the humidity-dependent O/P ratio under actual conditions by measuring and fitting the absorption lines of water vapor with different humidity values. The relative contents of *para* and *ortho* molecules in the water vapor were thus obtained with the ratio related to the humidity. Previous studies showed that the adsorption of *para* molecules always exists even with designed gas cells[11,12,16,19]. In our experiments, similar results were observed at lower humidity. The effect of *para* molecules adsorption was resulted from the divergence from normal 3:1 O/P ratio with a humidity 0.8%~20.0%. The dimers and clusters were formed from monomers of H_2O molecules at higher humidity. This decreases the concentration of *ortho* isomers. The effect of trapped *para* molecules is negligible on shorter time scales (a few minutes) under these circumstances. THz spectroscopy allows us to measure absorption features across a much wider frequency range from 0.1~1.5THz with spectral resolution below 6GHz. Here, we determined the humidity-dependent O/P ratio of the water vapor at room temperature. By measuring and fitting the absorption lines of water vapor with different humidity, values, the relative contents of *para* and *ortho* molecules in the water vapor were obtained; the ratio is related to the humidity. Thus, THz technology is a powerful tool to detect the spin isomers in H_2O molecules.

4 Conclusions

In summary, THz-TDS was employed to study the humidity-dependent O/P ratio of water vapor at room temperature. The relative contents of *para* and *ortho* molecules in water vapor were obtained by modeling the THz absorption lines of water vapor. and showed that the ratio is dependent on the humidity. The entire variation in the O/P ratio can be divided into two stages based on the changing tendency as well as the 3.0 reference line. The ratio basically remains unchanged at ~3.2 in the first stage with lower humidity (< 20%). In the second stage, there is a decrease in the concentration of the *ortho* isomers with increasing humidity (>20%) due to the formation of dimers and clusters as well as the spin selectivity; the ratio gradually decreases with rising humidity. Therefore, the THz technology remains a valuable tool for evaluating the spin isomers of water vapor.

Acknowledgements

This work was supported by the National Nature Science Foundation of China (Grant No. 11574401), the Science Foundation of China University of Petroleum, Beijing (Nos. 2462017YJRC029 and yjs2017019), and the Beijing Natural Science Foundation (No. 1184016).

References

[1] Turgeon P A, Ayotte P, Lisitsin E, et al. Preparation, isolation, storage, and spectroscopic characterization of water vapor enriched in the *ortho*-H_2O nuclear spin isomer. Physical Review A, 2012, 86(6): 29940-29948.

[2] Meier B, Mamone S, Concistrè M, et al. Electrical detection of *ortho-para* conversion in fullerene-encapsulated water. Nature Communications, 2015, 6: 8112.

[3] Mamone S, Concistrè M, Carignani E, et al. Nuclear spin conversion of water inside fullerene cages detected by low-temperature nuclear magnetic resonance. The Journal of Chemical Physics, 2014. 140(19): 194306.

[4] Goh K S K, Ruiz M J, Johnson M R, et al. Symmetry-breaking in the endofullerene $H_2O@C_{60}$ revealed in the quantum dynamics of *ortho* and *para*-water: A neutron scattering investigation. Physical Chemistry Chemical Physics, 2014. 16(39): 21330-21339.

[5] Hama T, Watanabe N. Surface processes on interstellar amorphous solid water: Adsorption, diffusion, tunneling reactions, andnuclear-spin conversion.Chemical Reviews, 2013, 113(12): 8783-8839.

[6] Van Dishoeck E F, Herbst E, Neufeld D A. Interstellar water chemistry: From laboratory to observations.Chemical Reviews, 2013, 113(12): 9043-9085.

[7] Kapralov P O, Artemov V G, Leskin A A, et al. On the possibility of sorting *ortho* and *para* water molecules during diffusion in nanopores. Bulletin of the Lebedev Physics Institure, 2008, 35(7): 221-223.

[8] Pershin S M. Coincidence of rotational energy of H_2O *ortho-para* molecules and translation energy near specific temperatures in water and ice. Physics of Wave Phenomena, 2008, 16(1): 15-25.

[9] Vigasin A A, Volkov A A, Tikhonov V I, et al. Spin-selective adsorption of water vapor. Doklady Physics, 2002, 47(12): 842-845.

[10] Hama T, Kouchi A, Watanabe N. Statistical *ortho*-to-*para* ratio of water desorbed from ice at 10 kelvin. Science, 2016, 351(6268): 65-67.

[11] Tikhonov V I, Makurenkov A M, Artemov V G, et al. Sorption experiments with water spin isomers in glycerol. Physics of Wave Phenomena, 2007, 15(2): 106-110.

[12] Stepanov E V, Zyryanov P V, Milyaev V A. Laser analysis of the relative content of *ortho*-and *para*-water molecules for the diagnostics of spin-selective processes in gaseous media. Physics of Wave Phenomena, 2010, 18(1): 33-43.

[13] Yang Y, Mandehgar M, Grischkowsky D R. Understanding THz pulse propagation in the atmosphere. IEEE Transactions on Terahertz Science and Technology, 2012, 2(4): 406-415.
[14] Tikhonov V I, Volkov A A. Separation of water into its ortho and para isomers. Science, 2002, 296(5577): 2363.
[15] Horke D A, Chang Y P, Długołęcki K, et al. Separating *para* and *ortho* water. Angewandte Chemie International Edition, 2014, 53(44): 11965-11968.
[16] Kapralov P O, Artemov V G, Makurenkov A M, et al. Deviations from the normal *ortho/para* ratio for water (3∶1) in the vapor phase in dynamic sorption. Russian Journal of Physical Chemistry A, 2009, 83(4): 663-669.
[17] Pershin S M, Bunkin A F. Temperature evolution of the relative concentration of the H_2O *ortho/para* spin isomers in water studied by four-photon laser spectroscopy. Laser Physics, 2009, 19(7): 1410-1414.
[18] Zhan H L, Sun S N, Zhao K, et al. Less than 6GHz resolution THz spectroscopy of water vapor. Science China: Technological sciences, 2015, 58(12): 2104-2109.
[19] Xin X, Altan H, Saint A, et al. Terahertz absorption spectrum of *para* and *ortho* water vapors at different humidities at room temperature. Journal of Applied Physics, 2006, 100(9): 094905.
[20] Miao X, Li H, Bao R, et al. Discriminating the mineralogical composition in drill cuttings based on absorption spectra in the terahertz range. Applied Spectroscopy, 2017, 71(2): 186-193.
[21] Xu J, Yuan T, Mickan S, et al. Limit of spectral resolution in terahertz time-domain spectroscopy. Chinese Physics Letters., 2003, 20(8): 1266-1268.
[22] Yu B L, Yang Y, Zeng F, et al. Terahertz absorption spectrum of D_2O vapor.Optics Communications, 2006, 258(2): 256-263.
[23] Yang Y, Mandehgar M, Grischkowsky D. Determination of the water vapor continuum absorption by THz-TDS and Molecular Response Theory. Optics Express, 2014, 22(4): 4388-4403.

Size effect on microparticle detection

Honglei Zhan[1]　Ru Chen[1]　Xinyang Miao[1]　Yizhan Li[1]　Kun Zhao[1]
Shijie Hao[2]　Xiaohong Chen[1]

(1. State Key Laboratory of Petroleum Resources and Prospecting and Beijing Key Laboratory of Optical Detection Technology for Oil and Gas, China University of Petroleum, Beijing 102249, China; 2. Department of Material Science and Engineering, China University of Petroleum, Beijing 102249, China)

Abstract: In this paper, terahertz (THz) spectroscopy was employed to analyze compacted sand particles. The dependence of attenuation coefficient spectra on thickness were obtained. Further analysis of the spectra indicated that as the particle size increased, the attenuation coefficient of the THz spectrum also increased. The smaller sand particle samples behaved as Rayleigh particles, their extinction spectrum was mainly affected by absorption of the THz waves, thus the attenuation was small. The extinction coefficient curve of the larger particle sand samples behaved as Mie particles and exhibited the strong characteristics of a Mie scattering curve. The THz wave attenuation in the larger sand particle samples was the result of the combined effect of absorption and scattering, thus the attenuation of the THz waves was larger.

Keywords: terahertz; microparticle detection; size effect; extinction coefficient

1 Introduction

Powders, droplets and bubbles with generally below 10^{-3} meters are referred to as particles. There are widespread particle problems in environmental science, life sciences and industrial production[1]. In industrial production, there are particle problems with powder particles as raw materials or products such as glass, ceramics, metallurgy, catalysts, cement, food, medicine and the like[2-5]. Characterization of particle properties is also of importance, resulting in a variety of methods of characterizing different parameters of the particle. Traditional methods of detecting particle characteristics often yield only a single physical characteristic, such as sieving, ultrasonic measurement, laser measurement, and electric measurement. For example, the particle size of the medium can be measured by measuring the attenuation of the ultrasonic waves passing through the measured medium[6]. However, the ultrasonic measurement needed the special working environment and the sensor demanding.

THz wavelength, which is located between millimeter wave and infrared light[7], overlaps the size range of the powders. The powders also respond to the THz waves. The substance absorbs and scatters the THz wave when the THz wave shines on the substance. Absorption is mainly related to the chemical composition of the material, and scattering is affected by the physical properties of the material, such as density, particle size, structure and so on. Scattering is mainly Rayleigh scattering when the size of the microparticles is small, following as[8]:

$$I_s = \frac{8\pi^4 \alpha^6}{r^2 \lambda^4}\left(\frac{m^2-1}{m^2+2}\right)I_0(1+\cos^2\theta) \qquad (1)$$

where I_s is scattering light intensity function, I_0 is the incident light intensity, λ is the wavelength of incident light, α is the particle size parameter, r is the distance from the scattering particle to the observation point, and m is the refractive index of the medium, respectively. The wavelength of the scattered light caused by the particles is the same as the wavelength of the incident light. However, the intensity of the light was changed. The intensity of the scattered light is proportional to the fourth power of the wavelength of the incident light, and the intensity of the scattering increases with the increase of the size of the particles.

As the particle size continues to increase beyond one THz wavelength, the scattering characteristics are calculated by Mie theory. Mie theory has been provided to deal with scattering of a single sphere. In Mie theory, the scattering cross section σ_s and the extinction cross section σ_t are[9]

$$\sigma_s = \frac{2\pi}{k^2}\sum_{n=1}^{\infty}(2n+1)\left(|a_n|^2+|b_n|^2\right) \qquad (2)$$

$$\sigma_t = \frac{2\pi}{k^2}\sum_{n=1}^{\infty}(2n+1)\mathrm{Re}\{a_n+b_n\} \qquad (3)$$

where a_n, b_n are the scattering coefficients, and $k = 2\pi/\lambda$ is the wave number in free space.

Chemical composition information and physical information of samples can be obtained simultaneously by THz technology, compared with other technologies that can only obtain the physical information. It is necessary to pretreat the sample when the terahertz technology is applied to the detection of microparticles. Solid samples may need to be ground into particles during the pretreatment. It is difficult for THz test to achieve in practice because thin samples can become brittle and fall apart. Therefore, the sample material is often mixed with a transparent binding material (such as polyethylene or Teflon). However, the particle size of a sample is often not uniform or it is simply a matter of screening a single size sample in most studies[10]. The impact of particle size on the THz wave is important either for the reflective terahertz or the transmissive terahertz tests. In this paper, the size effect of sand based was investigated on THz time-domain spectroscopy (THz-TDS).

2 Experimental methods

A schematic representation of the entire experimental process was shown in Figure 1. The sand sample was sequentially sorted by stacked sieves. Each of the sorted sand samples was mixed with polyethylene powder with a 0.8g : 0.8g mass ratio. The mixed powder was poured into a tablet press for 2min at 20MPa. The light direction used in this experiment was horizontal. Loose sands cannot be well placed in the laser focus area, and they need to be made into thin slices. It is difficult to achieve in practice because thin samples can become brittle and fall apart. The sample material is often mixed with a transparent binding material (such as polyethylene or Teflon). The grain sizes of the sample and binding material must be small enough to avoid scattering within the sample. The samples must also be pressed with sufficient pressure to reduce the size of air voids. The resulting compressed sheet had a diameter of 30mm and a thickness of 1.74mm±0.4mm. The reflection-type THz-TDS tablet was used to conduct a THz reflection spectrum test on at least 100 spots 2cm-diameter on the surface of the pressed sample. THz reflectance images of the sample's surface and the average reflectance of each sample were obtained.

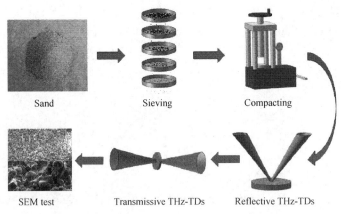

Figure 1　Schematic diagram of the entire experimental process

The sheet sample was subsequently analyzed using transmitted THz-TDS, which were positioned at the same location each time so that the THz radiation penetrated vertically through the center of the sample. The transmitted THz-TDS were conducted in a nitrogen environment to avoid the impact of water vapor on the experiment. For accuracy, all the THz-TDS measurements were performed by scanning the nitrogen blanket without a sample to obtain a reference signal, and then, the analyses were performed 3 times for each sample and the average value was used to calculate the optical parameters. A fast Fourier transform was applied to the time domain data to obtain the frequency-dependent spectra[11-13]. The extinction coefficient was calculated from the ratio and the relative phase difference between the sample and the reference power spectra.

Subsequently, we selected samples with a sand size of 74μm for scanning electron microscopy (SEM) analysis to investigate the sample's surface and bedding planes. According to Figure 2(a), the sand reflected a centralized distribution evenly over the surface. The bottom panels in Figure 2(a) are enlarged views of selected areas in the upper images. The upper image shows the secondary electron(SE) morphology of the sand sample at a scale of 200μm. The lower image shows the SE morphology of the same sample at a scale of 30μm. We also performed SEM analyses on the cross section of the sample and enlarged the area gradually, as shown in Figure 2(b). The lower image is an enlarged view of the selected area in the upper image. The upper image shows the SE morphology of the sand sample at a scale of 500μm. The lower image shows the SE morphology of the same sample at a scale of 50μm.

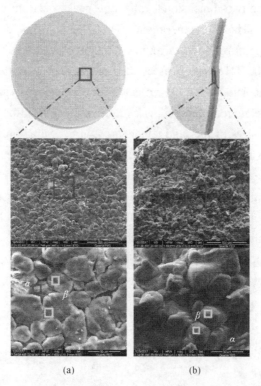

(a)   (b)

Figure 2   Surface and cross section SEM images of a sand sample measured by THz-TDS

In addition, two regions of the surface and cross section were selected for energy spectrum analysis. Two kinds of elemental distributions were obtained and are listed in Tables 1 and 2.

Table 1   Element distribution of area α

Elements	Mass fraction/%	Atom fraction/%
C	14.60	26.62
O	23.65	32.37
Mg	5.43	4.89
Al	7.41	6.01
Si	23.19	18.08
Ca	8.01	4.37
Fe	14.44	5.66

Table 2 Element distribution of area β

Elements	Mass fraction/%	Atom fraction/%
C	95.47	97.01
O	2.99	2.28

3 Results and discussion

Figure 3 display the two-dimensional reflection peak spectrum of the samples obtained from the reflection-type THz-TDS experiments. The reflection time-domain spectrometer that we used performed point-by-point scanning imaging with the center point polygon approaching a circle. The THz wave was emitted and reflected by the experimental sample. The reflectivity was calculated using the reflected light intensity divided by the emitted light intensity. The size and average reflection peak information (E_R) of the sand samples are listed in Table 3. The particle sizes of the sand tablets are 44μm, 55μm, 64μm, 106μm, 150μm, and 180μm. As can be seen from Figure 3, for the smallest particle size, the terahertz reflection peak value is large. The average reflection peak value is 0.04753V when the particle size is 44μm. In the case of largest particle size, the terahertz peak value is smaller, with an average reflection peak value of 0.03374V at a particle size of 180μm. The thicknesses of the experimental samples are not completely uniform due to the tablet making process, so the reflected image does not have a uniform reflection peak. However, as the particle size increases, the reflectivity of the particles decreases.

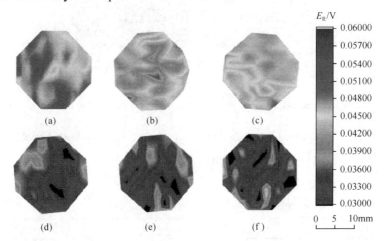

Figure 3 THz reflectance spectroscopic images of the different sand samples

Table 3 Reflectance parameters of six sand samples

	a	b	c	d	e	f
d/μm	44	55	64	106	150	180
E_R/V	0.04753	0.04286	0.04408	0.03427	0.03432	0.03374

Figure 4 display the selected transmission THz-TDS obtained by scanning the different sand samples in a nitrogen environment. Figure 4 only display the THz signal of the six samples. The frequency-dependent spectra were obtained using a complex Fourier transform as shown in the inset of Figure 4. The reference signal peak value is 0.19V at 5.8ps. The time delay of the sand tablet's signal was ~9ps because of the time delay caused by refraction of the THz wave through the sample. The signal peak of the sand samples decreased as the sand diameter increased. The peak value decreased from 0.145V to 0.062V with the particle size increasing from 34μm to 150μm in diameter. The signal attenuation in the sand sample was caused by absorption and scattering of the THz wave. It can be seen that the larger the sample size, the lower the cutoff frequency in the frequency domain spectrum.

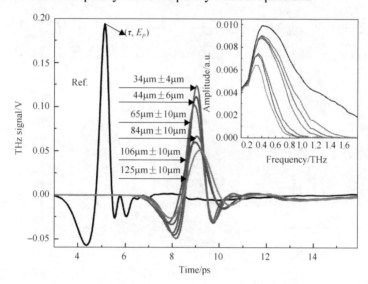

Figure 4　Terahertz spectroscopy observations of the reference and sand samples

Attenuation index β was calculated using $\beta = E_{\text{P-Sam}}/E_{\text{P-Ref}}$ to clear the size dependent attenuation coefficient, where $E_{\text{P-Sam}}$ is the sample signal peak value and $E_{\text{P-Ref}}$ is the reference signal peak value. As can be seen in Figure 5, as the sample size increases from 34μm to 180μm, the attenuation index of the unit thickness β/d changes from 0.8mm^{-1} to 3.1mm^{-1} linearly with an error of 5%, and the relation coefficient R of sand size dependent attenuation coefficient equaled 0.9928. The linear fitting formula is $\beta/d = 0.26529 + 0.01529s$, where d is the thickness of the tableting sample. The size of the sand sample was, and the corresponding attenuation index of the unit thickness is 0.9mm to 3.0mm.

An effective frequency range of 0.2~1.8THz was determined for the extinction spectra based on the amplitudes of the THz-frequency-domain spectroscopy (FDS)[14,15]. For the sake of clarity, only the extinction coefficients for the four samples are shown in the Figure 6. As can be seen from Figure 6, the extinction coefficients of all of the sizes occur before 0.2THz. The extinction coefficients are ~0.15. In the frequency range of 0.2THz to 1.0THz, the extinction

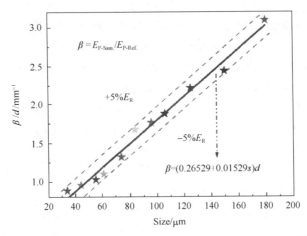

Figure 5 Sand size dependent attenuation coefficient. Error bars represent 5% fluctuation

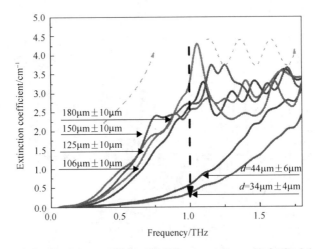

Figure 6 The calculated extinction coefficient of different sand samples in the 0.1~1.8 THz region

coefficient increases with increasing frequency, as shown by the dotted line. In addition, the larger the particle size of the sand, the larger the increase. After about 1.0THz, there was a clear difference in the extinction coefficients between the large particle sample and the small particle sample. The variation law of the extinction coefficient of the smaller particle sample is consistent with that before 1.0THz. The law of the extinction coefficient of the larger particle sample has a strong classical scattering resonance, as shown by the dotted line.

In order to understand the internal structure of the tablet and the composition of sand, SEM was employed to analyze sand samples. The SE technique was used to obtain the surface and cross section morphologies shown in Figure 2. Based on Figure 2, the conclusion that multiple polyethylene particles enclose a single sand particle was obtained by analyzing the elemental composition. It can be concluded that the smaller sand particle samples are densely distributed within the space of the compressed tablets, whether the surface or the cross section was observed. The distribution of the larger sand particle samples is relatively loose. Due to the dense distribution of the smaller sand particle samples on the tableting surface, the smaller sand particle samples are more likely to reflect THz waves than the larger sand particle samples. This results in more THz waves reflected by the smaller sand particle samples, i.e., a slightly higher reflectivity, than the larger sand particle samples. Analysis of the transmission THz spectroscopy experiments shows that the smaller sand particle samples exhibit Rayleigh scattering characteristics at all frequencies. The extinction is mainly dependent of absorption, and the scattering and absorption can be calculated using the following formula ($ka \ll 1$)[16]:

$$\sigma_s = \frac{8\pi}{3} k^4 a^6 |y|^2 \quad (4)$$

where $k = 2\pi/\lambda$, a is the linearity of particles, $y = (\varepsilon_s - \varepsilon)/(\varepsilon_s + 2\varepsilon)$, and ε is the permittivity of the medium, respectively. The larger sand particle sample's linearity is similar to that of a Mie particle. The extinction curve exhibits typical scattering resonance at 1.0THz, and the resonance curve exhibits a clear upward trend overall. This is consistent with the THz spectral characteristics of Mie particles, which are affected by both absorption and scattering with scattering as the dominant factor. This results in the larger particles altering the THz waves more than the smaller particles.

When actually selecting the samples for terahertz detection, small sand particles should be sifted out and used for THz testing in order to make the THz tests more accurate. The particle whose linearity is less than the THz wavelength is the most effective in that the attenuation of the THz wave is mainly due to absorption in the sand sample; however, that is simultaneously decided by absorption and scattering in terms of large particles. Therefore, the conclusions obtained by analyzing the experimental results will be more accurate.

4 Conclusion

In this paper, size effect was studied in THz range for the sand particles detection. Based on the extinction coefficient, it was concluded that smaller particles behave as Rayleigh particles in the THz band, while larger particles behave as Mie particles. The attenuation coefficient increases with increasing the size of the particles. Larger particles exhibit a higher THz reflectivity, a larger attenuation coefficient, and a greater extinction coefficient, which provided an effective means of terahertz detection in the future. With the constant development and improvement of THz-TDS, our results are of great significance for the exploration and development of terahertz detection.

Acknowledgements

We appreciate the National Nature Science Foundation of China (Grant No. 11574401), the Science Foundation of China University of Petroleum, Beijing (Nos. 2462017YJRC029,

2462018BJC005 and yjs2017019) and the Beijing Natural Science Foundation (No. 1184016) for the financial support of this work. We would like to thank LetPub (www.letpub.com) for providing linguistic assistance during the preparation of this manuscript.

References

[1] Lönnstedt O M, Peter E. Environmentally relevant concentrations of microplastic particles influence larval fish ecology. Science, 2016, 352(6290): 1213-1216.

[2] Kumar P, Morawska L, Birmili W, et al. Ultrafine particles in cities. Environment International, 2014, 66: 1-10.

[3] Joye I J, McClements D J. Biopolymer-based nanoparticles and microparticles: Fabrication, characterization, and application. Current Opinion in Colloid & Interface Science, 2014, 19.5: 417-427.

[4] Stone V, Miller M R, Clift M J, et al. Nanomaterials versus ambient ultrafine particles: An opportunity to exchange toxicology knowledge. Environmental Health Perspectives, 2017, 125(10): 106002.

[5] Samara C, Kantiranis N, Kollias P, et al. Spatial and seasonal variations of the chemical, mineralogical and morphological features of quasi-ultrafine particles (PM0.49) at urban sites. Science of the Total Environment, 2016, 553: 392-403.

[6] Stoik C D, Bohn M J, Blackshire J L. Nondestructive evaluation of aircraft composites using transmissive terahertz time domain spectroscopy. Optics Express, 2008, 16(21): 17039-17051.

[7] Baxter J B, Guglietta G W. Terahertz spectroscopy. Analytical chemistry, 2011: 83(12): 4342-4368.

[8] Kokkoniemi J, Lehtomäki J, Juntti M. Measurements on rough surface scattering in terahertz band. Antennas and Propagation (EuCAP) 2016 10th European Conference, 2006: 1-5.

[9] Fan X F, Zheng W T, Singh D J. Light scattering and surface plasmons on small spherical particles. Light: Science & Applications, 2014, 3(6): 179.

[10] Li Y Z, Wu S X, Yu X L, et al. Optimization of pyrolysis efficiency based on optical property of semicoke in terahertz region. Energy, 2017, 126: 202-207.

[11] Zhan H L, Xi J F, Zhao K, et al. A spectral-mathematical strategy for the identification of edible and swill-cooked dirty oils using terahertz spectroscopy. Food Control, 2016, 67: 114-118.

[12] Miao X Y, Zhan HL, Zhao K. Application of THz technology in oil and gas optics. Science China: Physics, Mechanics & Astronomy, 2016, 60(2): 024231.

[13] Zhu J L, Zhan H, Miao X, et al. Terahertz double-exponential model for adsorption of volatile organic compounds in active carbon. Journal of Physics D: Applied Physics, 2017, 50(23): 234103.

[14] Zhan H L, Kun Z, Xiao L Z. Spectral characterization of the key parameters and elements in coal using terahertz spectroscopy. Energy, 2015, 93: 1140-1145.

[15] Zhan H L, Wu S X, Bao R M, et al. Qualitative identification of crude oils from different oil fields using terahertz time-domain spectroscopy. Fuel, 2015, 143: 189-193.

[16] Peiponen K E, Zeitler A, Kuwata-Gonokami M. Terahertz Spectroscopy and Imaging. New York: Springer, 2012.

Surface phase-transition dynamics of ice probed by terahertz time-domain spectroscopy

Honglei Zhan[1,2,3]　Yan Wang[2]　Kun Zhao[1,2]　Xinyang Miao[2]　Jing Zhu[2]
Shijie Hao[3]　Wenzheng Yue[1]　Shangrong Wu[4]

(1. State Key Laboratory of Petroleum Resources and Prospecting, China University of Petroleum, Beijing 102249, China; 2. Beijing Key Laboratory of Optical Detection Technology for Oil and Gas, China University of Petroleum, Beijing 102249, China; 3. Department of Material Science and Engineering, China University of Petroleum, Beijing 102249, China; 4. Institute of Agricultural Resources and Regional Planning, Chinese Academy of Agricultural Sciences, Beijing 100081, China)

Abstract: In this paper, we present an investigation of the solid-liquid phase-transition process of ice based on terahertz time-domain spectroscopy(THz-TDS), which can precisely demonstrate the low-frequency vibrational and rotational modes in gaseous or condensed state materials. The intensity of the transmitted signal is used to characterize the interaction between the terahertz wave and the melting ice. The structure transition lead to the change of physical properties such as dielectric constant with the melting of the ice. When the temperature gradually rises, the H_2O molecules on ice surface overcome the hydrogen bond interaction and start to rotate under the driving of THz field which enhancing the interaction between water molecules and incident terahertz radiation. As H_2O is a polar molecule, a dipolar moment will be induced, which contributes to the dielectric polarization. An analysis of the intensities and dielectric constant of THz-TDS signal peaks reveals the dynamic and continuous melting process on the surface of ice at room temperature.

Keywords: THz time-domain spectroscopy; ice; melting; water layer

1 Introduction

Surface melting greatly affects material properties, and is an important factor in many phenomena such as friction and adhesion. Considerable attention has been focused on the surface melting of ice in particular since Faraday first proposed in 1859 that the extreme slipperiness of ice is caused by a thin film of liquid water formed between an incident force and the ice surface by regelation. Recently, this attention has been motivated by the significant role played by the surface melting of ice in many natural phenomena such as snow and ice disaster and melting glaciers. As a result, the solid-liquid phase transition of water has been the subject of numerous experimental and theoretical investigations [1-6].

Over the past few decades, Raman spectroscopy (RS) and infrared (IR) spectroscopy have been widely employed to characterize hydrogen bond stretching in liquid water and ice. Spectroscopic explorations of the hydrogen bond have been conducted extensively for various applications, but as yet the structural and dynamical properties of liquid water have long been are not fully understood[7-10]. However, with the recent development of generation and detection technologies for terahertz (THz) radiation[11-13]. THz spectroscopy has been well exploited and applied in many fields such as pharmaceuticals exploitation[14], security inspection[15], and material characterization[16,17]. By processing and analyzing the terahertz time-domain waveforms, the refractive index, extinction coefficient and real and imaginary parts of the dielectric function can be easily obtained. Therefore, terahertz radiation can extract valuable information, such as low-frequency vibrational motion and vibration of small gas molecules as well as rotation and twisting motion of small molecules[18-22]. It also has the potential to explore the dynamics of hydrogen bonding systems[18]. In particular, since the introduction of the terahertz spectrum, many researchers have studied water as an important liquid in biological and chemical systems to study the kinetics of water molecules through intermolecular hydrogen bonds. In addition, Takeya et al.[23-25] reveaed the molecular information of ice, tetrahydrofuran (THF) hydrate, propane hydrate, and sulfur hexafluoride (SF6) hydrate in terahertz region. The solute-induced change in the dynamics of water was probed directly by terahertz spectroscopy. THz-TDS was proven to be a very sensitive tool for hydrogen-bonded networks[26]. THz-TDS technology enhances clathrate hydrates in the THz region for several years.

Therefore, THz spectroscopy can be more favorably applied than other techniques, such as Raman spectroscopy and infrared spectroscopy, for probing the dynamics of water molecules. However, to the best of our knowledge, THz spectroscopy has not yet been applied for the analysis of the solid-liquid phase transitions of water. The spectral region ranging from 0.1THz to 3THz can provide valuable information regarding collective vibrational or torsional modes of hydrogen bonds[26]. The relaxation-like dielectric dispersion in the low-frequency portion can be attributed to the rotational motions of the coordinated water molecules in the hydrate crystal, which leads to dielectric polarization. The dielectric spectrum of liquid water in the sub-terahertz frequency region also shows the relaxation characteristic, which is related to the reorientation of water molecules through rotational motion under the driving of electric field[27-29]. Similar with the electric-field-induced reorientation of H_2O molecules in liquid water[30]. In hydrate crystals, the rotation motion will be more difficult due to the interaction between H_2O molecules. In this article, THz-TDS is experimentally applied to characterize the solid-liquid phase transition process of water by analyzing the transmission signal strength and dielectric constant under different melting conditions. We mainly explore the dynamic and continuous melting process on the surface of ice at room temperature.

2 Experimental methods

Ice samples were prepared by packing deionized water in 10cm×15cm×2.5cm transparent polypropylene containers, which have a strong permeability to THz radiation. The filled containers were then placed in on a rack of a conventional freezer at –20℃ for at least 12h. The ice samples were then taken from the freezer and immediately subjected to THz-TDS analysis, during which the ice was allowed to melt at room-temperature (22℃) and 43% humidity.

The experiments employed a conventional transmission-type THz-TDS system. In this system, a Ti-sapphire femtosecond (fs) laser beam with a central wavelength of 800nm, a pulse width of 100fs, and a repetition rate of 80MHz was split into a pump beam and a detection beam. The THz radiation employed in testing were in the range of 0.1~3THz which equivalent to 3.33~99.9cm^{-1}(wavenumbers), or 0.004~0.012eV photon energy. The spot size of the pump beam was 5mm. The detector was composed of ZnTe, and the detection apparatus used standard lock-in technology. All trials were performed under the discussed ambient conditions. We obtained THz-TDS spectra of both samples and references, and the time-domain data were transferred to a computer for further analysis.

3 Results and discussion

In order to investigate the interaction between the ice and the terahertz wave during melting, we measured the ice cube at room temperature 22℃ and collect the time-domain spectra. Figure 1 shows the intensity of the THz-TDS transmission signals of an ice sample obtained at time intervals over a period of several minutes. We note that the amplitude of the transmitted pulse (E_P) reduces, and the time delay (τ) increases as the melting process progresses, which indicates both the dielectric loss and dielectric constant change with the ice melting [31].

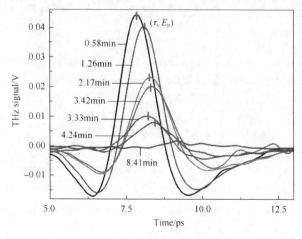

Figure 1 THz-TDS signals of an ice sample obtained at time intervals during the solid-to-liquid phase transition process of water. The time delay (τ) and peak signal amplitude (E_P) of each signal are marked

Figure 2 shows the values of E_P and τ for THz-TDS signals as a function of the measurement time. The peak intensity of the THz signal obtained at 0.58min exhibited a time delay (τ) of about 7.8ps and had an amplitude (E_P) of 44.3mV, whereas the peak signal intensity obtained at 8.41min exhibited a τ of about 9.1ps and an E_P of 1.8mV. The spectral changes are associated with the phase-transition process on the ice surface, and provide information regarding changes in the local environments of the water molecules. The structure transition lead to the change of physical properties such as dielectric constant. However, in this work, the driving force is the temperature and the THz electric field. Pauling pointed in a previous paper that the molecules in crystal may change from the oscillatory state to the rotation state if they have enough kinetic energy to overcome the lattice field potential. At low temperature, the molecules in crystal oscillate around the equilibrium position; as the temperature increases, the kinetic energy increases and binding force weakens; finally, a transition from oscillation to rotation occurs once the temperature reaches the critical value. The critical temperature of a diatomic molecular crystal can be estimated by the following expression:

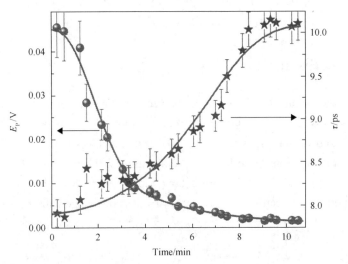

Figure 2 The values of E_P and τ for THz-TDS signals as a function of the measurement time, representing the dynamics of the solid-liquid phase transition process of water

$$T_c = \frac{2V_0}{k}$$

where T_c is the transition temperature, V_0 is a constant obtained from the heat data, which is related to the potential energy of crystal field, and k is the Boltzmann's constant.

Rectangular ice has six heat-absorbing surfaces with dimensions of length, width and thickness. Generally, it is assumed that the thickness of the cuboid is much smaller than the length and width, thus the total surface areas are replaced by the sum of the front and rear surface areas. The melting heat absorption rate of ice is the ratio of heat absorption to time, namely

$$\dot{Q} = \frac{dQ}{d\tau}$$

The amount of convective heat transfer between the outer surface and the environment per unit time is

$$Q_d = hS\Delta T$$

where h is the convective heat transfer coefficient of the ice surface, the unit is W/(m$^2\cdot$K), which is related to the cause of fluid flow, phase change, flow state, geometrical factors of heat transfer surface and physical properties of fluid. ΔT is the average temperature difference between the ambient temperature and the solid surface on the heat exchange surface. The convective heat transferring on the ice surface is completely absorbed by the ice and there is no heat loss. Consequently, the melting heat absorption rate of ice equals the ratio of heat absorption to time. Based on that, the relation between melting thickness and time can be described as

$$\tau = \frac{\rho \beta l}{h(T - 273.15)}$$

where τ is the melting time, l is the melting thickness, ρ represents the density of ice, β is the thermal parameter of melting, and T represents the ambient temperature in units of K.

Here, we note that the absorption of THz radiation by liquid water at low THz wavenumbers is much greater than that of ice which is related to the reorientation of water molecules through rotational motion under the driving of electric field. In hydrate crystals, the rotation motion will be more difficult due to the interaction between H_2O molecules.

As the temperature increases, the outermost water molecules overcome the lattice constraints, and the vibrational states of the water molecules transform from oscillation around equilibrium positions to rotations. And then under the driving of electric field, liquid water molecules on the surface of ice reoriented which leads to dielectric polarization. As a result, the absorption of the terahertz wave by the sample increases and the transmission decreases over time. Even though the liquid water layer at the surface is very thin (30~100μm), the thickness is sufficient to cause visible changes in the transmitted THz signal peak intensity. Thus, we can observe an obvious decrease in the THz signal peak intensity from Figure 2 with the appearance of the liquid water layer, which further decreases significantly as the thickness of the liquid water layer increases. Finally, once the water layer thickness reaches a critical value, the THz signals are nearly completely absorbed by the liquid water layer, and the transmitted signal gradually approaches zero. As such, the roughly sigmoidal shape of the plot of E_P versus measurement time in Figure 2 may be interpreted as a fairly abrupt weakening of the intermolecular hydrogen bonds in the top layers of the ice sample.

To understand the changes in the low frequency vibration mode of hydrogen bonds during the melting process better, we also obtained Raman spectra over the course of the melting process, and the results are presented in Figure 3. The temperature range of sample measurements was from −30℃ to 10℃, and the wavenumber range was from 50cm^{-1} to 4000cm^{-1}. We obtained two obvious hydrogen bond stretching modes in the wavenumber (ω) range of 3100cm^{-1} < ω < 3400cm^{-1} and ω < 300cm^{-1}, which is in agreement with the results of previous studies[32-35]. According to some previous works, the observable two groups of Raman bands between 90cm^{-1} and 4000 cm^{-1} can be assigned to lattice vibration as well as intramolecular vibrations of water molecules. Specifically, the lattice vibration bands are below 200 cm^{-1}. The Raman band below 300 cm^{-1} can be assigned to lattice vibration of ice. The inset of Figure 3 presents the shift in the ω position indicative of the O:H bond (R_{s1}) with respect to temperature in the range from 263.15K to 282.15K. Surprisingly, the shift in R_{s1} is not gradual with increasing temperature, but rather exhibits a rapid decrease from ~211.5cm^{-1} to ~68cm^{-1} at around 270.15K, indicating that the ice melted into liquid water. This change represents a weakening in the hydrogen bonds in the ice, while the interaction between the THz radiation and the water molecules is increased.

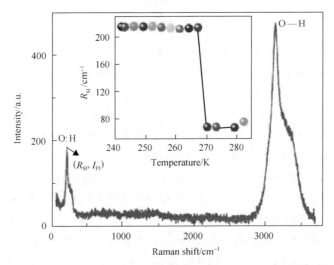

Figure 3 Raman spectrum indicative of the intermolecular van der Waals bond (O:H) and the intramolecular polar-covalent bond (O—H). The inset shows the temperature-dependent shifts in the wavenumber position (R_{s1}) of O:H bonds.

Figure 4 shows real and imaginary part of the dielectric spectra of the ice during the melting process. The thickness at room temperature is used when calculating the dielectric functions, so it will introduce some errors, especially for the real part, due to the expansion or contraction during melting. However, the temperature dependence of dielectric response can be generally observed. As can be seen, both the real part and imaginary part of the dielectric functions increase with the increasing temperature, which is consistent with the observation of the time-domain spectra. Besides, the relaxation like dielectric dispersion in the first two minutes is very weak. Thus, we may conclude that the rotation of H_2O molecules becomes difficult at low temperatures. As the ice begins to melt, the H_2O molecules in crystal overcome the hydrogen bond interaction and start to rotate under the driving of THz field. With the increase of THz frequency, the rotation motion of H_2O cannot catch up with the change of electric field, and therefore, a relaxation-like dielectric dispersion may appear. This molecular rotation mechanism well explains the low-frequency dielectric response shown in the last four minutes.

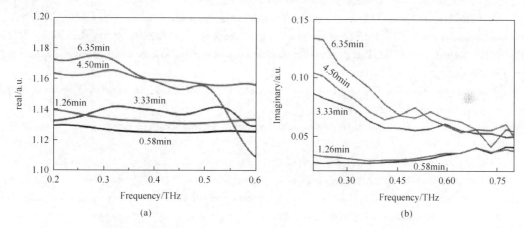

Figure 4　Real and imaginary part of the complex dielectric constants

4　Conclusions

In summary, we conducted an investigation of surface melting in the solid-liquid phase transformation process of water based on an analysis of THz-TDS signals and investigated the dynamic and continuous melting process on the surface of ice at room temperature. A remarkable attenuation in the THz time-domain spectra was observed. This feature was shown to indicate that the transformation of intermolecular hydrogen bonds on the ice surface marks. By analyzing the real and imaginary parts of the complex permittivity, we associated the signal change with the ice surface melting process. This melting took place at a thickness of 30~100μm from the ice surface and the melting rate is not positively correlated with time.

Acknowledgements

This work was supported by the National Nature Science Foundation of China (Grant No. 11574401), the Science Foundation of China University of Petroleum, Beijing (Nos. 2462017YJRC029, 2462018BJC005 and yjs2017019) and the Beijing Natural Science

Foundation (No. 1184016). We would like to thank LetPub (www.letpub.com) for providing linguistic assistance during the preparation of this manuscript. We would like to thank LetPub (www.letpub.com) for providing linguistic assistance during the preparation of this manuscript. Honglei Zhan and Yan Wang contributed equally to this work.

References

[1] Fecht H J. Defect-induced melting and solid-state amorphization. Nature, 1992, 356(6365): 133.

[2] Cahn R W. Melting from within. Nature, 2001, 413(6856): 582-583.

[3] Jin Z H, Gumbsch P H, Lu K H, et al. Melting mechanisms at the limit of superheating. Physical Review Letters, 2001, 87(5): 055703.

[4] Forsblom M, Grimvall G. How superheated crystals melt. Nature Materials, 2005, 4(5): 388-390.

[5] Iglev H, Schmeisser M, Simeonidis K, et al. Ultrafast superheating and melting of bulk ice. Nature, 2006, 439(7073):183.

[6] Lupi L, Hudait A, Peters B, et al. Role of stacking disorder in ice nucleation. Nature, 2017, 551(7679): 218-222.

[7] Butt H J, Döppenschmidt A, Hüttl G, et al. Analysis of plastic deformation in atomic force microscopy: Application to ice. The Journal of Chemical Physics, 2000, 113(3): 1194.

[8] DöPpenschmidt A, Butt, Hans-Jürgen. Measuring the thickness of the liquid-like layer on ice surfaces with atomic force microscopy. Langmuir, 2000, 16(16): 6709-6714.

[9] Dosch H, Lied A, Bilgram J H. Glancing-angle X-ray scattering studies of the premelting of ice surfaces. Surface Science, 1995, 327(1-2): 0-164.

[10] Goertz M P, Zhu X Y, Houston J E. Exploring the liquid-like layer on the ice surface. Langmuir, 2009, 25(12): 6905-6908.

[11] Baxter J B, Guglietta G W, Chem A. Terahertz spectroscopy. Analytical Chemistry, 2011, 83(12): 4342-4368.

[12] Berry C W, Wang N, Hashemi M R, et al. Significant performance enhancement in photoconductive terahertz optoelectronics by incorporating plasmonic contact electrodes. Nature Communications, 2013, 4(3): 1622.

[13] Beard M C, Turner G M, Schmuttenmaer C A. Terahertz spectroscopy. Journal of Physical Chemistry B, 2002, 106(29): 7146-7159.

[14] Liu H B, Zhang X C. Dehydration kinetics of D-glucose monohydrate studied using THz time-domain spectroscopy. Chemical Physics Letters, 2006, 429(1-3): 229-233.

[15] Miles R E, Zhang X C, Eisele H, et al. Terahertz frequency detection and identification of materials and objects. Springer Science & Business Media, 2007.

[16] Tsai T R, Chen S J, Chang C F, et al. Terahertz time-domain spectroscopy technique for characterizing Ytterbium-doped Yttrium Aluminum Garnet Crystals. Optical Society of America, 2006: JWD102.

[17] Ponomareva I, Bellaiche L, Ostapchuk T, et al. Terahertz dielectric response of cubic $BaTiO_3$. Physical Review B, 2008, 77(1): 012102.

[18] Harde H, Zhao J, Wolff M, et al. THz time-domain spectroscopy on ammonia. The Journal of Physical Chemistry A, 2001, 105(25): 6038-6047.

[19] Takeya K, Zhang C, Kawayama I, et al. Terahertz time domain spectroscopy for structure-II gas hydrates. Applied Physics Express, 2009, 2(12).

[20] Fu X, Wu H, Xi X, et al. Molecular rotation–vibration dynamics of low-symmetric hydrate crystal in the terahertz region. Journal of Physical Chemistry A, 2014, 118(2): 333-338.

[21] Kang H, Jung S, Koh D Y, et al. Physicochemical properties of semi-clathrate hydrates as revealed by terahertz time-domain spectroscopy. Chemical Physics Letters, 2013, 587: 14-19.

[22] Tan N Y, Li R, Bräuer P, et al. Probing hydrogen-bonding in binary liquid mixtures with terahertz time-domain spectroscopy: A comparison of Debye and absorption analysis. Physical Chemistry Chemical Physics, 2015, 17(8): 5999-6008.

[23] Masae T. Terahertz vibrations and hydrogen-bonded networks in crystals. Crystals, 2014, 4(2): 74-103.

[24] Takeya K, Zhang C, Kawayama I, et al. Terahertz time domain spectroscopy for structure-II gas hydrates. Applied Physics Express, 2009, 2(12).

[25] Liu H B, Zhong H, Karpowicz N, et al. Terahertz spectroscopy and imaging for defense and security applications. Proceedings of the IEEE, 2007, 95(8): 1514-1527.

[26] Glancy P, Beyermann W P. Dielectric properties of fully hydrated nucleotides in the terahertz frequency range. The Journal of Chemical Physics, 2010, 132(24): 245102.

[27] Fukasawa T, Sato T, Watanabe J, et al. Relation between dielectric and low-frequency Raman spectra of hydrogen-bond liquids. Physical Review Letters, 2005, 95(19): 197802.

[28] Choi D H, Son H, Jung S, et al. Dielectric relaxation change of water upon phase transition of a lipid bilayer probed by terahertz time domain spectroscopy. The Journal of Chemical Physics, 2012, 137(17): 175101.

[29] Ronne C, Thrane L, Astrand P O, et al. Investigation of the temperature dependence of dielectric relaxation in liquid water by THz. Journal of Chemical Physics, 1997, 107(14): 5319-5331.

[30] Cherkasova O, Nazarov M, Shkurinov A. Properties of aqueous solutions in THz frequency range. Journal of Physics: Conference Series. IOP Publishing, 2017, 793(1): 012005.

[31] Bernier M, Garet F, Kato E, et al. Comparative study of material parameter extraction using terahertz time-domain spectroscopy in transmission and in reflection. Journal of Infrared, Millimeter, and Terahertz Waves, 2018, 39(4): 349-366.

[32] Sun C Q, Zhang X, Fu X, et al. Density and phonon-stiffness anomalies of water and ice in the full temperature range. The Journal of Physical Chemistry Letters, 2013, 4(19): 3238-3244.

[33] Saito S, Ohmine I. Fifth-order two-dimensional Raman spectroscopy of liquid water, crystalline ice Ih and amorphous ices: Sensitivity to anharmonic dynamics and local hydrogen bond network structure. The Journal of Chemical Physics, 2006, 125(8): 084506.

[34] Li F, Skinner J L. Infrared and Raman line shapes for ice Ih. I. Dilute HOD in H_2O and D_2O. Journal of Chemical Physics, 2010, 133(24): 1758.

[35] Xue X, He Z Z, Liu J. Detection of water-ice phase transition based on Raman spectrum. Journal of Raman Spectroscopy, 2013, 44(7): 1045-1048.

第二篇　油气储层潜能的光学技术表征评价

Characterizing the rock geological time by terahertz spectrum

Wenxiu Leng[1] Qingyan Li[1] Rima Bao[1] Kun Zhao[1,2]
Xinyang Miao[1] Yizhang Li[1]

(1. Beijing Key Laboratory of Optical Detection Technology for Oil and Gas, China University of Petroleum, Beijing 102249, China; 2. State Key Laboratory of Petroleum Resources and Prospecting, China University of Petroleum, Beijing 102249, China)

In the field of geology, it is necessary to get the information about the rock geological time, which includes absolute geologic age and relative geologic age[1]. Tracing the evolutionary history of the earth is one of the research tasks of earth science. Time, space, material and motivation are the basic elements for studying geological processes and environmental evolution, and time is the basis for studying geological problems. A clear understanding of the interaction between the above elements will be able to clearly make sure the evolution of the earth and the conditions of rock formation. Nowadays, it is urgent to get the geochronological data of rocks for wide range comparison, so as to study the evolution trend and development rule of the earth's crust[2]. At the same time, exploring the geologic age of rocks is also of great significance to exploration and development of mineral deposits. Certain minerals are usually associated with a particular geological body, while the particular geological body is formed during a particular geological period, which requires explicit rock age data. Therefore, it is of great significance to identify the geological age of rocks. At present, the research methods of rock geological dating are method of lithological stratigraphy, paleontology method, paleomagnetic method, isotope method, fission track method, geophysical method, etc.[3]. The lithostratigraphy, palaeontology and paleomagnetism determine the relative geologic age of the rock; the absolute age of the rock is measured by the isotopic method. Isotope method is the most important and most widely used dating method in geologic dating, including stable isotope method, radioisotope method (U-Pb, K-Ar, etc.), cosmic nuclide method (^{10}Be, ^{36}Cl, etc.).

Owing to its unique advantages, the terahertz (THz) technique has received increasing attention in many fields[4-9]. Recently, the THz technique has also been applied in the geological field[10]. Here, we aim to characterize the relative geological time of rock by terahertz time-domain spectroscopy and principal component analysis. The relationship between the rock relative geological time and terahertz optical parameters is also investigated by analyzing the ingredients of the rock samples.

The clastic rock samples used here were collected from Liujiang basin, which located in Qinhuangdao, Hebei province. Figure 1 shows the composition and structure of the clastic rock samples by transmission polarization microscope. The sample details are shown in Table 1. As shown in Table 1, all samples are detritus particles and there are differences in composition and content of different samples. According to Table 1, six kinds of common ingredients were contained in the samples, and quartz, detritus and the argillaceous matrix are the main ingredients in all samples. The two samples have the same relative geologic age and are different from the other five. Before THz-TDS measurement, the rock samples were crushed into powder, then were sifted in a 200-mesh sieve to remove large particles and dried in a drying oven for 24h to remove the effect of moisture in the sample powder. Then they were mixed with polyethylene (PE) powder with rock powder/PE mass ratio of 1.3∶0.5, then pressed by a bead machine at the pressure of 20MPa for 2min. The mixture of polyethylene and rock powder was pressed into tablets with the thickness of approximately 1.750mm and diameter of 30mm.

Table 1 Quantitative analysis of components of the samples

Age /Ma	Geological time	Geological era symbol	Quartz /%	Detritus /%	Argillaceous matrix/%	Plagioclase /%	Glimmer /%	Silicone cement/%	Others /%
800~1000	Qingbaikouan period	Pt_3	96	1		2		1	
510~570	Middle cambrian	ϵ_2	60		15	13	7		5
290~362	Middle carboniferous	C_2	60	24	16				
290~362	Late carboniferous	C_3	48	30	20		2		
250~290	Early permian	P_1	58	32	10				
250~290	Early permian	P_1	70	20	10				
250~290	Late permian	P_2	69	21	10				

Figure 1 Polarized microscope pictures of clastic rocks and the images of the samples after preparation

Initially, the reference THz pulse was obtained after transmitting through the Nitrogen, and then the signals of the seven rock samples were collected. Terahertz time-domain spectra for reference and seven samples are shown in Figure 2(a). Time-domain waveforms of all the samples are variant from each other at the peak amplitude and delay time, indicating that THz waves could discriminate between the physical properties of the samples. Fast Fourier transform (FFT) was used for deriving the THz frequency domain spectra, and THz absorption coefficient spectrum and refractive index spectrum were calculated. Figure 2(b) shows the refractive indices of the samples are recorded over frequency intervals of 0.2~1.2THz, mainly due to the transmitted pulses had an acceptable SNR only such spectral windows. The refractive index of different samples was different from each other, ranging from 1.63 to 1.72. The curves of all the samples are flat in the entire frequency range. The result indicates that the effect of distinguishing rock geology era by the refraction index is not obvious.

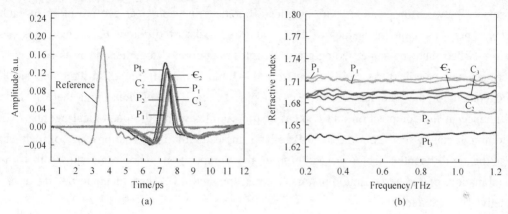

Figure 2 (a) Time dependent THz spectra of reference and the samples; (b) variations of THz refractive index with frequency from 0.2~1.2THz

Due to the complexity of rock composition, principal component analysis (PCA) was carried to establish the predictive models among the parameters obtained in this study. Principal component analysis, which has been successfully employed in THz-TDS to classify the different kinds of materials, is a common statistical method to decrease the dimensions and uses a dimension and uses an orthogonal transformation to convert the input data into a set of linearly uncorrelated principal components (PCs). The number of PCs is less than or equal to the number of the input data and the PC1 which has the highest variance and contribution rate reflects most information of input data[11-14]. In this study, PCA was done with the refractive index spectra of all samples ranging from 0.2THz to 1.2THz as input data. We put the selected data in the excel table. The first column of the table is the terahertz frequency, the second column is the first sample data, and the third column is second sample data, and so on. Then we put the excel table and the already compiled MATLAB program into the same file, and run the program to get the principal component analysis results[15]. Figure 3 is obtained by further optimizing the results. There are seven samples in Figure 3, and we use geological era symbol to represent each sample. For example, Pt_3 represents a sample of Qingbaikouan period in the relative geological age, and 800~1000Ma (million years) means that the sample started from 1 billion years to 800 million years ago. The geological age information of other samples is given in Table 1. There are two samples with the same geological age, both of which were Early Permian (P_1). As shown in Figure 3, PC1 accounts for 99.64% and PC2 accounts for a further 0.16% of the variance between the samples, with the total contribution rate equaling 99.80% in all deviation. Due to its large contribution, PC1 has vital significance for analyzing the relative geological era of rocks. Although the contribution rate of PC2 is not large, it is also very important for the analysis of geological age. We can find the distance between each sample based on the difference between the scores of PC1 and PC2 of each sample. The distances between the samples with large differences in geological age are far away, and the distances between the samples with smaller differences in geological age are close to each other. Then based on the values of PC1 and PC2, rocks of different geological ages are distinguished, and the same geologic age samples are grouped together (such as two Early Permian rock samples). As can be seen from the Table 1, the differences in the composition and content of the samples with large geologic age gaps are obvious, and there is little difference in the composition and content of the samples with little geological age gaps. In addition, it can be seen from the three Permian samples that the physical property of rock can also be an influential factor in addition to the composition. The result shows that the combination of terahertz time-domain spectroscopy and PCA can characterize the rocks relative geochronology.

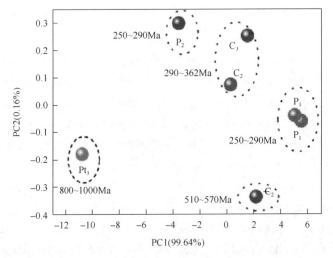

Figure 3　Two dimensional system of PC1 versus PC2 calculated from Figure 2 (b)

We aim to assess the relative geological time of rocks by measuring terahertz optical parameters, so the relationship between the relative geological time and the absorption coefficient is investigated. As shown in Figure 4(a), with the geological time increasing, the absorption coefficient of the samples at 0.7THz and 1.0THz tends to increase first and then decrease. The illustration in Figure 4(a) shows that the absorption coefficient increases from 240Ma to 310Ma. In order to analyze the tendency mentioned above, further study of the THz response to the percentage of each ingredient is essential. Figure 4(b)~(d) show the variations between each ingredient contents and the absorption coefficient. It is obvious that the variation tendency of the absorption coefficient is positively correlated with the detritus content and the argillaceous matrix content but negatively correlated with the quartz content. It could be speculated that detritus and the argillaceous matrix have a stronger absorption of the THz pulse than quartz. The outliers in Figure 4(b)~(d) are marked in different colors. In Figure 4(b) and (c), the outliers are the same sample, and the geologic age of the sample is P_1. P_1 and P_2 have the same composition and content, except for the geological age. It can be seen that there is a direct relationship between geological age and absorption coefficient. Due to the different tectonic movements experienced in different geological periods, the composition and content of rock formed in different geologic periods are different. That is why the absorption coefficient changes.

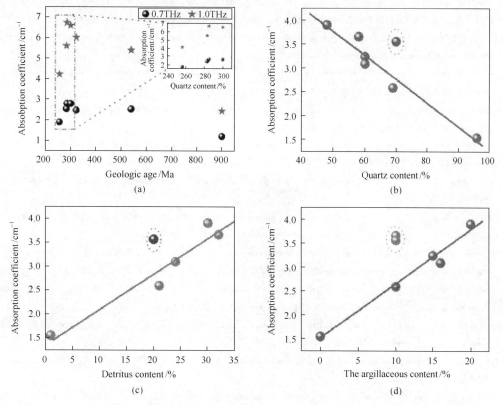

Figure 4 (a) Absorption coefficient of the samples at selected frequencies together with geologic age; (b) absorption coefficient versus quartz content; (c) absorption coefficient versus detritus content; (d) absorption coefficient versus the argillaceous content

In summary, clastic rocks of different geologic age in the same area were studied by using THz time domain spectroscopy. The research focused on reporting that the relative geological age of rock can be characterized by the combination of terahertz time-domain spectroscopy and PCA. Moreover, the terahertz absorption coefficient shows a trend of increasing first and then decreasing with the relative geological age of rocks. The effect of sample composition on the absorption coefficient was investigated with the identification of thin slices. The results prove that THz technology is a promising means for determining the relative geologic age of rocks, and it will be a significant supplementary method in geological survey fields.

Acknowledgements

This work was supported by the National Nature Science Foundation of China (Grant No. 61405259).

References

[1] Yan Q R, Wang Z Q, Yan Z, et al. Detailed dating of deformation/metamorphism of shear zones on the scale of orogen and its application. Earth Science Frontiers, 2001,8: 147.

[2] Guo G H, Han F. The overview of dating methods and the geophysical dating. Progress in Geophysics, 2007, 22(1): 87-94.

[3] Chen X H, Dong S W, Shi J.Brief review and prospect of geochronological development.Global Geology, 2009, 28(03): 384-396.

[4] Zhang Z W, Wang K J, Lei Y, et al. Non-destructive detection of pigments in oil painting by using terahertz tomography. Science China: Physics, Mechanics & Astronomy, 2015, 58(12): 124202.

[5] Bao R M, Li Y Z, Zhan H L, et al. Probing the oil content in oil shale with terahertz spectroscopy. Science China: Physics, Mechanics & Astronomy, 2015, 58(11): 114211.

[6] Wang D C, Huang Q, Qiu C W, et al. Selective excitation of resonances in gammadion metamaterials for terahertz wave manipulation. Science China: Physics, Mechanics & Astronomy, 2015, 58(8): 84201.

[7] Zhan H L, Li Q, Zhao K, et al. Evaluating $PM_{2.5}$ at a construction site using terahertz radiation. IEEE Transactions on Terahertz Science and Technology, 2015, 5(6): 1028-1034.

[8] Li Q, Zhao K, Zhang L W, et al. Probing $PM_{2.5}$ with terahertz wave. Science China: Physics, Mechanics & Astronomy, 2014, 57(12): 2354-2356.

[9] Zhan H L, Sun S N, Zhao K, et al. Less than 6 GHz resolution THz spectroscopy of water vapor. Science China: Technological Sciences, 2015, 58(12): 2104-2109.

[10] Miao X Y, Li H, Bao R M, et al. Discriminating the mineralogical composition in drill cuttings based on absorption spectra in the terahertz range. Applied spectroscopy, 2017, 71(2): 186-193.

[11] Wang X, Hu K X, Zhang L, et al. Characterization and classification of coals and rocks using terahertz time-domain spectroscopy. Journal of Infrared Millimeter & Terahertz Waves, 2016, 38(2): 1-13.

[12] Zhan H L, Wu S X, Bao R M, et al. Qualitative identification of crude oils from different oil fields using terahertz time-domain spectroscopy. Fuel, 2015, 143: 189-193.

[13] Ito H, Kono T. Quantitative analysis of organic additive content in a polymer by ToF-SIMS with PCA. Applied Surface Science, 2008, 255(4): 1044-1047.

[14] Kumar R, Kumar V, Sharma V. Discrimination of various paper types using diffuse reflectance Ultraviolet-Visible Near-Infrared(UV-VIS-NIR) spectroscopy: Forensic application to questioned documents. Applied Spectroscopy, 2015, 69(6): 714-720.

[15] Zhao K,Zhan H L,Terahertz Spectrum Analysis Technology. Beijing: Science Press, 2017.

Characterization of inclusions in evolution of sodium sulfate using terahertz time-domain spectroscopy

Rima Bao[1]　Zhikui Wu[1]　Hao Li[2]　Fang Wang[1]
Xinyang Miao[1]　Chengjing Feng[1]

(1. Beijing Key Laboratory of Optical Detection Technology for Oil and Gas, China University of Petroleum, Beijing 102249, China; 2. Petroleum Exploration & Production Research Institute, China Petroleum & Chemical Corporation (SINOPEC), Beijing 100083, China)

Abstract: The study of fluid inclusion is one of the important means to understand the evolution of mineral crystals, therefore can provide original information of mineral evolution. In the process of evolution, outside factors such as temperature and pressure, directly affect the number and size of inclusions, and thus are related to the properties of crystals. In this paper, terahertz time-domain spectroscopy (THz-TDS) was used to detect sodium sulfate crystals with different growth temperatures, and absorption coefficient spectra of the samples were obtained. It is suggested that the evolution of sodium sulfate could be divided into two stages, and 80℃ was the turning point. X-ray diffraction (XRD) and polarizing microscopy were used to support this conclusion. The research showed that THz-TDS could characterize the evolution of mineral crystals, and it had a unique advantage in terms of crystal evolution.

Keywords: inclusions; crystal growth; sodium sulfate; THz-TDS

1 Introduction

Making an important contribution to the study of the evolution of mineral crystals, researched on and applications of fluid inclusions have progressed tremendously over the last three decades[1,2]. Fluid inclusions are small volumes of ancient fluid trapped in minerals that provide indispensable information about geological processes, from high temperatures at depth towards low temperatures near the earth's surface[3,4]. The formation of inclusions is closely related to the mineral crystallization process, which exists throughout all geological processes. Fluid inclusions reveal geologically important information such as temperature, pressure, salinity, density and depth of trapping; and thereby providing direct information about the conditions at which given minerals and rocks are formed. So the study of the inclusions has great significance for clarifying the evolution of crystals.

The study of fluid inclusions has come a long way since their initial description by Sorby. In recent decades, the research of inclusions has made great progress, and many methods have been applied to the study of inclusions, such as homogenization method, burst method and

---

Originally published in *Analytical Science*, 2017, 33(9): 1077-1080.

quenching method for measuring temperature[5]; freezing method for the determination of salinity; laser Raman probes, chromatography, mass spectrometry, ultraviolet fluorescence spectroscopy for the determination of gas and liquid phase composition[6-8], and so on. These methods play an important role in the study of inclusions, and promote the development of research of the inclusions. However, these methods have their shortcomings, for example, invasion to samples, poor accuracy, poor reproducibility. Looking for new methods to study the inclusions is still important.

As a recently arisen spectroscopic method, THz-TDS uses terahertz wave that ranges from 0.1THz to 10THz, and bridges the gap between infrared and microwave as a light source[9]. Terahertz waves can penetrate many nonpolar materials, but they do not damage the sample because of the lower photon energy[10]. In addition, the waves can induce low frequency bond vibrations, crystal phonon vibration, hydrogen-bonding stretches, and torsion vibrations in materials[11]. In recent decades, because of these unique properties, THz-TDS has attracted more attention and has been employed in various fields, such as biochemistry, safety, medicine quality control, food, petroleum and geological evolution[12-16].

In this paper, THz-TDS was applied to measure sodium sulfate crystals with different growth temperatures. The change of fluid inclusion and its influence on the evolution of sodium sulfate were characterized by terahertz optical parameters, which was verified by XRD and polarizing microscopy. This paper indicated that THz-TDS could be used to characterize the fluid inclusions in the crystal growth process and provided a new method to characterize crystal evolution.

2 Experiment

Crystals of Na_2SO_4 were grown by evaporation crystallization. First, a saturated solution of Na_2SO_4 was prepared at room temperature. After filtering, it was placed in the drying oven that had been set temperature. Waiting for the solution to dry, sodium sulfate crystals were obtained. In order to get different samples, evaporation temperature was set at 20~150℃ at temperature intervals of 10℃. In order to eliminate the effect of the particle size, obtained crystals were ground and sifted, which made particle size within 120μm. Secondly, samples were made by tabletting with 25 MPa pressure for 2 min. This operation did not destroy the structure of the crystal so that it had no effect on the inclusions. Each sample weighed 2 g, and had an average thickness of 1.352mm. Finally, the samples were detected by THz-TDS.

In THz-TDS system, terahertz wave was produced by a femtosecond laser triggered photoconductive antenna. Femtosecond laser with center wavelength of 800nm, repetition frequency of 80MHz, and pulse width of 100fs was produced by titanium sapphire femtosecond mode-locked pulse laser. Laser pulse could be divided into two beams by a beam splitter, a beam as a pump, to motivate GaAs photoconductive antenna generate THz pulse, and the other as a probe light, to measure THz signal by detector. Finally, THz signal was gathered by phase-locked amplifier controlled by the chopper, and transmitted to the computer for processing. In order to reduce the absorption of water vapor in the air to the THz wave, dry air was used in the terahertz optical path, with humidity of less 1% and temperature of 298 K. The dry air in the system was measured by THz-TDS, and the results were taken as reference spectra. The absorption coefficient spectra were calculated automatically according to the samples and the reference spectra by the system.

3 Results and discussion

Growth conditions, such as temperature, pressure, humidity and so on, play a critical role of crystal growth, and are a major cause of different results in crystal growth. This experiment mainly studied the effect of temperature on crystal growth, in the temperature range of 20~150℃. Sodium sulfate crystals obtained by evaporating saturated solution at different temperatures were measured by THz-TDS system to obtain the terahertz time-domain spectra of samples. Figure 1 shows the three-dimensional time-domain waveforms of samples. As shown in Figure 1, there are different terahertz time-domain spectra of sodium sulfate crystals with different temperatures, especially the curve peak. The change of the peak value can be clearly seen from Figure 1. In order to show the phenomenon more clearly, we plotted a graph between square of the THz e-field (P_s) and growth temperatures, as shown in Figure 2. When the temperature rises from 20℃ to 80℃, P_s decrease linearly from $0.0091V^2$ to $0.0004V^2$. However, the change of P_S is various when the temperature continues to rise. As the temperature changes from 80℃ to 150℃, P_s linearly increases from $0.0004 V^2$ to $0.0084 V^2$. The change can be divided into two parts, and 80℃ is noted as a turning point.

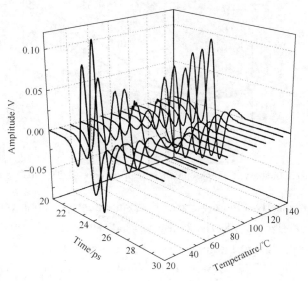

Figure 1 Terahertz three-dimensional time-domain spectrum of samples

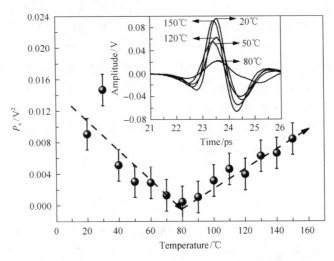

Figure 2 The relationship curve between the square of the THz e-field (P_s) and growth temperatures

The energy loss of the THz waves has a one-to-one correspondence with the absorption coefficient. Figure 3 shows a sample's absorption coefficient spectra within 0.2~1.0THz. There are absorption coefficient curves at growth temperatures of 20℃, 50℃, 80℃, 120℃ and 150℃, respectively. The absorption coefficients increase with the increase of frequency. In addition, samples with different growth temperature have disparate absorption coefficients. When the grown temperature is increased from 20℃ to 80℃, the absorption coefficients increase with grown a rise in temperature. On the contrary, absorption coefficients begin to decline when the grown temperature is greater than 80℃.

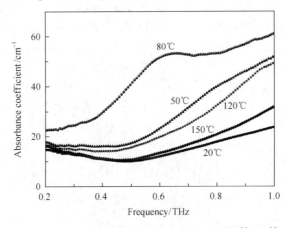

Figure 3 Absorbance coefficients of samples grown in temperatures of 20℃, 50℃, 80℃, 120℃ and 150℃

To exhibit more clearly, Figure 4 shows discontinuous absorption coefficient at 0.4THz, 0.5THz, 0.6THz, 0.7THz and 0.8THz, is used to indicate these objective law. At the same growth temperature, the higher the frequency are, the greater the absorption coefficients are, which can be seen in Figure 4. Within the temperature range of 20~80℃, absorption coefficients increase when the growth temperature increases. But when the temperature is greater than 80℃, it is just the opposite, absorption coefficients decrease with the increase of temperature, which is anastomotic with the P_s. However, there is a special case that absorption coefficient has reached the maximum at 70℃ rather than at 80℃ at 0.8THz. The reason for this phenomenon may be the error, and it also may be because the temperature is not divided carefully enough, leading to deviation of the extremum.

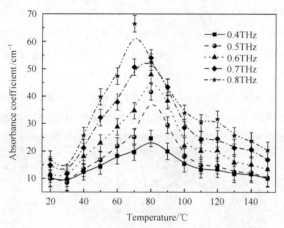

Figure 4　Absorbance coefficients of samples with different growth temperature in 0.4THz, 0.5THz, 0.6THz, 0.7THz and 0.8THz

In Figure 5, the X-ray diffractometer is used in order to explore the causes of this phenomenon, whose scanned range is from 5° to 90°. Five samples whose growth temperatures are respectively 20℃, 50℃, 80℃, 120℃ and 150℃ are selected to draw XRD profiles, and XRD curves are positioned perpendicularly in order to show a clear comparison. The result is that XRD peaks of samples are identical to the characteristic peaks of pure sodium sulfate, which shows that there is no crystal water inside sodium sulfate crystals. In other words, samples are Na_2SO_4 instead of $Na_2SO_4·10H_2O$ and $Na_2SO_4·7H_2O$, and the above phenomenon is not caused by crystal water.

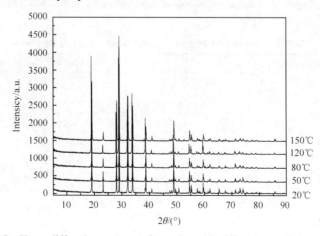

Figure 5　X-ray diffraction patterns of samples with different growth temperatures

The cause of this phenomenon may be fluid inclusions inside crystals. Microscope photos of inclusions inside sodium sulfate crystals are shown in Figure 6. Most of inclusions are gas-liquid inclusions. The number and size of them are disparate because of the different growth temperatures, but the size of them is within 50μm. The number of inclusions is less when the growth temperature is below 80℃, but when the growth temperature is higher than 80℃, it becomes more and more. This is due to growth temperature. As growth temperature increases, the crystallization rate becomes faster, and the possibility of crystal defect increases. So the number of inclusions becomes more and more. According to microscope photos, the number of inclusions' pixels is divided by the total number of pixels to get the proportion of the inclusions' area. It was measured in order to show more accurately differences of inclusions, and the result is shown in Table 1. It can be seen that proportions of inclusions' area

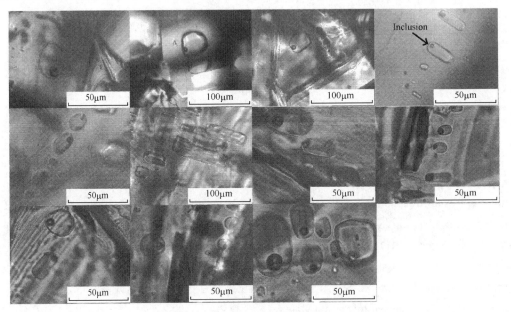

Figure 6 Microscope photos of fluid inclusions inside samples

Table 1 The proportion of inclusions' areas inside samples

Growth temperature/℃	The proportion of inclusions' areas/%
40	0.355
50	0.582
60	0.281
70	1.286
80	1.970
90	2.671
100	3.449
110	4.490
120	1.532
130	2.555
140	13.787

are mainly increased with the increase of growth temperature. When the growth temperature is below 80 ℃, the proportions of inclusions' area are minor, but when the growth temperature is higher than 80 ℃, they become more. So the growth temperature of the crystal has a great influence on inclusions.

Fluid inclusions contain water, and water has a strong absorption for terahertz waves. In order to measure the quantity of water in inclusion, 1g was extracted from every sample and dried for 24h in the drying oven whose temperature was set at 300 ℃, which ensured that inclusions were completely broken and water inside them was completely evaporated. Electronic balance was used to weigh the samples after drying, and the results were shown in Table 2. The quantity of water inside inclusions increased with the rise of growth temperature when the growth temperature was below 80 ℃. But at temperatures higher than 80 ℃, the quantity of water basically remained unchanged, which also indicated the content of the gas inside inclusions grew.

Table 2　Weights of samples before and after drying for 24 h in 300 ℃

Growth temperature/℃	Before drying/g	After drying/g	Change/g
20±0.1	1.0112±0.0001	1.0109±0.0001	0.0003±0.0002
30±0.1	1.0016±0.0001	1.0011±0.0001	0.0005±0.0002
40±0.1	1.0139±0.0001	1.0112±0.0001	0.0027±0.0002
50±0.1	1.0041±0.0001	1.0018±0.0001	0.0023±0.0002
60±0.1	1.0262±0.0001	1.0236±0.0001	0.0026±0.0002
70±0.1	1.0121±0.0001	1.0098±0.0001	0.0023±0.0002
80±0.1	1.0116±0.0001	1.0101±0.0001	0.0015±0.0002
90±0.1	1.0204±0.0001	1.0190±0.0001	0.0014±0.0002
100±0.1	1.0096±0.0001	1.0083±0.0001	0.0013±0.0002
110±0.1	1.0195±0.0001	1.0182±0.0001	0.0013±0.0002
120±0.1	0.9944±0.0001	0.9929±0.0001	0.0015±0.0002
130±0.1	0.9837±0.0001	0.9821±0.0001	0.0016±0.0002
140±0.1	1.0144±0.0001	1.0124±0.0001	0.0020±0.0002
150±0.1	0.9884±0.0001	0.9861±0.0001	0.0023±0.0002

In summary, the number and size of inclusions were small when the growth temperature was below 80 ℃, and the quantity of water inside inclusions increased with the rise of growth temperature. So absorption coefficients of samples had an increasing trend because of the absorption of water for terahertz. On the contrary, with growth temperatures above 80 ℃, the number of inclusions inside samples multiplied, and the proportions of inclusions' area were greater. But the quantity of water inside inclusions basically remained unchanged, which indicated the content of the gas grew with the increase of growth temperatures. When samples were irradiated by THz wave, the path that THz wave went through the gas gets longer.

Therefore, the absorption coefficient showed a trend of decline. This study showed that growth temperature had a larger influence on crystal growth, especially inclusions, and it also illustrated that THz-TDS could accurately and rapidly detect the change of the inclusions inside crystals.

4 Conclusions

Fluid inclusion is one of the important parameters for studying the evolution process of mineral crystal. In this paper, THz-TDS was used to study sodium sulfate crystals with different growth temperatures. The results showed that terahertz absorption coefficients of sodium sulfate crystals exhibited the tendency of increase first and then decrease with the increase of growth temperature, and the turning point was 80℃, which was closely related with fluid inclusions inside sodium sulfate crystals. It also showed that terahertz had a unique advantage in studying crystal growth and geological diagenetic and metallogenic evolution, and it would thus likely see greater applications in this area.

Acknowledgements

This research was supported by the National Natural Science Foundation of China (Grant No. 61405259).

References

[1] Chi G X, Zhou Y M, Lu H Z. An overview on current fluid-inclusion research and applications. Acta Petrologica Sinica, 2003, 19(2): 201-202.

[2] Xuan Z Q. Retrospection & prospection of China sail industry. Geology of Chemical Minerals, 1997, 19(3): 203-206.

[3] Lu H Z, Fan H R, Ni P, et al. Fluid inclusions (in Chinese). Beijing: Science Press, 2004.

[4] Zambito J J, Benison K C. Extremely high temperatures and paleoclimate trends recorded in Permian ephemeral lake halite. Geology, 2013, 41(5): 587-590.

[5] Johnson E L, Hollister L S. Syndeformational fluid trapping in quartz: Determining the pressure-temperature conditions of deformation from fluid inclusions and the formation of pure CO_2 fluid inclusions during grain-boundary migration. Journal of Metamorphic Geology, 2010, 13(2): 239-249.

[6] Fall A, Tattitch B, Bodnar R J. Combined microthermometric and Raman spectroscopic technique to determine the salinity of H_2O-CO_2-NaCl fluid inclusions based on clathrate melting. Geochimica Et Cosmochimica Acta, 2011, 75(4): 951-964.

[7] M. A. Pelch, M. S. Appold, P. Emsbo, et al. Constraints from fluid inclusion compositions on the origin of Mississippi Valley-type mineralization in the Illinois-Kentucky district. Economic Geology, 2015, 110(3): 787-808.

[8] Bourdet J, Burruss R C, Chou I M, et al. Evidence for a palaeo-oil column and alteration of residual oil in a gas-condensate field: Integrated oil inclusion and experimental results. Geochimica Et Cosmochimica Acta, 2014, 142(142): 362-385.

[9] Horiuchi N. Searching for terahertz waves. Nature Photonics, 2010, 4(9): 662.

[10] Nuss M C, Orenstein J. Terahertz time-domain spectroscopy. Berlin, Heidelberg: Springer, 1998: 7-50.

[11] Siegel P H. Terahertz technology in biology and medicine. Microwave Theory & Techniques IEEE Transactions on, 2004, 52(10): 2438-2447.

[12] Zhan H L, Wu S X, Bao R M, et al. Qualitative identification of crude oils from different oil fields using terahertz time-domain spectroscopy. Fuel, 2015, 143: 189-193.

[13] Chen J, Zhao K, Zhao L J, et al. Probing disaggregation of crude oil in a magnetic field with terahertz time-domain spectroscopy. Energy & Fuels, 2014, 28(1): 483-487.

[14] Jin W J, Zhao K, Yang C, et al. Experimental measurements of water content in crude oil emulsions by terahertz time-domain spectroscopy. Applied Geophysics, 2013, 10(4): 506-509.

[15] Zhao H L, Zhao K, Bao R M. Fuel property determination of biodiesel-diesel blends by terahertz spectrum. Journal of Infrared Millimeter & Terahertz Waves, 2012, 33(5): 522-528.

[16] Bao R M, Wu S X, Zhao K, et al. Applying terahertz time-domain spectroscopy to probe the evolution of kerogen in close pyrolysis systems. Science China: Physics, Mechanics & Astronomy, 2013, 56(8): 1603-1605.

CaCO₃, its reaction and carbonate rocks: Terahertz spectroscopy investigation

Honglei Zhan[1,2]   Shixiang Wu[3]   Kun Zhao[1,2]   Rima Bao[2]   Lizhi Xiao[1]

(1. State Key Laboratory of Petroleum Resources and Prospecting, China University of Petroleum, Beijing 102249, China; 2. Beijing Key Laboratory of Optical Detection Technology for Oil and Gas, China University of Petroleum, Beijing 102249, China; 3. Petroleum Exploration and Production Research Institute, China Petroleum and Chemical Corporation, Beijing 100083, China)

Abstract: Carbonate-rich rocks cover a primary part of the earth's petroleum geology reservoir. The study of carbonate has special significance and more effective study methods are now needed. In order to improve the availability of carbonate rock detection, terahertz (THz) spectroscopy was employed to investigate relevant materials in $Na_2CO_3+CaCl_2=CaCO_3+2NaCl$, which is often used to generate $CaCO_3$. By comparing the composited materials with different ions, it can be revealed that Ca^{2+}, CO_3^{2-}, Na^+ and Cl^- have respective absorption features at different frequencies. Furthermore, by utilizing a conservation equation it can be observed that the average refractive indices of Na_2CO_3 as well as $CaCl_2$ equal those of $CaCO_3$ and $NaCl$ in the entire range. Combining the absorption and refractive effect of the materials in the reaction can comprehensively characterize the different substances and reveal the inner interaction during the reaction. THz spectra can deduce the process of molecule rearrangement in the chemical reaction of long-term rock evolution. Besides, the absorption features of the real carbonate rock collected from the nearest town of Sinan county, Guizhou Province in Yunnan-Guizhou platea uvalidate the peaks' central frequencies of ions and the principal components of carbonates, which can be in agreement with the sanning electron microscope (SEM)-X-ray energy dispersive spectrum (EDS) analysis. This research will supply a spectral tool to identify the particles in the rock and deduce an evolution of petroleum carbonate reservoir.

Keywords: carbonate rocks; terahertz spectroscopy; chemical reaction; absorption; refractive

1 Introduction

For hundreds of years, petroleum geology has been of wide concern and in recent times has attracted the increasing attention of scientists. In petroleum geology, carbonate reservoirs are extremely significant in the global oil and gas industry. According to a previous report, carbonate reservoir had about 60% of the worldwide reserves of oil and 40% of the world's gas reserves[1]. In oil-gas reservoir detection, a primary work is to identify the lithology of the rocks. The composites of such rocks include heterogeneous fractured composites caused by great textural variation, which led to complex relationships between physical properties of the rock geophysical data[2, 3]. As a result of the physical properties of reservoir rocks belonged to the solid phase, especially the elemental composite and chemical structures[4]. In the evolution process of geology, a series of physical changes and chemical reactions happened due to long-term natural operation, such as high- and low-temperature switch. The characterization of the materials, which existed in the different stages of evolution, was useful and significant for the investigation of the reservoir rocks and for petroleum exploration[5].

Calcium carbonate ($CaCO_3$) was the primary component of the carbonate rocks in different petroleum geology. Due to the saturation of heavy $CaCO_3$ in nature, carbonate rocks were formed through the precipitation of water[6]. $CaCO_3$ was a kind of inorganic compound; it was neutral and not soluble in water, but soluble in hydrochloric acid. A common method to obtain $CaCO_3$ was to utilize a chemical reaction:

$$Na_2CO_3 + CaCl_2 = CaCO_3 + 2NaCl \tag{1}$$

where Na_2CO_3, $CaCl_2$, $CaCO_3$ and $NaCl$ were general compounds. The elemental and structural detection of these materials would promote the study of the chemical reaction and the simulated petroleum geology evolution. Herein, we introduced terahertz time-domain spectroscopy (THz-TDS), which was a newly developed spectral method. Owing to the development of ultra-short pulse lasers, semiconductors and sensitive detectors, THz technology advanced rapidly and has been widely used in various fields. THz-TDS, which bridged the gap between infrared and microwave, can provide rich information of intermolecular and intramolecular vibration modes, which cause little damage to carbonate, because of its low photon energy and it gives the amplitude and phase information of the sample simultaneously[7-11]. Therefore, THz technology is a promising tool for both qualitative and quantitative analysis for the reactants and products with the high signal-to-noise ratio (SNR). Some of our reports demonstrated that the THz spectrum was very sensitive to the oil-gas resources and pollutants[12-18]. In this research, we undertook an investigation on the reactants and products of a chemical reaction by using the THz-TDS method. The absorption effect revealed the close relationship between the different materials in this reaction. Relative ions had the respective absorption features in THz range. Meanwhile, the refractive index proved a conservation phenomenon where the average refractive index of the reactants equaled those of the products in the whole range. The results indicates that THz spectroscopy would supply a new selection process for geology evolution detection and really promote the detailed description of petroleum carbonate rock reservoir.

2 Experimental methods

The measurement setup is comprised of an atypical THz-TDS system with transmission mode and a femtosecond Ti-sapphire laser (MaiTai) from Spectral-Physics. The laser is diode-pump mode-locked with a center wavelength of 800nm, a pulse width of 100fs, and a repetition rate of 80MHz. The laser power is initially attenuated to make the average power equal ~100mW. The laser beam is then split by a splitter into two beams, which had relatively higher and lower power. The laser with the higher power is used as the pump beam and that with the lower power as the probe beam. The pump beam is used to generate THz radiation with a p-type GaAs with <100> orientation as the emitter. The generated THz pulse with diffusion state is focalized by a hyper-hemispherical len (HHSL1) and reflected by an indium titanium oxide (ITO1) to obtain collimated THz radiation. It will transmit into the detection system after being focalized by lenses (L2 and L3) and reflected by mirrors (M3 and M4). The probe beam comes from the splitter and transmits through the automatic delay stage, which can move to any set position. The probe beam then reaches the detection system, where the probe beam and THz pulse are collinear and pass through hyper spherical lens (HSL); the hybrid beams are focalized onto a 2.8mm thick <110> ZnTe, whose index ellipsoid will be changed by the THz electric field. In this setup, the THz signal is detected and amplified by a lock-in amplifier, which can greatly improve the SNR. A computer was used to control the measurement process and collect the signal data[19].

Sodium carbonate (Na_2CO_3), calcium chloride ($CaCl_2$), calcium carbonate ($CaCO_3$) and sodium chloride (NaCl) were measured, which belonged to the reactants and products of the reaction (1), where the relative molecular mass equaled 105.99, 110.98, 100 and 58.5, respectively. In this experiment, Na_2CO_3 and $CaCl_2$ particles have the purity of 99.9%. We initially weighed 10.599g of Na_2CO_3 and 11.098g of $CaCl_2$, and then mixed each with 100ml of deionized water so that both the Na_2CO_3 and $CaCl_2$ solutions with a content of 1mol/L were obtained. By blending them together, turbid liquid with the $CaCO_3$ precipitation and NaCl solution can be obtained. It was then filtered by two conventional filter papers. $CaCO_3$ was collected on the paper and NaCl solution flowed into a clean glass. To avoid residues of NaCl on the $CaCO_3$, the $CaCO_3$ was washed and filtered with water three times. Herein, the Na_2CO_3 and $CaCl_2$ solutions were re-prepared in a similar way to that mentioned above. By being heated in an oven at ~90℃, the Na_2CO_3, $CaCl_2$ and NaCl solutions would be volatile and crystallize gradually. Then, the solid Na_2CO_3, $CaCl_2$ and NaCl crystals were collected. Finally, the four materials including Na_2CO_3, $CaCl_2$, $CaCO_3$ and NaCl were placed in an oven and heated at ~80℃ for 12h, after which the fine powders of the four materials were obtained for processing. The covered spectral range in the THz region would indeed be broader if thinner pressed discs could be used. For this reason, the size of the Na_2CO_3, $CaCl_2$ and NaCl pressed discs should be smaller than that of $CaCO_3$ because the $CaCO_3$ has the smallest absorption coefficient in the THz range. Such work would confirm similar ranges of effective spectra in the THz region among the four materials. Therefore, we weighed the dry material powders of Na_2CO_3, $CaCl_2$, $CaCO_3$ and NaCl, and the mass equaled 1.2g, 1.2g, 1.5g, and 1.2g, respectively. All the substances with corresponding mass were pressed by a bead machine. The pressure and the time length in the pressing process were 25MPa and 3min, respectively. After being manufactured, they were measured by a Vernier caliper to obtain their diameters and depths. The diameters were all 30mm each, while the thickness equaled 0.92mm, 1.04mm, 1.46mm, and 0.91mm, respectively. In addition, in order to discuss the relation between pure salts and real rock, limestone was collected from the nearest town of Sinan county, Guizhou Province in Yunnan-Guizhou plateau, which has beautiful Karst landform in southwest China. The block of rock was first cut into slices with a thickness of ~3.1mm; the slice was then rubbed down to ensure a smooth surface.

The THz spectra of the samples are obtained by scanning the Na_2CO_3, $CaCl_2$, $CaCO_3$ and NaCl pressed discs one by one, and the reference spectrum is obtained by scanning nitrogen. To minimize the absorption of water vapor and enhance the SNR, the THz beam path was purged with dry nitrogen at room temperature with keeping the humidity below 0.5%. Similarly the THz-TDS of the rock and its reference were obtained several weeks later. The THz parameters, such as refractive index (n), absorption coefficient (α) and the dielectric constant (r: real part, i: imaginary part) reflect the dispersion and absorption characteristics. After fast Fourier transform (FFT) of the reference and samples' THz-TDS, the THz parameters can be calculated by the formulas below[20]:

$$n(\omega) = \varphi(\omega)\frac{c}{\omega d} + 1 \qquad (2)$$

$$\alpha(\omega) = \frac{2k(\omega)\omega}{c} = \frac{2}{d}\ln\left\{\frac{4n(\omega)}{\rho(\omega)[n(\omega)+1]^2}\right\} \qquad (3)$$

where ω is the frequency, c equals the velocity of light in vacuum, d is the thickness of the sample, $k(\omega)$ represents the dispersion index, $\rho(\omega)$ and $\varphi(\omega)$ are the amplitude ratio and phase difference of the reference and sample signal.

3 Results and discussion

A basic investigation was initially performed of the THz dielectric effect of reactants and products of the Reaction (1). By using the THz setup shown in Figure 1, the THz-TDS of four materials was obtained. Figure 2 shows the THz field signal as a function of time after the transmission of the THz pulses through the nitrogen (reference) and Na_2CO_3, $CaCl_2$, $CaCO_3$ as well as the NaCl pressed discs. Combined with the reference spectra, there is an obvious attenuation of the peak intensity in the THz-TDS, indicating an evident absorption of the substances in the THz range. Meanwhile, larger delay time lengths of the four materials can also be observed. The effective optical lengths of THz through the samples are much larger than those through nitrogen. Comparing the samples with each other, Na_2CO_3 had the minimum peak intensity and time delay, followed by $CaCl_2$, NaCl and $CaCO_3$. The reactants and products had different elemental composition and chemical structure, indicating different vibration modes of the molecules. Some of the vibration modes were caused by intermolecular and intramolecular interaction. The vibrations on the ps timescale were detected due to the sensitivity of the THz-TDS.

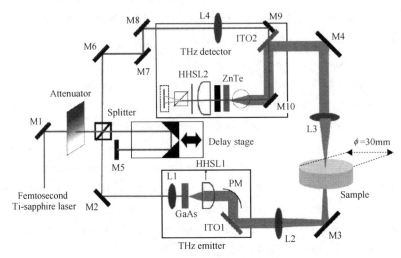

Figure 1　The measurement setup (THz-TDS system)

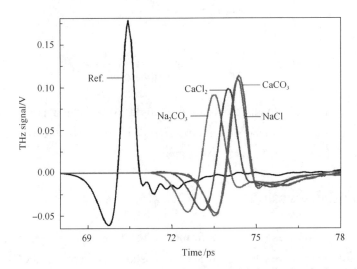

Figure 2　THz-TDS of the reference and four materials in the chemical reaction

According to the measured THz-TDS of the reference and four materials, the absorption coefficient α spectra can be obtained by using Formulas (2) and (3). As shown in Figure 3, the α spectra of Na_2CO_3, $CaCl_2$, $CaCO_3$ as well as NaCl were plotted from 0.2~1.9THz. It is obvious that there was little difference among the four materials in less than 0.5THz. With the increasing of frequency, the α was gradually augmented with different slopes, which can be described as dy/dx; in 0.5~1.3THz range. There were no obvious absorption peaks and the absorption of Na_2CO_3 ranked largest, followed by $CaCl_2$, NaCl and $CaCO_3$, while in the 1.3~1.9THz range, absorption features were observed in all four spectra, comparing the four spectra we can easily identify Na_2CO_3 with the 1.40THz, 1.51THz, 1.60THz, 1.73THz and 1.82THz absorption peaks. Similarly, $CaCl_2$ can be distinguished by the 1.71THz and 1.87THz absorption peaks. Besides, the 1.72THz peak position can be observed in the absorption spectra of $CaCO_3$ tablet. In addition, we can identify NaCl with the 1.62THz and 1.88THz absorption peaks, and with an additional 1.39THz peak, which is not so sharp. Based on the analysis of the central frequencies of the absorption peaks, it can be found that several peak location were very close to each other among Na_2CO_3, $CaCl_2$, $CaCO_3$ as well as NaCl. For instance, ~1.4THz and ~1.6THz belonged to the peak frequencies of both Na_2CO_3 and NaCl; ~1.73THz corresponded to those of Na_2CO_3, $CaCl_2$ and $CaCO_3$; and ~1.87THz belonged to those of $CaCl_2$ and NaCl. Although random noises in the frequency band may have caused some deviations of peak frequencies, several common central frequencies existed in which two materials had similar molecular vibration modes. On the one hand, the four materials can be directly identified according to the respective absorption features in this frequency range, which was in agreement with previous reports[21,22]. On the other hand, the absorption peaks and their central frequencies indicated that the relationship between the reactants and products in this chemical reaction can be reflected in the THz spectroscopy.

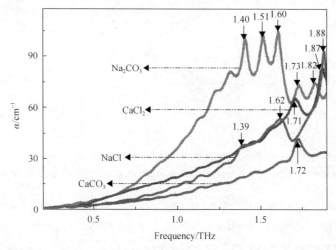

Figure 3　The frequency dependence of absorption coefficient spectra in the frequency range from 0.2THz to 1.9THz

Comparing the THz-TDS of the four materials in Figure 2, we find the time delays were different from each other, indicating variant refractive index among them. According to the THz-TDS and formula (2), the frequency dependence of refractive index n spectra can be obtained and are plotted in Figure 4, where the frequency ranged from 0.2~1.9THz. Four kinds of materials exhibited distinctive n. The refractive effect retained the maximum of ~2.29 for the NaCl compared with the others over the entire range, and subsequently ranked $CaCl_2$ (~2.02), Na_2CO_3 (~1.99) and $CaCO_3$ (~1.81) in decreasing order. The refractive effect of every material remained constant and there was no overlapping phenomenon between the materials. According to Figure 4, the products in the chemical reaction of $Na_2CO_3+CaCl_2=CaCO_3+2NaCl$ had the largest as well as the smallest refractive index, and the reactants had middle values. In order to compare the refractive information before and after the reaction, we calculated the average index of the reactants and products at all frequencies in a selected range. Then, the relative error E_R of refractive index was obtained in the entire range by using the deviation being divided by average n of four materials. The detailed E_R were depicted in the inset of Figure 4 over the entire range. The maximum and minimum E_R equaled ~1.7% and ~5.8%, and the average value was ~2.8% over the entire range. Such small error proved that n was unchanged before and after the chemical reaction. Therefore, the n conservation phenomenon was validated in the THz range, indicating that the reaction information was reflected by the THz spectra.

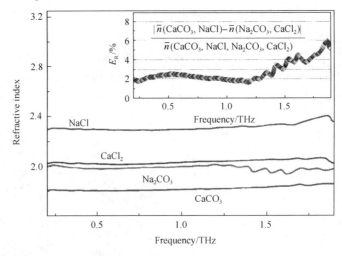

Figure 4 The frequency dependent refractive index spectra from 0.2THz to 1.9THz. The inset reflected the relative refractive error of the deviation after and before the reaction over the whole range

The study suggests that THz spectroscopy is very sensitive to the materials that exist before and after the chemical reaction. Figures 3 and 4 reflect the absorption and refractive response of the reactants and products of reaction (1) in the THz range. Absorption peaks were found among the four materials and there were several peaks of the reactants and products at the same frequencies. The central frequencies were related to the characteristic vibration, which was caused by the atomic compositions and the structure. Chemical reaction referred to a process during which the molecules were broken down and new substances were generated by atomic rearrangement. All the materials discussed were ionic compounds and composed of

Na^+, CO_3^{2-}, Ca^{2+} or Cl^-. By comparing the central frequencies of the absorption peaks and ions in their composition, some vibration information can be revealed among the four materials. The absorption features at 1.40THz and 1.60THz belonged to Na^+ vibration due to the common frequencies of Na_2CO_3 and NaCl. Cl^- had a characteristic absorption peak at 1.87THz because of the common features of $CaCl_2$ and NaCl at this frequency. 1.51THz, 1.63THz and 1.82THz belonged to the central frequencies of CO_3^{2-} vibration and 1.71THz was assigned to Ca^{2+}. It can be observed that $CaCO_3$ had an absorption peak at 1.72THz, which was different from the central frequencies of CO_3^{2-} and Ca^{2+}. Here, $CaCO_3$ had the amorphous state due to the powder precipitation. The waveform around the absorption peak exhibited the lower peak height and wider peak width. The ionic bond between CO_3^{2-} and Ca^{2+} was relatively strong and the properties of $CaCO_3$ were relatively stable. On the one hand, the absorption coefficient of $CaCO_3$ was smaller than the others in the THz range. On the other hand, the absorption features of single Ca^{2+} and CO_3^{2-} ions discussed above would not reflect in the absorption spectra of $CaCO_3$. In terms of refractive index, the four materials reflected an interesting phenomenon. NaCl crystal had the largest refractive index, followed by $CaCl_2$, Na_2CO_3 and $CaCO_3$. The refractive index intensities of the four materials in the THz range were different from those in the visible spectra, proving the strong frequency scattering effect of these substances at different frequencies. When passing the inner medium of the materials, THz radiation would interact with the atomic internal electronic system so that the velocity of propagation would slow down. In spite of the difference between the materials, the average refractive index of the reactants and products were unchanged in the different stages of the chemical reaction. The results indicated that the rearrangement of atoms in the four materials had little influence on the scattering effect of the ions in the THz range.

Figures 2~4 shows the detailed response of pure binary salts of reaction (1) in the THz range. In terms of real geological rocks, they are composed of different compounds and have various structures. Carbonate rock often has the characteristics of crystal superposition and orientation. The correct detection of crystal classifications is of great importance. Herein, a real limestone which belonged to a typical carbonate rock was used and discussed. In order to confirm the morphology and chemical composition of the carbonate rock, we initially used scanning electron microscopy (SEM) and energy dispersive spectroscopy (EDS) methods. Spray-gold pretreatment was used for the electric conduction of the carbonate surface in the SEM and EDS measurement. Figure 5 showed the EDS spectrum of carbonate rock at the surface, which was shown in the SEM image. According to the EDS spectrum, the mass fraction and atom fraction of the chemical elements were measured and listed in the table. The atom fractions (at%) were pressed in order from large to small as follows: O 44.81%, C 18.14%, Ca 12.52%, Mg 11.12%, Si 5.65%, Al 2.25%, K 1.12%, Fe 0.69%, in addition to Au, which was introduced by the spray-gold treatment. The real rock was then measured by the THz-TDS system. As shown in the inset of Figure 6, the THz-TDS peak intensity E_P of rock has an

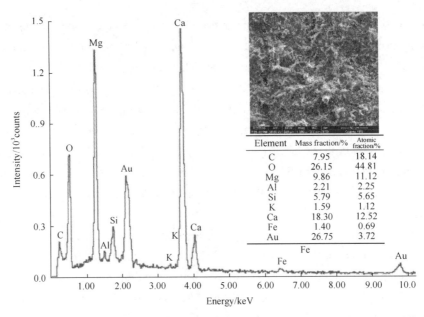

Figure 5　SEM-EDS analysis of limestone collected from the nearest town of Sinan county, Guizhou province in Yunnan-Guizhou plateau

obvious attenuation compared with that of the reference. In order to detect THz response, the absorption coefficient spectra were calculated in the range from 0.2~1.9THz. The real carbonate rock has several significant absorption features, whose central frequencies included 1.53THz, 1.64THz, 1.72THz and 1.82THz, respectively. By comparing Figure 3 and Figure 6, it can be revealed that the principal components of the rock included $CaCO_3$ not only because the absorption peak at 1.72THz was assigned as the vibration of $CaCO_3$, but because the absorption coefficients of rock varied from $2cm^{-1}$ to $30cm^{-1}$, which were close to those of $CaCO_3$ in Figure 3. In addition, according to the discussion above, CO_3^- had characteristic vibration frequencies such as 1.51THz, 1.63THz and 1.82THz. In Figure 6, absorption features exist at 1.53THz, 1.64THz and 1.82THz, which were corresponded with 1.51THz, 1.63THz and 1.82THz mentioned in Figure 3. The EDS results demonstrated that the rock was mainly composed of $CaCO_3$ crystal, other carbonates such as $CaMg(CO_3)_2$ and some clay minerals like silicates mineral with Al and Mg elements; Meanwhile, the absorption peaks in the THz frequency range reflected the principal carbonates in limestone. Due to different ions of ionic compounds in rocks, the absorption features in Figure 5 also validated the description of the peaks' central frequencies of cations and anions in their composition. The results of THz and EDS measurements were in agreement with the actual phenomenon of carbonate rocks[4].

Geological science has attracted the attention of scientists for a long time. Geology mainly refers to the earth's geological structure, material composition and development history. Its evolution was related to sphere differentiation, physical properties, chemical properties, rock properties, mineral composition, rock as well as rock occurrence, contact relationships, the history of the earth's tectonic development, biology evolution, climate change and the

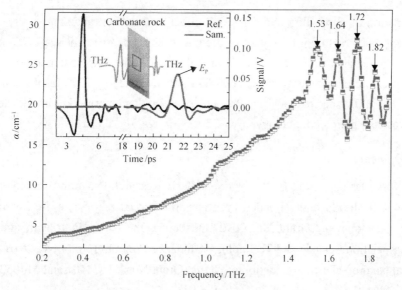

Figure 6   Absorption coefficient α spectra of a typical carbonate rock collected from the nearest town of Sinan county, Guizhou Province in Yunnan-Guizhou plateau, China. The inset picture represents the real carbonate and the measurement

condition and distribution of the occurring mineral resources[23]. The results in this study proved the sensitivity of the THz wave to the heteropolar compound and the ion vibration. The chemical reaction information can be revealed in the THz spectra according to the absorption and refractive index. The common carbonate rocks reflect a sensitive response and significant features in the THz range. According to the ion vibration, the principal components of rocks can be rapidly determined, which is agreement with the analysis of the SEM-EDS. Therefore, the detection of carbonate rocks and their evolution can be realized by THz spectroscopy due to the absorption features and the conservation equation. Such study is worth continuing by combining the spectrum and geology information[24].

4   Conclusions

In summary, the response of THz radiation was discussed regarding the measurement of four materials in the chemical reaction of reaction (1) and a real limestone collected from Yunnan-Guizhou plateau in China. The absorption and refractive effect were investigated separately, based on THz-TDS. In the selected THz frequency range, absorption and refractive spectra reflected different information regarding these materials. The ions, including Ca^{2+}, CO_3^{2-}, Na^+ as well as Cl^-, had respective absorption intensities and features at different frequencies. Such features can be used to identify the materials. Furthermore, they revealed the inner relationship of the elements among the substances, which was useful to analyze the reaction process. In addition, the refractive index spectra validated a common conservation relation that the average index of the reactants equaled those of the products in the whole range.

The combination of absorption and refractive effect in the THz range can really extract the inner element and structure information of the materials in the different stages. By comparing the absorption features of pure materials and limestone, the principal components of real geological rock were characterized by THz-TDS and validated by EDS results. This research suggested that THz spectroscopy would act as a supplementary tool in the detection of carbonate rocks in petroleum geology.

Acknowledgments

This work was supported by the National Basic Research Program of China (Grant No. 2014CB744302), the Specially Funded Program on National Key Scientific Instruments and Equipment Development (Grant No. 2012YQ140005), the China Petroleum and Chemical Industry Association Science and Technology Guidance Program (Grant No. 2016-01-07) and the National Nature Science Foundation of China (Grant Nos. 11574401 and 61405259).

References

[1] Ceia M A R, Misságia R M, Neto I L et al. Relationship between the consolidation parameter, and aspect ratio in microporous carbonate rocks. J. Appl. Geophys., 2015, 122:111-121.

[2] Vanorio T, Scotellaro C, Mavko G. The effect of chemical and physical processes on the acoustic properties of carbonate rocks. Lead. Edge, 2008, 27: 1040-1048.

[3] Silva A A, Neto I A L, Misságia R M, et al. Artificial neural networks to support petrographic classification of carbonate-siliciclastic rocks using well logs and textural information. J. Appl. Geophys., 2015, 117: 118-125.

[4] Nabawy B S. Impacts of the pore-and petro-fabrics on porosity exponent and lithology factor of Archie's equation for carbonate rocks. J. Afr. Earth Sci., 2015, 108: 101-114.

[5] Li D F, Chen H Y, Zhang L, et al. Ore geology and fluid evolution of the giant Caixiashan carbonate-hosted Zn-Pb deposit in the Eastern Tianshan, NW China. Ore Geol. Rev., 2016, 72: 355-372.

[6] Wilford J, De Caritat P, Bui E. Modelling the abundance of soil calcium carbonate across Australia using geochemical survey data and environmental predictors. Geoderma, 2015, 259-260: 81-92.

[7] Horiuchi N, Zhang X C. Searching for terahertz waves. Nat. Photonics, 2010, 4: 662.

[8] Bao R M, Wu S X, Zhao K, et al. Applying terahertz time-domain spectroscopy to probe the evolution of kerogen in close pyrolysis systems. Sci. China: Phys. Mech. Astron., 2013, 56: 1603-1605.

[9] Siegel P H. Terahertz technology in biology and medicine. IEEE Trans. Microw. Techn. 2004, 52(10): 2438-2447.

[10] Reklaitis A. Monte Carlo study of pulsed terahertz emission from multilayer GaAs/AlGaAs structure. J. Phys. D: Appl. Phys., 2013, 46: 145107.

[11] Gu L, Zhou T, Tan Z Y et al. Computed tomography using a terahertz quantum cascade laser and quantum well photo-detector. J. Opt., 2013, 15: 105701.

[12] Feng X, Wu S X, Zhao K, et al. Pattern transitions of oil-water two-phase flow with low water content in rectangular horizontal pipes probed by terahertz spectrum. Opt. Express, 2015, 23: 1693-1699.

[13] Leng W X, Zhan H L, Ge L N, et al. Rapidly determinating the principal components of natural gas distilled from shale with terahertz spectroscopy. Fuel, 2015, 159: 84-88.

[14] Zhan H L, Wu S X, Bao R M, et al. Qualitative identification of crude oils from different oil fields using terahertz time-domain spectroscopy. Fuel, 2015, 143: 189-193.

[15] Ge L N, Zhan H L, Leng W X, et al. Optical characterization of the principal hydrocarbon components in natural gas using terahertz spectroscopy. Energy & Fuels, 2015, 29: 1622-1627.

[16] Zhan H L, Wu S X, Bao R M, et al. Water adsorption dynamics into active carbon probed by terahertz spectroscopy. RSC Adv., 2015, 5: 14389-14392.

[17] Zhan H L, Zhao K, Xiao L Z. Spectral characterization of the key parameters and elements in coal using terahertz spectroscopy. Energy, 2015, 93: 1140-1145.

[18] Zhan H L, Li Q, Zhao K, et al. Evaluating $PM_{2.5}$ at a construction site using terahertz radiation. IEEE T. THz Sci. Techn., 2015, 5: 1-6.

[19] Zhan H L, Sun S N, Zhao K, et al. Less than 6GHz resolution THz spectroscopy of water vapor. Sci China Tech Sci, 2015, 58: 2104-2109.

[20] Dorney T D, Baraniuk R G, Mittleman D M. Material parameter estimation with terahertz time-domain spectroscopy. J. Opt. Soc. Am. A, 2001, 18: 1562-1571.

[21] Mizuno M, Fukunaga K, Hosako I. Terahertz spectroscopy and its applications to dielectric and electric insulating materials. International Symposium on Electrical Insulating Materials, 2008: 613-616.

[22] Rungsawang R, Ueno Y, Ajito K. Detecting a sodium chloride ion pair in ice using terahertz time-domain spectroscopy. Anal. Sci., 2007, 23: 917-920.

[23] Maier A C, Cates N L, Trail D et al. Geology, age and field relations of Hadean zircon-bearing supracrustal rocks from Quad Creek, eastern Beartooth Mountains (Montana and Wyoming, USA). Chem. Geol., 2012, 312-313: 47-57.

[24] Koo T, Jang N, Kogure T, et al. Structural and chemical modification of nontronite associated with microbial Fe (III) reduction: Indicators of "illitization". Chem. Geol., 2014, 377: 87-95.

Spectral characterization of the key parameters and elements in coal using terahertz spectroscopy

Honglei Zhan[1,2] Kun Zhao[2] Lizhi Xiao[1]

(1. State Key Laboratory of Petroleum Resources and Prospecting, China University of Petroleum, Beijing 102249, China; 2. Beijing Key Laboratory of Optical Detection Technology for Oil and Gas, China University of Petroleum, Beijing 102249, China)

Abstract: The augmenting necessity of energy saving and environmental protection has led to the increasing technical requirements about on-line monitoring of key parameters and elements in coal. In this study, terahertz spectroscopy combined with principal components analysis (PCA) was employed to analyze nine kinds of coal materials. Due to the strong absorption of the organics with the relatively high C/H ratio, such as aromatic compounds, coals with lower hydrogen content show higher absorption effects in terahertz range. Based on PC1 score calculated by PCA, the anthracite and bituminous coal can be clearly classified, and the clean as well as meagre coals were also distinguished by opposite trends. All the critical points in PCA system were in agreement with those (volatile matter and ash, respectively) classified in international standard of coal classification. In addition, significant elements, including carbon, hydrogen, nitrogen and sulfur, can be directly characterized using PC1 score with linear and non-linear models. This research indicates that the terahertz spectral analysis of key parameters and elements of coal is a promising tool for improving the detection method and advancing the technical innovation in coal processing industry.

Keywords: terahertz spectroscopy; principal component analysis; coal; volatile matter; ash; elemental characterization

1 Introduction

Coal, always known as the "black gold" on earth, has been one of the most important energy resources since eighteenth century. However, the combustion of coal also leads to a series of environmental problems. Most countries are turning to the optimization of the industrial structure. Technological innovation is the most significant way to improve the benefit of coal processing[1-4]. The monitoring of key parameters and elements is an important part in processing and utilization of coal[5]. One of the key parameters is ash, which is the solid residue after combustion. The larger ash content is, the lower combustion efficiency reflects. Volatile matter is another key parameter to characterize the classification and quality. Coal quality and rank often augment with the dwindling of volatile. Coal with high volatile matter produces unburned carbon particles under combustion and leads to more carbon monoxide, polycyclic aromatic hydrocarbons, aldehydes pollutants et al., especially when the combustion conditions are not appropriate[6,7]. In the process of combustion, carbon and hydrogen are the principal elements to generate heat. Nitrogen hardly generates energy, but transmits to nitrogen oxides and ammonia[8]. Moreover, sulfur leads to the largest harm to the health of animals, plants and human, and corrodes the equipment[9].

A series of researches focused on the dynamics behavior and interaction between the elemental components in gasification, pyrolysis and combustion process[10-13]. Conventional methods such as derivative thermo-gravimetric and designed reactor represented clear features of special composites in pyrolysis. Moreover, some new technique such as computer controlled scanning electron microscopy and numerical simulation also presented advantages on the potential of various inorganic and organic components. Based on the data of dynamics behavior and interaction in pyrolysis or combustion, the key parameters such as volatile and ash can be revealed and the elemental components including carbon, hydrogen, oxygen et al. can be determined for further processing. In terms of direct characterization of the components in coal without pyrolysis, a series of techniques, such as proximate analysis, ultimate analysis, gross calorific value determination, chemical analysis, X-ray diffraction et al., are often required to determine multiple parameters[14].

In the last decades, optical technologies, especially spectral methods, were used to realize the on-line characterization of coal because of actual technical requirements in present coal industry[15-17]. Terahertz (THz) spectroscopy, a developing spectral technique bridging the gap between microwave and infrared spectroscopy, is becoming a hot research because of the unique advantages[18-20]. A series of studies reported the practicability of THz technique to apply in energy field such as petroleum industry[21-25]. Due to the strong absorption effect of water in THz range, water content in coal can be qualitatively and quantitatively determined[26]. In this research, nine kinds of coal materials were studied by THz time domain spectroscopy (THz-TDS). PCA was then employed and PC1 scores were extracted to analyze in detail the relationship between volatile matter as well as ash and THz radiation. In terms of anthracite versus bituminous and clean versus meagre coal, the different trends were reflected in volatile matter and ash dependent PC1 scores, respectively. In addition, based on the PC1 scores, the contents of carbon, hydrogen, nitrogen and sulfur can be determined according to linear and non-linear models.

2 Experimental methods

2.1 THz-TDS setup

The experimental setup used in this research is comprised of a transmission THz-TDS system and a mode-locked femtosecond Ti-sapphire laser. As shown in Figure 1, the femtosecond laser with the central wavelength of 800 nm is initially attenuated by an attenuator to obtain an input beam with an appropriate power ~100mW. The input laser is then split into two beams. The probe beam has the lower power. After transmitting through the delay stage, the probe beam is reflected by a mirror (M3) to reach detector. Another laser with the larger power is the pump beam to generate THz radiation through an emitter composed of photo conductive antenna. The generated THz pulses are then focused onto the sample by optical lens as well as mirror (M5). The sample-information-carried THz beam is focused as a collimated light and then reflected by mirror (M6) to meet the probe beam at the ZnTe crystal. The electro-optic effect (signal) is detected and amplified by a lock-in amplifier. The amplified information will be sent to a computer and processed by a special software program. Our previous reports also mentioned the principal measurement of this setup[19]. Because of the significant absorption features of water vapor in THz range, the setup was covered with dry nitrogen at room temperature (294.1±0.3K). The signal-to-noise ratio equaled ~1500 during the measurement.

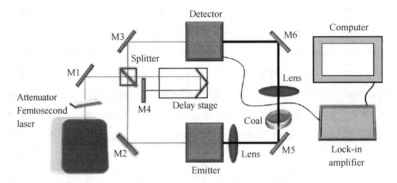

Figure 1　Experimental setup of THz-TDS measurement

2.2　Sample preparation

Nine kinds of reference coal materials were from the National Center of Coal Quality Supervision and Inspection, China. The standard coal materials were manufactured as following: raw coals from coal mine were dried to remove the water; the dry coals were then ground into fine particles with the size of less than 200μm; finally, the manufactured coals were sealed and stored for next processing. The significant properties of coals such as carbon, hydrogen, nitrogen, sulfur, ash and volatile matter content were tested and confirmed in Table 1 by the National Center of Coal Quality Supervision and Inspection. Polytene(PE) powder, which has the little absorption effect in THz range, was employed to blend the coals because of the too large THz absorption effect of coal. The coal materials was mixed by PE with the coal/PE mass ratio equaling 1∶2, and then pressed by a bead machine. The pressure and the time length in the pressing process were 15MPa and 5min, respectively. Finally, nine coal tables with the thickness of 1.5mm and the diameter of 13mm were obtained for THz measurement.

Some detailed tendency of relationship between the properties can be observed according to the given data in Table 1. Figure 2 shows the relationship between volatile matter or ash and the elemental contents. Hydrogen and nitrogen increase monotonically when the volatile matter increases. In Figure 2 (b), ash dependent carbon reflects that carbon content keeps unchanged in the ash content range of 0%~15% and decreases in 15%~30%; on the contrary, sulfur content remains firstly unchanged and then increases.

Table 1 Contents of the physical properties of nine kinds of coal

Number	Mass content/%					
	Carbon	Hydrogen	Nitrogen	Sulfur	Volatile matter	Ash
GBW 11101p	77.03	4.7	1.4	0.51	33.97	8.48
GBW 11101q	79.21	4.21	1.4	0.49	22.19	9.63
GBW 11103h	78.73	2.66	0.98	0.4	9.99	13.16
GBW 11104g	79.77	2.14	0.92	1.19	5.58	13.8
GBW 11109g	56.98	3.91	0.94	2.81	29.43	28.78
GBW 11110g	61.14	2.98	0.96	4.43	18.34	26.31
GBW 11111f	64.08	3.97	1.11	1.77	29.96	20.36
GBW 11113e	64.84	2.87	1.06	3.05	10.96	24.45
GBW 11126a	80	0.95	0.24	0.26	5.82	14.56

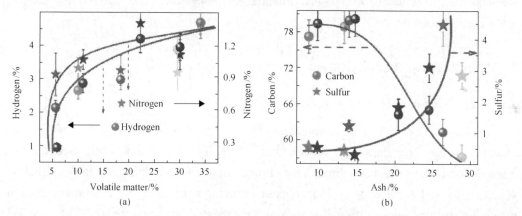

Figure 2 The relationship between volatile matter or ash and elemental content of nine coal materials: (a) volatile matter dependent hydrogen as well as nitrogen content; (b) ash dependence of carbon as well as sulfur

2.3 PCA

PCA is a mathematical statistical method reducing the number of dimensions within the data while retaining as much of the overall variations as possible based on uncorrelated projections. It can be employed to identify the underlying structure of the large data set and to identify groups within the data while at the same time removing any contribution from noise. The calculation with PCA will result in a series of variables which are known as principal components (PCs). PCs remain a set of new maximized variables and are uncorrelated and expressed as linear combination of original variables. These PCs contain information about the samples and spectral variables which are called as scores and loadings, respectively. They can be calculated via eigenvalue decomposition of the following scatter matrix S:

$$S = \sum_m (d_m - \eta)(d_m - \eta)^T \tag{1}$$

where d_m is the m-th input pattern and η is the average value of d_m. If we let D be a diagonal matrix of eigenvalues in a descending order, and E be an orthogonal matrix whose columns are the corresponding eigenvectors, the principal components P_m can be obtained as following:

$$S = EDE^T \tag{2}$$

$$P_m = E^T d_m \tag{3}$$

In terms of PC1, the eigenvector corresponding to the maximum eigenvalue is the direction of the maximum variance distribution of the data. Similarly, the variance of the data points along the direction of the PC2 has the second largest change and this eigenvector is orthogonal with the first. Dimension reduction is achieved by discarding the unimportant elements of d_m. The number of retained principal components would be determined according to the classification results[27-29].

3 Results and discussion

3.1 Statistical analysis of THz spectra

Initially, we perform a basic characterization of the THz dielectric effect of nine coal materials. Figure 3(a) shows the THz field signal amplitude as a function of time after the transmission of the THz pulses through the nitrogen (reference) and nine coal tablets. The THz signal peaks and relative delay time remain different from each other, showing that the physical properties can be differentiated using THz spectra. Based on the THz-TDS, the THz frequency-domain spectroscopy (THz-FDS) can be also calculated. According to ratio of the amplitude spectra of a reference measurement (nitrogen) and a sample in the THz-TDS, frequency dependent absorbance spectra are obtained as shown in Figure 3(b)[30]. Due to the strong absorption, the effective frequency range is reduced to 0.1~1.5THz. Differences can be observed in the absorbance spectra. GuoBiaoWu (GBW) 11126a and 11104g lead all the samples over the whole frequency range. Besides, the deviation of absorption among the other 7 coals is small so that the qualitative identification and determination are difficult to realize. Also, none of absorption features can be observed over the whole frequency range in Figure 3(b). Thus present spectra are not sufficient to directly reveal the relationship between the components and THz spectra.

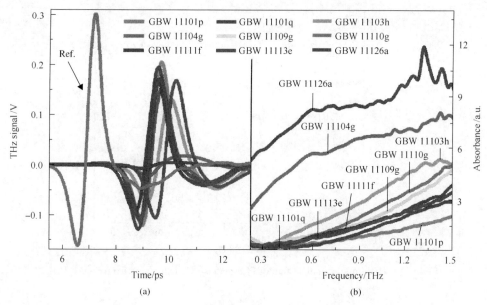

Figure 3 (a) Time dependence of THz signal spectra of reference and nine samples; (b) the frequency dependent absorbance spectra in 0.1~1.5THz

To build a more precise model between coals' parameters and THz spectra, PCA was used with an input of the absorbance spectra from 0.1THz to 1.5THz. In PCA, PC1, whose variance was maximized, reflected most information according to the largest contribution rate which can be used to describe the importance of PCs to the samples. PC2 reflected second most information. By plotting the scores of the early PCs against each other, a two- or three-dimensional score space could be obtained, where samples which were closely related would cluster together and unrelated samples were outliers with each other. As shown in Figure 4, a two-dimensional space indicates the two principal components (PC1 and PC2) and their contribution rates. Concretely, the first two of the data set are found to describe 98.1% and 1.5% of the variance within the data, with the total contribution rate equaling 99.6% in all deviations. For simplicity the samples' number were simplified in next statement and figures (i.e., GBW 11101p—101p). Due to the large contribution, PC1 is of great importance to analyze the components and absorption effect. All the samples can be reordered by the ascending PC1 score as following: 101p, 101q, 113e, 111f, 109g, 110g, 103h, 104g and 126a. Similar to absorbance values at any one frequency, 126a and 104g lead the PC1 score beyond other 7 coals, where 101p is the smallest. Such tendency was actually caused by the inner compositions and structure among the coals, and this phenomenon would be discussed later. In the following discussion, only PC1 will be exacted to characterize coals' features because of its high contribution rate (98.1%). On account of the input data which was selected over the whole range, PC1 can represent the general absorption effect in the selected frequency range. Such selection can reduce the randomness caused by the absorbance only at one frequency, especially when the absorption peaks are not existed.

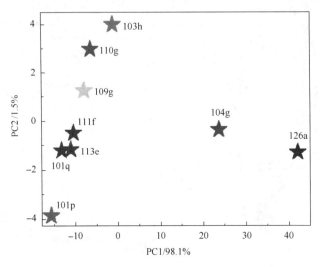

Figure 4 Two dimensional system of PC1 versus PC2 calculated from Figure 3(b)

3.2 Key parameters monitoring

Volatile matter refers to the gas product, such as H_2, C_mH_n, CO, CO_2 et al. escaped from the organics and minerals when heated to a certain temperature under anoxic conditions. It is a main index in coal quality evaluation system and has significant function before the way of processing and the condition of utilization are determined. Anthracite coal, a common kind of coal used in citizens' family, has volatile matter less than 10% according to the classification standard of China coal. Bituminous coal is another famous material with a large volatile matter range from 10% to 37%[31,32]. Differences of absorption effect can be observed between anthracite and bituminous coals in Figure 3 and 4. Figure 5 shows the volatile matter content dependent PC1 score, which represents the general absorption effect in the whole frequency

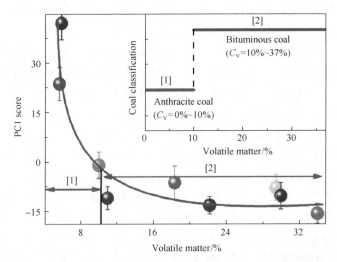

Figure 5 Plot of the PC1 versus volatile matter content of nine coal materials. The inset picture indicates the standard classification of coal based on volatile matter. C_V represents the content of volatile matter in coal

range, of the all the nine coals. A significant tendency can be fitted and observed that the PC1 decreases with the increasing of volatile matter, and the trend is similar to a hyperbola curve, indicating that the gradient is a monotone decreasing function. Moreover, it should be noted that the PC1 scores are larger than 0 among the coals whose volatile contents are less than 10%; on the contrary, negative PC1 scores can be found among the coals which have larger volatile. Therefore, a significant agreement is proved that the anthracite coal and bituminous coal can be qualitatively classified based on the combination of THz spectra and principal components analysis.

The tendency of PC1 score versus volatile in Figure 5 reflects the absorption effects of coals with different coalification in THz range. It is of interest that coal with a special coalification is related to a special hydrogen or nitrogen content from Figure 2(a). In terms of the coal 126a, it has the largest coalification and the lowest volatile matter. H is an element with smallest mass among all the elements; also, always exists in the form of H_2, CH_4 et al., which are pyrolyzed from aromatic and aliphatic hydrocarbons, respectively. Generally, they always have a C/H ratio because of the cycle-type structure; for example, the C/H ratio of benzene (C_6H_6) equals 1∶1. The fewer volatile matter is, the larger percentage of H_2 and CH_4 are, and the bigger ratio of aromatic and aliphatic hydrocarbons are existed in coal. In terms of aromatic compounds, they have strong absorption effects and features in THz range, so the coal with smaller volatile matter has larger PC1 score[33]. Consequently, PC1 score is an effective parameter to the coal classification and THz technique should be a promising selection as a supplementary tool for the detection and classification of anthracite and bituminous coals.

Another parameter of coal classification is the ash content, which is defined as the residue after calcination. Ash content belongs to important index of characterization and utilization research. Generally, mineral matter in coal is not conductive to the production and use; the lower ash content is, the better combustion characteristic coal has. One of the significant features of Chinese coal is the low content of ash and sulfur; most crude coals have a small ash content less than 15%; moreover, the ash contents of coals from some coal fields such as Shenfu and Dongsheng vary from 3% to 5%. So clean coal refers to coal with less than 15% and meagre coal larger than 15% but less than 30%[6]. Figure 6 shows the ash content dependence of PC1 score and the inset picture indicate the normal classification of clean and meagre coal in China. It is of great interest that ~15% of ash content is existed as a cut-off point, left of which indicates that PC1 score increases gradually and right of which PC1 is a decreasing function. In the first stage, the absorption increases due to the augmenting ash content including some carbonates and oxides. On the contrary, the decreasing trend of PC1 score in second stage reflect the decreasing aromatic compounds. This can be proved by the decreasing trend of ash dependent carbon in Figure 2(b). The ash content related to critical point is in good agreement with the actual transition position of clean and meagre coal in Chinese coal industry. Therefore, Figure 6 indicates that THz technique can be a new way to determine and characterize the coal classified by ash.

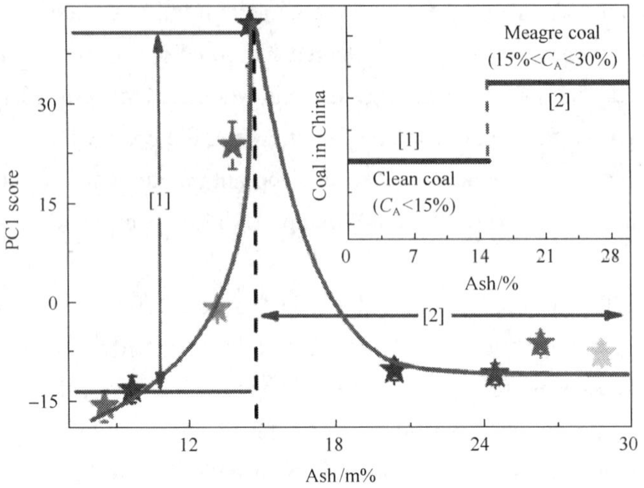

Figure 6 PC1 score versus ash content of nine coal materials. The inset picture reflects the normal classification of clean coal and meagre coal in China. C_A represents the content of ash in coal

3.3 Elemental characterization

Further study of the quantitation about each element mass percentage is important for component monitoring and quality control in coal processing industry. Environmental protection department should exercise strict control over the pollution elements like S, N et al.. In this research, each element was quantitatively determined by extracting the PC1 score of all coals. Different quantitative relationship, including linear and non-linear trend, between elemental contents and PC1 score are depicted in Figure 7, where x-axis indicates the PC1 score and y-axes are the elemental contents. It is of significant interest that every element has relative tendency with the augmenting of PC1 score. Concretely, single linear regression models can be built between PC1 and hydrogen as well as nitrogen contents, which reflect the negative growth in the linear models; non-linear models are existed in carbon and sulfur contents' characterization. However, a given PC1 score, which can be calculated by PCA based on the absorbance spectra, reveals detailed elemental contents, including carbon, hydrogen, nitrogen and sulfur. These models have significant referring value to the elemental analysis and determination in actual coal industry.

Different responses in THz range were observed in absorbance spectra concerning nine kinds of coals including anthracite versus bituminous and clean versus meagre coal. PCA was employed to calculate PC1 score with the highest contribution rate to discuss the absorption effects of coal with the different classification. The results in Figure 5~7 were based on the PC1 scores and the physical properties of coals. PC1 scores were calculated by PCA with the input of absorbance spectra over the whole range. The calculation process referred to Formulas (1)~(3).

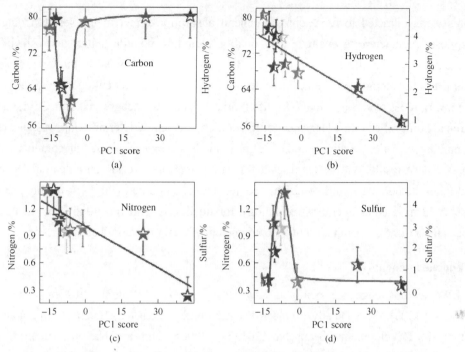

Figure 7 PC1 score dependent (a) carbon, (b) hydrogen, (c) nitrogen and (d) sulfur content of nine coal materials

The coal materials were standard coals and the physical properties were measured by National Center of Coal Quality Supervision and Inspection in China. Therefore, the calculation and the properties are correct, and their combination would result in correct results. In terms of the key parameters in coal, such as volatile matter and ash content, PC1 scores based on THz absorption spectra can distinguish the coals directly. In addition, the linear and nonlinear models can be built between elements and PC1 score. The classification about volatile matter as well as ash and the quantitative models of elements can be used as an important reference to the coal determination in further research. Actually, kinds of coal materials and new mathematical methods can be employed and further investigation can be performed in the future. The more coals, especially the common species, are used to analyze, the more abundant models can be built and the higher accuracy models reflect to predict the unknown coals. Based on the results in this research, more and more coal materials can be measured by THz setup and then combined with PCA to obtain an abundant data base, which could be acted as one of the identification standards to monitor the key parameters and elements.

4 Conclusions

In summary, THz technique was employed in this research to determine nine coal materials. PCA was used with the input of absorbance spectra over the whole range and PC1 scores were extracted to represent the general absorption effect in the selected range. The significant parameters to evaluate coal quality, including volatile matter and ash, indicated different trend with the increase of PC1. The PC1 dependent volatile matter revealed that anthracite and bituminous coals have clear demarcation point equaling zero of PC1 score. Moreover, 15%, which is a mark to differentiate clean and meagre coals in China, acted as extremal point reflected in THz spectra. In addition, elemental determination was also realized by building a series of linear and non-linear models between elemental contents and PC1 scores. These results will be significant for parameters' monitoring and elemental control in coal fields. According to this research, THz technique, combined with PCA method, should be considered as a supplementary technology for qualitative and quantitative characterization, especially on-line determination of key parameters of coal in coal industry.

Acknowledgements

We acknowledge the National Key Basic Research Program of China (Grant No. 2014CB744302), the Specially Founded Program on National Key Scientific Instruments and Equipment Development (Grant No. 2012YQ140005) and the National Natural Science Foundation of China (Grant No. 11574401) for the financial support of this work.

References

[1] Yan L, He B, Pei X, et al. Energy and exergy analyses of a zero emission coal system. Energy, 2013, 55: 1094-1103.

[2] Man Y, Yang S, Zhang J, et al. Conceptual design of coke-oven gas assisted coal to olefins process for high energy efficiency and low CO_2 emission. Appl. Energy, 2014, 133: 197-205.

[3] Prabu V, Geeta K. CO_2 enhanced in-situ oxy-coal gasification based carbon-neutral conventional power generating systems. Energy, 2015, 84: 672-683.

[4] Cormos C C. Economic evaluations of coal-based combustion and gasification power plants with post-combustion CO_2 capture using calcium looping cycle. Energy, 2014, 78: 665-673.

[5] Dai S, Hower J C, Ward C, et al. Elements and phosphorus minerals in the middle Jurassic inertinite-rich coals of the Muli Coalfield on the Tibetan Plateau. Int. J. Coal Geol., 2015, 144: 23-47.

[6] Howaniec N, Smoliń ski A. Influence of fuel blend ash components on steam co-gasification of coal and biomass – Chemometric study. Energy, 2014, 78: 814-825.

[7] Lee B H, Song J H, Kim R G, et al. Simulation of the influence of the coal volatile matter content on fuel NO emissions in a drop-tube furnace. Energy & Fuels, 2010, 24: 4333-4340.

[8] Yao H M, Vuthaluru H B, Tadé M O, et al. Artificial neural network-based prediction of hydrogen content of coal in power station boilers. Fuel, 2005, 84: 1535-1542.

[9] Mesroghli S H, Yperman J, Jorjani E, et al. Changes and removal of different sulfur forms after chemical desulfurization by peroxyacetic acid on microwave treated coals. Fuel, 2015, 154: 59-70.

[10] Wen C, Xu M, Zhou K, et al. The melting potential of various ash components generated from coal combustion: Indicated by the circularity of individual particles using CCSEM technology. Fuel Process. Technol., 2015, 133: 128-136.

[11] Zhang Y, Wang M, Qin Z, et al. Effect of the interactions between volatiles and char on sulfur transformation during brown coal upgrade by pyrolysis. Fuel, 2013, 103: 915-922.

[12] Wijayanta A T, Alam M S, Nakaso K, et al. Numerical investigation on combustion of coal volatiles under various O_2/CO_2 mixtures using a detailed mechanism with soot formation. Fuel, 2012, 93: 670-676.

[13] Frau C, Ferrara F, Orsini A, et al. Characterization of several kinds of coal biomass for pyrolysis and gasification. Fuel, 2015, 152: 138-145.

[14] Chakravarty S, Mohanty A, Banerjee A, et al. Composition, mineral matter characteristics and ash fusion behavior of some Indian coals. Fuel, 2015, 150: 96-101.

[15] Mahoney S A, Rufford T E, Rudolph V, et al. Creation of microchannels in Bowen Basin coals using laser and reactive ion etching. Int. J. Coal Geol., 2015, 144-155: 78-57.

[16] Morga R, Jelonek I, Kruszewska K. Relationship between coking coal quality and its micro-Raman spectral characteristics. Int. J. Coal Geol., 2014, 134-135: 17-23.

[17] Li Z, Fredericks P M, Rintoul L, et al. Application of attenuated total reflectance micro-Fourier transform infrared (ATR-FTIR) spectroscopy to the study of coal macerals: Examples from the Bowen Basin. Australia. Int. J. Coal Geol. 2007, 70: 87-94.

[18] Horiuchi N, Zhang XC. Searching for terahertz waves. Nat. Photonics, 2010, 4: 662-662.

[19] Bao R M, Wu S X, Zhao K, et al. Applying terahertz time-domain spectroscopy to probe the evolution of kerogen in close pyrolysis systems. Sci. China-Phys. Mech. Astron., 2013, 56: 1603-1605.

[20] Siegel P H. Terahertz technology in biology and medicine. IEEE Trans. Microw. Theory Techn., 2004, 52(10): 2438-2447.

[21] Zhan H L, Wu S X, Bao R M, et al. Water adsorption dynamics in active carbon probed by terahertz spectroscopy. RSC Adv., 2015, 5: 14389-14392.

[22] Jin W J, Zhao K, Yang C, et al. Experimental measurements of water content in crude oil emulsions by terahertz time-domain spectroscopy. Appl. Geophys., 2013, 10: 506-509.

[23] Jiang C, Zhao K, Zhao L J, et al. Probing disaggregation of crude oil in magnetic field with terahertz time-domain spectroscopy. Energy & Fuels, 2014, 28: 483-487.

[24] Zhan H L, Wu S X, Bao R M, et al. Qualitative identification of crude oils from different oil fields using terahertz time-domain spectroscopy. Fuel, 2015, 143: 189-193.

[25] Ge L N, Zhan H L, Leng W X, et al. Optical characterization of the principal hydrocarbon components in natural gas using terahertz spectroscopy. Energy & Fuels, 2015, 29: 1622-1627.

[26] Takenori T, Takahiro O, Ikumi K, et al. Estimation of water content in coal using terahertz spectroscopy. Fuel, 2013, 105: 769-770.

[27] Burnett A D, Fan W, Upadhya P C, et al. Broadband terahertz time-domain spectroscopy of drug-of-abuse and the use of principal component analysis. Analyst, 2009, 134: 1658-1668.

[28] Hwang J, Choi N, Park A, et al. Fast and sensitive recognition of various explosive compounds using Raman spectroscopy and principal component analysis. J. Mol. Struct., 2013, 1039: 130-136.

[29] Shao R P, Hu W T, Wang Y Y, et al. The fault feature extraction and classification of gear using principal components analysis and kernel principal component analysis based on the wavelet packet transform. Measurement, 2014, 54: 118-132.

[30] Dorney T D, Baraniuk R G, Mittleman D M. Material parameter estimation with terahertz time-domain spectroscopy. J. Opt. Soc. Am. A, 2001, 18(7): 1562-1571.

[31] Eom S, Ahn S, Rhie Y, et al. Influence of devolatilized gases composition from raw coal fuel in the lab scale DCFC (direct carbon fuel cell) system. Energy, 2014, 74: 734-740.

[32] Hashimoto N, Hiromi S. Numerical simulation of sub-bituminous coal and bituminous coal mixed combustion employing tabulated-devolatilization-process model. Energy, 2014, 71: 399-413.

[33] Cataldo F, Angelini G, García-Hernández D A, et al. Far infrared (terahertz) spectroscopy of a series of polycyclic aromatic hydrocarbons and application to structure interpretation of asphaltenes and related compounds. Spectrochimica Acta Part A: Mol. and Biomol. Spectrosc., 2013, 111: 68-79.

Applying terahertz time-domain spectroscopy to probe the evolution of kerogen in close pyrolysis systems

Rima Bao[1] Shixiang Wu[2] Kun Zhao[1] Lunju Zheng[3] Changhong Xu[1]

(1. College of Science, China University of Petroleum, Beijing 102249, China; 2. Petroleum Exploration and Production Research Institute, China Petroleum and Chemical Corporation, 100083, China; 3. Exploration and Production Research Institute, China Petroleum and Chemical Corporation, Wuxi 214151, China)

Abstract: Terahertz time-domain spectroscopy (THz-TDS) has been used to probe the evolutionary paths of kerogen in selected black mudstone. The evolutionary regime of kerogens (for instance, the immaturity, early maturity, middle maturity, late maturity, and catagenesis stages) can be indicated by the absorption coefficient in the THz region. The present study of identification based on THz-TDS was in good agreement with programmed pyrolysis experiments and suggested that THz technology can act as a nondestructive, contact-free tool for probing the ability to generate hydrocarbons from kerogens.

Keywords: evolution; kerogen; terahertz spectroscopy

Kerogens are complex mixtures of organic components, primarily the result of the decomposition of algae deposited in fine-grained sediments under anaerobic conditions. Kerogens cannot be given a specific chemical formula because the distinctive chemical composition from sample to sample varies[1]. Therefore, kerogens are still not characterized sufficiently to allow scientists to construct accurate fundamental predictive models of thermal cracking to generate petroleum[2]. In sedimentary rock ~95% of the organic matter is composed of kerogen, the key intermediate in the formation of oil and gas. The kerogen types and maturity levels has an important role in determining the characteristics of the products that will be generated from a specific kerogen[3]. The significance for understanding the ability to generate hydrocarbons derives from two primary observations: (i) kerogen can serve as a significant energy source (for instance, oil shales and coals), where recoverable shale oil and coal reserves far exceed the remaining reserves of petroleum, and (ii) kerogens possess significant sorption capacity for organic compounds[4]. Previous research work has revealed that kerogens are primarily composed of alicyclics, aromatics, and other functional groups. The ability to generate oil and gas from kerogens is determined primarily by composition and structure[5]. There are several techniques for studying kerogen structure, which are divided into two major categories: Direct (physical, such as electron microscopy, electron diffraction, X-ray, infrared absorption spectroscopy, and NMR)[6,7] and indirect (chemical, such as oxidative degradation, hydrogenolysis, and pyrolysis) methods[2]. Each of these methods has its advantages and unavoidable deficiencies in providing specific parameters of kerogen. For instance, oxidation degradation can decompose kerogen completely, but the sample cannot be used repeatedly, and methodology is complex and time-consuming. Spectrum methods such as infrared absorption spectroscopy can provide information about the vibration and rotation of different functional groups in kerogen from the molecular vibration generated by interaction between electromagnetic radiation and materials.

THz-TDS is a method of determining the real and imaginary parts of the dielectric function, or the absorption coefficient and refractive index[8,9]. Kerogen in mudstone is non-polar and has low dielectric permittivity values at frequencies greater than 10 MHz[10]. Furthermore, this method has been used in a wide range of research fields, including studies of carrier dynamics in semiconductors[11], low-frequency vibration modes in molecular crystals[12], and low-energy interactions in polar and non-polar liquids[13]. Remarkable efforts have been made recently to employ THz-TDS in the petroleum industry[14-19].

According to modern hydrocarbon generation theory, the hydrocarbon evolution paths of sedimentary organics such as kerogens can be divided into three stages: diagenesis, catagenesis, and metagenesis[3,20]. The reflectance of vitrinite is defined as the proportion of normal incident light reflected by a plane polished surface of vitrinite, which alters according to the level of maturation. Therefore, the vitrinite reflectance (R_o) is typically used to describe the maturity stage of kerogens. In different evolutionary stages, the composition and structure of kerogen alters greatly with the generation of oil and gas. The pyrolysis hydrocarbon evolutionary paths of kerogen in the long geological history can be simulated by increasing the temperature and pressure artificially in the laboratory.

To meet the demands of oil and gas optics, we applied THz-TDS as a nondestructive, contact-free tool for identifying the evolutionary paths and hydrocarbon generation ability of kerogen. The absorption coefficient at different temperatures and pressures indicated the evolutionary regime of the kerogen, which was in good agreement with the results of programmed pyrolysis experiments. We hope to provide further discussions on whether THz technology will establish a lead in petroleum resource and prospecting research.

The THz spectra in this study were measured by the THz-TDS technique[15]. The selected sample used was low-maturity black mudstone (from 1271 m deep well at Eh3 geological age). The sample was crushed and ground to fine powders (20~60 mesh) and then extracted with chloroform for 72h to remove soluble organic molecules. After the samples were dried, they were placed into an airtight reactor to pyrolyze. The initial temperature was set to 275℃, and the final temperature was 500℃. The process was divided into different pyrolysis sections according to the heating temperature. The pyrolyzed products from different temperature ranges were demineralized with HCl-HF treatments to remove most of the carbonates, sulfates, and silicates. Kerogens of various maturity levels were obtained after washing by distilled water and were identified as Ⅱ 1-type kerogen.

As shown in Figure 1, the different maturity levels of the kerogens were identified at different simulated temperatures (T_s): (i) immature (IM) stage, in which kerogen generally cannot generate oil and gas ($R_o<0.5\%$, $T_s<T_1\sim320℃$); (ii) early mature (EM) stage, heavy oil zone ($0.5\%\leqslant R_o<0.7\%$, $T_1\leqslant T_s<T_2\sim345℃$); (iii) middle mature (MM) stage, primaryzone of crude oil generation, also referred to as the oil window ($0.7\%\leqslant R_o<1.2\%$, $T_2\leqslant T_s<T_3\sim370℃$); (iv) late mature (LM) stage, zone of light oil and natural gas ($1.2\%\leqslant R_o<2.0\%$, $T_3\leqslant T_s<T_4\sim395℃$); (v) over-mature (OM) stage, in which methane remains as the only hydrocarbon (dry gas zone)($R_o\geqslant2.0\%$, $T_s\geqslant T_4$).

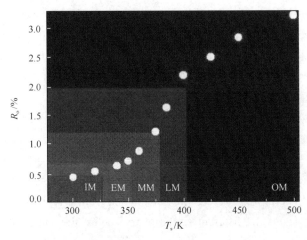

Figure 1 Simulated temperature (T_s) dependence of R_o (vitrinite reflectance)

To avoid the effects of moisture in kerogen during THz measurement, the samples were dried at 105℃ for 4h prior to the experimental procedure. Because polystyrene has low dispersion and absorption in the THz region[21], the sample cells were made of polystyrene, and the optical path in the sample was given at 2mm.

Figure 2 shows the R_o dependence of α at fixed THz frequencies. By comparing the kerogen THz curves under different T_s (or R_o) and the evolutionary characteristics of hydrocarbons, we can conclude that a relationship exists between the kerogen THz optical constants and the hydrocarbon evolutionary stages. The THz optical constant curves at the same frequency can be divided into several sections denoted by IM, EM, MM, LM, and OM. When $T_s<T_1$ in the IM stage ($R_o\leqslant0.5\%$), the kerogens cannot generate any significant amount of oil or gas. Because the functional groups and characteristics do not alter, the THz optical constants exhibit little change. In the main oil generation zone ($0.7\%<R_o<1.2\%$, $T_2<T_s<T_3$), the functional groups of methyl, methylene, aromatic hydrocarbon, oxygen, and nitrogen are separated from kerogen and began to generate oil and gas. Residual kerogen is condensed to macromolecules with aromatic nuclei. With the change in the structures and features from those of kerogen, the first peak values of the THz absorption coefficient and real part of the

relative dielectric permittivity curves appear as the oil-generating peak of kerogen. As the simulated temperature increased, at a more mature evolutionary stage ($R_o \geqslant 1.2\%$, $T_s \geqslant T_3$), alkyls in aromatic groups separated from kerogen and initiate to generate hydrocarbons in the primary gas zone.

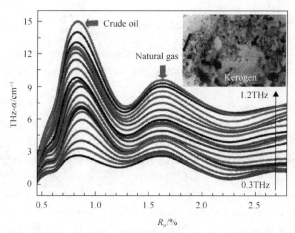

Figure 2　R_o (vitrinite reflectance) dependence of α (absorption coefficients) of kerogen of different maturities at selected frequencies

As the temperature and pressure continued to increase, more bonds were broken, such as those of the esters contained in kerogens. The predominance of molecules with odd carbon number weakened and were absorbed because n-alkanes were diluted by the mature oil. Concurrently, the number of carbons and cylinders in naphthenes and arenes continued to decrease, resulting in a change in the kerogen structure. With increasing temperature, pyrolysis occurred in petroleum and the residual kerogens. The liquid hydrocarbons decreased rapidly, whereas the light hydrocarbons increased, particularly those of C1 to C8. In addition, colloids and bituminous components tapered or disappeared with the increase in alkanes and light hydrocarbons. Therefore, the oil density decreased, and gases were generated under proper conditions.

In conclusion, THz-TDS technology was applied to probe the evolution of kerogen in close pyrolysis systems. According to the absorption coefficient we can obtain the evolutionary regime of kerogen and a set of conditions in which to describe the characteristics of kerogen at different oil and gas generation points. THz technology can be effective as a nondestructive, contactless technique for determining the kerogen evolutionary path and hydrocarbon generation ability in the oil industry.

Acknowledgements

This work was supported by the National Key Basic Research Program of China (Grant No. 2013CB328706), the Specially Funded Program on National Key Scientific Instruments and Equipment Development (Grant No. 2012YQ14005), the Beijing National Science Foundation (Grant No. 4122064), and the Science Foundation of the China University of Petroeum (Beijing) (Grant Nos. QZDX-2010-01 and KYJJ2012-06-27). BAO RiMa and XU ChangHong contributed equally to this work.

References

[1] Scales J A, Batzle M. Millimeter wave analysis of the dielectric properties of oil shale. Appl. Phys. Lett., 2006, 89: 024102.

[2] Peters K E, Fowler M G. Application of petroleum geochemistry to exploration and reservoir management. Org. Geochem., 2002, 33: 5-36.

[3] Tissot B P, Welte D H. Petroleum Formation and Occurrence. New York: Springer-Verlag, 1984: 160-198.

[4] Zhang L, LeBoeuf E J. A molecular dynamics study of natural organic matter. 1: Lignin, kerogen and soot. Org. Geochem., 2009, 40: 1132-1142.

[5] Faulon J L, Vandenbroucke M, Drappier J M, et al. 3D chemical model for geological macromolecules. Org. Geochem., 1990, 16: 981-993.

[6] Tong J H, Han X X, Wang S, et al. Evaluation of structural characteristics of Huadian oil shale kerogen using direct techniques (solid-state 13C NMR, XPS, FT-IR, and XRD). Energy & Fuels, 2011, 25: 4006-4013.

[7] Kelemen S R, Afeworki M, Gorbaty M L, et al. Direct characterization of kerogen by X-ray and solid-state 13C nuclear magnetic resonance methods. Energy & Fuels, 2007, 21: 1548-1561.

[8] Hirakawa Y, Ohno Y, Gondoh T, et al. Nondestructive evaluation of rubber compounds by terahertz time-domain spectroscopy. J. Infrared Millim. Terahertz Waves, 2011, 32: 1457-1463.

[9] Ikeda T, Matsushita A, Tatsuno M, et al. Investigation of inflammable liquids by terahertz spectroscopy. Appl. Phys. Lett., 2005, 87: 034105.

[10] Sweeney J J, Roberts J J, Harben P E. Study of dielectric properties of dry and saturated Green River oil shale. Energy & Fuels, 2007, 21: 2769-2777.

[11] Ulbricht R, Hendry E, Shan J, et al. Carrier dynamics in semiconductors studied with time-resolved terahertz spectroscopy. Rev. Mod. Phys., 2011, 83: 543-586.

[12] Franz M, Fischer B M, Walther M. Probing structure and phase-transitions in molecular crystals by terahertz time-domain spectroscopy. J. Mol. Struct., 2011, 1006: 34-40.

[13] Kindt J T, Schmuttenmaer C A. Far-infrared dielectric properties of polar liquids probed by femtosecond terahertz pulse spectroscopy. J. Phys. Chem., 1996, 100: 10373-10379.

[14] Ikeda T, Matsushita A, Tatsuno M, et al. Investigation of inflammable liquids by terahertz spectroscopy. Appl. Phys. Lett., 2005, 87: 034105.

[15] Al-Douseri F M, Chen Y Q, Zhang X C. THz wave sensing for petroleum industrial applications. Int. J. Infrared Millim. Waves, 2006, 27: 481-503.

[16] Jin Y S, Kim G J, Shon C H. Analysis of petroleum products and their mixtures by using terahertz time domain spectroscopy. J. Korean Chem. Soc., 2008, 53: 1879-1885.

[17] Tian L, Zhou Q L, Jin B, et al. Optical property and spectroscopy studies on the selected lubricating oil in the terahertz range. Sci. China Ser. G-Phys. Mech. Astron., 2009, 39: 1938-1943.

[18] Zhao H, Zhao K, Bao R M. Fuel properties determination of biodiesel-diesel blends by terahertz spectrum. J. Infrared Millim. Terahertz Waves, 2012, 33: 522-528.

[19] Zhao H, Zhao K, Tian L, et al. Spectrum features of commercial derv fuel oils in terahertz region. Sci. China: Phys. Mech. Astron., 2012, 55: 195-198.

[20] Durand B. A history of organic geochemistry. Oil Gas Sci. Technol., 2003, 58: 203-231.

[21] Jin Y S, Kim G J, Jeon S G. Terahertz dielectric properties of polymers. J. Korean Chem. Soc., 2006, 49: 513-517.

Discriminating the mineralogical composition in drill cuttings based on absorption spectra in terahertz range

Xinyang Miao[1,3] Hao Li[2] Rima Bao[3] Chengjing Feng[3] Hang Wu[1,4]
Honglei Zhan[3] Yizhang Li[3] Kun Zhao[1,3]

(1. State Key Laboratory of Petroleum Resources and Prospecting, China University of Petroleum, Beijing 102249, China; 2. Petroleum Exploration & Production Research Institute, China Petroleum & Chemical Corporation(SINOPEC), Beijing 100083, China; 3. Beijing Key Laboratory of Optical Detection Technology for Oil and Gas, China University of Petroleum, Beijing 102249, China; 4. Faculty of Earth Sciences, China University of Petroleum, Beijing 102249, China)

Abstract: Understanding the geological units of a reservoir is essential to the development and management of the resource. In this paper, drill cuttings at several depths from oil field were studied using THz time-domain spectroscopy (THz-TDS). Cluster analysis (CA) and principal component analysis (PCA) were employed to classify and analyze the cuttings. The cuttings were clearly classified based on CA and PCA methods, and the results performed agreement with the lithology. Moreover, calcite and dolomite have stronger absorption of THz pulse than any other minerals, based on analysis of PC1 scores. Quantitative analyses of minor minerals were also realized by building a series of linear and non-linear models between contents and PC2 scores. The results prove THz technology to be a promising means for determining reservoir lithology as well as other properties, which will be a significant supplementary method in oil fields.

Keywords: drill cuttings; terahertz spectroscopy; cluster analysis; principal component analysis

1 Introduction

The economic viability of an oil field is dependent on the quality and accuracy of lithology distribution prediction. Understanding the geological units of a reservoir is essential to the development and management of the resource[1-3]. Recently, the rise in unconventional resource prospecting and the increasing complexity of conventional methods have made accurate lithology prediction more critical[4]. Usually, a lithological characterization is performed by direct measurements of cores. However, the recovery of cores is expensive, and adequate interpretation depends on the experience of a geologist[5,6]. This fact motivates the development of widely used, efficient and low-cost methodologies for the lithological description of wells, such as well-log analysis and cutting logging technique[7-9].

Drill cuttings have been proven to be useful for developing a broad understanding of the rock strata for lithology and many other properties[10-12]. For rapid recognition and classification of minerals in cuttings, the optical properties of minerals such as color and birefringence in the visible region are important. Such properties are widely used in optical mineralogy for the recognition of minerals in order to identify rocks[13]. Nevertheless, recognition and correlation of units from fine-grained cuttings are difficult, particularly when trying to distinguish units with similar lithology. To solve this problem, optical and spectroscopic techniques for element identification are used, such as natural gamma radiation[14], X-ray fluorescence spectral scanning[15], laser-induced breakdown spectroscopy[16,17], as well as Raman spectroscopy[18].

Owing to the unique advantages, terahertz (THz) technique has received increasing attention in many fields[19-24]. Recently, THz technique has been also applied in petroleum and coal industry[25-29]. As a natural material, rocks with different compositions are expected to present different dielectric properties, which can be detected by THz waves. To date, THz spectroscopy has been applied for the investigation of optical properties of marble, sandstone and limestone[30]. The optical constants and dielectric properties of montmorillonite clay[31], micaceous clay[32], kaolinite, halloysite[33] and clay based polyamide nanocomposite films[34] were measured in THz region. Nevertheless, those studies were focused on optical characteristics of rocks and minerals, and the relationship between rock properties and THz parameters had not been analyzed. In our previous study, the particle sizes and porosities of simulated pore structures were probed by THz-TDS[35]. In this paper, authentic specimens were used as research subjects. Drill cuttings at several depths from oil field were tested using THz-TDS. We aim to analyze in detail the relationship between the mineral content and THz radiation. In addition, CA and PCA were employed to classify and analyze the cuttings.

2 Materials and methods

2.1 Characteristics of the cuttings

The drill cuttings used in the experiments were achieved from Liaohe Oil Field in China. Twelve samples were obtained at the depth from 1944m to 2415m. Those samples were decontaminated by means of vibrating screens and hydro cyclones to remove mud and drilling fluid. The oil content of the samples was measured in the oil field using fluorescence spectroscopy. The results showed that there was no hydrocarbon contained in the samples. X-ray diffraction analyses were performed on the samples to determine the mineral compositions. The results were shown in Table 1.

According to Table 1, eight kinds of common minerals were contained in the samples. On the basis of major element analysis, the samples could be divided into four kinds of lithology: Quartzite breccia, clay-bearing quartzite breccia, dolomitic limestone and mudstone. Samples

Table 1 Quantitative analysis of mineral compositions of the samples

Serial number	Identifier	Measured depth/m	Quartz /%	Feldspar /%	Plagioclase /%	Calcite /%	Dolomite /%	Clay /%	Iron pyrite/%	Siderite /%	Lithology
1	S1944	1944	96.9	2.2	—	1	—	—	—	—	Quartzite breccia
2	S1946	1946	97.47	—	—	0.9	—	—	—	—	Quartzite breccia
3	S1954	1954	78.9	—	—	2.3	—	18.9	—	—	Clay - bearing quartzite breccia
4	S1956	1956	94.5	3.2	—	2.3	—	—	—	—	Quartzite breccia
5	S1958	1958	95.4	2.9	—	1.7	—	—	—	—	Quartzite breccia
6	S2160	2160	6.9	—	—	48.2	44.9	—	—	—	Dolomitic limestone
7	S2166	2166	18.2	—	—	52.9	29	—	—	—	Dolomitic limestone
8	S2170	2170	3	0.8	—	48.8	47.3	—	—	—	Dolomitic limestone
9	S2330	2330	42.9	5.2	9.1	5.6	7	28	2.3	3.4	Mudstone
10	S2365	2365	39.2	17.5	27.1	7.1	4.2	—	1.5	8.9	Mudstone
11	S2385	2385	32.5	—	7.7	5.5	9.6	30.8	5	—	Mudstone
12	S2415	2415	24.7	9.2	33.5	2.5	9.4	18.1	2.7	—	Mudstone

with the quartz content of 78.9%~97.47% occur over the interval of 1944~1958m, while clay appears firstly at a depth of 1954m. The main mineral from 2160m to 2170m is carbonate (calcite and dolomite) and accounts for ~90% totally. All the cuttings described above (from 1944m to 2170m) contained fewer mineral species, while the composition over 2330~2415m was more complex. By contrasting the variation of mineral contents with measured depth of the samples, we found that the changing tendency of all the minerals differs from each others. Generally, the content of quartz decreases in the range of 1944~2160m and goes up in 2170~2330m, while the calcite and dolomite show opposite variation trends. The samples contained only small amounts of the other minerals such as clay, feldspar, plagioclase, iron pyrite, and siderite from 1944m to 2170m (except clay at 1954m). As the depth increased (2330~2415m), a non-monotonic change of the element (except quartz and plagioclase) content occurs, probably owing to the growing number of species and larger spacing.

2.2 Terahertz time-domain spectroscopy experiment

The experimental setup used in this research is comprised of a transmission THz-TDS system and a femtosecond Ti-sapphire laser (MaiTai) from Spectral-Physics[24-28]. As shown in Figure 1, a femtosecond laser beam (800nm) was split into two beams. The pump beam was to generate THz radiation through an emitter composed of a photo-conductive antenna, and the probe beam was for electro-optic detection. To prevent vapor absorption in the air and enhance the signal-to-noise ratio (SNR), the setup was covered with dry nitrogen at room temperature.

Before each measurement, the samples were sifted in an 80-mesh sieve to remove larger particles. They were then mixed with polyethylene (PE) powder with a cuttings/PE mass ratio of 1∶2, then pressed by a bead machine at a pressure of 25MPa for 2min.

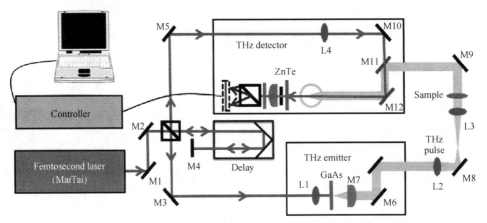

Figure 1　Experimental setup of THz-TDS measurement

3　Results and discussions

3.1　Terahertz spectra of the samples

Terahertz time-domain spectra for reference and twelve samples are shown in Figure 2(a). Time-domain waveforms of all the samples are variant from each other at the peak amplitude and delay time, indicating that THz waves could discriminate between the physical properties of the samples. Fast Fourier transform (FFT) was used for deriving the THz frequency domain spectra and THz absorbance spectra were calculated. Figure 2(b) shows the frequency-dependent absorption spectra of the samples at various depths. The absorbance spectra of the samples were recorded over the frequency intervals of 0.2~1.2THz, mainly because the transmitted pulses had acceptable SNR only in such spectral windows. Generally, the absorbance was observed to be proportional to the frequency. The disparities were not significant at lower frequencies (below 0.3THz); however, different grow rates could be observed with the frequencies increasing. Overlapping phenomenon was also found at several frequencies. In general, the absorbance spectra of S2160, S2166 and S2170 led across most of the frequency range; while those of S1954, S2330, S2365, S2385 and S2415 were distributed in the middle of the range; and S1944, S1946, S1956 and S1958 are at the bottom of the range. The dependence of absorbance at selected frequencies on depth was displayed in Figure 2(c) and the range of absorbance values gradually grows with the increase in frequency. In addition, the absorbance values of most samples at adjacent depth resembled each other, signifying similar optical characteristics across the THz band. Therefore, THz technique was expected to be a tool for the easy detection of cuttings from different depth due to their special responses in THz range. CA and PCA were used to achieve the confirmation of cuttings with different compositions.

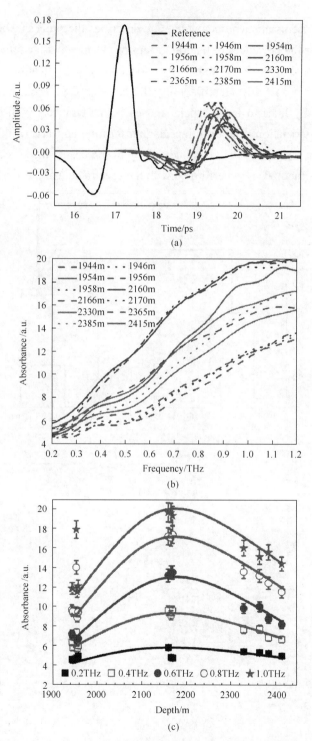

Figure 2 (a) Time dependent THz spectra of reference and the samples; (b) variations of THz absorbance with frequency from 0.2~1.2THz; (c) absorbance of the samples at selected frequencies together with depth, error bars indicating 5% fluctuation are shown for all samples

As commonly used statistical technique, CA describes the process whereby multivariate data are analyzed for the presences of natural groups or clusters that possess certain properties. It provides an objective method to automate the procedure of classification, and quantify group associations that are not readily identifiable by the naked eye. In CA, smallest distances are connected together by lines to build a pair, and each line connects with another at a node, showing a cluster association. The resulting diagram has the appearance of the branches of a tree[36-38]. According to the dispersion and adjacency of the absorbance spectra in Figure 2(b), calculation with CA methods was performed with the spectra data as shown in Figure 3(a).

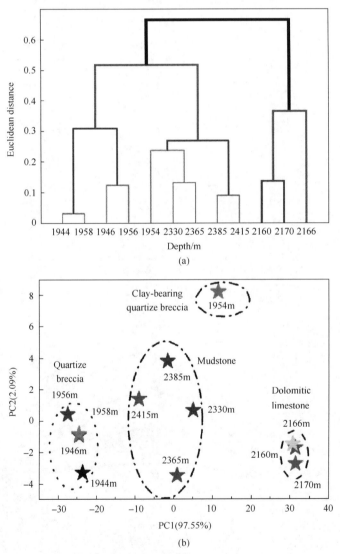

Figure 3 (a) The Euclidean distance dendrogram calculated with absorbance data; (b) two dimensional system of PC1 versus PC2 calculated from Figure 2 (b)

In Figure 3(a), all the samples are clustered into one class with 11 steps, and the samples were divided into three categories: The first kind is The quartz breccia samples S1944, S1946, S1956 and S1958; the second is clay-bearing quartz breccia sample S1954 and mudstone samples S2330, S2365, S2385 and S2415; the third category is dolomitic limestone samples S2160, S2166 and S2170. The results of classification are in agreement with the lithology and the depth distribution except that of S1954. The Euclidean distance of S1944 and S1958 are the least, and several samples are close to each other although not in the same relative position as in the dendrogram. This proximity is owing to the similar constituent of the samples, as shown in Table 1, such as the similar quartz and feldspar contents of S1944 and S1958, as well as the carbonate content (calcite and dolomite) of S2160 and S2170. It is noteworthy that S1954 is grouped with the mudstone samples, due to each of them containing clay. Generally, it is surmised that the samples are grouped by the THz absorbance, which is influenced by 4~5 constituents: the major element of quartz breccia samples is quartz. Dolomitic limestone samples mainly composed of two kinds of carbonate rocks, and clay and plagioclase appeared only in the rest of the samples.

3.2 Statistical analysis of THz spectra

In order to analyze the response of THz waves to different components, PCA was carried out to establish the predictive models among the parameters obtained in this study, and allow a discussion of the absorption effects of cuttings with different classifications. PCA is a classical method that provides a sequence of the best linear approximations to a given high-dimensional observation, which is used abundantly in all forms of analysis as a simple, nonparametric method to extract relevant information from confusing data sets. It starts with the correlation matrix describing the dispersion of the original variables and extracting the eigenvalues and eigenvectors. With minimal additional effort, PCA provides a roadmap on how to reduce a complex data set to a lower dimension[39-41]. In this research, each spectrum was extracted as a single point plotted in PC space, with similar samples were plotted close together. The scores plot for all samples is shown in Figure 3(b). PC1 accounts for 97.55% and PC2 accounts for a further 2.09% of the variance between the samples, with the total contribution rate equaling 99.64% in all deviations. Due to the large contribution, PC1 has vital significance to analyze the components and absorption effect. Based on the value of PC1 and PC2, the scatters were mainly divided into four gatherings. The cluster results performed agreement with the lithology of the samples.

3.3 Mineral characterization

Further study on THz response to the percentage of each ingredient is essential for lithology identification and quantification. Therefore, the variations between mineral content and PC1 score with the identifier are depicted in Figure 4. It was obvious that most of the content had relative tendency with PC1 score. Overall, the variation tendency of calcite and dolomite content roughly conformed to that of PC1 (Figures 4(b) and (c)), whereas that of quartz is just the opposite. It could be speculated that carbonate (including calcite and dolomite) has stronger absorption of THz pulse than any other minerals. In addition, the content of clay and PC1 match well in the samples from 1~5, indicating that clay is the main factor causing the anomalies absorption of THz wave in S1954. However, this trend is not consistent with the mudstone samples, which is probably caused by the complex components of those samples, and the relationship between the mineral content and THz response would be confused. Furthermore, as clay is a mixture of many components, similarities and differences existed at the same time, and divergence was primarily displayed by the PC1 scores. Likewise, PC2 values of the samples were plotted with the minor mineral content in order to figure out the effects of those elements (Figure 5). A noticeable tendency can be observed in Figure 5 by connecting the content scatters into lines. For the PC2 scores of the last four samples, the changing trend is in contrast with the varying content of plagioclase and feldspar, and in accordance with that of clay and iron pyrite. A different quantitative relationship between mineral contents and PC1/PC2 scores were depicted in Figure 6, where x-axis indicated the elemental contents and y-axes was the PC1/PC2 score.

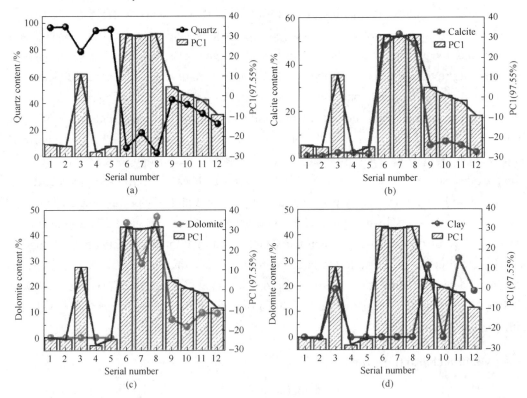

Figure 4   The content of minerals ((a) quartz, (b) calcite, (c) dolomite and (d) clay) and PC1 score of the samples numbered 1~12

第二篇 油气储层潜能的光学技术表征评价 ·103·

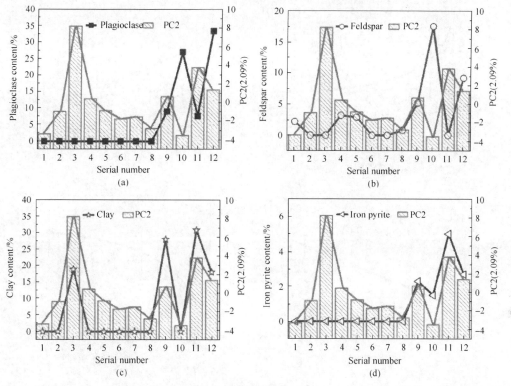

Figure 5 The content of minerals ((a) plagioclase, (b) feldspar, (c) clay and (d) iron pyrite) and PC2 score of the samples numbered 1~12

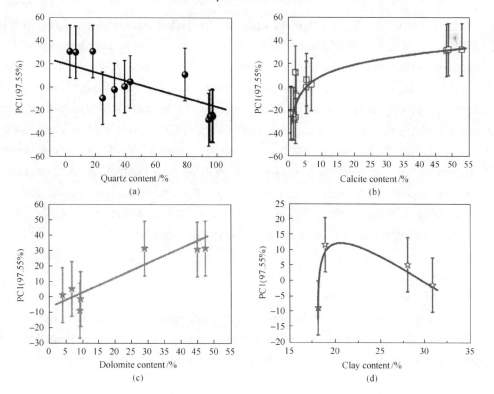

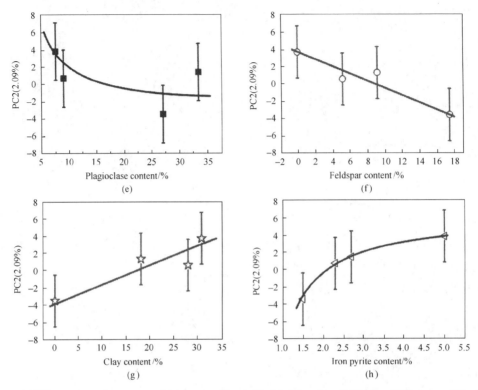

Figure 6. PC1 scores vs. the content of (a) quartz, (b) calcite, (c) dolomite and (d) clay; PC2 scores of the last four samples vs. the content of (e) plagioclase, (f) feldspar, (g) clay and (h) iron pyrite

It is clear from the figures that the content of selected minerals correlates with the PC1/PC2 scores. To be specific, single linear regression models can be built between PC1 and the content of quartz and dolomite, as well as PC2 and the content of feldspar and clay. Here, PC1 is proportionate to dolomite content and inversely related to that of quartz, while PC2 is in proportion to clay content and inversely proportional to that of feldspar. Meanwhile, non-linear models exist for calcite and clay contents with PC1, as well as plagioclase and iron pyrite contents with PC2. The PC1/PC2 scores vary monotonically with the element contents in those models with the exception of clay with PC1. Mineral content can be estimated by PC1/PC2 score for quartz, calcite dolomite, plagioclase, feldspar, iron pyrite, and clay. Therefore, THz technology with CA and PCA has a significant reference value for mineral analysis and lithological identification in geological exploration.

4 Conclusion

In summary, in this research drill cuttings at several depths from an oil field were studied using THz time domain spectroscopy. Cluster analysis and principal component analysis were employed to classify and analyze the cuttings. The cuttings were clearly classified based on

CA and PCA methods, and the results were in agreement with the lithology. Furthermore, calcite and dolomite have stronger absorption of a THz pulse than the other minerals in the samples, based on analysis of PC1. Quantitative analyses of minerals were also realized by building a series of linear and non-linear models between the results and the PC1/PC2 scores. The results prove that THz technology is a promising means for determining reservoir lithology as well as other properties, and will be a significant supplementary method in oil fields.

Acknowledgements

This work was supported by the National Basic Research Program of China (Grant No. 2014CB744302), the Specially Funded Program on National Key Scientific Instruments and Equipment Development (Grant No. 2012YQ140005) and the National Natural Science Foundation of China (Grant Nos. 61405259 and 11574401).

References

[1] Shorttle O, Maclennan J, Lambart S. Quantifying lithological variability in the mantle. Earth. Planet. Sc. Lett., 2014, 395: 24-40.

[2] Hartmann J, Moosdorf N, Lauerwald R, et al. Global chemical weathering and associated P-release—The role of lithology, temperature and soil properties. Chem. Geol., 2014, 363: 145-163.

[3] Herzberg C. Identification of source lithology in the Hawaiian and Canary Islands: Implications for origins. J. Petrol., 2011, 52(1): 113-146.

[4] Grana D, Schlanser K, Campbell-Stone E. Petroelastic and geomechanical classification of lithologic facies in the Marcellus Shale. Interpretation, 2014, 3(1): SA51-SA63.

[5] Ross P S, Bourke A, Fresia B. Improving lithological discrimination in exploration drill-cores using portable X-ray fluorescence measurements: (1) Testing three Olympus Innov-X analysers on unprepared cores. Geochem. Explor. Env. A., 2014, 2012: 163.

[6] Bradbury K K, Evans J P, Chester J S, et al. Lithology and internal structure of the San Andreas fault at depth based on characterization of Phase 3 whole-rock core in the San Andreas Fault Observatory at Depth (SAFOD) borehole. Earth. Planet. Sc. Lett., 2011, 310(1): 131-144.

[7] Perez-Muñoz T, Velasco-Hernandez J, Hernandez-Martinez E. Wavelet transform analysis for lithological characteristics identification in siliciclastic oil fields. J. Appl. Geophys., 2013, 98: 298-308.

[8] Li X, Li H. A new method of identification of complex lithologies and reservoirs: Task-driven data mining. J. Petrol. Sci. Eng., 2013, 109: 241-249.

[9] Ajayi O, Torres-Verdín C, Preeg W E. Fast numerical simulation of logging-while-drilling gamma-ray spectroscopy measurements. Geophysics, 2015, 80(5): D501-D523.

[10] Petri I, Pereira M S, dos Santos J M, et al. Microwave remediation of oil well drill cuttings. J. Petrol. Sci. Eng., 2015, 134: 23-29.

[11] Mazidi S M, Haftani M, Bohloli B, et al. Measurement of uniaxial compressive strength of rocks using reconstructed cores from rock cuttings. J. Petrol. Sci. Eng., 2012, 86: 39-43.

[12] Phan T T, Capo R C, Stewart B W, et al. Trace metal distribution and mobility in drill cuttings and produced waters from Marcellus Shale gas extraction: Uranium, arsenic, barium. Appl. Geochem., 2015, 60: 89-103.

[13] Baykan N A, Yılmaz N. Mineral identification using color spaces and artificial neural networks. Comput. & Geosci-UK, 2010, 36(1): 91-97.

[14] Ramola R C, Choubey V M, Prasad Y, et al. Variation in radon concentration and terrestrial gamma radiation dose rates in relation to the lithology in southern part of Kumaon Himalaya, India. Radiat. Meas., 2006, 41(6): 714-720.

[15] Vekshin N L. Increasing the sensitivity of fluorescent analysis by coating a preparation on a plate diagonally cutting the cuvette. Instrum. Exp. Tech, 2008, 51(2): 315-317.

[16] Washburn K E. Rapid geochemical and mineralogical characterization of shale by laser-induced breakdown spectroscopy. Org. Geochem., 2015, 83: 114-117.

[17] Kuhn K, Meima J A, Rammlmair D, et al. Chemical mapping of mine waste drill cores with laser-induced breakdown spectroscopy (LIBS) and energy dispersive X-ray fluorescence (EDXRF) for mineral resource exploration. J. Geochem. Explor., 2016, 161. 72-84.

[18] Nibourel L, Herman F, Cox S C, et al. Provenance analysis using Raman spectroscopy of carbonaceous material: A case study in the Southern Alps of New Zealand. J. Geophys. Res. Earth. Surf., 2015, 120(10): 2056-2079.

[19] Ferguson B, Zhang X C. Materials for terahertz science and technology. Nat. Mater., 2002, 1(1): 26-33.

[20] Wimmer L, Herink G, Solli D R, et al. Terahertz control of nanotip photoemission. Nat. Phys., 2014, 10(6): 432-436.

[21] Parrott E P, Zeitler J A. Terahertz time-domain and low-frequency Raman spectroscopy of organic materials". Appl. Spectrosc., 2015, 69(1): 1-25.

[22] Lepodise L M, Horvat J, Lewis R A. Terahertz (6-15THz) spectroscopy and numerical modeling of intermolecular vibrations in benzoic acid and its derivatives. Appl. Spectrosc., 2015, 69(5): 590-596.

[23] Özer Z, Gök S, Altan H, Severcan F. Concentration-based measurement studies of L-tryptophan using terahertz time-domain spectroscopy (THz-TDS). Appl. Spectrosc., 2014, 68(1): 95-100.

[24] Zhan H, Li Q, Zhao K, et al. Evaluating $PM_{2.5}$ at a construction site using terahertz radiation. IEEE T. THz. Sci. Techn., 2015, 5(6): 1028-1034.

[25] Bao R, Wu S, Zhao K, et al. Applying terahertz time-domain spectroscopy to probe the evolution of kerogen in close pyrolysis systems. Sci. China: Phys. Mech. Astron., 2013, 56(8): 1603-1605.

[26] Zhan H, Zhao K, Xiao L. Spectral characterization of the key parameters and elements in coal using terahertz spectroscopy. Energy, 2015, 93: 1140-1145.

[27] Jin W J, Zhao K, Yang C, et al. Experimental measurements of water content in crude oil emulsions by terahertz time-domain spectroscopy. Appl. Geophys., 2013, 10(4): 506-509.

[28] Jiang C, Zhao K, Zhao L J, et al. Probing disaggregation of crude oil in a magnetic field with terahertz time-domain spectroscopy. Energ. Fuel., 2014, 28(1): 483-487.

[29] Ge L N, Zhan H L, Leng W X, et al. Optical characterization of the principal hydrocarbon components in natural gas using terahertz spectroscopy. Energy & Fuel, 2015, 29(3): 1622-1627.

[30] Schwerdtfeger M, Castro-Camus E, Krügener K, et al. Beating the wavelength limit: Three-dimensional imaging of buried subwavelength fractures in sculpture and construction materials by terahertz time-domain reflection spectroscopy. Appl. Optics., 2013, 52(3): 375-380.

[31] Wilke I, Ramanathan V, LaChance J, et al. Characterization of the terahertz frequency optical constants of montmorillonite. Appl. Clay Sci., 2014, 87(87): 61-65.

[32] Janek M, Matejdes M, Szöcs V, et al. Dielectric properties of micaceous clays determined by terahertz time-domain spectroscopy. Philos. Mag., 2010, 90(17-18): 2399-2413.

[33] Zich D, Zacher T, Darmo J, et al. Far-infrared investigation of kaolinite and halloysite intercalates using terahertz time-domain spectroscopy. Vib. Spectrosc., 2013, 69: 1-7.

[34] Nagai N, Imai T, Fukasawa R, et al. Analysis of the intermolecular interaction of nanocomposites by THz spectroscopy. Appl. Phys. Lett., 2004, 85(18): 4010-4012.

[35] Dong C, Bao R, Zhao K, et al. Terahertz time-domain spectroscopy of a simulated pore structure to probe particle size and porosity of porous rock. Chinese. Phys. B, 2014, 23(12): 127802.

[36] Jerram D A, Cheadle M J. On the cluster analysis of grains and crystals in rocks. Am. Mineral., 2000, 85(1): 47-67.
[37] Al-Baldawi B A. Applying the cluster analysis technique in logfacies determination for Mishrif Formation, Amara oil field, South Eastern Iraq. Arabian. J. Geosci-Czech., 2014, 8(6): 3767-3776.
[38] Giorio C, Tapparo A, Dall'Osto M, et al. Local and regional components of aerosol in a heavily trafficked street canyon in central London derived from PMF and cluster analysis of single-particle ATOFMS spectra. Environ. Sci. Technol., 2015, 49(6): 3330-3340.
[39] Ito H, Kono T. Quantitative analysis of organic additive content in a polymer by ToF-SIMS with PCA. Appl. Surf. Sci., 2008, 255(4): 1044-1047.
[40] Fontalvo-Gómez M, Colucci J A, Velez N, et al. In-line near-infrared (NIR) and Raman spectroscopy coupled with principal component analysis (PCA) for *in situ* evaluation of the transesterification reaction. Appl. Spectrosc., 2013, 67(10): 1142-1149.
[41] Kumar R, Kumar V, Sharma V. Discrimination of various paper types using diffuse reflectance ultraviolet-visible near-infrared (UV-Vis-NIR) spectroscopy: Forensic application to questioned documents. Appl. Spectrosc., 2015, 69(6): 714-720.

Oil yield characterization by anisotropy in optical parameters of the oil shale

Xinyang Miao   Hongli Zhan   Kun Zhao   Yizhang Li   Qi Sun   Rima Bao

(Beijing Key Laboratory of Optical Detection Technology for Oil and Gas, China University of Petroleum, Beijing 102249, China)

Abstract: Oil yield is an important indicator for oil shale to optimize the comprehensive utilization. Generally, oil shale is highly anisotropic owing to the combined effects similar to shale. In this paper, terahertz time-domain spectroscopy(THz-TDS) was employed to investigate the anisotropic response of oil shale samples from Longkou, Yaojie, and Barkol with different oil yields. All the samples had significant anisotropy of the refractive index (n) and absorption coefficient (α) with symmetries at a location of 180°, which corresponded with the bedding plane and the partial alignment of particles. Additionally, the D values of experiment n in the vertical and parallel directions of the bedding plane were calculated as $\Delta n' = n_\perp - n_\parallel$, and samples from Beipiao and Huadian were also tested in the horizontal and vertical directions for a sufficient number of terahertz (THz) parameters. Linear regression was built between the $\Delta n'$ of the samples from five regions and the oil yield, described as $y = 60.86x + 3.72$ for oil yield (y) and $\Delta n'(x)$, with the correlation coefficient R equaling 0.9866 and the residual sum of squares was 1.182, indicating THz technology could be an effective selection for evaluating the oil yield of oil shale.

Keywords: oil yield, oil shale, anisotropy, refractive index

1 Introduction

Shale formations have been found in most of the global sedimentary basins, acting as the cap rock of conventional reservoirs. Moreover, energy demands have motivated the development of shale formations as significant unconventional reservoirs. Oil shale, a finely grained sedimentary rock with kerogen contained, has been gradually developed in China since the 1920s[1]. Numerous oil and gas products as fuels and raw materials in petrochemical industries can be yielded by pyrogenation of kerogen[2-3]. In order to optimize the comprehensive utilization of oil shale, a series of methods have been carried out for oil content and oil yield evaluation, including pyrolysis[4], thermogravimetry analysis (TGA)[5], well logging[6], and X-ray diffraction as well as diffuse reflectance infrared Fourier transforms spectroscopy (DRIFTS)[7].

Generally, shales as well as oil shales are often highly anisotropic owing to the combined effect of partial alignment of platy clay particles, layering, microcracks, low-aspect-ratio pores and kerogen inclusions[8-10]. Ultrasonic measurements have demonstrated that the elastic properties are isotropic in the directions parallel to the bedding, while anisotropic in other directions[11-13]. Additionally, solid organic matters are usually more compliant than other minerals in shales. Organics' shapes and distribution exhibit some elongation parallel with bedding direction[14], which makes organic materials a strong source of anisotropy in shales. The organic matter content has impact on the anisotropy degree in shales; thus, anisotropy can be employed to assess the oil content theoretically. Actually, precisely quantitative relations between oil content and anisotropy were seldom built in relative reports[15-18].

In the past decades, optics have been proved to be the ultimate means of sending information to and from the interior structure of materials[19]. Birefringence, defined as the division of a light ray into two rays when it passes through an optically anisotropic material, is dependent on the polarization of the light. It derives from the electrical anisotropy of a material, and is applied for characterizing the internal material properties[20]. THz radiation, which is located between far-infrared (IR) and millimeter-wave bands of the spectrum, spans the transition range from radio-electronics to photonics[21]. At present, analysis with THz radiation has attracted much attention owing to the unique advantages[22-27]. Various crystals (e.g., quartz[28], sapphire[29], $LiNbO_3$[30] and ZnO[31]) have been studied to exhibit THz birefringence. Alternating structures with two different dielectric materials on the submillimeter scale were also investigated to show form birefringence, such as multilayered polymers[32] and stacks of silicon wafers with air gaps[33]. In addition, THz birefringence was observed for fibrous materials (e.g., textiles[34], leaf and wood[35]), which exhibits an orientation arrangement of the fiber during the production process.

Recently, THz-TDS has been utilized for the investigation of dielectric properties of reservoir rocks. Optical parameters in the THz range were sensitive to the mineral and structures in rocks, e.g., marble, limestone, sandstone, and clay as well as mudstone[36-38]. Spectral features of organics were also studied, and THz-TDS was proved efficiently in probing the evolution of kerogen with different maturity and disaggregation of crude oil under magnetic fields[23,25]. Additionally, combination of THz technology and stoichiometry had provided a practical means to analyze crude oils[26] as well as fuel oils blended with various additives[39]. In our previous study, oil yields of oil shales were studied with laser-induced voltage (LIV), showing correlations between oil yields and the LIV parameters[40]. In this paper, the anisotropic responses of oil shales were initially studied by THz-TDS. The anisotropy of experimental n was then calculated in order to investigate organic matters content based on the anisotropy in THz parameters.

2 Experimental methods

The oil shales used in the experiments were obtained from three districts in China, including Longkou, Yaojie, and Barkol. The oil yields of the samples from these regions were measured as ~14.16%, ~9.05%, and ~5.66%, respectively[40]. From each shale block, we cut slices at different angles (θ_0) relative to the bedding plane (Table 1). The size of each slice equaled ~8×8mm², and the thickness was measured by micrometer caliper one by one. Therefore, totals of 17 slices were prepared. In order to avoid the influence of water, all the slices were dried in a vacuum at 90℃ for 10h before testing.

Table 1 Cutting and testing angles of the slices from three kinds of blocks

Region	Longkou	Yaojie	Barkol
Cutting angles of the slices (θ_0)/(°)	0, 15, 30, 45, 60, 75, 90	0, 30, 45, 60, 90	0, 30, 45, 60, 90
Testing angles of the slices (θ)/(°)	0, 15, 30, 45, 60, 75, 90, 105, 120, 135, 150, 165, 180, 195, 210, 225, 240, 255, 270, 285, 300, 315, 330, 345, 360	0, 30, 45, 60, 90, 120, 135, 150, 180, 210, 225, 240, 270, 300, 315, 330, 360	0, 30, 45, 60, 90, 120, 135, 150, 180, 210, 225, 240, 270, 300, 315, 330, 360

The experiments were performed with a transmission THz-TDS system (Figure 1) under air atmosphere with a temperature of 294.1K and a humidity of 30%. The signal-to-noise ratio (SNR) of the setup was ~1500. A detailed description and schematic drawing of the THz system based on photoconductive antenna and electro-optical sampling has been explained in our previous study[26]. All the slices were tested, and then the refractive index (n) and absorption coefficient (α) anisotropy were calculated. As shown in Figure 1, the measurement process was divided into four steps: (i) each slice was placed in the THz-TDS equipment at the focus position, with the angles between the input orientation of terahertz waves and the normal direction of the bedding plane (θ); (ii) revolved 180° of the slice around the THz-propagation direction (define as x-axis) with $\theta'=180°-\theta$; then the slice was turned 180° around the (iii) z-axis and (iv) x-axis, with $\theta''=360°-\theta$ and $\theta'''=180°+\theta$, respectively.

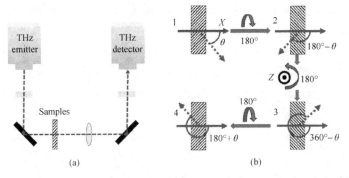

Figure 1 Top view of (a) experimental setup for the detection of slices with transmission THz-TDS, (b) schematic diagram of the four steps during the test

3 Results and discussion

THz-TDS of the reference and the slices were initially measured. Fast Fourier transform

(FFT) was used for deriving the THz frequency domain spectra (THz-FDS), and the *n* as well as *α* spectra were then calculated[25]. Owing to the intense absorption of the slices, the spectra were not only composed of the samples' characteristic features, but also many kinds of noises. Therefore, weaker signals at some frequencies were covered by the noises, and the effective frequency range was reduced to 0.4~1.0THz for acceptable signal-to-noise ratio. Frequency dependent *α* and *n* spectra of slices with the testing angle 0° and 90° were exhibited in Figure 2.

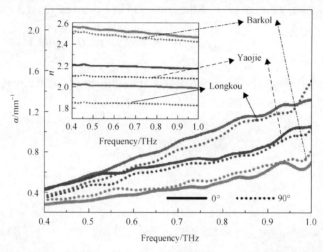

Figure 2 Frequency dependent spectra of the absorption coefficient (*α*) and refractive index (*n*) measured with the propagate direction parallel (dotted line, *θ* = 90°) and vertical (solid line, *θ* = 0°) to the bedding plane at 0.4~1.0THz

As shown in Figure 2, the values of *n* were almost unchanged in the range of 0.4~1.0THz, while those of *α* augment with the increasing of the frequency. Owing to the divergence in mineral compositions and oil yields, the values of *n* and *α* in the three regions differed from each other. Generally, oil shales collected from Barkol have the maximum *n*, followed by Yaojie and Longkou. On the contrary, the *α* spectra of oil shales from Barkol were the minimum and those samples from Longkou led across most of the frequency range. Divergence was also found in the spectra of *n* and *α* for the samples from the same place with 0° and 90°, respectively.

Then, we investigated the *θ* dependences of *n* and *α* at various angles ranging from 0° to 360°. Expressions of *n* and *α* variation with the angle were obtained first. Oil shale was assumed as a uniaxial birefringent medium with transverse isotropy, in which the optical axis was perpendicular to the bedding plane. The dielectric constant of the oil shale was denoted by ε_0 in the direction of the optical axis, as well as ε_{90} along the layer. The dielectric tensor was expressed as a tensor in the principal axes, i.e.

$$\boldsymbol{\varepsilon} = \begin{bmatrix} \varepsilon_{90} & 0 & 0 \\ 0 & \varepsilon_{90} & 0 \\ 0 & 0 & \varepsilon_0 \end{bmatrix} \qquad (1)$$

Then, we defined the propagation velocities of electromagnetic wave in x, y, and z directions, with 0 and 90 representing the transmission direction parallel and vertical to the optical axis, respectively.

$$v_x = v_y = v_{90}, \quad v_z = v_0 \tag{2}$$

According to the theory of light propagation in an anisotropy medium[41], the relationship between the propagation velocity (v) and the angle (θ) in a uniaxial crystal was given by

$$v_\theta = \sqrt{v_0^2 \cos^2\theta + v_{90}^2 \sin^2\theta} \tag{3}$$

where θ represented the angle between the propagation direction and the optical axis. Thus, the refractive index (n) could be calculated by

$$n_\theta = \frac{1}{\sqrt{\dfrac{\cos^2\theta}{n_0^2} + \dfrac{\sin^2\theta}{n_{90}^2}}} \tag{4}$$

As n_0 and n_{90} were constants associated with the properties of samples, n_θ varied periodically as a function of θ with the cycle 180° in the case of $n_0 \neq n_{90}$.

Additionally, the angle dependent attenuation of the THz field was also calculated. The output field was expressed as a function of θ, and then α was obtained based on that. Considering the diagram shown in Figure 3, the output fields of the shale parallel and perpendicular to the layer were expressed as

$$E_\perp^{out} = E^{in} \cos\theta \exp(i\Gamma_\perp) \exp(-\alpha_\perp d)$$

$$E_\parallel^{out} = E^{in} \sin\theta \exp(i\Gamma_\parallel) \exp(-\alpha_\parallel d) \tag{5}$$

in which E^{in} was the incident electric field amplitude, $\Gamma_\perp$ and $\Gamma_\parallel$ were the phase retardances parallel and vertical to the layer, respectively, d was the thickness of the oil shales, $\alpha_\perp$ and $\alpha_\parallel$ were defined as the absorption coefficients along the two directions.

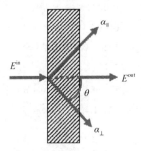

Figure 3 Schematic diagram of the shale sample for the input THz fields. Directions $\alpha_\perp$ and $\alpha_\parallel$ were parallel and perpendicular to the layer, respectively

In Figure 3, the output fields in directions perpendicular and parallel to the layer recombined and the output THz field was given by

$$E^{out} = E_{\perp}^{out} \cos\theta + E_{\parallel}^{out} \sin\theta \tag{6}$$

Then, angular dependent E^{out} was obtained as

$$E^{out} = E^{in} \exp(-\alpha_{\parallel}d) | \cos^2\theta \exp(-\Delta\alpha d) \exp(i\Delta\Gamma) + \sin^2\theta \tag{7}$$

where $\Delta\alpha$ was the difference value of $\alpha_{\parallel}$ and $\alpha_{\perp}$, and $\Delta\Gamma$ represented the phase retardance which could be expressed by

$$\Delta\Gamma = \frac{2\pi d}{\lambda}\Delta n \tag{8}$$

with the birefringence (Δn) and the wavelength (λ)[27].

Then, the relation of α could be expressed as $-\ln(E^{out}/E^{in})/d$.

$$\alpha_{\theta} = \alpha_{\parallel} - \frac{1}{d}\ln|\cos^2\theta \exp(-\Delta\alpha d)\exp(i\Delta\Gamma) + \sin^2\theta \tag{9}$$

Using Equation (8) in Equation (9), we obtained the variation of α with θ, d as well as λ:

$$\alpha_{\theta} = \alpha_{\parallel} - \frac{1}{2d}\ln\left(\cos^4\theta\, e^{-2\Delta\alpha d} + 2\cos^2\theta e^{-\Delta\alpha d}\cos\frac{2\pi l}{\lambda}\Delta n + \sin^4\theta\right) \tag{10}$$

Figure 4 shows the angular dependent n and α (at 0.8THz) of the samples from Longkou, Yaojie, and Barkol. It is clear from Figure 4 (a) that all the oil shale samples from three regions has significant anisotropy of n, and there was data symmetry at the location of 180°. Moreover, the data also showed a significant attenuation of n in the THz propagation direction parallel to the layering compared with that perpendicular to the layering. Then, calculated n were obtained and plotted (the dotted lines in Figure 4(a) using Equation (4) with the experimental n_0 and n_{90}. Slightly larger differences existed between the calculated and observed n for Longkou and Barkol samples compared to that of Yaojie but followed the same overall trend. Likewise, the measured (solid curves) and calculated (dotted curves) α_{θ} as a function of the angle were shown in Figure 4(b). For all the three samples, the peak values of measured α were obtained at ~45° in the range from 0° to 90°, which remained coincide with the calculated values. The experimental data met the expectations in Equation(10), and the results clearly demonstrated the anisotropy behavior of the measured oil shale samples. Various studies have shown that the observed THz anisotropy of materials with natural fibers are contributed by their intrinsic structure[34, 35]. Similarly, for the oil shale samples, the anisotropy in THz parameters are caused by the preferential orientation of basic structural units. The data symmetries at the location of 180° in the experimental data as well as the calculated lines are corresponded with the bedding plane and the partial alignment of particles in oil shales.

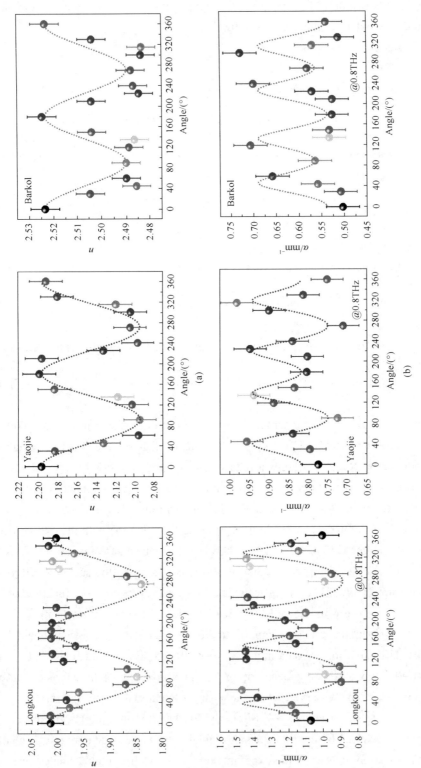

Figure 4 (a) θ dependent measured n values of the samples (dots) and calculated n (dotted curve) from Longkou, Yaojie, and Barkol, respectively; (b) experimental value of α at 0.8THz (dots) as a function of angle along with the expected angular dependence of α (dotted curve) calculated by Equation (10). Error bars represent 0.5 fluctuation of standard deviation of n and α

In addition, the presence of alternating layers with varying organic matter content has influence on the degree of anisotropy of elastic wave velocity and attenuation[14]. The organic matter appeared to have little intrinsic birefringence, nevertheless, the effect of organic matter on the anisotropy was speculated to be strongly related to the texture of shales[15-19]. Organic matter such as kerogen has always exhibited some elongation parallel with the bedding, which is caused by lithostatic overburden and deformation, and by the original orientation when kerogen first deposits[42]. The organic matter content resulted in the variation of Δn with the oil yields in this research. Herein, we calculated the D value of $n_\perp$ and $n_\parallel$ as $\Delta n'$, in which $n_\perp$ and $n_\parallel$ represented the average value of experiment n in the vertical and parallel directions of the bedding plane. Organic matter has a very low refractive index in THz range[36,43], in addition, owing to the effect of oriented organic matters, $n_\parallel$ decreased more than $n_\perp$ with the increase in organic matter content. The possibility of using $\Delta n'$ to infer oil yield prompted us to test two more samples with different oil yields. Oil shale samples from Beipiao and Huadian were tested in the horizontal and vertical directions, with the oil yields ~5.00 and ~9.96, respectively. $\Delta n'$ of the samples from five places was calculated and plotted with oil yield in Figure 5. Significant positive correlation was shown between the anisotropy parameters and the oil yield, indicating that the increase in organic matter has promoted the anisotropy in the THz range. Therefore, THz technology was expected to be valid in evaluating the oil yields of oil shales.

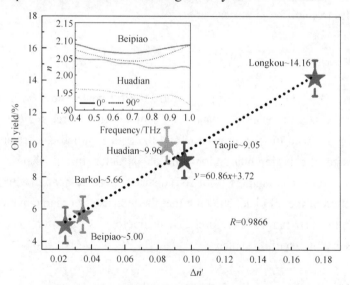

Figure 5 Linear fit of oil yield as a function of $\Delta n'$, with the correlation coefficient R equal to 0.9866 and the residual sum of squares equal to 1.182. The inset is the n spectra of Beipiao and Huadian samples with the propagate direction parallel and vertical to the bedding plane. Error bars represent 0.5 fluctuation of standard deviation of the oil yields

Anisotropic degree of n in horizontal and vertical directions have provided alternative parameters for oil yield characterization, with a small amount of samples required. According to previous research, a variety of methods can be implemented to evaluate the organic matter content in oil shales, mostly assessed by heating and combusting[44]. Fisher assay has been extensively applied in oil yield characterization, which is time-consuming, destructive, and expensive with large amounts of samples required[7]. TGA measurements have been utilized to determine the temperature affected pyrolyzation and kinetic parameters of oil shales with the heating rates 5~40 ℃/min and final temperature of 950 ℃ [5,7]. Other methods such as well-logging provide the total organic carbon content (TOC) by relations between the TOC and $\Delta \lg R$, however, the lg curve is always influenced by the well environment and cannot reflect the information on undisturbed formation accurately[6]. According to our previous studies, positively related relationships were observed between the photoinduced voltage (Δv) and the oil yield, in which Δv was directly affected by organic matter content[39]. In this paper, we determined the content of organic matter by its promotion of the anisotropic degree. In terms of the anisotropy in THz parameter, by measuring n of oil shale slices with two bedding plane direction, n_0 and n_{90} could be obtained based on relations formulated in Equation(4), and $\Delta n'$ can be quantitatively related to the oil yield. Linear correlation has been shown with the results of the samples from three places, described as $y = 60.86x+3.72$ for oil yield (y) and $\Delta n'$ (x), with the correlation coefficient R equaled 0.9866 and the residual sum of squares (RSS) being 1.182. Therefore, THz technology has provided a promising means in nondestructive detection of oil yields in oil shales from different places.

4 Conlusions

In summary, THz-TDS was utilized to study the oil shale slices with different angles between the surface and the bedding plane. The samples from three regions had significant anisotropy of n and α with the variation of θ, and the data symmetries corresponded with the bedding plane and the preferential alignment of particles in oil shales. In addition, n anisotropies of the oil shale samples from five districts were compared with each other and plotted as a function of the oil yield, showing that the increase of organic matter has promoted the anisotropy in the THz range. Therefore, THz technology is supposed to be a valuable tool for the oil yields evaluation of oil shales.

Acknowledgements

This work was supported by the National Nature Science Foundation of China (Grant Nos. 61405259 and 11574401), National Basic Research Program of China (Grant No. 2014CB744302), the Specially Founded Program on National Key Scientific Instruments and Equipment Development, China (Grant No. 2012YQ140005) and the China Petroleum and Chemical Industry Association Science and Technology Guidance Program (Grant No. 2016-01-07).

References

[1] Dyni J R. Geology and resources of some world oil-shale deposits. Oil Shale, 2003, 20: 193-252.

[2] Niu M T, Wang S, Han X X, et al. Yield and characteristics of shale oil from the retorting of oil shale and fine oil-shale ash mixtures. Appl. Energ., 2013, 111: 234-239.

[3] Reinik J, Irha N, Steinnes E, et al. Characterization of water extracts of oil shale retorting residues form gaseous and solid heat carrier processes. Fuel Process. Technol., 2015, 131: 443-451.

[4] Bai F T, Wei G, Lü X S, et al. Kinetic study on the pyrolysis behavior of Huadian oil shale via non-isothermal thermogravimetric data. Fuel, 2015, 146: 111-118.

[5] Wang W, Li L Y, Ma Y, et al. Pyrolysis kinetic of North-Korean oil shale. Oil Shale, 2014, 31: 250-265.

[6] Jia J L, Liu Z J, Meng Q T, et al. Quantitative evaluation of oil shale based on well log and 3-d seismic technique in the Songliao basin, northeast China. Oil Shale, 2012, 29: 128-150.

[7] Bhargava S, Awaja F, Subasinghe N D. Characterisation of some Australian oil shale using thermal, X-ray and IR techniques. Fuel, 2005, 84: 707-715.

[8] Zhubayev A, Houben M E, Smeulders D M J, et al. Ultrasonic velocity and attenuation anisotropy of shales, Whitby, United Kingdom. Geophysics, 2016, 81: D45-D56.

[9] Blum T E, Adam L, Wijk K V. Noncontacting benchtop measurements of the elastic properties of shales. Geophysics, 2013, 78: C25-C31.

[10] Barkia H, Belkbir L, Gerard N, et al. Investigation of the anisotropy of Timahdit oil shale (morocco) by dilatohetry. Thermochim. Acta, 1989, 152: 197-202.

[11] Stanley D, Christensen N I. Attenuation anisotropy in shale at elevated confining pressures. Int. J. of Rock. Mech. Min., 2001, 38: 1047-1056.

[12] Bhandari A R, Flemings P B, Polito P J, et al. Anisotropy and stress dependence of permeability in the Barnett shale. Transport. Porous. Med, 2015, 108: 393-411.

[13] Allan A M, Kanitpanyacharoen W, Vanorio T. A multiscale methodology for the analysis of velocity anisotropy in organic-rich shale. Geophysics, 2015, 80: C73-C88.

[14] Sayers C M, Guo S, Silva J. Sensitivity of the elastic anisotropy and seismic reflection amplitude of the eagle ford shale to the presence of kerogen. Geophys. Prospect, 2014, 63: 151-165.

[15] Vernik L, Nur A. Ultrasonic velocity and anisotropy of hydrocarbon source rocks. Geophysics, 1992, 57: 758-759.

[16] Vernik L, Landis C. Elastic anisotropy of source rocks: Implications for hydrocarbon generation and primary migration. AAPG Bull., 1996, 80: 531-544.

[17] Vernik L, Liu X. Velocity anisotropy in shales: A petrophysical study. Geophysics, 1997, 62: 521-532.

[18] Sayers C M. The effect of kerogen on the elastic anisotropy of organic-rich shales. Geophysics, 2013, 78: D65-D74.

[19] Shalaev V M. Optical negative-index metamaterials. Nat. Photonics, 2007, 1: 41-48.

[20] Turchette Q A, Hood C J, Lange W, et al. Measurement of conditional phase shifts for quantum logic. Phys. Rev. Lett., 1995, 75: 4710-4713.

[21] Tonouchi M. Cutting-edge terahertz technology. Nat. Photonics, 2007, 1: 97-105.

[22] Feng X, Wu S X, Zhao K, et al. Pattern transitions of oil-water two-phase flow with low water content in rectangular horizontal pipes probed by terahertz spectrum. Opt. Express, 2015, 23: A1693-A1699.

[23] Bao R M, Wu S X, Zhao K, et al. Applying terahertz time-domain spectroscopy to probe the evolution of kerogen in close pyrolysis systems. Sci. China: Phys. Mech. Astron, 2013, 56: 1603-1605.

[24] Zhan H L, Zhao K, Xiao L Z. Spectral characterization of the key parameters and elements in coal using terahertz spectroscopy. Energy, 2015, 93: 1140-1145.

[25] Jiang C, Zhao K, Zhao L J, et al. Probing disaggregation of crude oil in a magnetic field with terahertz time-domain spectroscopy. Energy & Fuels, 2014, 28: 483-487.

[26] Zhan H L, Wu S X, Bao R M, et al. Qualitative identification of crude oils from different oil fields using terahertz time-domain spectroscopy. Fuel, 2015, 43: 189-193.

[27] Jin Z M, Mics Z, Ma G H, et al. Single-pulse terahertz coherent control of spin resonance in the canted antiferromagnet $YFeO_3$, mediated by dielectric anisotropy. Phys. Rev. B, 2013, 87: 269-275.

[28] Kaveev A K, Kropotov G I, Tsygankova E V, et al. Terahertz polarization conversion with quartz waveplate sets. Appl. Optics, 2013, 52: B60-B69.

[29] Kim Y, Yi M, Kim B G, et al. Investigation of THz birefringence measurement and calculation in Al_2O_3 and $LiNbO_3$. Appl. Optics, 2011, 50(18): 2906-2910.

[30] Li D, Ma G, Ge J, et al. Terahertz pulse shaping via birefringence in lithium niobate crystal. Appl. Phys. B-Lasers O., 2009, 94: 623-628.

[31] Kim Y, Ahn J, Kim B G, et al. Terahertz birefringence in zinc oxide. Jpn. J. Appl. Phys., 2011, 50(3): 3437-3442.

[32] Saha S C, Ma Y, Grant J P, et al. Low-loss terahertz artificial dielectric birefringent quarter-wave plates. IEEE Photonic. Tech. L., 2010, 22: 79-81.

[33] Scherger B, Scheller M, Vieweg N, et al. Paper terahertz wave plates. Opt. Express, 2011, 19: 24884-24889.

[34] Hirota Y, Hattori R, Tanim M, et al. Polarization modulation of terahertz electromagnetic radiation by four-contact photoconductive antenna. Opt. Express, 2006, 14: 4486-4493.

[35] Todoruk T M, Hartley I D, Reid M E. Origin of birefringence in wood at terahertz frequencies. IEEE T. THz. Sci. Techn., 2012, 2: 123-130.

[36] Schwerdtfeger M, Castrocamus E, Krügener K, et al. Beating the wavelength limit: Three-dimensional imaging of buried subwavelength fractures in sculpture and construction materials by terahertz time-domain reflection spectroscopy. Appl. Optics, 2013, 52: 375-380.

[37] Wilke I, Ramanathan V, Lachance J, et al. Characterization of the terahertz frequency optical constants of montmorillonite. Appl. Clay Sci., 2014, 87: 61-65.

[38] Janek M, Matejdes M, Szöcs V, et al. Dielectric properties of micaceous clays determined by terahertz time-domain spectroscopy. Philos. Mag., 2010, 90: 2399-2413.

[39] Zhan H L, Zhao K, Zhao H, et al. The spectral analysis of fuel oils using terahertz radiation and chemometric methods. J. Phys. D: Appl. Phys, 2016, 49: 395101.

[40] Lu Z Q, Sun Q, Zhao K, et al. Laser-induced voltage of oil shale for characterizing the oil yield. Energy & Fuels, 2015, 29: 4936-4940.

[41] Sharma K K. Optics-Principles and Applications. Burlington, San Diego, London: Elsevier, 2006.

[42] Shitrit O, Hatzor Y H, Feinstein S, et al. Effect of kerogen on rock physics of immature organic-rich chalks. Mar. Petrol. Geol., 2016, 73: 392-404.

[43] Washburn K, Birdwell J E, Foster M, et al. Detailed description of oil shale organic and mineralogical heterogeneity via FTIR-microscopy. Energy & Fuels, 2015, 29: 4264-4271.

[44] Arik E, Altan H, Esenturk O. Dielectric properties of diesel and gasoline by terahertz spectroscopy. J. Infrared Millim. Te., 2014, 35: 759-769.

Optimization of pyrolysis efficiency based on optical property of semicoke in terahertz region

Yizhang Li[1]　Shixiang Wu[2]　Xiaolu Yu[3]　Rima Bao[1]　Zhikui Wu[1]　Wei Wang[1]
Honglei Zhan[1]　Kun Zhao[1]　Yue Ma[1]　Jianxun Wu[1]　Shaohua Liu[1]　Shuyuan Li[1]

(1.Beijing Key Laboratory of Optical Detection Technology for Oil and Gas, China University of Petroleum, Beijing 102249, China; 2.Petroleum Exploration and Production Research Institute, China Petroleum and Chemical Corporation, 100083, China; 3.Institute of Petroleum Geology, Sinopec Oil Exploration and Development Research Institute, Wuxi 214126, China)

Abstract: The correlation between pyrolysis conditions and fuel production has been extensively studied. Terahertz parameters, instead of thermal kinetic parameters, were investigated to reveal such interior correlation in this work. Different ventilating rate (V), heating rate (β) and final temperature (T) were controlled in several pyrolysis experiments of oil shale to prepare semicoke with varied amount of oil content, respectively. It was observed that $V = 0.6$L/min, $\beta = 15$℃/min and $T=550$℃ were the expected conditions to produce fuel. The critical points in terahertz parameter corresponded well with that in oil yield. Intervals where increasing V, β and T contributed to oil yield best can be determined by comparing the trend of absorption index at 0.4THz, 0.6THz, 0.8THz, 1.0THz. Therefore terahertz parameter of semicoke may characterize the organic matter in semicoke that may convert to additional fuel and terahertz method can be applied to optimize pyrolysis efficiency.

Keywords: optimization; pyrolysis; oil shale; semicoke; terahertz time-domain spectroscopy; absorption index

1 Introduction

A number of countries attach importance on developing alternative sources of conventional energy[1-7]. As an alternative energy, shale oil is prepared through an oxygen-free pyrolysis process with the conversion of kerogen in oil shale. Enhancing production efficiency is concerned by investigators to reduce costs of extracting fuel from oil shale. Thus the motivation generates for seeking methods to optimize the pyrolysis conditions.

Thermal parameters are gradually concerned to interpret the process and play a great role in thermal analyzing. Experiments based on thermal gravity (TG), differential thermal analysis (DTA) or X-ray diffraction (XRD) have been widely reported where impacts of mineral matrix and organic matter to pyrolysis have been noted[8-13]. The effects of mineral matrix in shale char on pyrolysis and combustion of organic matter have been explored by testing shale char and its organic matter with TG-Fourier transform infrared (FTIR)[3]. The catalytic effect of minerals in oil shale on pyrolysis was investigated through the TG-FTIR test on original oil shale and p-kerogen [12]. Similar views form that major components of oil shale consist of organic system and nonorganic system. Additional, spectroscopy method is also helpful in evaluating the effects that organic system deviates from the nonorganic system. In other words, optical parameters may reflect the potential for obtaining additional fuel.

Terahertz spectrum has been employed to obtain information of materials by many investigators and efforts were made to apply terahertz approach to detecting oil shale in this work [14-24]. The response of object matter to terahertz pulse reflects internal information about molecule distribution. Therefore, in-lab or even commercial THz techniques such as terahertz time-domain spectroscopy (THz-TDS) are desired and gradually developed in recent years. Studies on material identification using THz-TDS have been reported. However, energy investigators seldom note the impacts of varied pyrolysis process on terahertz spectrum of semicoke. The difference in optical properties of semicoke within terahertz range ought to be addressed to enhance cognition of pyrolysis conditions. THz-TDS may provide information regarding semicoke as a means of assessing the pyrolysis of this material. As such, THz studies are very likely to contribute to the evaluation of oil shale and optimization of pyrolysis efficiency.

In the present study, oil shale from Longkou, a large reserve of oil shale in China, was collected to carry out experiments. Then the change in the THz absorption related to the pyrolysis conditions was determined. It was observed that the critical points of THz absorption correspond well with that of oil yield. The analysis indicated that the organic matter remained in semicoke, which may convert to additional fuel, can be characterized with THz-TDS.

2 Experimental section

2.1 Pyrolysis experimental procedure

In Figure 1, the oil shale was preprocessed by pulverizing and sieving sequentially. Typically, oil shale powder prepared in this manner is assessed through pyrolysis to obtain related thermal dynamics. In the present study, however, THz analysis of the oil shale semicoke was conducted. As is known, change in ventilating rate (V), heating rate (β) and final temperature (T) can all have a significant effect on the products of pyrolysis experiments. As such, semicoke with different oil content was prepared by adjusting V, β and T solely.

Oil shale from Longkou, China was obtained as raw materials. To carry out pyrolysis experiments, rocks with individual shape and size were all crushed in a pulverizer. The output was fragment of oil shale with diameter less than 3mm to eliminate size effect in pyrolysis. The pyrolysis was carried out in Fischer assay under nitrogen atmosphere and semicoke was produced simultaneously. During the assay, the heating rate β and final temperature T were programmed. Ventilating rate V was controlled according to the real-time data from a flowmeter. To determine impacts brought about by single variable, only one sort of condition was adjusted between experiments. Several groups of samples were assessed according to Table 1. In group 1, β and T were set at 15℃/min and 550℃ respectively while V was adjusted from 0 to 1L/min. In group 2, V and T were set at 0.6L/min and 550℃ respectively while β was adjusted from 5℃/min to 25℃/min. Finally, in group 3, V and β were set at 0.6L/min and 15℃/min respectively while T was adjusted from 400℃ to 550℃.

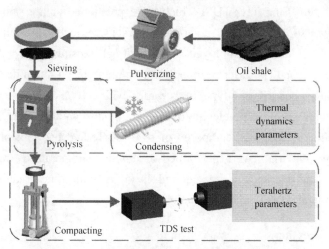

Figure 1　Schematic representation of the entire experimental process. Mesh box above briefly shows how conventional method based on thermal analysis works, whereas mesh box below briefly shows how terahertz method in this paper works. Semicoke samples with different amount of organic matter were prepared by pyrolysis and detected in Terahertz TDS

Table 1　Pyrolysis parameters of three groups of semicoke samples

Pyrolysis parameter	V/(L/min)	β/(℃/min)	T/℃
Group 1	0 0.3 0.6 0.8 1.0	15	550
Group 2	0.6	5 10 15 20 25	550
Group 3	0.6	15	400 430 460 490 520 550

2.2 THz experimental procedure

Semicoke powder was produced during pyrolysis of shale oil. To conduct THz experiment, the powder was compacted under 20MPa for 5min to allow it to be mechanically sectioned. The diameter of slice was 30mm and the thickness was around 2mm. Sliced samples were subsequently analyzed using THz-TDS, which were positioned at the same position in the equipment each time so that the THz radiation penetrated vertically through the center of the sample. The experimental setup employed a conventional transmission-type THz-TDS[18]. In this system, a femtosecond (fs) laser beam split into a pump beam and a detection beam. The THz pulse was generated by a p-type InAs wafer with <100> orientation, pumped by a Ti-sapphire laser with a central wavelength of 800 nm, a pulse width of 100 fs and a repetition rate of 80MHz. The detector was composed of <110>ZnTe and the apparatus used standard lock-in technology. All trials were performed at ambient temperature. Initially, THz-TDS spectra of samples and references were acquired. Data involving time-domain information are transferred to computer so as to make further analysis.

2.3 THz data processing

THz-TDS was exploited to obtain time-resolved data that reflected time-dependence of transient electric field. Fast Fourier transform (FFT) was applied to calculate the THz-FDS. In THz-FDS, signal components with different frequencies and phases were shown. Based on these spectra, the THz absorption index of semicoke was calculated as $\alpha = \ln[A_0(\omega)/A(\omega)]/d$, where $A_0(\omega)$ and $A(\omega)$ are the spectrum functions of the reference and sample respectively, and d is the thickness of the slice. An effective frequency range of 0.2~1.0THz was determined for the absorption spectra based on the amplitudes of the THz-FDS. Moreover, the absorption index at 0.4THz, 0.6THz, 0.8THz and 1.0THz were picked up and depicted along with variables. Thus the effects of increasing V, β and T can be determined by comparing optical parameters between different samples.

3 Results and discussion

3.1 Condition-dependent pyrolysis results

Oil yield in each experiment was diverse due to specific pyrolysis conditions. The condition-dependent oil yield in three groups is shown in Figure 2. Owing to the change of V from 0 to 0.6L/min, oil yield rose from 14.16% to 17.46%. Nevertheless, further rise of V from 0.6L/min to 1.0L/min led to decrease of oil yield from 17.46% to 16.11%. Similarly, increasing β from 5℃/min to 15℃/min worked in improving oil yield from 14.45% to 17.46%. On the contrary, increasing β from 15℃/min to 25℃/min may cause decline in oil yield from 17.46% to 13.95%. It was much easier to extract fuel from oil shale by increasing T from 400℃ to

460℃, which led to increase in oil yield from 6.03% to 14.29% while increasing T from 460℃ to 550℃ only led to increase in oil yield from 14.29% to 17.46%.

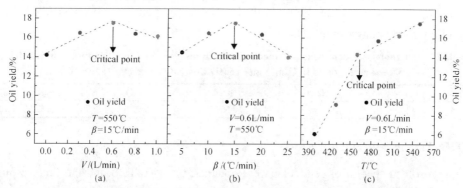

Figure 2 Condition-dependent oil yield: (a) V from 0 to 1L/min at β=15℃/min and T=550℃; (b) β from 5 to 25℃/min at V=0.6L/min and T=550℃; (c) T from 400℃ to 550℃ at V=0.6L/min and β=15℃/min

We also get condition-dependent pyrolysate composition including water content, semicoke, oil yield and dry distillation gas. According to the result shown in Figure 3, water content remains almost constant in all conditions and thus remaining water content within semicoke exerts equivalent effects on absorption of terahertz wave. Besides, the mass conservation of shale oil, gas and semicoke manifest the transference and conversion of organic matter. When semicoke reduces but oil yield rises, kerogen in oil shale decomposes to a greater degree. On the contrast, descent of semicoke and oil yield accompanied by ascent of gas indicates a great chance that secondary cracking takes place. Some of shale oil with higher molecular weight involves reactions to generate hydrocarbons with lower molecular weight.

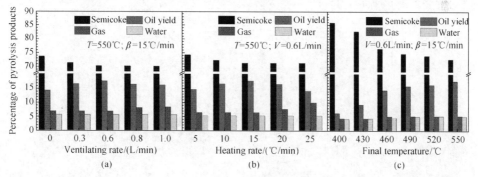

Figure 3 Condition-dependent pyrolysis products: (a) V from 0 to 1L/min at β=15℃/min and T=550℃; (b) β from 5 to 25℃/min at V=0.6L/min and T=550℃; (c) T from 400℃ to 550℃ at V=0.6L/min and β=15℃/min

3.2 Semicoke ash analysis

Semicoke ash analysis for one case is conducted to get insight into composition. Table 2 shows the specific value for all sorts of ash component including Fe_2O_3, Al_2O_3, SiO_2, Na_2O, K_2O, TiO_2 MnO_2, P_2O_5, CaO, MgO and SO_3. As we can see, SiO_2, Al_2O_3, Fe_2O_3 are predominant since these are common in mineral matrix of oil shale. Another 4% of component is not clear due to constraints of experimental conditions. These components involve elements such as carbon, sulfur and etc. Particularly, the listed components are fairly stable in temperature range below 600℃ and change little after pyrolysis. These components are consistent with the situation for mineral matrix in oil shale.

Table 2 Ash analysis of semicoke produced at conditions that final temperature equals 550℃, heat rate equals 15℃/min and ventilating rate equals 0.6L/min

Fe_2O_3	Al_2O_3	SiO_2	Na_2O	K_2O	TiO_2	MnO_2	P_2O_5	CaO	MgO	SO_3
6.78	12.17	62.49	1.98	1.24	0.55	0.07	0.38	7.76	1.52	1.03

3.3 THz-TDS measurement results

Time-resolved spectra of semicoke were obtained via THz-TDS and the time resolution was in picosecond scale. The waveforms of reference and samples in time domain are shown in Figure 4. The equipment emitted pulse whose width was less than 3 ps and amplitude was around 0.27 V. Due to semicoke absorption of terahertz rays, the amplitude was attenuated as much as 0.09 V. Since velocity of light reduced in solid media, the peak time changed from 17 ps to 23 ps approximately. It turned out that slight difference in shape was seldom noted except for attenuation and time delay.

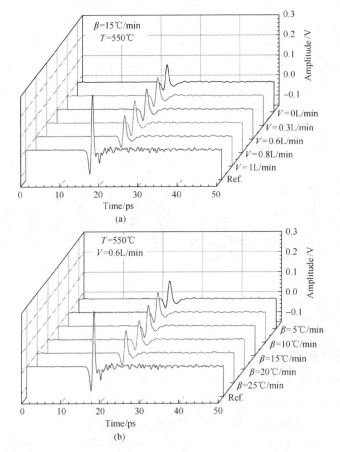

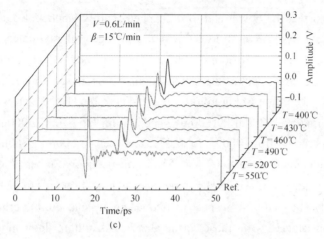

Figure 4   Time dependence of THz signal spectra of reference and semicoke samples: (a) V from 0 to 1L/min at β =15℃/min and T=550℃; (b) β from 5 to 25℃/min at V=0.6L/min and T=550℃; (c) T from 400℃ to 550℃ at V=0.6L/min and β=15℃/min

Absorption index values (α) for frequencies ranging from 0.2THz to 1.0THz were calculated for each sample, as shown in Figure 5. In general, α were observed to increase with increasing frequency for all samples. At approximately 0.2THz α varied between 150m^{-1} and 600m^{-1}, while at approximately 1.0THz α ranged from 1600m^{-1} to 3500m^{-1}. Besides, the difference between samples in terahertz absorption increased with increasing frequency.

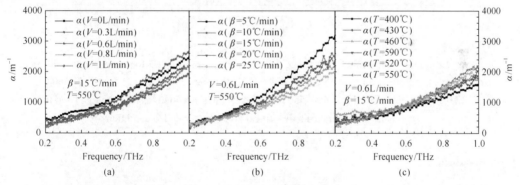

Figure 5   Absorption index of semicoke samples: (a) V=0~1L/min, β=15℃/min, T=550℃; (b) V=0.6L/min, β=5~25℃/min, T=550℃; (c) V=0.6L/min, β=15℃/min, T=400~550℃

Some valuable information can be gathered by comparing absorption index values at different conditions and frequencies. In Figure 6, absorption indexes α at 0.4THz, 0.6THz, 0.8THz and 1.0THz were depictured with the variables of pyrolysis conditions. Figure 6(a) also indicates that α at 0.4THz, 0.6THz, 0.8THz and 1.0THz decreased approximately from ~750m^{-1}, 1200m^{-1}, 1900m^{-1} and 2600m^{-1} to 530m^{-1}, 900m^{-1}, 1600m^{-1} and 2300m^{-1}, respectively when V increased from 0 to 0.6L/min. However, when V rose from 0.6L/min to 1.0L/min, α at 0.4THz, 0.6THz, 0.8THz, 1.0THz turned to 600m^{-1}, 850m^{-1}, 1400m^{-1}, 2000m^{-1} respectively. On account of errors existed, the downtrend of α changed sharply when $V > 0.6$L/min. In Figure 6(b), when β changed from 5℃/min to 15℃/min, α changed approximately from ~ 660m^{-1}, 1280m^{-1}, 2050m^{-1} and 3200m^{-1} to 500m^{-1}, 900m^{-1}, 1560m^{-1} and 2300m^{-1} at 0.4THz, 0.6THz, 0.8THz, 1.0THz, respectively. But α fluctuated and almost unchanged in terahertz region when β rose to 25℃/min. According to Figure 6(c), α at 0.4THz, 0.6THz, 0.8THz, 1.0THz changed approximately from 480m^{-1}, 780m^{-1}, 1200m^{-1}, and 1600m^{-1} to 670m^{-1}, 900m^{-1}, 1360m^{-1} and 1800m^{-1} when T increased from 400℃ to 460℃, and then to 500m^{-1}, 860m^{-1}, 1500m^{-1} and 2400m^{-1} when T increased from 460℃ to 550℃, respectively. When T exceeded 460℃, consistent upgrade tendency of α in terahertz region was less observed. Actually, the ascending trend of α gradually changed with respect to frequency. To sum up, the critical points in oil yield were in consistence with those in terahertz absorption, indicating that terahertz absorption was dependent on pyrolysis condition. By obtaining terahertz parameter of semicoke using THz-TDS, optimal values for certain pyrolysis conditions can be ascertained.

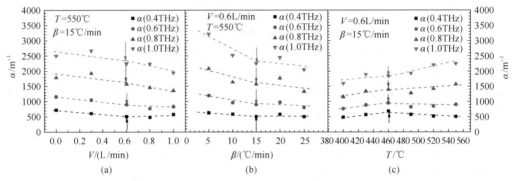

Figure 6 Absorption index values at 0.4THz, 0.6THz, 0.8THz and 1.0THz for samples under different conditions: (a) V=0~1L/min, β=15℃/min, T=550℃; (b) V=0.6L/min, β=5~25℃/min, T=550℃; (c) V=0.6L/min, β=15℃/min, T=400~550℃. The dash lines are fitted to show the trend and the chromatic arrows indicate the critical points under varied conditions

3.4 Discussion

The pyrolysis conditions we choosed made a difference in energy transferring and, above all, fuel production. Flowing gas carries both substance and heat away that generates in the process, which brings about dual consequences. The detachment of generated alkanes promotes better transferring of heat from the surface to the core of oil shale grains. Besides, products of initial reactions are also heated to cause secondary pyrolysis and in this case some of shale oil generated is converted to gas. The time when products stay in the reaction chamber and condenser actually plays a role in regulating oil yield. Venting is adverse to secondary pyrolysis since flow shortens the residence time and initial products are carried into condenser before secondary cracking. However, if venting flow is excessive, little time is left for effective condensation to collect shale oil. Thus the ideal value of V ought to be determined to seek for best production. On account of difference between the heat received by reactants and the heat that exotic environment provides, moderate heating rate is essential for optimal fuel production. For one thing, adequate heating rate ensures temperature gradient to boost heat transference and escape of products. For another, higher heating rate stands for shorter period for pyrolysis. The contradiction in diffusion and generation of products also leads to prolonging of residence time and conversion of shale oil to gas due to secondary cracking. Accordingly, an inappropriate heating rate may result in a loss of oil yield. Final temperature works on the degree of kerogen's artificial maturation to influence oil yield. According to our study, change of final temperature from 400℃ to 550℃ leads to change of oil yield around 12% and the maximum oil yield equals almost three times of the minimum one. In other words, the final temperature is of overriding importance to the production of shale oil. Higher final temperature ensures that the final fuel production is next to the expected one.

As is known, terahertz parameter is sensitive to the change of component and structure. In general, the photons whose energies are in the vicinity of energy levels corresponding to vibration or rotation modes of molecules are more likely to be absorbed. The organic matter surrounded by continuous mineral matrix suffers from great changes that take place in pyrolysis. During the process, reactions with lower activation energies are triggered first at lower temperature and weak bonds are destroyed. As temperature continues rising, impediment to reactions from higher activation energy is gradually overcome and more reactions are involved. Conversions of kerogen to shale oil and gas lead to notable change in component of semicoke. Such change can be observed indirectly via terahertz TDS. Besides, mineral matrix is little influenced in the temperature range that we focus. As ash within semicoke applies similar impacts on terahertz experiments, we conclude that the change of kerogen results in varied response of semicoke to terahertz wave.

4 Conclusions

In this study, components of oil shale change partially during the process of pyrolysis. With the rising of temperature, the energy of organic matter exceeds the threshold to initiate related reactions. The mineral matrix, which is relatively stable in pyrolysis, remains and changes less compared to organic matter. Parallel and sequential reactions occur, resulting in change of component ratio, typically the ratio of organic matter to inorganic one. Therefore, mineral matrix predominates in determining the terahertz parameter with respect to semicoke. The change in terahertz parameter reflects the facts that organic opponent changes, which is dependent on pyrolysis conditions. In the cases that shifting pyrolysis conditions supply sufficient energy to initiate the reaction, like what happens in rising T, step increment of T gradually activates reactions in need of further heat each time so that organic matter deviates from the mixture to greater extend with the rising of T. Thus the output is increasing as expected, corresponding to persistent drop of oil content contained by semicoke.

By comparing changes of absorption index at different frequency in terahertz range, we found that transition in amount of fuel production can be confirmed, which was helpful in improving pyrolysis efficiency. In addition, we realized that enhanced performance of THz-TDS was essential for tests. Terahertz equipment with fine dynamic range and high resolution is prerequisite for this method since information involving entire frequency range needs to be acquired. We prepared semicoke with varied amount of organic matter and conducted uniform preprocesses so as to prepare samples with sufficient difference, which is suitable for primitive study on oil shale. In this work, observed values deviated from the fitted lines due to reasons involving geometry errors as well as slight difference between individual oil shale. Hence specially-designed THz-TDS is in demand for rapid and accurate test aimed at original oil shale.

Acknowledgements

This work was supported by the Specially Funded Program on National Key Scientific Instruments and Equipment Development (Grant No. 2012YQ140005), the National Basic Research Program of China (Grant No. 2014CB744302), and the National Nature Science Foundation of China (Grant Nos. 61405259 and 11574401).

References

[1] Jiang X M, Han X X, Cui Z G. New technology for the comprehensive utilization of Chinese oil shale resources. Energy, 2007, 32: 772-777.

[2] Raji M, Gröcke D R, Greenwell H C, et al. The effect of interbedding on shale reservoir properties. Mar. Pet. Geol., 2015, 67: 154-169.

[3] Fan C, Yan J, Huang Y, et al. XRD and TG-FTIR study of the effect of mineral matrix on the pyrolysis and combustion of organic matter in shale char. Fuel, 2015, 139: 502-510.

[4] Wang W, Li S, Yue C, et al. Multistep pyrolysis kinetics of North Korean oil shale. J. Therm. Anal. Calorim., 2015, 119: 643-649.

[5] Na J G, Im C H, Chung S H, et al. Effect of oil shale retorting temperature on shale oil yield and properties. Fuel, 2012, 95: 131-135.

[6] Han X, Kulaots I, Jiang X, et al. Review of oil shale semicoke and its combustion utilization. Fuel, 2014, 126: 143-161.

[7] Guo H, Lin J, Yang Y, et al. Effect of minerals on the self-heating retorting of oil shale: Self-heating effect and shale-oil production. Fuel, 2014, 118: 186-193.

[8] Kok M V, Pokol G, Keskin C, et al. Combustion characteristics of lignite and oil shale samples by thermal analysis techniques. J. Therm. Anal. Calorim., 2004, 76: 247-254.

[9] Wright M C, Court R W, Kafantaris F C A, et al. A new rapid method for shale oil and shale gas assessment. Fuel, 2015, 153: 231-239.

[10] Guo H, Lin J, Yang Y, et al. Effect of minerals on the self-heating retorting of oil shale: Self-heating effect and shale-oil production. Fuel, 2014, 118: 186-193.

[11] Alstadt K N, Katti D R, Katti K S. An in situ FTIR step-scan photoacoustic investigation of kerogen and minerals in oil shale. Spectrochim. Acta. Part A Mol. Biomol. Spectrosc., 2012, 89: 105-113.

[12] Yan J, Jiang X, Han X, et al. A TG-FTIR investigation to the catalytic effect of mineral matrix in oil shale on the pyrolysis and combustion of kerogen. Fuel, 2013, 104: 307-317.

[13] Li B, Lv W, Zhang Q, et al. Pyrolysis and catalytic pyrolysis of industrial lignins by TG-FTIR: Kinetics and products. J. Anal. Appl. Pyrolysis., 2014, 108: 295-300.

[14] Siegel P H. Terahertz technology in biology and medicine. IEEE. Trans. Microw. Theory Tech., 2004, 52(10): 2438-2447.

[15] Ikeda T, Matsushita A, Tatsuno M, et al. Investigation of inflammable liquids by terahertz spectroscopy. Appl. Phys. Lett., 2005, 87: 7-10.

[16] Löffler T, Hahn T, Thomson M, et al. Large-area electro-optic ZnTe terahertz emitters. Opt. Express, 2005, 13: 53-62.

[17] Adbul-Munaim A M, Reuter M, Koch M, et al. Distinguishing gasoline engine oils of different viscosities using terahertz time-domain spectroscopy. J. Infrared, Millimeter, Terahertz Waves, 2015, 36: 687-696.

[18] Jiang C, Zhao K, Zhao L J, et al. Probing disaggregation of crude oil in a magnetic field with terahertz time-domain spectroscopy. Energy & Fuels, 2014, 28: 483-487.

[19] Bao R, Wu S, Zhao K, et al. Applying terahertz time-domain spectroscopy to probe the evolution of kerogen in close pyrolysis systems. Sci. China: Physics, Mech. Astron., 2013, 56: 1603-1605.

[20] Ren L, Xia D, Xu Y, et al. Research on pyrolysis mechanism of Huadian oil shale. Energy Procedia, 2015, 66: 13-16.

[21] Jin W J, Zhao K, Yang C, et al. Experimental measurements of water content in crude oil emulsions by terahertz time-domain spectroscopy. Appl. Geophys., 2013, 10: 506-509.

[22] Bao R, Li Y, Zhan H, et al. Probing the oil content in oil shale with terahertz spectroscopy. Sci. China: Physics, Mech. Astron., 2015, 58: 114211.

[23] Zhan H, Wu S, Bao R, et al. Qualitative identification of crude oils from different oil fields using terahertz time-domain spectroscopy. Fuel, 2015, 143: 189-198.

[24] Zhan H L, Wu S X, Bao R M, et al. Water adsorption dynamics in active carbon probed by terahertz spectroscopy. RSC Advances, 2015, 143: 89-92.

The mechanism of the terahertz spectroscopy for oil shale detection

Honglei Zhan[1,2,3] Mengxi Chen[2] Kun Zhao[1,2,3] Yizhang Li[2] Xinyang Miao[2]
Haimu Ye[3] Yue Ma[2] Shijie Hao[3] Hongfang Li[3] Wenzheng Yue[1]

(1.State Key Laboratory of Petroleum Resources and Prospecting, China University of Petroleum, Beijing 102249, China; 2.Beijing Key Laboratory of Optical Detection Technology for Oil and Gas, China University of Petroleum, Beijing 102249, China; 3.Department of Material Science and Engineering, China University of Petroleum, Beijing 102249, China)

Abstract: Terahertz time-domain spectroscopy (THz-TDS) can directly detect oil shales. The absorption coefficient is related to the oil content in the rock. This value can be compared across regions to measuring the oil content in the oil shale. Here we studied three regions and included scanning electron microscope (SEM) and thermogravimetric analysis (TGA) to verify the amount of kerogen within oil shale prior to the constitution of the mineral matrix via the THz response. Aromatic and the aliphatic compounds contribute to the absorption of shales in the THz range due to the relatively high intramolecular interactions—this reveals the mechanism of THz radiation penetration through shales as reported in previous reports. The differences in quantum structure of a molecule between organic and inorganic materials suggest that THz-TDS can be applied to geophysical prospecting and improve the effectiveness of the detection of organics in oil shale.

Keywords: oil shale; terahertz spectroscopy; aliphatic compounds; kerogen

1 Introduction

With continuous worldwide declines in conventional oil and gas production and increasing demand for energy resources, unconventional oil and gas have become a viable supplement due to technological breakthroughs. One is oil shale due to its enormous reserves and valuable chemical composition[1,2]. Oil shale is a sedimentary rock that contains an organic complex called kerogen within its mineral matrix. Oil shale retorting can process kerogen into vapor products. This practice has a long history, and different technologies have been developed for such processing—this offers energy and economic benefits[3].

Pyrolysis or retorting of oil shale processes up to 90% of kerogen into shale oils. These are raw materials for the production of boiler and motor fuels as well as valuable chemicals[4,5]. Many investigations have studied the oil contents of oil shale in recent years. Many techniques have been traditionally used to characterize oil contents such as X-ray diffractometry, scanning electron microscopy, fluorescence microscopy, gas chromatography, etc. [6]. Measuring the oil content in the oil shale is a significant challenge that has not yet been satisfactorily addressed; further research is needed to develop a more reliable technique that provides *in situ* information with minimal sample perturbation and destruction[7].

Recently, terahertz (THz) waves have emerged as a new tool in science and technology especially in the field of oil-gas industry. This was created as a new instrumental analysis tool[8]. THz spectroscopy can detect various intermolecular, lattice, and skeletal vibrations especially organic. Terahertz absorption spectroscopy in addition to conventional instrumental analysis methods is useful for characterizing oil contents. THz wave is sensitive to the polar compounds, especially those with hydrogen bond. The hydrogen bond collective network formed by organic molecules changes on a picosecond time scale, thus causing the THz spectrum to be sensitive to fluctuations in the dipole moment. There have been many experimental studies on evaluating the potential for oil shale production[9]. The data reported here suggest that analysis based on the THz method is likely to contribute to the evaluation of oil shale. Evaluating potential oil content in oil shale using THz spectroscopy is thus possible. Another paper reported that THz spectra reflected the organic matter remaining in the semicoke, and this facilitated pyrolysis results. The critical points of THz absorption correspond well with the oil yield. The analysis indicated that the organic matter remained in the semicoke, and it may convert to additional fuel that can be characterized with THz spectroscopy[10].

Previous work has demonstrated that THz can characterize the lithology and oil content of oil shale. This technique offers clear results for oil shale, semi-coke, and kerogen. It also offers different forms of the parameters including the absorption coefficient and THz anisotropy parameters. However, the problem is that the corresponding theoretical model of the oil shale THz spectrum is not clear, and the oil shale has a complicated compositional structure. Based on the model and mechanism of the THz spectrum response of an oil shale in a certain area, it is difficult to apply the oil to the oil extracted from other areas. It is very difficult to apply the model and mechanism of the THz spectrum collected in one area to other areas. Thus, we selected oil shales from different regions to obtain the various THz response characteristics. The key components that determine the THz spectral response characteristics of oil shale were studied in depth. The problem was solved stepwise via SEM and TGA/DTA.

In this paper, the absorption coefficients of shales from different districts are discussed, and a possible mechanism is presented. The results suggest that oil shale absorption is very complex. The absorption is a combination of several components in the oil shales. Ash has little effect on THz absorption. The content of kerogen determines the THz absorption coefficient, which is highly correlated with oil content of oil shales. Thus, this study indicates that using THz-TDS to evaluate the oil content in oil shale is an effective method for resource exploration.

2 Experimental section

Oil shale samples in this study were obtained from three regions of China including Beipiao city in Liaoning, Huadian city in Jilin, and Kazakhstan Barkol county in Xinjiang. They were mining in the depth of ~320m, ~360m and ~290m, respectively. We sliced these with uniform pressure and time. The actual photographs of the measured oil shales are shown in Figure 1. Their thicknesses are 1.231mm, 1.420mm and 2.370mm, respectively. The relative thickness errors at different locations are less than 5%; thus, the measured spectra can nicely represent the physical information of the shale samples.

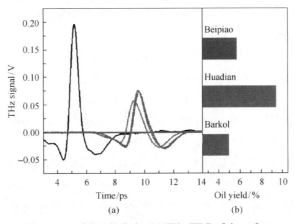

Figure 1   THz-TDS and oil content of the oil shale. (a) THz-TDS of the reference and oil shale samples; (b) oil content of the oil shale selected from Beipiao, Huadian, and Barkol

The measurement setup is comprised of an atypical THz-TDS system with transmission mode and a femtosecond Ti-sapphire laser (MaiTai) from Spectra-Physics. The laser is a diode-pump mode-locked with a center wavelength of 800 nm, a pulse width of 100 fs, and a repetition rate of 80MHz[11]. The laser power is initially attenuated to make the average power equal to 100mW[12]. The THz-TDS system is shown in Figure 2. All tests were carried out in air at 21℃.

TGA is a conventional method widely used in the material and energy sciences as well as geology to determine the changes in weight of a material[13]. It observes mineral decompositions or reactions graphically through peaks or valleys. Sample pretreatment was done before the experiments. Crude samples were crushed and sieved to a size range of 0.04~0.16mm to obtain the experimental oil shale samples. To determine the effects of temperature and heating rate on the thermal degradation of the samples, a TGA with differential thermal analysis (DTA) was used from 30℃ to 1000℃ with a heating rate of 20℃/min under an N_2 atmosphere. TGA-DTA has been extensively used to study weight loss and acquire kinetic parameters during the thermal treatment of fuels. SEM is another technique used to directly characterize the microstructure of minerals. The surface morphology of oil shale samples was recorded using SEM including morphology and energy-dispersive X-ray spectroscopy test[14].

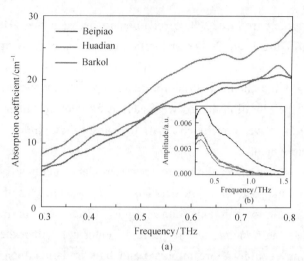

Figure 2 The frequency dependence of absorption coefficient spectra from 0.2~0.8THz. (a) Frequency domain spectroscopy of the reference and the oil shales; (b) the frequency dependence of absorption coefficient spectra from 0.2~0.8THz

SEM characterized the petrographic composition of the different lithologies including the textural, structural and anisotropic features of the rock[15]. The chip samples were coated with gold to make them conductive prior to imaging. These analytical techniques and imaging methods were used to characterize the oil shales to better understand the internal microstructure and chemical composition.

3 Results and discussion

Figure 1(a) shows the THz field signal as a function of time after the transmission of the THz pulses through the air (reference) and oil shales selected from Beipiao, Huadian, as well as the Barkol, respectively. The oil contents of the oil shale from Beipiao, Huadian, and Barkol are 5.00%, 9.95%, and 5.66% (Figure 1(b))[16]. Combined with the reference spectra, there is an obvious attenuation of the peak intensity in the THz-TDS indicating absorption in the THz range[17]. Meanwhile, larger delay time lengths were also observed for the three oil shale samples. The effective optical lengths of the THz through the samples were much larger than those through nitrogen. The THz signal peaks and relative delay time are different from each other showing that physical properties can be identified using THz spectra.

After a fast Fourier transmission, THz frequency-domain spectra (THz-FDS) of reference and samples can be calculated and plotted in the inset of Figure 2. The absorption coefficient α spectra were then obtained by the THz-FDS of the reference and oil shale samples. As shown in Figure 2, the α spectra of oil shale samples were plotted from 0.2THz to 0.8THz. The curves are smooth and have no obvious peaks. In general, α increased with increasing frequency. Besides, the difference between samples in the THz absorption increased with increasing frequency. The relationship between the THz response and oil content is shown in Figure 1 and 2. The absorption effect increased with oil contents in the shale and α was related to the oil shales from different districts. The component had different elemental composition and chemical structure indicating different vibrational modes for the molecules. Some of the vibration modes were caused by intermolecular and intramolecular interactions. Therefore, we concluded that the absorption index was a means of characterizing the oil content.

Organic matter is seen under the microscope (circular structures; Figure 3(a)). The oil shale was a medium/dark gray color. The detailed distribution analysis of organics in shale is plotted in Figure 3. These different colors suggest differences in these area oil shale compositions because the minerals usually have a lighter color than the organic matter[18]. The gray fraction may contain more minerals and thus is darker than the black fraction. Under this scenario, these oil shales may be separated into two general fractions: A black fraction with more organic matter and a gray fraction with more inorganic minerals. The analysis were 62.69wt% (mass fraction) of carbon in the black fraction and 14.17wt% of carbon in the gray fraction. This means that the black fraction contains more shale oil than the gray fraction. We think that the black fraction is kerogen. Kerogen is a mixture of various macromolecular moieties encompassing a wide range of compositions and molecular weights. Its changes with increasing maturity coupled with differences in chemical and structural compositions during geochemical processing.

The mineralogy most commonly found in oil shale includes clay and quartz minerals. SEM-EDS analysis of oil shale offers the same results. As shown in Figure 3(b), the oil shale contains several mineral layers. A microporosity was also found.

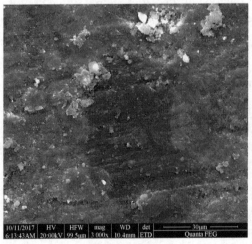
(a)

(b)

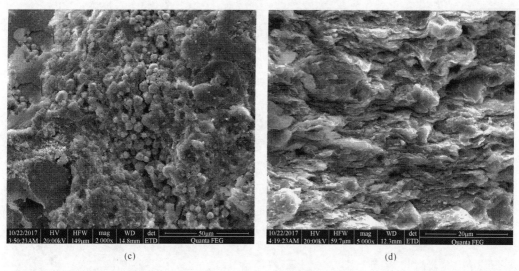

Figure 3 Micrograph characterization of the oil shale. (a) Organic matter is shown under the microscope; (b) the surface of the oil shale selected from Barkol; (c) Huadian oil shale material with the presence of framboidal pyrite; (d) cross-section of oil shale from Beipiao

Another observation was the presence of framboidal pyrite between quartz aggregates in the samples. A detailed distribution analysis of mineral composites in shales is plotted in Figure 4. This mineral was seen only in the upper layer of the oil shale, which was parallel to the sedimentary bedding of the rock[19]. Pyrite distributes unevenly on the surface of oil shale and often bunches together like grapes. Figure 3(c) shows the presence of pyrite in the oil shale. Pyrite was identified in a large number of the oil shale samples, which meant oil shale is a complex mixture containing a variety of minerals[20].

In Figure 3(d), the silt and clay were deposited in a plane-parallel bedding structure of thin strata less than 1mm thick intercalated with organic matter (kerogen). The detailed analysis of anisotropy in shales is plotted in Figure 5. Beipiao oil shale thin sections examined by optical microscopy reveal a fabric that is well-defined by micro-laminations (Figure 3(d)). The laminations are due to alternating layers of rich and lean organic matter. The organic-rich laminae—primarily in the form of kerogen—are 10mm thick. These alternate with thicker layers of predominantly carbonate minerals in the form of dolomite and calcite. The fine kerogen laminae are both continuous and discontinuous laterally with the occasional presence of wavy laminations.

The detailed elemental analysis of the oil shale samples is shown in Figure 4. Some common elements could be observed including C, N, O, Si, Fe, and S. The organic and inorganic compositions had a difference in oil shales from the Beipiao, Huadian, and Barkol districts in China due to the differences in the degree of coalification, local climate, geologic ages, etc. (Figure 4). However, they still resembled the chief constituents, e.g., aliphatic, poly-aromatic, and long chain hydrocar- bons in organic matter as well as illite-muscovite, gypsum, and quartz in the inorganic part. Calcite and dolomite are the main minerals in the carbonate constituents of the shale[21].

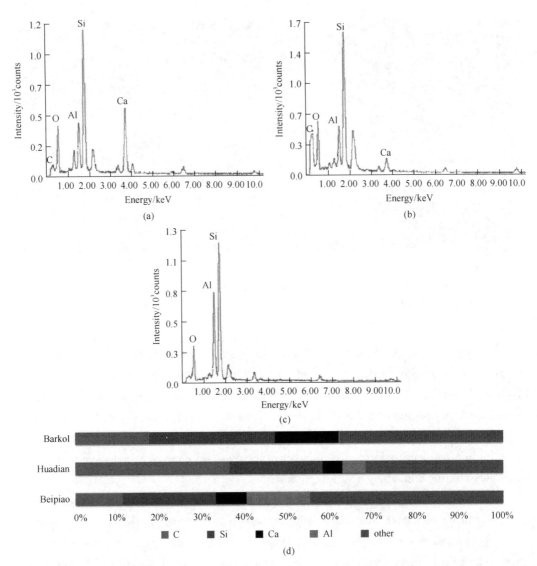

Figure 4 (a) (b) (c) SEM-EDS analysis of the oil shale selected from Beipiao, Huadian, and Barkol; (d) elemental analysis of oil shale samples

Oil shale has several mineral layers. In terms of chemical composition, the oil shales have high levels of SiO_2 and $CaCO_3$ to a lesser degree. Silica (SiO_2), aluminum (Al_2O_3), and iron (Fe_2O_3) were the main oxides[22]. A comparison of the chemical composition of the three oil shale samples showed that C content was lower in the Beipiao and Barkol samples than the Huadian. The Al_2O_3 content was relatively high in the Huadian and Barkol. The high S content of oil shales was consistent with the presence of the FeS[23].

The TGA/DTG curves obtained by heating individual fuel samples at 20 ℃/min are depicted in Figure 5. TGA has been extensively used to investigate the effect of process conditions on the pyrolysis characteristics of oil shales by measuring weight change of samples[24]. The first weight loss of oil shale samples decreased to less than 5wt% and ended at 160℃. This is associated to the separation of the physically absorbed moisture. The loss of labile volatiles led to a DTG peak at 350℃ in the pyrolysis of oil shale samples. Depolymerization and decomposition mainly occurred in this range of temperatures. This led to gas and tar precipitation and char formation[25]. The temperature of the DTG peak maxima (T_{max}) of oil shales from the Beipiao, Huadian, and Barkol entirely overlapped at 350℃. There was a remarkable sharp right-side shoulder peak at 450℃ and 480℃ in the oil shale pyrolysis. The DTG peaks were found at 775℃ in oil shales from the Huadian pyrolysis. This is likely caused by the decomposition of minerals. The residual char of each sample was 80.0wt% (Beipiao), 61.7wt% (Huadian), and 74.7wt% (Barkol). These values matched nicely the ratio of ash and fixed carbon.

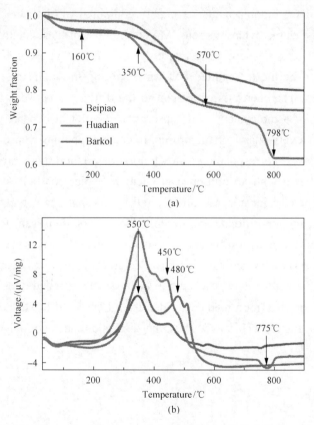

Figure 5 Thermogravimetric analysis of oil shale

The TG/DTG results suggest that pyrolysis of fossil fuel under inert atmosphere is dominated by the decomposition and release of organic compounds (kerogen)[26]. Oil shale pyrolysis involves complex physical and chemical reactions. During the pyrolysis of oil shale, it undergoes the following four successive variations or partial overlap stages: (Ⅰ) moisture evaporation, (Ⅱ) DE volatilization and combustion of volatile matter, (Ⅲ) char combustion, and (Ⅳ) carbonate decomposition[27]. Chemical bonds are broken contenting lighter hydrocarbon molecules in stage 2[28]. The organic matter (kerogen) in oil shale decomposes to produce oil (shale oil or tar), gas, and char/coke. This is the heart of the decomposition-based conversion process. The pyrolysis of kerogen determined the possibly maximal oil yield. In this method, the oil content of oil shale was reflected by changes in the absorption index within the THz range.

Spectroscopy is connected to the quantum structure of a molecule. Kerogen is the insoluble macromolecular organic matter (OM) in sedimentary rocks and is by far the most abundant form of OM on Earth. It is difficult to characterize the specific structure, and it has no defined molecular weight[29]. There are three structures compose most of oil shale kerogens: aromatic structure, aliphatic structure, and heteroatom species (O, N and S). All the carbon skeletal structure of the organic matter is mainly composed of aliphatic carbons, and these carbons mainly exist as methylene carbons. Moreover, most of these methylene carbons exist as straight chains. Bonds in the organic matrix have an effect on absorption. The molecular absorption is caused by the transmitted EM wave shifting the molecules in the medium to higher energy states. The energy is equivalent to the difference between the higher and the lower energy state of a molecule[30]. The absorption coefficient depends on frequency and gives the THz band a very characteristic profile. In general, photons whose energies are in the vicinity of energy levels corresponding to a vibration or rotation mode are more likely to be absorbed[31]. THz has a high absorption coefficient for polar substances especially organic matter such as kerogen. Major minerals present in these oil shales are quartz, calcite, kaolinite, dolomite, analcime, montmorillonite and pyrite, which is not significant for THz absorption. The oil yields of oil shale with different regions are different, which mainly depend on the content of kerogen. The kerogen can be converted into oil, gas and semi-coke by thermal degradation when the temperature reached 500 ℃ in isolate air condition. There is no doubt that kerogen detection is a good method to identify a high yield of shale oil. Organics have significant absorption in the THz wave. Thus, the absorption index strongly depends on the oil yields of oil shale[32].

4 Conclusions

In this project, THz-TDS was employed to obtain important information about oil shale samples. The absorption coefficient strongly depends on the properties of the oil shales and increases as the oil yield increases from 5.00% to 5.66%~9.96% for Beipiao, Barkol, and Huadian, respectively. The mechanism was proposed via comparative analysis with SEM and TGA. The figures emerging in the THz region are related to the kerogen, which is composed of complex organic matter. Aromatic and the aliphatic compounds contribute to the absorption of shales in the THz range due to the relatively high intramolecular interactions. Slices of oil shale were used to directly measure the oil content and did not require other treatments such as semi-coke. The absorption intensity was related to the content of oil. With the development of THz technique, the emitter and detector will be more powerful and the detection of oil shales can overcome the technical challenges. THz spectroscopy can directly detect oil shale without any pyrolysis and its mechanism is revealed. Our data suggests that analysis based on the THz method could directly determine the oil shale as an effective method.

Acknowledgements

This work was supported by the National Nature Science Foundation of China (Grant No. 11574401), the Science Foundation of China University of Petroleum, Beijing (Grant Nos. 2462017YJRC029, 2462018BJC005 and yjs2017019) and the Beijing Natural Science Foundation (Grant No. 1184016). We would like to thank LetPub (www.letpub.com) for providing linguistic assistance during the preparation of this manuscript.

References

[1] Berge J, Norling M, Vorobiev A, et al. Yield and characteristics of shale oil from the retorting of oil shale and fine oil-shale ash mixtures. Appl. Energy, 2013, 111: 234-239.

[2] Saif T, Lin Q, Bijeljic B, et al. Microstructural imaging and characterization of oil shale before and after pyrolysis. Fuel, 2017, 197: 562-574.

[3] Nazzal J M. Influence of heating rate on the pyrolysis of Jordan oil shale. Anal. Appl. Pyrol., 2002, 62: 225-238.

[4] Wang S, Jiang X, Han X, et al. Investigation of Chinese oil shale resources comprehensive utilization performance. Energy, 2012, 42: 224-232.

[5] Williams P T, Chishti H M. Influence of residence time and catalyst regeneration on the pyrolysis-zeolite catalysis of oil shale. Anal. Appl. Pyrol., 2001, 60: 187-203.

[6] Saif T, Lin Q, Butcher A R, et al. Multi-scale multi-dimensional microstructure imaging of oil shale pyrolysis using X-ray microtomography, automated ultra-high resolution SEM, MAPS Mineralogy and FIB-SEM. Appl. Energy, 2017, 202: 628-647.

[7] Wang J, Feng L, Steve M, et al. China's unconventional oil: A review of its resources and outlook for long-term production. Energy, 2015, 82: 31-42.

[8] Li Y, Wu S, Yu X, et al. Optimization of pyrolysis efficiency based on optical property of semicoke in terahertz region. Energy, 2017, 126: 202-207.

[9] Tong J, Han X, Wang S, et al. Evaluation of structural characteristics of huadian oil shale kerogen using direct techniques. Energy & Fuels, 2011, 25: 4006-4013.

[10] Zhan H, Wu S, Bao R, et al. Qualitative identification of crude oils from different oil fields using terahertz time-domain spectroscopy. Fuel, 2015, 143: 189-193.

[11] Zhan H, Zhao K, Xiao L, et al. Spectral characterization of the key parameters and elements in coal using terahertz spectroscopy. Energy, 2015, 93: 1140-1145.

[12] Ge L, Zhan H, Leng W, et al. Optical characterization of the principal hydrocarbon components in natural gas using terahertz spectroscopy. Energy & Fuels, 2015, 29: 1622-1627.

[13] Huang Y, Zhang M, Lyu J, et al. Modeling study on effects of intraparticle mass transfer and secondary reactions on oil shale pyrolysis. Fuel, 2018, 221: 240-248.

[14] Kelly S, El-Sobky H, Torres-Verdin C, et al. Assessing the utility of FIBSEM images for shale digital rock physics. Adv. Water Resour, 2015, 95: 302-316.

[15] Saif T, Lin Q, Butcher A R, et al. Multi-scale multi-dimensional microstructure imaging of oil shale pyrolysis using X-ray microtomography, automated ultra-high resolution SEM, MAPS Mineralogy and FIB-SEM. Appl. Energy, 2017, 202: 628-647.

[16] Bai F, Guo W, Lü X, et al. Kinetic study on the pyrolysis behavior of Huadian oil shale via non-isothermal thermogravimetric data. Fuel, 2015, 146: 111-118.

[17] Zhan H, Wu S, Bao R, et al. Water adsorption dynamics in active carbon probed by terahertz spectroscopy. RSC Adv., 2015, 5: 14389-14392.

[18] Li D, Li R, Zhu Z, et al. Origin of organic matter and paleo-sedimentary environment reconstruction of the Triassic oil shale in Tongchuan City, southern Ordos Basin (China). Fuel, 2017, 208: 223-235.

[19] Na J G, Im C H, Chung S H, et al. Effect of oil shale retorting temperature on shale oil yield and properties. Fuel, 2012, 95: 131-135.

[20] Ribas L, Neto J M D R, França A B, et al. The behavior of Irati oil shale before and after the pyrolysis process. J. Petrol. Sci. Eng., 2017, 152: 156-164.

[21] Al-Otoom A Y, Shawabkeh R A, Al-Harahsheh A M, et al. The chemistry of minerals obtained from the combustion of Jordanian oil shale. Energy, 2005, 30: 611-619.

[22] Guo H, Lin J, Yang Y, et al. Effect of minerals on the self-heating retorting of oil shale: Self-heating effect and shale-oil production. Fuel, 2014, 118: 186-193.

[23] Al-Makhadmeh L A, Maier J, Batiha M A, et al. Oxyfuel technology: Oil shale desulfurization behavior during staged combustion. Fuel, 2016, 190: 229-236.

[24] Lahti P, Heikinheimo A, Johansson T, et al. A TGA/DTA-MS investigation to the influence of process conditions on the pyrolysis of Jimsar oil shale. Energy, 2015, 86: 749-757.

[25] Bai F, Sun Y, Liu Y, et al. Evaluation of the porous structure of Huadian oil shale during pyrolysis using multiple approaches. Fuel, 2017, 187: 1-8.

[26] Li S, Ma X, Liu G, et al. A TG-FTIR investigation to the co-pyrolysis of oil shale with coal. Anal. Appl. Pyrol., 2016, 120: 540-548.

[27] Moine E, Groune K, El A, et al. Multistep process kinetics of the non-isothermal pyrolysis of Moroccan Rif oil shale. Energy, 2016, 115: 931-941.

[28] Li S, Yue C. Study of different kinetic models for oil shale pyrolysis. Fuel Process Technol, 2003, 85: 51-61.

[29] Wang Q, Hou Y, Wu W, et al. The relationship between the humic degree of oil shale kerogens and their structural characteristics. Fuel, 2017, 209: 35-42.

[30] Kokkoniemi J, Lehtomäki J, Juntti M. A discussion on molecular absorption noise in the terahertz band. Nano. Commun. Networks, 2016, 8: 35-45.

[31] Wang S, Jiang X, Han X, et al. Investigation of Chinese oil shale resources comprehensive utilization performance. Energy, 2012, 42: 224-232.

[32] Liu H, Bai W, Feng J, et al. The synthesis of large-diameter ZnTe crystal for THz emitting and detection. Journal of crystal Growth, 2017, 475: 115-120.

Layer caused an anisotropic terahertz response of a 3D-printed simulative shale core

Xinyang Miao[1,2] Limei Guan[2] Rima Bao[1,2] Yizhang Li[2] Honglei Zhan[2]
Kun Zhao[1,2] Xiaodong Wang[3] Fan Xu[3]

(1.State Key Laboratory of Petroleum Resources and Prospecting, China University of Petroleum, Beijing 102249, China; 2.Beijing Key Laboratory of Optical Detection Technology for Oil and Gas, China University of Petroleum, Beijing 102249, China; 3.School of Physics and Optoelectronic Engineering, Yangtze University, Jingzhou 434023, China)

Abstract: Energy demands have motivated the development of shale formations as significant unconventional reservoirs. The anisotropy of shales plays a significant role in both the mechanical behavior and engineering activities. Alternating layers presented in shales affect the propagation of waves, causing anisotropy at various frequencies. Simplifying the complicated interior structures of shales is conducive to characterize the anisotropic properties. Therefore, simulative shale core samples were designed and fabricated using additive manufacturing processes, and layer-caused dielectric anisotropy was investigated by terahertz time-domain spectroscopy (THz-TDS). On the basis of effective medium theory, the change of the optical length caused by refraction of rays was discussed and modeled. It is believed that the refraction of rays at the interfaces is the source of terahertz (THz) propagation anisotropy in the multilayered structure, and the anisotropy degree is mainly influenced by the layer thickness as well as the refractive index.

Keywords: anisotropy; shale; terahertz; additive manufacturing

1 Introduction

Shales, acting as cap rocks of conventional reservoirs formations, have been found in most global sedimentary basins. Most recently, energy demands have motivated the development of shale formations as significant unconventional reservoirs. Layers, one of the most common structural features of sedimentary rocks, have been the focus of numerous studies by both geologists and engineers[1]. As the most common of sedimentary rocks, shales have been estimated to form ~75% of all sedimentary rocks on earth[2]. They are fine-grained sedimentary rocks composed primarily of clay, organics, as well as hard minerals such as quartz, feldspar, and pyrite[3]. Owing to various combined effects of partial alignment of platy clay particles, layering, micro-cracks, low-aspect-ratio pores, and kerogen inclusions, shales are often highly anisotropic[4]. The anisotropy of shales plays a significant role in both the mechanical behavior and engineering activities. For hydraulic fracturing, the noted anisotropy affects drilling, borehole stability as well as perforation orientation, and also controls the initiation and propagation of hydraulic fractures, and further impacts on the flow and leakage of the fracturing fluid[5]. Thus, the influences in the anisotropy of shales must be taken into account for traditional and advanced geophysical exploration technologies. Ultrasonic systems have been widely used to study shale anisotropy in the laboratory[6,7]. These studies have mainly focused on the elastic properties of shales, and have revealed that the preferential alignment of clay minerals and cracks is the major factor leading to the intrinsic anisotropy of shales. Organics' shapes and distribution exhibit some elongation parallel with bedding direction, making the organics in shales an important source of elastic anisotropy[8]. Besides, the scanning electron microscopy (SEM) technique has been commonly used for the microstructure determination of shales[9]. The planes perpendicular and parallel to the bedding were analyzed by SEM, showing some patchy kaolinite and acicular illite with preferred orientations parallel to the bedding. An oblique-incidence reflectivity difference (OIRD) technique was utilized to characterize the layered distribution of the dielectric properties in shales, indicating that directional changes of the dielectric and surface properties resulted due to micro-cracks and particle orientation[10]. The presence of alternating layers in shales also affects the propagation of waves, causing anisotropy at various frequencies[11]. Shales have complex layer structures consisting of various mineral components and kerogen contents, and simplifying the complicated interior structure of shale is conducive to characterizing the anisotropic properties[12]. In recent years, additive manufacturing (AM) technology has become a powerful tool for direct digital manufacturing[13]. Being well known as "3D printing", the technique possesses the advantages of accuracy and cost-efficiency. Based on the data from three-dimensional models, the materials can be rapidly solidified to fabricate physical objects. AM also provides an innovative approach to study rock[14]. This technique has been used to fabricate test specimens of complicated rock structures with cracks and joints to investigate the stress changes in the process of mining, showing a promising way to quantify and visualize the complex fracture structures and their influences on the stress distribution of underground rocks[15].

Since the layering in shales was on the sub-millimeter scale, THz wave, with wavelengths of 30μm~3mm and frequencies of 0.1~10THz, it is a sensitive tool for studying the effect of alternating layers on the degree of anisotropy. Owing to the unique advantages, this technique is expected to be a "silver bullet" in various aspects of oil and gas resources[16], especially in reservoirs and rocks[17]. At present, THz-TDS has been applied to investigating the optical properties of marble, sandstone, and limestone[18]. The dielectric properties of montmorillonite clay[19], micaceous clay[20], kaolinite, halloysite[21], and clay-based polyamide nanocomposite films were measured at the THz region[22]. Meanwhile, the THz technique has also been employed to study the anisotropic response of materials. Various crystals and alternating structures have been studied to exhibit THz birefringence[23]. In our previous work[24], anisotropic responses of oil shales were initially studied by THz time domain spectroscopy (THz-TDS). The refractive index anisotropies of the oil shales were compared with each other and plotted as a function of the oil yield, showing that the increase of organic matter has promoted anisotropy in the THz range. In this article, we consider the anisotropic response of 3D-printed simulated "shale core" by THz-TDS. The refraction of the ray at the interfaces was considered and modeled to figure out the source of THz propagation anisotropy in the multilayered structure.

2 Experiments

Actual shales are characterized by breaks along thin laminaes or parallel layering less than 1cm in thickness, with layers composed of different minerals or liquids[2,5]. Here, in this research, we designed cylindrical structures consisting of alternating layers made of solid and air gaps to simulate the shale core. We focused on the anisotropic response caused by the layer in this study, herein, the regular structures of the 3D printed samples were beneficial to the theoretical analysis and helpful to exclude other factors. Simulative samples were designed and fabricated using the AM processes started from a software model illustrated in Figure 1(a). Three cylindrical structures (numbered S_1, S_2 and S_3) consisting of alternating layers made of solid and air gaps were designed with the diameter 21mm and the wall thickness 1mm. The number of the layers in them were uniform but the thicknesses varied. 3D printing of the models was achieved by a XYZ Printing^{TM} 3D printer (da Vinci 2.0) based on a technique of curing polymer named fused deposition modeling technology (FDM). Acrylonitrile-butadiene-styrene (ABS), which is a kind of thermoplastic material commonly employed in FDM, was used as the raw material of the samples. The material was firstly heated and melted, then extruded and solidified quickly.

We used a THz-TDS setup with the frequency range of 0.1~3THz based on photoconductive antenna and electro-optical sampling to study the layer-caused dielectric anisotropy. Figure 1(b) appears in a schematic drawing of the experimental setup. An femtosecond laser beam (800 nm) was split into two beams, the pump beam was focused onto the surface of a biased GaAs photo-conductive antenna for THz generation and the probe beam for electro-optic detection. THz pulses were focused onto a sample by an optical lens, and met the probe laser beam at the ZnTe crystal in the THz detector. The 3D-printed model was installed in the center of a manually rotational stage, and the emitter and detector beams were aligned on the antipodes for a transmission experiment. The experiments were performed under air atmosphere at a temperature of 294.1 K and a humidity of ~10% with a signal-to-noise ratio of ~1000 during the measurement. The setup provides a direct measurement of THz wave propagation along the diameter of the circle, and hence the angle dependent THz parameters can be obtained. The rotation angles (θ) were defined as the angles between the input orientation of THz ray and the layer. THz-TDS of the air was initially measured as the reference signal, and the samples with different d_1 and d_2 were rotated and tested one by one.

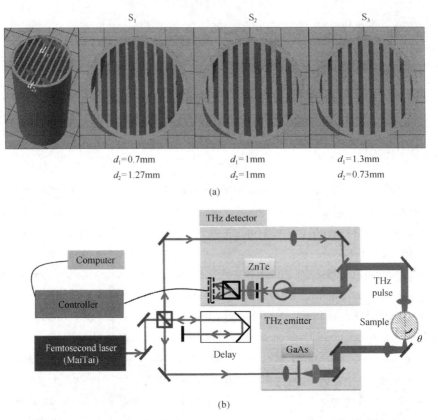

Figure 1 (a) 3D models of the simulated "shale core"; (b) top-view, schematic of the experimental setup, in which θ is defined as the angle between the input orientation of terahertz ray and the layer

3 Results and discussion

As an example of the raw THz data, time-domain spectra of S_1 with varies θ from 0° to 90° are presented in Figure 2. Waveforms of the sample with different θ are variant from each other. The observed THz wave signals correspond to direct waves with different optical distances owing to the circular structure followed by the scattered waves arriving at later times. The delay time (t_1) of the first peak is located (marked by the short vertical line in Figure 2) and extracted. It is obvious that the time delay of the THz wave is strongly dependent upon θ for the arrival of the first THz signal. t_1 gradually increases with rotation from 0° to 90°. To be specific, t_1 of S_1 at the first signal is 12.06 ps, 14.05 ps, 16.55 ps, 18.72 ps, 19.56 ps, 20.43 ps and 20.28 ps with θ equals 0°, 15°, 30°, 45°, 60°, 75° and 90° in Figure 2, respectively. The same measurements have also been performed for S_2 and S_3; the group angle dependent t_1 is presented in Figure 3. We then calculated the effective refractive index (n) by t_1 as well as the reference (t_0),

$$n = 1 + \frac{c(t_1 - t_0)}{D_0} \qquad (1)$$

where c is the speed of light, and D_0 is the external diameter of the circle. According to Equation (1), n is related to t_1; thus the variation trends of n are consistent with t_1, as shown in Figure 3. It is clear from Figure 3 that all the samples have significant t_1 and n anisotropy, meanwhile, there is a data symmetry every 180° owing to the geometry of the layer consisted samples. In an angular period of 180°, both t_1 and n basically increase in the first half from 0° to 60°, and the fall is symmetric from 120° to 180°. At angles of 60°~120°, the values are basically invariant, especially for S_2 and S_3. Besides, both of them are approximately the same for all three samples with the transmission direction parallel to the layer (at 0°, 180° and 360°). The average values of t_1 are 12.29 ps, 12.64 ps and 12.66 ps, as well as n 1.10, 1.11 and 1.11, respectively. However, different variation curves can be observed among the samples. The maximum t_1 and n raised with increasing d_1.

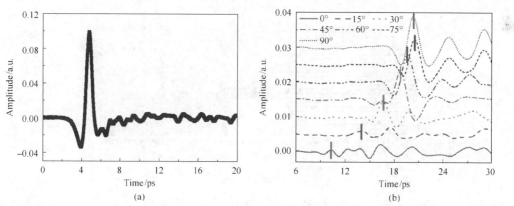

Figure 2　Time-domain spectra of the (a) reference and (b) S_1 with varies angles from 0° to 90°. The short vertical bar marks the first THz waveform signal for each angle

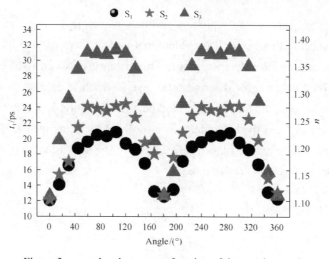

Figure 3　t_1 and n change as a function of the rotation angle

In order to verify the experimental result, we conduct an analysis of the THz ray propagation. Given that the ray is subject to lateral movement after going through the solid part, the directions of rays intro or inter the layers are consistent respectively. Thus we propose an imaginary solid layer and an air layer to simulate the layered 3D-printed models, whose thicknesses are the sum of solid and air layers that THz ray penetrates, as shown in Figure 4. In this equivalent model, the original optical length separated in same sort of layers are aligned without any change in the total optical length or incident angle. A rectangle, whose diagonal is the diameter of the circumscribed circular intercept, is depicted to subsume the trace of the ray within the sample. The ray enters this rectangle along its diagonal at first, and then it refracts when traveling through the interface due to different n values on both sides of the interface. We managed to figure out the optical length in two equivalent layers, and then applied effective medium theories to calculate n. As we suppose, the calculated n should be close to the measured one.

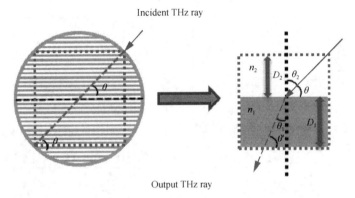

Figure 4 Schematic diagram of the model based on the effective medium theory, in which n_1 and n_2 as well as D_1 and D_2 represent the refractive index as well as the effective thickness of the solid layers and air layers, respectively

We define a reference plane which is parallel to any layer and the entire sample is symmetry with respect to this plane. It is evident that two surfaces of both the air layer and the solid layer are parallel to the reference plane. The distance between two surfaces of a solid layer is the effective thickness for discrete solid layers which is defined as D_1. Similarly, the effective thickness for discrete air layers is marked as D_2 (Figure 4). On account of the angle between the incident ray and the reference plane (defined as θ), as well as the angle between the refracted ray and the reference plane (defined as θ'), the actual length of the ray within the solid layer (L_1) and air layer (L_2) ought to be

$$L_1 = \frac{D_1}{|\sin\theta'|} \tag{2}$$

and

$$L_2 = \frac{D_2}{|\sin\theta|} \tag{3}$$

respectively. According to the refractive index of the ABS material and the air (defined as n_1 and n_2), a relationship was built based on the equivalent optical length:

$$n_1 L_1 + n_2 L_2 + 2n_1 D' = n(L_1 + L_2 + 2D') \tag{4}$$

where D' is the thickness of the wall. On the basis of the Snell's law as well as the refracting angle θ_1 and the incident angle θ_2, we obtained

$$n_1 \sin\theta_1 = n_2 \sin\theta_2 \tag{5}$$

Then, according to the relationship between the angles shown in Figure 4,

$$\sin\theta' = \sqrt{1 - \left(\frac{n_2}{n_1}\right)^2 \cos^2\theta} \tag{6}$$

by applying Equations (2), (3) and (6) to Equation (4), we obtained

$$n = \frac{\dfrac{n_2 D_2}{|\sin\theta|} + \dfrac{n_1 D_1}{\sqrt{1 - \left(\dfrac{n_2}{n_1}\right)^2 \cos^2\theta}} + 2n_1 D'}{\dfrac{D_2}{|\sin\theta|} + \dfrac{D_1}{\sqrt{1 - \left(\dfrac{n_2}{n_1}\right)^2 \cos^2\theta}} + 2D'} \tag{7}$$

The theoretical results of n at varies angles of 15°, 30°, 45°, 60°, 75° and 90° were calculated by Equation (7) with the THz-TDS measured refractive index of air (n_2=1) and ABS material (n_1≈1.56) used in this experiment. As plotted in Figure 5, the calculated values show a good agreement with that of the experiment. The maximum relative errors between the calculated and experimental results are no larger than 3%. Besides, in the cases of $\theta = 0°$, part of the rays penetrate air layers and approach the detector at first, followed by those travelling through solid layers; n is approximately calculated by the length of air layer and the thickness of the wall. Actually, owing to a slight deflection of the rays, the measured n is larger than the calculated one under this circumstance.

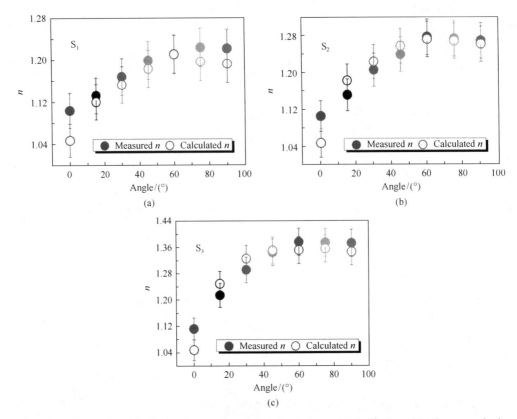

Figure 5   Measured n and calculated n as a function of the angle for (a) S_1, (b) S_2 and (c) S_3, respectively.
The solid and hollow circles represent the experiment and simulation results, respectively.
Error bars represent 3% of the data values

It is obvious that the presence of alternating layers in the model has a strong influence on the degree of dielectric anisotropy. The results have also shown an attenuation of n in the THz propagation direction parallel to the layering compared with that perpendicular to the layering. According to an analysis of THz ray propagation as well as experimental data, we consider that the change of optical length, which results refraction of the ray at the interfaces, leads to the n anisotropy of the models. As described in Equations (2) and (6), the deflection of THz ray shortens the actual optical length within the solid layers, and such deflection diminishes with the increase of θ in the range of 0°~90°, while the proportion of the optical length within the solid raises. Since n_1 is larger than n_2, n augments gradually with the increase of θ from 0° to 90°.

With a lack of thorough understanding of anisotropy theoretically, the anisotropic properties of shales have been of long-standing interest, and the sources of elastic anisotropy are inherently complex and difficult to discriminate[5]. Ultrasonic measurements under variable saturation and pressure conditions in the laboratory have revealed that anisotropy of elastic wave propagation is caused mainly by the preferential alignment of mineral grains and thin layers[6,7]. Near cracks, the effective density and thickness are clearly different from those at other locations. Such heterogeneity is minimized at high confining pressures upon the closure of cracks[8,10]. Moreover, solid organic matter is also a strong source of anisotropy in shales owing to the elongation distribution of organic matter parallel with the bedding[2,5]. The organic matter appears to have little intrinsic birefringence; however, organic matter, such as kerogen, has always exhibited some elongation parallel with the bedding, which is caused by lithostatic overburden and deformation, and by the original orientation when kerogen first deposits[8]. For electromagnetic waves, the presence of alternating layers in shales has a strong influence on the degree of dielectric anisotropy[11]. In our previous study, we modeled the oil shale samples as a uniaxial birefringent medium, and discussed the organic matter-dependent n anisotropy[23]. Here, in this work, the effect of the layer structure is investigated. The refraction of the ray at the interfaces is believed to be the source of anisotropy, which was never considered in previous studies. In this circumstance, the anisotropy degree is mainly influenced by the layer thickness and the refractive index. In addition, for shales consisting of kerogen-rich and kerogen-poor layers, the refractive index of the layers is mainly decided based on the organic matter content. For a better guidance of the development of shale, the relationship between the anisotropy caused by refraction and the organic matter content should be investigated in the future to quantify the anisotropic properties of shales.

4 Conclusions

In conclusion, THz-TDS was employed to investigate the anisotropic response of 3D-printed multilayered samples. On the basis of the effective medium theory, the change of the optical length caused by refraction of the ray was considered and modeled. It is believed that the refraction of the ray at the interfaces is the source of THz propagation anisotropy in the multilayered structure, and the anisotropy degree is mainly influenced by the layer thickness and the refractive index. The results prove that THz-TDS combined with the AM technique is willing to be an effective means for researching the anisotropy of shales.

Acknowledgements

This work was supported by the National Nature Science Foundation of China (Grant Nos. 61405259 and 11574401).

References

[1] Wild P, Crampin S. The range of effects of azimuthal isotropy and EDA anisotropy in sedimentary basins. Geophysical Journal International, 1991, 107(3): 513-529.

[2] Sayers C M. The elastic anisotropy of shales. Journal of Geophysical Research Solid Earth, 1994, 99(B1): 767-774.

[3] Han S, Zhang J, Li Y, et al. Evaluation of lower Cambrian shale in northern Guizhou Province, south China:Implications for shale gas potential. Energy & Fuels, 2013, 27(6): 2933-2941.

[4] Zhubayev A, Houben M E, Smeulders D M J, et al. Ultrasonic velocity and attenuation anisotropy of shales, Whitby, United Kingdom. Geophysics, 2016, 81(1): 45-56.

[5] Vernik L, Liu X. Velocity anisotropy in shales: A petrophysical study. Geophysics, 1997, 62(2): 521-532.

[6] Blum T E, Adam L, Van Wijk K. Noncontacting benchtop measurements of the elastic properties of shales. Geophysics, 2013, 78(3): C25-C31.

[7] Piane C D, Dewhurst D N, Siggins A F, et al. Stress-induced anisotropy in brine saturated shale. Geophysical Journal International, 2011, 184(2): 897-906.

[8] Allan A M, Kanitpanyacharoen W, Vanorio T. A multiscale methodology for the analysis of velocity anisotropy in organic-rich shale. Geophysics, 2015, 80(4): C73-C88.

[9] Lin B, Cerato A B. Applications of SEM and ESEM in microstructural investigation of shale-weathered expansive soils along swelling-shrinkage cycles. Engineering Geology, 2014, 177: 66-74.

[10] Zhan H L, Wang J, Zhao K, et al. Real-time detection of dielectric anisotropy or isotropy in unconventional oil-gas reservoir rocks supported by the oblique-incidence reflectivity difference technique. Scientific Reports, 2016, 6(1): 39306.

[11] Scales J A, Batzle M. Millimeter wave analysis of the dielectric properties of oil shales. Applied Physics Letters, 2006, 89(2): 024102.

[12] Liu W Q, Wang D N, Su Q. Dual media model of shale layer with anisotropy involved and its simulation on gas migration. Natural Gas Geoscience, 2016 27(8): 1374-1379.

[13] Gross B C, Erkal J L, Lockwood S Y, et al. Evaluation of 3D printing and its potential impact on biotechnology and the chemical sciences. Analytical Chemistry, 2014, 86(7): 3240-3253.

[14] Jiang C, Zhao G F. A preliminary study of 3D printing on rock mechanics. Rock Mechanics & Rock Engineering, 2015, 48(3): 1041-1050.

[15] Bennett K C, Berla L A, Nix W D, et al. Instrumented nanoindentation and 3D mechanistic modeling of a shale at multiple scales. Acta Geotechnica, 2015, 10(1): 1-14.

[16] Miao X Y, Zhan H L, Zhao K. Application of THz technology in oil and gas optics. Science China: Physics, Mechanics & Astronomy, 2017(2): 88-90.

[17] Bao R M, Wu Z K, Li H, et al. Characterization of inclusions in evolution of sodium sulfate using terahertz time-domain spectroscopy. Analytical Sciences the International Journal of the Japan Society for Analytical Chemistry, 2017, 33(9): 1077-1080.

[18] Schwerdtfeger M, Castro-Camus E, Krügener K, et al. Beating the wavelength limit:Three-dimensional imaging of buried subwavelength fractures in sculpture and construction materials by terahertz time-domain reflection spectroscopy. Applied Optics, 2013, 52(3): 375-380.

[19] Wilke I, Ramanathan V, Lachance J, et al. Effects of the November 2012 flood event on the mobilization of Hg from the Mount Amiata Mining District to the sediments of the Paglia River Basin. Applied Clay Science, 2014, (87): 61-65.

[20] Janek M, Matejdes M, Szőcs V, et al. Dielectric properties of micaceous clays determined by terahertz time-domain spectroscopy. Philosophical Magazine, 2010, 90(17-18): 2399-2413.

[21] Zich D, Zacher T, Darmo J, et al. Far-infrared investigation of kaolinite and halloysite intercalates using terahertz time-domain spectroscopy. Vibrational Spectroscopy, 2013, 69(6): 1-7.

[22] Nagai N, Imai T, Fukasawa R, et al. Analysis of the intermolecular interaction of nanocomposites by THz spectroscopy. Applied Physics Letters, 2004, 85(18): 4010-4012.

[23] Arikawa T, Zhang Q, Ren L, et al. Review of anisotropic terahertz material response. Journal of Infrared Millimeter & Terahertz Waves, 2013, 34(11): 724-739.

[24] Miao X Y, Zhan H L, Zhao K, et al. Oil yield characterization by anisotropy in optical parameters of the oil shale. Energy & Fuels, 2016, 30(12): 10365-10370.

Characterizing the oil and water distribution in low permeability core by reconstruction of terahertz images

Rima Bao[1] Xinyang Miao[1] Chengjing Feng[1] Yingzhi Zhang[2]
Honglei Zhan[1] Kun Zhao[1] Maorong Wang[3] Jianquan Yao[3]

(1.Beijing Key Laboratory of Optical Detection Technology for Oil and Gas, China University of Petroleum, Beijing 102249, China; 2.Qingxin Oil Field Development Company Limited, Anda 151400, China; 3.College of Precision Instrument and Opto-electronics Engineering, Tianjin University, Tianjin 300072, China)

Exploitation of low permeability oilfield has become an interest of the world due to the decline of conventional oil reserves. Flooding has been observed to be able to enhance oil recovery in tight sandstone reservoir. The pore throat diameter of the sandstone described above is in nano/micro scale, and remaining oil and irreducible water exist simultaneously. Therefore, better understanding of reservoir behavior is essential to predict future performance of water flooding process[1-4]. Terahertz (THz) radiation (wavelengths from 30 μm to 3mm) has recently emerged as an efficient tool in many fields such as oil and gas characterization, cultural heritage investigation, air pollution detection and design of metamaterials[5-8]. Applications of THz techniques are promising such as time domain spectroscopy (TDS) in solid-state physics and aqueous chemistry[9]. Response of THz wave is intrinsically associated with low energy events such as molecular torsion or vibration as well as inter and intramolecular hydrogen-bonding, causing the strong attenuation of THz waves by water[10]. Therefore, THz spectroscopy was considered as an effective means for detection of water content and distribution[11]. In this letter, we investigated the distribution of oil and water in low permeability core via reconstruction of THz images.

The argillaceous sandstone reservoir cores with low porosity (from 12% to 14%) and low permeability (1.0mD, 1D=0.986923×10^{-12}m^2) were employed as the object in this study. The core has a diameter of 25mm. Before each measurement, cores were firstly deoiled and saturated with water, then flooded with kerosene to an irreducible water state. All the cores were immersed into the solution blended by kerosene and water to simulate a real situation underground. One of the cores was sliced into 10 pieces. The thickness of each wafer was about 2mm, and all of them were numbered from 1 to 10 according to the original positions.

The THz spectroscopy was measured with a reflection THz-TDS. The THz wave source was introduced by the Cherenkov type phase matching and optical rectification in a non-linear optical crystal (LiNbO$_3$). The effective frequency range was 0.1~4THz and the minimum spatial resolution of the system was below 0.3mm. Scanning was achieved via mechanical motion of object so as to obtain information to characterize the entire slice. Parameters such as peak amplitude and delay time were extracted directly from the spectrum for imaging. The center of the wafer was selected as the starting point of scanning. The sweep range referred to an 8mm radius circle, and the step interval was 0.4mm.

The reflective spectra of a selected region (2mm×2mm) on the first slice (numbered 1) was shown in Figure 1(a). The amplitudes of the spectra were different from each other due to the component changes of the rock at different positions. Since water had a much stronger interaction with THz waves compared with oil and kerosene[12], the attenuation of the radiation can be related to the water content in the porous media. In order to display more clearly, the peak amplitude of the THz spectrum at each position was selected for imaging, and the levels of the intensity were represented by various colors. Figure 1(c) presents the THz amplitude images of the samples numbered 1, 3, 6 and 10, respectively. Since the oil and water distribution was not absolutely homogenous, regions with different compositions were identified in the spatial distribution map. Colors of the images displayed in Figure 1(c) varied gradually from red to azure, and changed back to red at last among these images, indicating that the content of water rose first and then declined. As shown in Figure 1(b) the THz images of all the pieces were rebuilt according to the original position in order to figure out the oil and water distribution within the core. On the basis of analysis above, we could predict the oil-water distribution via the reconstructed diagram. By the changes of the colors, kerosene mainly existed in the upper area, and remaining oil failed to be flooded and occupied the lower regions of the core while the irreducible and free water was distributed in the center of the core. This result is consistent with the flooding processes in low-permeability cores. It is noteworthy that the color changed suddenly at various positions due to a few number of slices, and the accurate evaluation of water content will be studied with the achievement of thinner slices in the future.

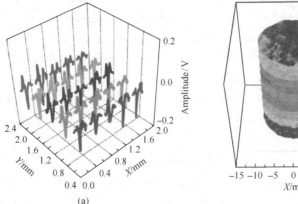
(a)

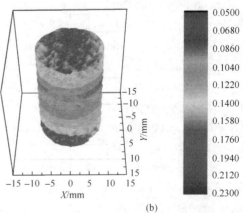
(b)

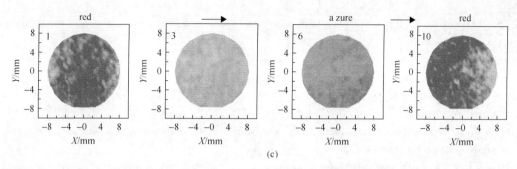

(c)

Figure 1 (a) THz spectra at the selected regions of the sample 1; (b) reconstruction of all the 10 samples by order; (c) THz images of the samples numbered 1, 3, 6 and 10 obtained by exacting the peak amplitude of each point, and mixed colors were selected to represent the level of intensity visually

In this letter, THz imaging technology was used to predict the occurrence state of oil and water in handled sandstone cores. THz amplitude images were performed to characterize the distribution of oil and water in each piece. The existence of kerosene, water and remaining oil could be clearly identified by THz three-dimension (THz-3D) reconstruction. The results prove THz to be a promising means for detecting reservoir properties. Furthermore, the image reconstruction method used in this letter may provide a new idea for THz tomography technique and application.

Acknowledgements

This work was supported by the National Key Basic Research Program of China (Grant No. 2014CB744302), the Specially Funded Program on National Key Scientific Instruments and Equipment Development (Grant No. 2012YQ140005) and the National Natural Science Foundation of China (Grant nos. 61405259 and 11574401). The authors also thank the testing service provided from the Advantest Corporation.

References

[1] Yuan D Y, Hou J R, Song Z J, et al. Residual oil distribution characteristic of fractured-cavity carbonate reservoir after water flooding and enhanced oil recovery by N_2 flooding of fractured-cavity carbonate reservoir. Journal of Petroleum Science & Engineering, 2015, 129: 15-22.

[2] Zhao D W, Gates I D. On hot water flooding strategies for thin heavy oil reservoirs. Fuel, 2015, 153: 559-568.

[3] Shojaei M J, Ghazanfari M H, Masihi M. Relative permeability and capillary pressure curves for low salinity water flooding in sandstone rocks. Journal of Natural Gas Science & Engineering, 2015, 25: 30-38.

[4] Torabi F, Mosavat N, Zarivnyy O. Predicting heavy oil/water relative permeability using modified Corey-based correlations. Fuel, 2016, 163: 196-204.

[5] Zhang Z W, Wang K J, Lei Y, et al. Non-destructive detection of pigments in oil painting by using terahertz tomography. Science China: Physics, Mechanics & Astronomy, 2015, 58(12): 124202.

[6] Li Q, Zhao K, Zhang L W, et al. Probing $PM_{2.5}$ with terahertz wave. Science China: Physics, Mechanics & Astronomy, 2014, 57(12): 2354-2356.

[7] Bao R M, Li Y Z, Zhan H L, et al. Probing the oil content in oil shale with terahertz spectroscopy. Science China: Physics, Mechanics & Astronomy, 2015, 58(11): 114211.

[8] Wang D C, Huang Q, Qiu C W, et al. Selective excitation of resonances in gammadion metamaterials for terahertz wave manipulation. Science China: Physics, Mechanics & Astronomy, 2015, 58(8): 84201.

[9] Jepsen P U, Cooke D G, Koch M. Terahertz spectroscopy and imaging-Modern techniques and applications. Laser & Photonics Reviews, 2011, 5(1): 124-166.

[10] Zhan H L, Wu S X, Bao R M, et al. Water adsorption dynamics in active carbon probed by terahertz spectroscopy. RSC Advances, 2015, 5(19): 14389-14392.

[11] Jin W J, Zhao K, Yang C, et al. Experimental measurements of water content in crude oil emulsions by terahertz time-domain spectroscopy. Applied Geophysics, 2013, 10(4): 506-509.

[12] Jin Y S, Kim G J, Shon C H, et al. Analysis of petroleum products and their mixtures by using terahertz time domain spectroscopy. Journal of Korean Physical Society, 2008, 53: 1879-1885.

Laser-induced voltage of oil shale for characterizing the oil yield

Zhiqing Lu[1,2] Qi Sun[1,2] Kun Zhao[2] Hao Liu[3] Xin Feng[2]
Jin Wang[2] Lizhi Xiao[1]

(1. State Key Laboratory of Petroleum Resources and Prospecting, China University of Petroleum, Beijing 102249, China; 2. Beijing Key Laboratory of Optical Detection Technology for Oil and Gas, China University of Petroleum, Beijing 102249, China; 3. School of Science, China University of Geosciences, Beijing 100083, China)

Abstract: The oil yields of oil shales from the Longkou, Yaojie, and Barkol districts in China were characterized with laser-induced voltage, which was monitored by a voltage tester or a digital oscilloscope when oil shale was irradiated directly by a continuum solid-state laser with a wavelength of 532 nm without an applied bias or at a bias of 9 V. The laser-induced voltage strongly depends upon the properties of the oil shales and exhibits an increasing trend as the oil yield increases from ~5.66%, ~9.05%, to ~14.16% for Barkol, Yaojie, and Longkou, respectively. In addition, the rise time constant increases with the decrease of the oil yield, while the fall time constant declines as the oil yield decreases. A possible mechanism was proposed by the comparative analysis of the experimental results. The experimental results provide useful optical information on oil shale, suggesting the laser-induced voltage can be used to characterize the oil yield of oil shale.

1 Introduction

The gradual decrease in the reserves of conventional energy resources has motivated many countries to use oil shale as an alternative energy. Oil shale is a natural, impermeable, fine-grained, laminated black or brown combustible material and consists of complex organic material of high molecular weight called kerogen[1]. Kerogen, a complex combination of carbon, hydrogen, sulfur, and oxygen, cannot be extracted with ordinary solvents. Liquid hydrocarbons and combustible shale gas can be obtained from insoluble kerogen after heat treatment[2,3].

The evaluation of oil shale plays a vital role in industrial exploitation. The oil yield is an important indicator to evaluate oil shale. As an energy source, oil shale pyrolysis has been used through heating the oil shale to a temperature that the kerogen in the oil shale decomposes into gas, oil, and coke[4,5]. The pyrolysis is a complicated thermochemical conversion process involving extremely complex reactions[6,7] and is generally applicable to the evaluation of potential oil yields. Other methods of thermal analysis give an estimate of the oil yield by the collection of shale oil and measurement of the parameters that vary with the oil yield, such as the response of a flame ionization detector on pyrolysis of the oil shale[8-10] and the weight loss of the sample (thermogravimetry)[11-16]. In addition, some indirect methods have been developed, e.g., well-logging[17], X-ray diffraction[18-21], and nuclear magnetic resonance spectroscopy[22-25].

Optical analysis provides additional information on the nature of oil shales. Fourier transform infrared spectroscopy evaluates the oil-yielding potential of raw oil shale by giving a quantitative measure of aliphatic, aromatic, and hydroxyl hydrogen[26]. Environmental scanning electron microscopy and confocal laser scanning microscopy can contribute to the study of laminated organic-rich sediments, allowing for the identification of different organic and inorganic components and their spatial arrangement and the characterization of microfacies[27]. Fluorescence microscopy promotes the development of organic petrography. The basis of fluorescence microscopy is that the organic matter absorbs primary wavelengths and re-emits longer wavelengths within the visible light spectrum. A total of 80 samples from the deposit have been recognized using fluorescence microscopy for the included organic matter[28]. Macroscopic examination of oil shale beds during ultraviolet excitation may thus be used as a fast screening method to assess oil shale quality[29].

In this paper, the photoelectric properties of oil shales have been studied under the irradiation of a 532 nm laser. The laser-induced voltage (LIV) was monitored by a voltage tester without an applied bias and by a digital oscilloscope at a bias of 9 V. The relationship between the LIV and oil yield was determined, and a possible mechanism was presented. The present results suggested that the LIV could be used to evaluate the oil yield of oil shale.

2 Experimental section

2.1 Oil shale sample preparation

Oil shale samples in this study were obtained from three regions of China, Longkou city in Shandong, Yaojie street in Lanzhou, and Kazakhstan Barkol county in Xinjiang. Shale oil of oil shales produced through the pyrolysis process was collected and analyzed off-line by gas chromatography (GC). The results showed that CH_4, H_2, CO_2, C_2H_4, CO, and C_2H_6 were the major components. CH_4 is the most abundant component in Yaojie and Barkol, while Longkou is rich in CO_2. The GC results for the shale oil indicated strongly the aliphatic and aromatic hydrocarbon characteristics of the shale oil, and their total contents were 60wt%~80 wt%. The rest of the organic matter contained small amounts of phenols, olefin compounds, nitrogen-containing compounds, etc.

The pyrolysis process is applicable to the evaluation of potential oil yields. Crushed and sieved oil shale particles with the size of less than 200 mesh were put in a stainless-steel vessel and heated to 510℃ at the heating rate of 15℃/min with the ventilating rate of 0.6L/min under a nitrogen atmosphere. When the vessel was heated according to mentioned conditions, reactions occurred to convert solid organic matter to liquid or gas, which were exhausted and condensed. The oil yield was calculated by the ratio between the mass of liquid condensation and gross mass of oil shale. Thus, the oil yields of the oil shale from Longkou, Yaojie, and Barkol can be ~14.16%, ~9.05%, and~5.66%, respectively.

The oil samples for the LIV measurement, which have the size of $8\times 8mm^2$ with a thickness of 1mm, are mechanically fabricated by a rock-cutting machine and then polished by sandpaper. The roughness of the samples detected by a surface roughness tester is 1.5~2.0μm. Our current-voltage curve measurements confirmed that the contact between the silver electrode and the sample is a good Schottky contact. Because of the high resistance of the organic compound, silver interdigitated electrodes were fabricated on one surface of the oil shale sample, as shown in Figure 1, to obtain high responsivity[30]. Silver film with a thickness of ~500 nm was deposited onto the oil shale samples by magnetron sputtering at a power of 100 W and a pressure of 0.6~0.7Pa for 10min, and standard lithography and etching were performed to define the interdigitated contact pattern. The finger width w of the interdigitated electrode was equal to its separated spacing $s = 100$μm. The sizes of the outer electrode and interdigitated electrode were $2 \times 2mm^2$ and $6.4 \times 7mm^2$, respectively.

Figure 1 Schematic diagram of the interdigitated electrode configuration on the oil shale sample surface, where w and s represent the width and interspacing of the electrodes, and $w = s = 100$μm

2.2 LIV measurement

The oil shale sample surfaces were irradiated directly by a continuum solid-state laser with a wavelength of 532 nm. The laser power was 138mW, and the irradiation area was approximately $9mm^2$. The schematic circuit of the measurement is displayed in Figure 2(a), where a 9 V bias was applied, and a 350MHz digital oscilloscope with 1MΩ input impedance was used to record the photoresponse signal. The circuit of a typical open-circuit photovoltage is shown in Figure 2(b), where the voltage signal was monitored by a voltage tester. All of the measurements were performed at room temperature in the atmosphere.

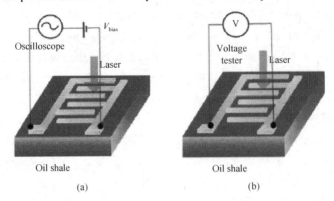

Figure 2 (a) Schematic circuit of the oil shale sample measurement. V_{bias} is the bias voltage. (b) Circuit of a typical open-circuit photovoltage of oil shale samples

3 Results and discussion

The inset of Figure 3 displays the steady-state photovoltaic responses of oil shales from the three regions under the excitation of a 532 nm laser at a bias of 9 V. A sudden increase in voltage occurs when applying a bias of 9 V and then gradually becomes stable. Under the laser irradiation, the voltage signals decline obviously and then become stable. Recovery is obtained using the light-off condition for 20min. Short arrows specify the time for the laser turning on and off. The baseline V_1 for the laser-off state, which is caused by the external bias, was recorded by the oscilloscope and can be fitted by $V_1 \propto \exp(-t/\tau)$, where the time constant τ is ~7.5s, 600s, and 45 s for the oil shales from Longkou, Yaojie, and Barkol, respectively. V_2 is the actual experimental value under illumination. The photovoltaic signal was defined by V_2-V_1 and was plotted as a function of the time in Figure 3. Three curves of different color represent the LIV characterization curves of the three regions. The photoinduced voltage ΔU of the oil shale was recorded as $\Delta U_1 \approx 467\text{mV}$, $\Delta U_2 \approx 341\text{mV}$, and $\Delta U_3 \approx 266\text{mV}$ for Longkou, Yaojie, and Barkol, respectively. Figure 3 also displays the different response times of the oil shale. The response time can be extracted from the enlarged voltage-time curve, and the rise time and fall time are ~4.69min and 12.73min, ~ 5.13min and 12.43min, and ~5.38min and 10.99min for Longkou, Yaojie, and Barkol, respectively. The oil yields of the oil shale from Longkou, Yaojie, and Barkol are ~14.16%, ~9.05%, and ~5.66%. Thus, the relationship between the LIV and oil yield is built as shown in Figure 4, where the LIV and fall time show an increasing trend as the oil yield increases, whereas the rise time decreases with an increasing oil yield.

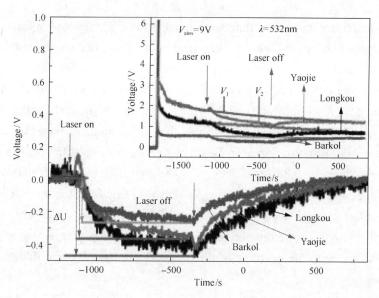

Figure 3 LIV as a function of time for oil shale samples under 532nm laser irradiation at a bias of 9V. The short arrows specify the time for the laser turning on and off. The inset shows the fitted curves and the actual experimental curves of the oil shales

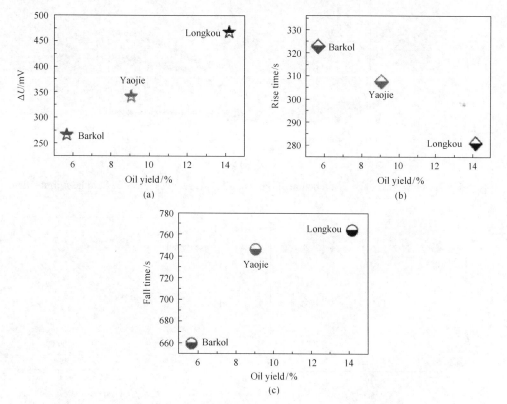

Figure 4 (a) Photoinduced voltage ΔU as a function of the oil yield of oil shales from the three regions; (b) rise time dependence of the oil yield of the oil shales from three regions; (c) fall time dependence of the oil yield of oil shales from three regions

The open-circuit voltage as a function of time with a 532nm He-Ne laser on or off at room temperature is shown in Figure 5. When the light is switched on, the voltages decrease suddenly at the beginning and then the signals become stable with increasing irradiation time. The open-circuit photoinduced voltage signals $\Delta U'$ are measured directly by a voltage tester and are ~65mV, ~33mV, and ~5.2mV for Longkou, Yaojie, and Barkol, respectively. Recovery is obtained using the light-off condition for 20min, and the voltage response signals increase. The response times of the oil shale samples are different. The rise and fall times are approximately 0.88s and 799.2s, 3.08s and 598.8s, and 59.94s and 99s for Longkou, Yaojie, and Barkol, respectively. As shown in Figure 6, the open-circuit voltage and response time strongly depend upon the oil yields of the oil shales from the different regions, which show the same relation as that under a bias in Figure 4, where the LIV and falling time decline with the decrease of the oil yield and the rise time increases as the oil yield decreases.

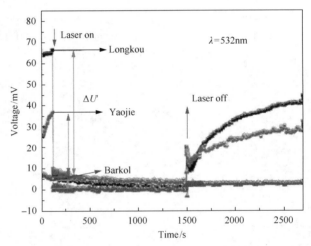

Figure 5 Open-circuit voltage as a function of time for oil shale samples under 532 nm laser irradiation

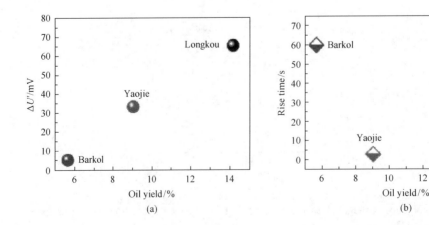

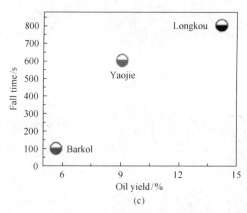

Figure 6 (a) Photoinduced voltage $\Delta U'$ as a function of the oil yield of oil shales from three regions; (b) rise time dependence of the oil yield of oil shales from the three regions; (c) fall time dependence of the oil yield of oil shales from the three regions

The fluorescence properties of oil shale have been widely studied[31,32]. The fluorescence phenomenon of organic matter is due to the emission of photons by fluorophores when excited by electromagnetic radiation. Fluorophores can absorb a part of the incident energy, which promotes electrons from a ground state to an excited state. The fluorescence of organic molecules is due to energy transition phenomena in the π orbitals of C=C bonds. The aromatic and polyaromatic compounds are designated the main source of the fluorescence properties of sedimentary organic matter[32]. In general, organic electronic material, specifically organic semiconductors, contain small organic molecules (polycyclic aromatic compounds, etc.) and conjugated polymers. The important physics of conjugated systems come from the π electrons. When the photon energy is higher than the optical gap of organic molecules, many photons are absorbed into the molecules and change the molecules from the ground state to the excited state. The electrons of the molecule are excited to the lowest unoccupied molecular orbital (LUMO) level from the highest occupied molecular orbital (HOMO) level, which induces the electron-hole pairs called excitons[33]. Then, carriers follow a hopping transport mechanism in organic semiconductors, which is different from band transport in inorganic semiconductors[34].

On the basis of the photoelectric tests of the samples, a possible mechanism is proposed and illustrated in Figure 7.

Under visible light irradiation, the oil shale samples absorb photons and excite electron-hole pairs, which diffuse to a Schottky junction before recombination and are dissociated into holes and electrons by charge separation at the metal/oil shale interface. The surface science investigations of organic semiconductor interfaces were interpreted in the case of accepted models of inorganic semiconductor interfaces[35]. The interface electron and hole barriers are not simply defined by the difference between the work functions of the metals and organic solids. The range of interface Fermi level positions is material-dependent, and dipole

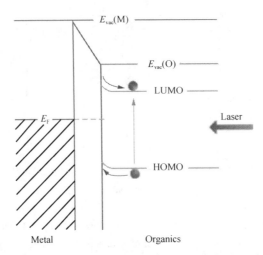

Figure 7 Charge transfer process in the oil shale under laser irradiation

barriers are present at all of these interfaces[36]. Typical Schottky models are described briefly and later applied to organic materials[37-39]. Especially, the Schottky junction barrier between metal electrode and organic semiconductor plays an important role in organic electronic devices[40-42]. Because of the different work functions of silver electrodes and complex organic components, electrons may flow from the complex organic components to silver to achieve thermal equilibrium in the junction until the Fermi levels aligned, resulting in energy band bending. When the metal/oil shale interface was irradiated by visible light [i.e., the 532 nm laser (~2.3 eV)], the electrons in the HOMO energy levels of the complex organic components were excited to the LUMO energy levels to form electron-hole pairs. Then, these pairs were separated by the built-in electric field (pointing from the complex organic components to silver) induced by the Schottky junction, leading to an increase in the Fermi level of the complex organic components. During the photoelectric tests of the oil shale, carriers were collected by the electrodes and driven into the external circuit. Figure 3 shows the combined action of the Schottky junction and the polarization effect at the metal/oil shale interface. In Figure 5, the voltages tend to zero when the oil shales are irradiated by a laser without bias. The LIV of the oil shales from different regions are different as a result of the various organic component contents in the oil shales. In Figure 4 (a) and 6 (a), LIV has a proportional relationship with the shale oil yield. This result may be interpreted that the oil shale sample, which contains different contents of organic matter (such as aliphatic and aromatic hydrocarbons), generated different values of excitons when they are irradiated by laser. The oil shale from Longkou shows the highest LIV, indicating the efficient photogenerated charge transfer, which may result from an increase in the mobile carrier generation efficiency and the mobile carrier lifetime. In oil shales, both electrons and holes are mobile, with the charge transport proceeding by way of an intermolecular hopping process. The transport of the delocalized electrons is limited by phonon scattering[33], impurities, and traps, which can influence the LIV values. The rise time of the Longkou oil shale shows the fastest value in Figure 4 (b) and 6 (b), which is attributed to the fastest generating process of photogenerated electron-hole pairs in the oil shale. The results illustrate that the higher the oil yield of the oil shale, the more easily the carriers are generated. In addition, the oil shale from Barkol has the shortest falling time in Figure 4 (c) and 6 (c), illustrating that the fastest carrier recombination occurs in the oil shale from this region.

It is noted that the relaxation time is in the hundreds of seconds order, as shown in Figure 4 (c) and 6 (c), indicating that the voltage response was also thermal in origin, resulting from the heating by the laser pulse. With increasing temperature as a result of laser illumination, the spontaneous electric polarization results in the variation of the surface charge and the associated flow of a thermoelectric current. A voltage results from the thermoelectric current through equivalent impedance of the sample and system recording the signal. The present typical voltage responses from the chopped laser illumination can be fitted to $A\left[\exp\left(-\dfrac{t}{\tau_e}\right)-\exp\left(-\dfrac{t}{\tau_{th}}\right)\right]$, yielding distinct time constants for the fall and rise of the voltage signal. The rise time constant was observed to decrease with increasing oil yield, while the fall time constant increased. The rise time constant is interpreted to be the electrical time constant τ_e determined by the equivalent circuit of the sample and measurement system, while the fall time constant is interpreted as the thermal time constant τ_{th}. Thus, the thermal excitation may be another factor responsible for the present LIV behavior. Therefore, the nature of the LIV behavior is believed to be the cooperation of photoexcitation and thermal excitation.

4 Conclusions

LIV was demonstrated to be efficient to characterize the oil yield of oil shale samples from different regions. There were obvious differences among the three types of oil shale samples with respect to the photoelectric properties. The results obtained in this study suggest that the higher oil yield of the oil shale, the larger the LIV that occurs when the oil shale is irradiated by a laser. There is a certain relationship between the oil yield and response time. The rise time of Longkou oil shale with the highest oil yield shows the fastest value, which indicates that the higher the oil yield of the oil shale, the more easily carriers are generated. The fall time of the Barkol oil shale with the lowest oil yield is the shortest, which shows the fastest carrier recombination. A possible mechanism of the LIV of oil shale was proposed by the comparative analysis of the experimental results. Therefore, LIV may provide useful photoelectric property information on the oil shale to characterize the oil yield.

Acknowledgements

This work was supported by the Specially Funded Program on National Key Scientific Instruments and Equipment Development (Grant No. 2012YQ140005), the National Key Basic Research Program of China (Grant No. 2014CB744302), and the Higher Education Young Elite Teacher Project of China University of Petroleum (Beijing) (Grant No. 2462015YQ0603). The authors are also thankful for the help from Prof. Shuyuan Li, Dr. Yue Ma, and Wei Wang.

References

[1] Dyni J R. Geology and resources of some world oil-shale deposits. Oil Shale, 2003, 20: 193-252.

[2] Na J G, Im C H, Chung S H, et al. Effect of oil shale retorting temperature on shale oil yield and properties. Fuel, 2012, 95: 131-135.

[3] Martins M F, Salvador S, Thovert J F, et al. Co-current combustion of oil shale-Part 1: Characterization of the solid and gaseous products. Fuel, 2010, 89: 144-151.

[4] Wilson M A, Lambert D E, Colin P J, et al. Chemical transformations during pyrolysis of Rundle oil shale. Fuel, 1985, 64: 1647-1654.

[5] Burnham A K, Happe J A. On the mechanism of kerogen pyrolysis. Fuel, 1984, 63: 1353-1356.

[6] Tiwari P, Deo M. Detailed kinetic analysis of oil shale pyrolysis TGA data. AIChE J., 2012, 58: 505-515.

[7] Wang Q, et al. Study on pyrolysis characteristics of Huadian oil shale with isoconversional method. Oil Shale, 2009, 26: 148-162.

[8] Korth J, Hutton A C, Ellis J. A comparison of microscale pyrolysis and organic petrography for the estimation of yield from oil shale. Org. Geochem., 1990, 15: 477-483.

[9] Charlesworth J M. Monitoring the products and kinetics of oil shale pyrolysis using simultaneous nitrogen specific and flame ionization detection. Fuel, 1986, 65: 979-986.

[10] del Río J C, García-Mollá J, González-Vila F J, et al. Flash pyrolysis-gas chromatography of the kerogen and asphaltene fractions isolated from a sequence of oil shales. J. Chromatogr. A, 1993, 657: 119-122.

[11] Jaber J O, Probert S D. Non-isothermal thermogravimetry and decomposition kinetics of two Jordanian oil shales under different processing conditions. Fuel Process. Technol., 2000, 63: 57-70.

[12] Jiang X M, Han X X, Cui Z G. Progress and recent utilization trends in combustion of Chinese oil shale. Prog. Energy Combust. Sci., 2007, 33: 552-579.

[13] Skala D, Sokic M, Kopsch H. Oil shale pyrolysis—A new approach to the kinetic investigation of different kerogen sample types. Thermochim. Acta, 1988, 134: 353-358.

[14] Dogan O M, Uysal B Z. Non-isothermal pyrolysis kinetics of three Turkish oil-shales. Fuel, 1996, 75: 1424-1428.

[15] Williams P F V. Thermogravimetry and decomposition kinetics of British Kimmeridge clay oil-shales. Fuel, 1985, 64: 540-545.

[16] Rajeshwar K. The kinetics of the thermal decomposition of Green River oil-shale kerogen by non-isothermal thermogravimetry. Thermochim. Acta, 1981, 45: 253-263.

[17] Bardsley S R, Algermissen S T. Evaluating oil shale by log analysis. JPT, J. Pet. Technol., 1963, 15: 81-84.

[18] Kumar R, Bansal V, Badhe R M, et al. Characterization of Indian origin oil shale using advanced analytical techniques. Fuel, 2013, 113: 610-616.

[19] Cogo S L, Brinatti A M, Saab S C, et al. Characterization of oil shale residue and rejects from Irati formation by electron Paramagnetic resonance. Braz. J. Phys., 2009, 39: 31-34.

[20] Al-Otoom A Y, Shawabkeh R A, Al-Harahsheh A M, et al. The chemistry of minerals obtained from the combustion of Jordanian oil shale. Energy, 2005, 30: 611-619.

[21] Muhammad A F, El Salmawy M S, Abdelaala A M, et al. El-nakheil Oil shale: Material characterization and effect of acid leaching. Oil Shale, 2011, 28: 528-547.

[22] Bartuska V J, Maciel G, Schaefer J, et al. Prospects for carbon-13 nuclear magnetic resonance analysis of solid fossil-fuel materials. Fuel, 1977, 56: 354-358.

[23] Feng Y, Doan T V L, Pomerantz A E, et al. The chemical composition of bitumen in pyrolyzed Green River oil shale: Characterization by ^{13}C NMR spectroscopy. Energy & Fuels, 2013, 27: 7314-7323.

[24] Cao X Y, Yang J, Mao J D. Characterization of kerogen using solid-state nuclear magnetic resonance spectroscopy: A review. Int. J. Coal Geol., 2013, 108: 83-90.

[25] Solum M S, Mayne C L, Orendt A M, et al. Characterization of macromolecular structure elements from a Green River oil shale, I. Extracts. Energy & Fuels, 2014, 28: 453-465.

[26] Solomon P R, Miknis F P. Use of Fourier transform infrared spectroscopy for determining oil shale properties. Fuel, 1980, 59: 893-896.

[27] Nix T, Feist-Burkhardt S. New methods applied to the microstructure analysis of Messel oil shale: Confocal laser scanning microscopy (CLSM) and environmental scanning electron microscopy (ESEM). Geol. Mag., 2003, 140: 469-478.

[28] Henstridge D A, Hutton A C. Geology and organic petrography of the Nagoorin oil shale deposit. Fuel, 1987, 66: 301-304.

[29] Kalkreuth W, Naylor R, Pratt K, et al. Fluorescence properties of alginite-rich oil shales from the Stellarton Basin, Canada. Fuel, 1990, 69: 139-144.

[30] Jeong J W, Huh J W, Lee J I, et al. Interdigitated electrode geometry effects on the performance of organic photoconductors for optical sensor applications. Thin Solid Films, 2010, 518: 6343-6347.

[31] Bertrand P, Pittion J L, Bernaud C. Fluorescence of sedimentary organic matter in relation to its chemical composition. Org. Geochem., 1986, 10: 641-647.

[32] Pradier B, Largeau C, Derenne S, et al. Chemical basis of fluorescence alteration of crude oils and kerogens—I. Microfluorimetry of an oil and its isolated fractions; relationships with chemical structure. Org. Geochem., 1990, 16: 451-460.

[33] Brazovskii S, Kirova N. Physical theory of excitons in conducting polymers. Chem. Soc. Rev., 2010, 39: 2453-2465.

[34] Karl N. Charge carrier transport in organic semiconductors. Synth. Met., 2003, 133-134: 649-657.

[35] Hill I G, Milliron D, Schwartz J, et al. Organic semiconductor interfaces: Electronic structure and transport properties. Appl. Surf. Sci., 2000, 166: 354-362.

[36] Hill I G, Rajagopal A, Kahn A, et al. Molecular level alignment at organic semiconductor-metal interfaces. Appl. Phys. Lett., 1998, 73: 662-664.

[37] Yang B, Xiao Z, Huang J, et al. Polymer aggregation correlated transition from Schottky-junction to bulk heterojunction organic solar cells. Appl. Phys. Lett., 2014, 104: 1-5.

[38] Zhu Y W, Tao S, Zhang F, et al. Efficient organic-inorganic hybrid Schottky solar cell: The role of built-in potential. Appl. Phys. Lett., 2013, 102: 113504-113507.

[39] Matsuki N, Irokawa Y, Nakano Y, et al. π-conjugated polymer/GaN Schottky solar cells. Sol. Energy Mater. Sol. Cells, 2011, 95: 284-287.

[40] Tang C W. Two-layer organic photovoltaic cell. Appl. Phys. Lett., 1986, 48: 183-185.

[41] Campbell I H, Smith D L. Schottky energy barriers and charge injection in metal/Alq/metal structures. Appl. Phys. Lett., 1999, 74: 561-563.

[42] Li Y F, Yang W, Tu Z Q, et al. Schottky junction solar cells based on graphene with different numbers of layers. Appl. Phys. Lett., 2014, 104: 043903.

Ultraviolet laser-induced lateral photovoltaic response in anisotropic black shale

Xinyang Miao[1,2] Jing Zhu[2] Kun Zhao[1,2] Wenzheng Yue[1]

(1.State Key Laboratory of Petroleum Resources and Prospecting, China University of Petroleum, Beijing 102249, China; 2.Beijing Key Laboratory of Optical Detection Technology for Oil and Gas, China University of Petroleum, Beijing 102249, China)

Abstract: The anisotropy of shale has significant impact on oil and gas exploration and engineering. In this paper, a 248nm ultraviolet laser was employed to assess the anisotropic lateral photovoltaic (LPV) response of shale. Anisotropic angle-depending voltage signals were observed with different peak amplitudes (V_p) and decay times. We employed exponential models to explain the charge carrier transport in horizontal and vertical directions. Dependences of the laser-induced LPV on the laser spot position were observed. Owing to the Dember effect and the layered structure of shale, V_p shows an approximately linear dependence with the laser irradiated position for the 0° shale sample but nonlinearity for the 45° and 90° ones. The results demonstrate that the laser-induced voltage method is very sensitive to the structure of materials, and thus has great potential in oil and gas reservoir characterization.

Keywords: laser-induced voltage; shale; anisotropy; position detector

1 Introduction

Shales are cap rocks in many petroleum deposits and are of great interest in geochemical and geological investigations because of the hydrocarbons generated and hosted in them[1]. Increasing demand for fossil fuels and depletion of traditional resources has driven the development of shale formations as significant unconventional oil and gas reservoirs. Shales are fine-grained sedimentary rocks composed primarily of clay, organics, and hard minerals such as quartz, feldspar, and pyrite. They are often highly anisotropic because of the combined effects of partially aligned platy clay particles, layering, micro-cracks, low-aspect-ratio pores, and kerogen inclusions[2]. The anisotropy affects both mechanical behaviors and engineering activities, and numerous studies have been performed to understand the structural, mechanical, acoustic, electrical, thermal, and geochemical properties of shales at various length scales[3]. Increases in oil and gas drilling depths and environmental conditions have challenged the development of those resources. Development has included optical methods, which are insensitive to harsh physical environments. For example, distributed sensing with optical fibers has been successfully employed in shale gas exploration in China and worldwide. Terahertz spectroscopy and oblique-incidence reflectivity difference techniques have been used in the characterization of reservoirs, recognition of oil-water fluids in boreholes, as well as evaluation of petrochemicals and pollutants[4,5]. Since its discovery by Schottky in 1930, Literal Photovoltaic LPV effects have been utilized in a variety of optical transducers and sensors, including biomedical devices, robotics, process control, and position information systems[6]. Position-sensitive detector devices are made possible because the output voltage changes linearly with the light spot position[7]. This effect has been found in many different materials, and is sensitive to their interior structures[8-11].

Black shale found in the earth's upper crust has been investigated by electrical logging methods, showing low electrical resistivity anomalies[12,13]. The chemical composition of organic carbon in black shale was modified by high temperatures (>400°C), which converted it into an electrically conducting phase. Therefore, it is feasible to study its electrical properties in the laboratory by laser-induced voltages. Here, a 248nm ultraviolet (UV) laser was used to assess its anisotropic LPV response. Anisotropic angle-depending voltage signals at three angles were observed with different peak amplitudes V_p and decay times. Exponential models were employed to provide explanations for the carrier transportation process different directions. In addition, the laser-induced LPV was measured as a function of spot position. Overall, the results demonstrate that laser-induced voltage is very sensitive to the structure of materials, and has great potential in oil and gas reservoir characterization.

2 Experimental methods

Black shale samples were obtained from a vertically cored shale formation. Component and structural information of the shale samples were examined with scanning electron microscopy (SEM) at planes parallel and perpendicular to the bedding plane, as shown in Figures 1(a) and (b). Various morphologies were observed, with more micro-pores in the surface perpendicular to the bedding plane. In addition, black areas in Figure 1(b) exhibited parallel elongation, indicating a preferred orientation of the minerals (e.g., quartz, feldspar, and clay). According to energy dispersive spectroscopy (EDS), the main elements in the shale are O, Na, Al, Si, and Ca.

From each shale block, slices were cut at three different angles (0°, 45° and 90°), relative to the bedding plane. Each slice was $10\times8mm^2$ and 2mm thick. The surface for light incidence was polished and two $1\times8mm^2$ silver electrodes for LIV detection were separated on the surface by 2mm. A 248nm (5eV), 20ns KrF pulsed laser was used to irradiate the surface. Scanning via lateral movement of the slices was used to obtain light-spot position-dependent LPV. Time-dependent voltages between indium electrodes were measured and recorded at room temperature by an oscilloscope having a 350MHz bandwidth and a 1MΩ input impedance.

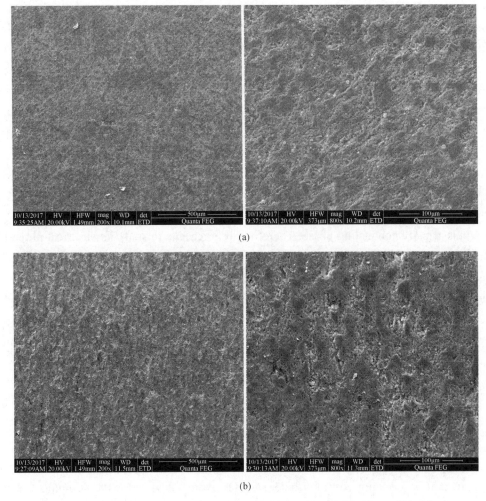

Figure 1 SEM surface morphology of a shale sample at planes (a) parallel and (b) perpendicular to the bedding plane

3 Results and discussion

Figure 2 plots bias-voltage-free laser-induced voltage waveforms of black shale samples at three cutting angles with respect to the incident radiation spot close to the measurement electrodes. All the pulses initially rose rapidly and then gradually decreased after reaching maximum values at ~2μs. Because of the electrical anisotropy, the time-dependent voltage for different angles varied at V_p and the full width at half maximum (FWHM). Specifically, V_p at 0° had a maximum value of ~106.2mV, followed by ~51.3mV at 45°, and ~38.5mV at 90°. The FWHM at 0° was ~4.02 μs, which was faster than that for the 45° (~7.6 μs) and 90° (~12.8 μs) samples. Moreover, there were different relaxation times for the voltage to reach zero after the laser pulse, indicating anisotropic carrier mobilities.

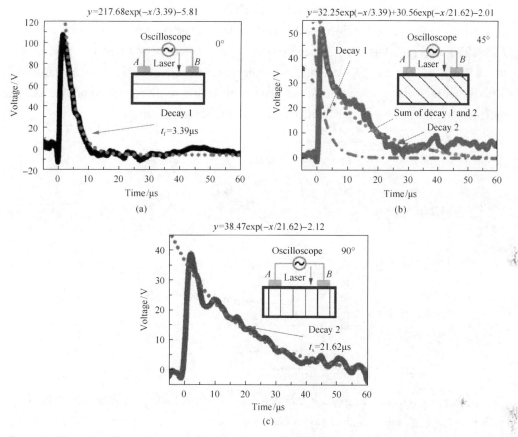

Figure 2 Time-dependent measured and fitted voltages of black shale samples in response to 248nm laser illumination. Waveforms of samples at different cutting angles between the incident surface and the bedding plane: (a) 0°, (b) 45°, and (c) 90°

To achieve further insight into the anisotropic electrical properties affected by carrier recombination, the voltage curves were fitted (short dotted lines) during the gradual decrease of the waveform following the laser pulse for the different shale samples. At 0° and 90°, the measured voltage waveforms were well-characterized by a single exponential decay $A\exp(-x/\tau)$ with different decay times. Here, τ_f =3.39 μs and τ_s=21.62 μs were the resistivity-affected decay times of the faster (decay 1, in Figure 1(a)) and slower decay components (decay 2, in Figure 1(c)). The fast and slow processes were caused by carrier transportation along and across the shale bedding plane, respectively. In addition, for the sample with the 45° cutting angle, the initial voltage exhibited a quick decay, after which the decay process slowed down significantly. Both the faster and slower processes contributed to the decrease. Thus the decay was fit by a double-exponential function $A_s \exp(-t/\tau_s)+A_f \exp(-t/\tau_f)$, where A_f =32.25 and A_s=30.96 were the coefficients of decay 1 and decay 2, respectively. For the 45° cutting angle sample, the horizontal and vertical carrier transportation distances were equal, and the proportions of decay 1 and decay 2 at 45° were similar. Hence, the voltage decay anisotropy was indeed related to carrier transportation in the shale.

Figure 3 indicated that V_p increased linearly with increasing energy over the range of 24~39.9mJ. The corresponding laser-induced voltaic sensitivities were ~4.20mV/mJ, ~1.57mV/mJ, and ~0.74mV/mJ for the 0°, 45°, and 90° samples, respectively. The linear relationships revealed that the laser intensity made the main contribution to the signal amplitude for each irradiation angle. The energy sensitivities also indicated shale anisotropy.

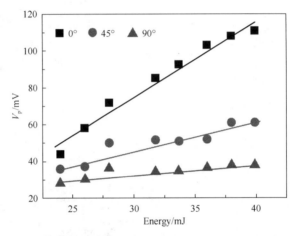

Figure 3   Laser-energy dependence of V_p for shale at three cutting angles

Another anisotropic observation was the laser-induced LPV as a function of laser spot position relative to the center of the electrodes. The charge carriers and the photovoltaic responses were generated by 248nm, 36.0mJ laser pulses. For the 0° shale sample depicted in Figure 4(a), V_p exhibited an approximately linear dependence with a 105.7mV/mm position sensitivity as the spot moved in the region between the electrodes, along the horizontal direction of the surface. A monotonic drop in V_p was observed when the laser spot moved beyond the electrodes. Furthermore, the signal became stronger when the light spot was closer to the measurement electrode; a maximum V_p=106.5mV was observed when the incident radiation spot was closest to the electrode B shown in the Figure 4(a). V_p underwent a sign reversal when the laser spot traveled from one electrode to the other, and became zero at the center of the electrodes. This laser-induced LPV effect of the 0° shale sample was based on the Dember effect, which is shown schematically in Figure 5(a). Because of the high laser pulse energy, the number of laser-induced carriers was much larger than that of the majority carriers in the shale. Therefore, the Dember process was dominated by the lateral voltage induced by the laser. The mobility difference between holes and electrons, which resulted in the transient distribution of electrons far away from the irradiation spot and holes remaining close to the spot, created a higher electric potential in the region close to the laser spot[14,15].

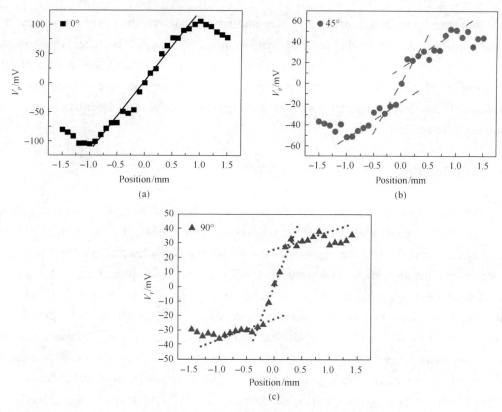

Figure 4 V_p vs. laser position on shale samples at various cutting angles: (a) 0°, (b) 45° and (c) 90°

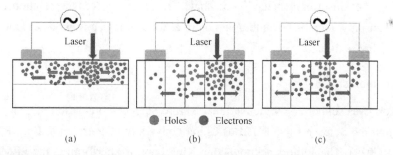

Figure 5 Schematic of holes and electrons diffusing in (a) 0° and (b) 90° shale samples with the laser spot located between one electrode and the boundary, or (c) between two boundaries

Additionally, for the 45° and 90° shale samples, the V_p sensitivities varied with laser position. In Figures 4(b) and (c), V_p variation for both samples could be divided into three sequential stages by the changing range of laser position, with 0.3mm and −0.3mm demarcation points. This nonlinearity of LPV caused by the intrinsic microstructure of the sediments in the shale, had not been previously reported. Generally, owing to the effects of preferentially aligned fractures, pores, and grains, the electrical conductivity of shale exhibited distinct differences in various directions. Most of the micro-cracks existed near the boundaries of adjacent bedding planes, causing higher porosity and lower electrical conductivity[16].

Therefore, the layers and boundaries at sub-millimeter scales resulted in the nonlinear V_p response to position for the 45° and 90° shale samples. Specifically, boundaries with many more defects and traps restricted the movement of electrons, weakening the Dember effect. Therefore, V_p changed little when the laser spot was located between one electrode and the boundary [Figure 5(b)], while the variation rate increased rapidly as the laser passed over the boundary [Figure 5(c)].

4 Conclusions

The laser-induced anisotropic voltaic response of black shale was examined. Voltage signals of shale samples at three cutting angles were observed under illumination of a UV laser, showing anisotropy both at the peak amplitude V_p and the decay time. Carrier transportation processes in different directions were modeled with exponential functions. Laser-induced LPV as a function of spot position were discovered for the first time in shale. Specifically, V_p exhibited an approximately linear dependence with position for the 0° cut shale sample, but nonlinearity for the 45° and 90° cuts, which was understood in terms of the Dember effect and the layered shale structure. With the increase in oil and gas drilling depths, the development of oil and gas resources will become increasingly difficult; therefore, characterization of the oil-gas reservoirs is of great significance. Because it is very sensitive to the structure of materials, laser-induced voltage method should be a very promising and practical method for characterizing the anisotropic behavior of shale. In general, given its relationship between optics and electricity, there is considerable potential for a multi-physical description of oil and gas reservoirs in the future.

Acknowledgements

This work was supported by the National Nature Science Foundation of China (Grant No. 11574401) and the Science Foundation of China University of Petroleum, Beijing (Grant No. 2462017YJRC029). The authors acknowledge Xing Fang for performing the SEM test shown in Figure 1.

References

[1] Zhubayev A, Houben M E, Smeulders D M J, et al. Ultrasonic velocity and attenuation anisotropy of shales, Whitby, United Kingdom. Geophysics, 2016, 81(1): 45-56.

[2] Vernik Lev, Liu Xingzhou. Velocity anisotropy in shales: A petrophysical study. Geophysics, 1997, 62(2): 521-532.

[3] Blum T E, Adam L, Van Wijk K. Noncontacting benchtop measurements of the elastic properties of shales. Geophysics, 2013, 78(3): C25-C31.

[4] Zhan H L, Wang J, Zhao K, et al. Real-time detection of dielectric anisotropy or isotropy in unconventional oil-gas reservoir rocks supported by the oblique-incidence reflectivity difference technique. Scientific Reports, 2016, 6(1): 39306.

[5] Miao X Y, Zhan H L, Zhao K. Application of THz technology in oil and gas optics. Science China: Physics, Mechanics & Astronomy, 2017(2): 88-90.

[6] QiaoS, Liu Y N, Liu J H, et al. The reverse lateral photovoltaic effect in boron-diffused Si p-n junction structure. IEEE Electron Device Letters, 2016, 37(2): 201-204.

[7] Toda K, Yoshida S. Pb_2CrO_5 photovoltaic device for the detecting light-beam position. Applied Physics B, 1988, 45(2): 65-69.

[8] Zhao K, Jin K J, Lu H B, et al. Transient lateral photovoltaic effect in p-n heterojunctions of $La_{0.7}Sr_{0.3}MnO_3$ and Si. Applied Physics Letters, 2006, 88(14): 141914.

[9] Gan Z K, Zhou P Q, Huang X, et al. Using electric pulse and laser to trigger a sharp and nonvolatile change of lateral photovoltage in nano-carbon film. Applied Physics Letters, 2016, 108(13): 666.

[10] Qiao S, Chen J, Liu J, et al. Large lateral photovoltaic effect in μc-SiOx:H/a-Si:H/c-Si p–i–n structure. Applied Physics Express, 2016, 9(3): 031301.

[11] Moon I K, Ki B, Yoon S, et al. Corrigendum:Lateral photovoltaic effect in flexible free-standing reduced graphene oxide film for self-powered position-sensitive detection. Scientific Reports, 2016, 6: 36800.

[12] Duba A, Huengest E, Nover G, et al. Impedance of black shale from Münsterland 1 borehole:an anomalously good conductor?. Geophysical Journal of the Royal Astronomical Society, 1988, 94(3): 413-419.

[13] Adão F, Ritter O, Spangenberg E. The electrical conductivity of Posidonia black shales-from magnetotelluric exploration to rock samples. Geophysical Prospecting, 2016, 64(2): 469-488.

[14] Jin K J, Zhao K, Lu H B, et al. Dember effect induced photovoltage in perovskite p-n heterojunctions. Applied Physics Letters, 2007, 91(8): 081906.

[15] Liao L, Jin K, Ge C, et al. A theoretical study on the dynamic process of the lateral photovoltage in perovskite oxide heterostructures. Applied Physics Letters, 2010, 96(6): 062116.

[16] Dewhurst D N, Siggins A F. Impact of fabric, microcracks and stress field on shale anisotropy. Geophysical Journal of the Royal Astronomical Society, 2010, 165(1): 135-148.

Ultraviolet laser-induced voltage in anisotropic shale

Xinyang Miao[1,2]   Jing Zhu[1,2]   Yizhang Li[2]   Kun Zhao[1,2]
Honglei Zhan[2]   Wenzheng Yue[1]

(1.State Key Laboratory of Petroleum Resources and Prospecting, China University of Petroleum, Beijing 102249, China; 2.Beijing Key Laboratory of Optical Detection Technology for Oil and Gas, China University of Petroleum, Beijing 102249, China)

Abstract: The anisotropy of shales plays a significant role in oil and gas exploration and engineering. Owing to various problems and limitations, anisotropic properties were seldom investigated by direct current resistivity methods. Here in this work, a 248nm ultraviolet laser was employed to assess the anisotropic electrical response of a dielectric shale. Angular dependence of laser-induced voltages (V_p) were obtained, with a data symmetry at the location of 180° and a ~62.2% V_p anisotropy of the sample. The double-exponential functions have provided an explanation for the electrical field controlled carrier transportation process in horizontal and vertical directions. The results demonstrate that the combination of optics and electrical logging analysis (opti-electrical logging) is a promising technology for the investigation of unconventional reservoirs.

Keywords: laser-induced voltage; shale; anisotropy; opti-electrical logging

1 Introduction

The need for more accessible energy resources has made shale formations increasingly important as promising alternative source rocks for unconventional oil and gas[1]. Unlike conventional reservoirs that have been characterized through numerous methodologies, shale formations have not yet undergone advanced characterization, and much remains to be done[2]. Due to preferential orientation of mineral particles and finely laminated bedding planes, several of the macroscopic physical properties of shales are anisotropic[3,4]. Such anisotropy has important impacts on shale energy exploration, wellbore stability as well as interpretation of micro-seismic monitoring, and many studies have been taken to understand the mechanical, acoustic, electrical, thermal and geochemical properties of shales at various scales for inferring the structural characteristics[5-7]. Electromagnetic (EM) techniques and direct current (DC) resistivity methods were widely used for hydrocarbon exploration and reservoir characterization[8,9]. Various studies of anisotropic formations and rocks have been taken in the laboratory to provide a basis to interpret the EM logging data[10,11]. However, for DC resistivity methods, anisotropy is seldom incorporated into practical investigations owing to various reasons such as problems of electrical equivalence, the limitation in resolution capability of the geoelectric field, and the increase of parameters to be recovered from the data[12]. Moreover, due to the insulation characteristics caused by pores, cracks as well as the complex composition, it is difficult to measure the anisotropic electrical properties of shales directly by using a conventional DC power in the laboratory. Usually, dielectrics with large band-gap exhibit extremely low conductivity at low applied fields, however, recent studies have shown that currents and charge in insulator could be induced by optical fields[13]. The irradiation of light strongly distorted the electronic band structure to make the insulator instantaneously transforms into a conductor[14]; therefore, optical fields can be employed to assess the transient conductivity of dielectrics theoretically. Laser-induced voltaic response has been found in many different materials, and is sensitive to their interior structures. A built-in potential can be observed in tilting $SrTiO_3$ single crystal under the irradiation of an ultraviolet (UV) laser pulse owing to the Seebeck effect, with the voltage dependent on the tilting angle[15]. Anisotropy of the Seebeck coefficient (ΔS) has also been observed in thin films of $YBa_2Cu_3O_{7-\delta}$, $La_{1-x}Ca_xMnO_3$, $La_{1-x}Sr_xMnO_3$ and various cobalt oxides[16-19], showing great potential applications in the detection of the laser radiations for a wide spectral range. Manganites with doped Ag show significantly enhanced electrical and magnetic transport properties by increasing the amount of Mn^{4+} through Ag-ion substitution. Obviously, ions in substitution, changing the doping level and the lattice mismatch contribute to enhanced initial laser-induced voltage (LIV) signals[20, 21].

With the advantages of non-interference with the earth magnetic fields, optical measurements have been proved effectively to evaluate petroleum resources[22], especially for properties of unconventional reservoirs[23,24]. In our previous study, the oblique-incidence reflectivity difference (OIRD) technique was initially employed to characterize the layered distribution of the dielectric properties in shales, showing that directional changes of the dielectric and surface properties were resulted by the micro-cracks and particle orientation[25]. Dielectric responses of anisotropic oil shales in the terahertz (THz) range were studied by teranertz time-domain spectroscopy (THz-TDS), and the refractive index anisotropies of the oil shales were plotted as a function of the oil yield, indicating that the increase of organic matter had promoted the degree of anisotropy[26]. Here in this research, the anisotropic electrical response of a shale core has been studied under the UV laser irradiation. We have monitored the LIV signals at various angles of the shale. The experimental results suggest that the LIV measurement is willing to build a new platform for the combination of optics and electrical logging analysis (opti-electrical logging), which is helpful to provide basic research for geophysical prospecting of unconventional oil and gas resources.

2 Experimental methods

The shale sample used in the LIV measurement was cored in the horizontal direction with a cylindrical plug from a whole specimen of a shale formation at the depth of 3325.4m. The diameter and height of the core was 25mm and 60mm, respectively. Optical microscopy images and SEM pictures of the core sample used here were shown in our previous study[25], in which the orientational arrangement of the particles and crevices could be observed by the secondary electron (SE) morphology contrast and BSE images. Besides, the layered distribution of the dielectric properties in the shale sample were also identified, and the width of the adjacent main beddings was measured as 100~300μm, as well as the secondary laminations 30~50μm.

A KrF pulse laser at a wavelength of 248 nm (5eV photon energy) and the pulse width 20 ns was used as the source. The waveforms of laser-induced voltages were recorded by an oscilloscope with a 350MHz bandwidth and 1MΩ input impedance. A Keithley 2400 sourcemeter was adopted as the bias voltage source for the sake of a significant photo-response measurement. Before the LIV measurement, the core sample was polished for a better contact between the electrode and the sample. As shown in Figure 1, twelve colloidal silver electrodes with 2×8mm^2 separated by 30° were equidistantly arrayed around the cylindrical sample. Two adjacent electrodes were connected in turn, being used for both applied bias injection and voltage probing. The 248nm UV laser was employed to irradiate at the core surface at the center of the two electrodes. All measurements were performed at room temperature.

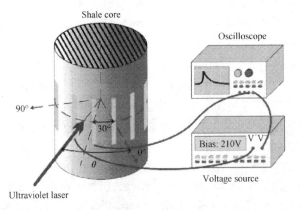

Figure 1 Schematic diagram of the measurement system

3 Results and discussion

Figure 2 displays the measured LIV waveforms of the shale at three irradiation angles (θ) with respect of the bedding plane under a bias of 210V. The response of each measured waveform is composed of a fast rise and a slow decay. All the pulses rise rapidly at the beginning (the inset of Figure 2), and then the increases start to slow up. After reaching the maximum values at~80μs, the LIV signals decrease gradually. The full width at half-maximum (FWHM) of ~290μs and rapidly rising time of ~ 3.3μs were measured by using the digital oscilloscope with 1MΩ input impedance. Besides, the peak amplitude of the voltage (V_p) are variant from each other at different angles. To be specific, V_p at 75° have the maximum value of ~1.52V under a 210V bias, followed by 45° and 15°, with the V_p equals ~1.21V and ~1.07V, respectively.

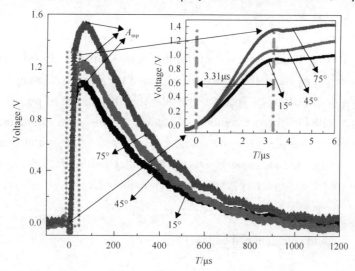

Figure 2 Time dependent voltages of the shale in response to the laser illumination at 248 nm applied with a 210V bias. Lines with various colors represent the waveforms of different angles between the incident laser and the bedding plane. The inset displays the rapidly rising period of the waveform

The nature of the LIV behavior is believed to be the cooperation of photoexcitation and thermal excitation. Consisting of clay and tiny fragments of minerals such as calcite, pyrite, silica, and etc., the shale has exhibited a dielectric property with large bandgap. Because of the reaction of dielectrics to applied fields[13], the valence band and conduction band levels become close to each other with the applied bias (210V in this study). According to previous studies, both thermoelectric and photovoltaic effect originate the LIV signals[16,17,27,29]. The UV laser photon energy E_{photon} exceeds the bandgap energy E_g of the shale, which induces interband transition and generation of photo-excited carriers. In addition, the transverse thermoelectric effect was also considered as the origin of the LIV signal. With increasing temperature as a result of laser illumination, the spontaneous electric polarization results in the variation of the surface charge and the associated flow of a thermoelectric current.

Figure 3(a) exhibits V_p as a function of applied bias voltage (V_b). For each angle, V_p increases linearly with the increasing V_b from 80 V to 210 V, and V_p is also found to depend linearly on the on-sample energy from 28mJ to 42.5mJ at 210 V bias (Figure 3(b)), with the corresponding laser-induced voltaic sensitivities of ~0.036V/mJ, ~0.042V/mJ and ~0.058V/mJ for 15°, 45° and 75°, respectively. Such results indicate that the built-in field as well as the intensity of laser makes the main contribution to the amplitude of the signals for each irradiation angle, and owing to the anisotropy in electrical properties of the shale, the values of V_p in the three angles differ from each other. Actually, due to the configuration of the measurement setup presented in Figure 1, V_p is determined by

$$V_p = \frac{V_b R_T}{R_s + R_T} \approx \frac{V_b R_T}{R_s} = \frac{N}{R_s} \tag{1}$$

where $R_T = 1 M\Omega$ is the impedance at the input channel of the oscilloscope, and $V_b R_T$ is defined as a constant N. R_s is the irradiated resistance of the sample between electrodes with the value far more than R_T owing to the dielectric characteristics of the shale. According to Equation (1), V_p is inversely proportional to R_s. Thus we consider that the change of resistance at different angles, which is resulted by the preferential orientation of shale's basic structural units, led to the V_p anisotropy of the sample. Then, we take the anisotropy of resistivity into account, as $\rho_{90} \neq \rho_0$, in which ρ_{90} and ρ_0 are the surface resistivity parallel and vertical to the bedding plane, respectively, and we have the surface resistivity ρ_s variation with θ[28]

$$\rho_s = \rho_0 \cos^2\theta + \rho_{90} \sin^2\theta \tag{2}$$

Therefore, V_p could be calculated by

$$V_p = \frac{N}{\rho_s d} = \frac{N}{\rho_0 d \cos^2\theta + \rho_{90} d \sin^2\theta} = \frac{N}{R_0 \cos^2\theta + R_{90}\sin^2\theta} = \frac{1}{\dfrac{\cos^2\theta}{V_0} + \dfrac{\sin^2\theta}{V_{90}}} \tag{3}$$

where d is the linear distance between the electrodes. As V_0 and V_{90} are peak amplitude of the voltage vertical and parallel to the bedding plane, V_p varies periodically as a function of θ with the cycle 180° in the case of $V_0 \neq V_{90}$. The possibility of using V_0 and V_{90} to calculate V_p prompted us to test the significant angles parallel and perpendicular to the bedding plane, and V_p at 0°, 90°, 180°, and 270° were measured with the re-prepared electrodes.

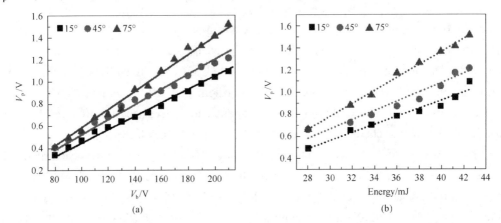

Figure 3 (a) Biases and (b) laser energy dependence of V_p for the shale at three irradiation angles. The solid and dotted lines represent linear fits to the data

Figure 4(a) shows the angular dependent V_p from 0° to 360°, where the shale sample has a significant anisotropy of V_p, and there is a data symmetry at the location of 180°. The results have demonstrated the anisotropy behavior of the shale, with the data symmetries and anisotropy degrees in V_p correspond with the bedding plane and the partial alignment of particles in the shale. Then, V_p was fitted and plotted (the dotted curves in Figure 4) using Equation (3) with the measured V_0 and V_{90}. As depicted in Figure 4, V_p reaches a maximum (V_{pmax}) and minimum (V_{pmin}) of 1.59V and 0.98V at the parallel and perpendicular direction. Therefore, A ~62.2% V_p anisotropy (calculated by [($V_{pmax} - V_{pmin}$)/V_{pmin}] could be observed in the sample. The electrical conductivity of shale exhibited distinct differences in various directions as a result of preferentially aligned fractures, pores, and grains. Most of the micro-cracks existed near the boundaries of adjacent bedding planes, causing higher porosity and lower electrical conductivity. Specifically, boundaries with many more defects and traps restricted the movement of carriers, causing a significant decrement of V_p in the direction perpendicular to the layering compared with that parallel to the layering.

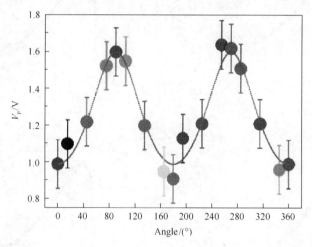

Figure 4　Irradiation angle dependent measured (dots) and fitted V_p (dotted curve) of the sample between 0°~360°. Error bars represent 0.5-fluctuation of standard deviation of V_p

Such anisotropic electrical properties also have an effect on the process of carrier recombination. In Figure 5, we have shown the measured and the fitted voltage curve at the decrease stage of the waveform with different θ. The measured voltage waveforms are characterized well by the double-exponential function:

$$V = A_f \exp(-T/\tau_f) + A_s \exp(-T/\tau_s) + V_0 \qquad (4)$$

Here, τ_f =238 μs and τ_s=387 μs are the resistivity influenced decay times of the faster (decay 1, in Figure 5) and slower decay components (decay 2, in Figure 5), in which the faster process and the slower process are caused by the electrical field controlled carrier transportation along and across the bedding plane, respectively. At 0° and 90°, the waveforms are characterized by a single-exponential decay, corresponding with the slower process and faster process, respectively. A_f and A_s are the coefficients of decay 1 and decay 2, implying the contribution of each component to the whole process. The proportions of decay 1 and decay 2 at various θ are different from each other as shown by the relative position of the curves in Figure 5.

With the augment of θ from 0° to 90°, the carrier transportation distance in horizontal direction increases, and decay 1, with the curve raising in Figure 5, contributes more to the entire voltage attenuation stage. Then we calculated the values of A_f / A_s and the ratio of A_f or A_s to the entire voltage signal, as depicted in Figure 6, the agreements between the calculated values and the fitted trigonometric functions indicate that the coefficients of decay 1 and decay 2 are related to the carrier transportation distance in horizontal and vertical directions, respectively.

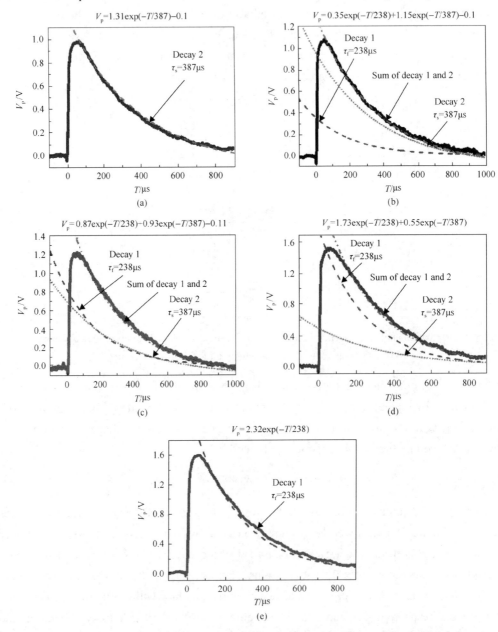

Figure 5 Measured and double-exponential fitted voltage waveforms of the shale sample with θ equals (a) 0°, (b) 15°, (c) 45°, (d) 75°, and (e) 90° respectively

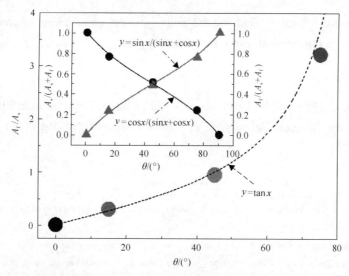

Figure 6 Nonlinear fit of the relation between A_f and A_s at 0°, 15°, 45°, 75°, and 90° as a function of θ from 0° to 90°

The anisotropy of shale has significant impact on oil and gas exploration, which affects both mechanical behaviors and engineering activities (e.g., drilling efficiency and borehole deviation). With the increase of the unconventional oil and gas drilling depth, the environmental conditions become challenging, and the development of oil and gas resources will become increasingly difficult. Optical methods, which are insensitive to harsh physical environment, can be served as an appropriate method to promote the development of unconventional oil-gas industry. In our previous study[25], the anisotropy of shale was clearly observed in the OIRD images, and such method was proved promising in surface anisotropy detection in rock. Here in this work, LIV is employed to study the electrical anisotropy of shale, and periodic variation of V_p was observed with the angle. As it is very sensitive to the structure of materials, LIV method should be a very promising and practical method for characterizing the anisotropic behavior of shale.

4 Conclusions

In summary, the electrical property of shale have been characterized using LIV method. Owing to the inversely proportional relationship between V_p and R_s, angular dependence of V_p is resulted by the preferential orientation caused resistance variation, and the anisotropy of V_p was observed by plotting the values with θ. V_p varied periodically as a function of θ, showing a ~62.2% anisotropy. In addition, the double-exponential functions have provided an explanation for the electrical field controlled carrier transportation process in horizontal and vertical directions. The introduction of optical fields in this study can be broadly applied for a better understanding of shale's electrical behavior, and the opti-electrical logging technique is promising to provide a basis to describe the multi-physical connections between geophysical measurements of unconventional reservoirs in the future.

Acknowledgements

This work was supported by the National Nature Science Foundation of China (Grant No. 11574401) and the Science Foundation of China University of Petroleum, Beijing (No. 2462017YJRC029).

References

[1] Zhubayev A, Houben M E, Smeulders D M J, et al. Ultrasonic velocity and attenuation anisotropy of shales, whitby, united kingdom. Geophysics, 2016, 81: D45-D56.

[2] Vernik L, Liu X. Velocity anisotropy in shales: A petrophysical study. Geophysics, 1997, 62: 521-532.

[3] Sayers C M, Guo S, Silva J. Sensitivity of the elastic anisotropy and seismic reflection amplitude of the eagle ford shale to the presence of kerogen. Geophys Prospect, 2014, 63: 151-165.

[4] Blum T E, Adam L, Wijk K V. Noncontacting benchtop measurements of the elastic properties of shales. Geophysics, 2013, 78: C25-C31.

[5] Vernik L, Nur A. Ultrasonic velocity and anisotropy of hydrocarbon source rocks. Geophysics, 1992, 57: 758-759.

[6] Lin B T, Cerato A B. Application of SEM and ESEM in microstructural investigation of shale-weathered expansive soils along swelling-shrinkage cycles. Eng. Geol., 2014, 177: 66-74.

[7] Allan A M, Kanitpanyacharoen W, Vanorio T. A multiscale methodology for the analysis of velocity anisotropy in organic-rich shale. Geophysics, 2015, 80: C73-C88.

[8] Woodruff W F, Revil A, Prasad M, et al. Measurements of elastic and electrical properties of an unconventional organic shale under differential loading. Geophysics, 2015, 80: D363-D383.

[9] Zhou J, Revil A, Karaoulis M, et al. Image-guided inversion of electrical resistivity data. Geophys J Int, 2014, 197: 292-309.

[10] López A A, Mousatov A, Markov M, et al. Modeling and inversion of elastic wave velocities and electrical conductivity in clastic formations with structural and dispersed shales. J. Appl. Geophys., 2015, 116: 28-42.

[11] Wei W, Cai J, Hu X, et al. An electrical conductivity model for fractal porous media. Geophys. Res. Lett., 2015, 42: 4833-4840.

[12] Wiese T, Greenhalgh S, Zhou B, et al. Resistivity inversion in 2-D anisotropic media: numerical experiments. Geophys. J. Int., 2015, 201: 247-266.

[13] Schultze M, Bothschafter E M, Sommer A, et al. Controlling dielectrics with the electric field of light. Nature, 2013, 493: 75-78.

[14] Schiffrin A, Paasch-Colberg T, Karpowicz N, et al. Optical-field-induced current in dielectrics. Nature, 2013, 493: 70-74.

[15] Zhao K, Jin K J, Huang Y H, et al. Ultraviolet fast-response photoelectric effect in tilted orientation $SrTiO_3$ single crystals. Appl. Phys. Lett., 2006, 89: 173507.

[16] Lengfellner H, Kremb G, Schnellbögl A, et al. Giant voltages upon surface heating in normal $YBa_2Cu_3O_{7-\delta}$ films suggesting an atomic layer thermopile. Appl. Phys. Lett., 1992, 60: 501-503.

[17] Zhang P X, Lee W K, Zhang G Y. Time dependence of laser-induced thermoelectric voltages in $La_{1-x}Ca_xMnO_3$ and $YBa_2Cu_3O_{7-\delta}$ thin films. Appl. Phys. Lett., 2002, 81: 4026-4028.

[18] Wang S, Chen S, Liu F, et al. Laser-induced voltage effects in c-axis inclined Na_xCoO_2 thin films. Appl. Surf. Sci., 2012, 258: 7330-7333.

[19] Liu X, Yan Y Z, Chen Q M, et al. Laser-induced voltage (LIV) enhancement of $La_{2/3}Sr_{1/3}MnO_3$ films with Ag addition. Appl. Phys. A., 2014, 115: 1371-1374.

[20] Liu X, Yin X, Chen Q, et al. Effect of annealing oxygen pressure on the enhancement of laser-induced voltage in $La_{2/3}Ca_{1/3}MnO_3$: $Ag_{0.04}$ films. Mater. Sci. Eng. B-ADV., 2014, 185: 105-108.

[21] Liu X, Zhang M, Cao M G. Enhancement of laser-induced voltage (LIV) in $La_{2/3}Ca_{1/3}MnO_3$:$Ag_{0.04}$ films: $Ag_{0.04}$ films. Applied Physics A, 2017, 123(1): 36.

[22] Zhan H L, Zhao K, Zhao H, et al. The spectral analysis of fuel oils using terahertz radiation and chemometric methods. J. Phys. D Appl. Phys., 2016, 49: 395101.

[23] Li Y Z, Wu S X, Yu X L, et al. Optimization of pyrolysis efficiency based on optical property of semicoke in terahertz region. Energy, 2017, 126: 202-207.

[24] Lu Z Q, Sun Q, Zhao K, et al. Laser-induced voltage of oil shale for characterizing the oil yield. Energy & Fuels, 2015, 29(8): 4936-4940.

[25] Zhan H, Wang J, Zhao K, et al. Real-time detection of dielectric anisotropy or isotropy in unconventional oil-gas reservoir rocks supported by the oblique-incidence reflectivity difference technique. Sci. Rep., 2016, 6: 39306.

[26] Miao X Y, Zhan H L, Zhao K, et al. Oil yield characterization by anisotropy in optical parameters of the oil shale. Energ & Fuels, 2016, 30: 10365-10370.

[27] Takahashi K, Kanno T, Sakai A, et al. Influence of interband transition on the laser-induced voltage in thermoelectric Ca_xCoO_2 thin films. Phys. Rev. B, 2011, 83: 115107.

[28] Koren G, Cohen D, Polturak E. ab-Plane anisotropy of transport properties in unidirectionally twinned $YBa_2Cu_3O_{7-\delta}$ films. Phys. Rev. Lett., 1996, 77: 3913-3916.

[29] Liu H, Zhao K, Zhao S Q, et al. Ultrafast and spectrally broadband photovoltaic response in quartz single crystals. J. Phys. D Appl. Phys., 2009, 42: 075104.

Transient laser-induced voltaic response in a partially illuminated dielectric core

Xinyang Miao[1,2]   Jing Zhu[2]   Yizhang Li[2]   Honglei Zhan[2]
Kun Zhao[1,2]   Wenzheng Yue[1]

(1.State Key Laboratory of Petroleum Resources and Prospecting, China University of Petroleum, Beijing 102249, China; 2.Beijing Key Laboratory of Optical Detection Technology for Oil and Gas, China University of Petroleum, Beijing 102249, China)

Abstract: Ultraviolet laser-induced voltage (LIV) measurements were performed on dielectric rock cores. LIV response in the dielectric core samples were measured with the laser spot partially irradiated at the surfaces under a bias of 210 V, and the accumulation of interface charges was considered resulting in the transient voltage response. The voltage waveforms at the shale and sandstone were different in maximum values (V_p) and full width at half maximum in accordance with the components and structures. V_p, as a function of spot position, depended on the distance between the laser spot and the electrode planes. In addition, ablation was observed in the rock surface irradiated by the pulsed laser, which was associated with rapid buildup of electrons and generation of intrinsic defects in the cores. Therefore, LIV method is willing to promote understanding of the interaction between intense laser and dielectric materials.

Keywords: dielectric; core; laser-induced voltage

1 Introduction

Over the past years, a number of methods have been developed for measuring the charge distributions in solid insulating materials owing to their potential applications in sensors and equipments (e.g., pressure pulses, electron beams, thermal pulses and so on)[1]. Being used as the source of pressure pulses and currents, laser interaction with solids has been the subject of numerous studies[2]. Many researchers focused on laser-induced pressure pulse and plasma[3], electron density[4], laser-matter interaction by laser-induced breakdown spectroscopy[5] as well as time of flight measurement[6-10]. LIV response has been found utilized in a variety of optical transducers and sensors, which is sensitive to their interior structures[11]. Position-sensitive laser detector devices are made possible because the output voltage changes linearly with the light spot position[12,13]. In contrast to the conventional view of laser-material interactions, the processes of material response to intense laser irradiation are far more complex, especially for dielectrics[14-16]. Recent studies have shown that charge and current in insulator could be induced by optical fields[15-17]. The irradiation of ultraviolet (UV) laser strongly distorted the electronic band structure to make the insulator instantaneously transforms into a conductor; thus, UV laser could be employed to assess the transient conductivity of dielectrics.

In the petroleum industry, application of advanced high power laser technology in well drilling has attracted significant research interests[18]. With the increase of the unconventional oil and gas drilling depth, the environmental conditions become challenging; thus, the development of oil and gas resources will become increasingly difficult[19]. Lasers, which are insensitive to harsh physical environment, can be served as an appropriate method to promote the development of unconventional oilgas industry. In our previous studies, dependences of the voltages on the laser spot position were observed in electrically conducting black shale owing to the Dember effect and the layered structure[20,21], and the anisotropic electrical response of a dielectric shale core were also characterized under the UV laser irradiation[22]. Experimental study of the destruction of carbonate rock was also performed using a KrF laser beam monitored by LIV method[23]. Here in this study, LIV response in the dielectric shale and sandstone cores were measured on condition that UV laser spot partially irradiated at the surfaces. The transient LIV signals produced by the accumulated and displacement of polarization charges were found depending on the laser position. Ablation was also observed in the rock surface irradiated by the pulsed laser owing to the generation of intrinsic defects. Our results prove that the LIV method is effective for oil-gas reservoirs characterization, and has great potential for the insulator based laser detectors.

2　Materials and methods

Cylindrical shale and sandstone cores were used in this research. Each core is a cylinder of 25mm diameter and 50mm long. Shale was cored in the horizontal direction; thus, the petrophysical layers are perpendicular to the upper and lower surfaces of the core. Sandstone is the main reservoir of oil, gas and groundwater, which can be considered as isotropic because of the absence of striations and lack of orientation in the particle arrangement. Due to the insulation characteristics caused by pores, cracks as well as the complex composition contained insulating minerals, the electrical resistivity were extremely high and up to $\sim 4.2 \times 10^7$ Ω·cm for the shale and $\sim 9.1 \times 10^7$ Ω·cm for the sandstone at room temperature.

Experimentally, a 248nm excimer laser was employed to irradiate the core surface with the full width at half maximum (FWHM) of 20 ns, laser spot of $\sim 1 \times 2 mm^2$ and energy up to 300mJ at a repetition rate of 1Hz. To study the transfer of the laser-excited carriers in rock cores, the test electrodes need to be prepared, and the shale and sandstone cores were polished to ensure better contact between the electrodes and the samples. As shown in Figure 1, a set of electrodes with two $2 \times 8 mm^2$ colloidal silver separated by 60° was equidistantly arrayed on the cylindrical surface of the sandstone core (labeled "sandstone"). Similarly, two sets of electrodes are prepared around the shale, with the electric field parallel (labeled "shale-1") and perpendicular (labeled "shale-2") to the shale bedding plane, respectively. The signals from the electrodes, being the voltage response of the laser-excited samples at the focal spots, were recorded by means of the 1MΩ input of a DOP-4032 digital oscilloscope (Tektronix) with a 350MHz bandwidth. A Keithley 2400 sourcemeter was adopted as the bias voltage source for the sake of a significant photo-response measurement[20]. All measurements were performed at room temperature 24℃.

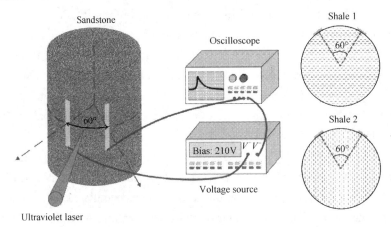

Figure 1   Schematic diagram of the measurement system

3   Results and discussion

The LIV method was employed to monitor the transient generation and transportation of the electronic excitations in the cores. As shown in Figure 2, voltage waveforms in the dielectric core samples were measured with the laser spot partially illuminated close to an electrode under a bias of 210 V. All the pulses initially rose rapidly and then gradually decreased after reaching maximum values (V_p), indicating that the carriers transferred inside the dielectric cores could be captured by the silver electrodes and then conducted to the measuring oscilloscope. The experimental fact of the measured LIV signals shows that both the irradiated- and unirradiated-areas have participated in transmitting carriers in the cores. In our case, the core sample is insulating and the 210V-bias-generated electric field can induce polarized local charges inside the cores.

Heterogeneous distributed mineral components, pores and cracks in shale and sandstone have leaded to inhomogeneous polarization charges inside the cores. Therefore, at the moment of bias importing, a transient charging current will be produced inside the dielectric core. This happens on a time scale of a few microseconds. When the laser pulse irradiates the surface of a core, it induces microscopic displacement of the bound charges, forming oscillating electric dipoles that add up to the macroscopic polarization. Accordingly, interface charges of opposite sign are accumulated, resulting in a transient voltage response[17]. In addition, the different

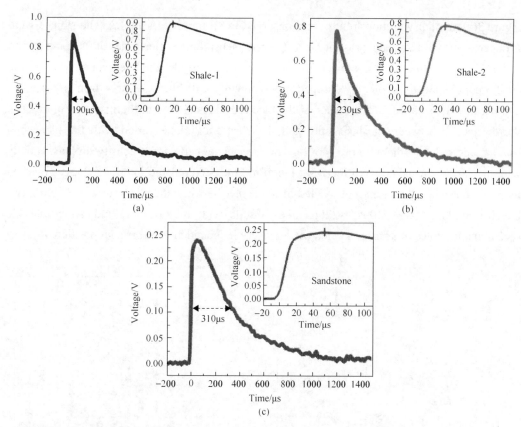

Figure 2 Time dependent voltages of the rock cores in response to the partial laser illumination applied with a 210 V bias. The inset displays the rapidly rising period of the waveform

voltage response can be observed for different samples, which varied at the rise time (τ_s), V_p and the FWHM in related with the composition and structure of the rocks. τ_s and V_p were ~15.7μs and 0.89V, ~23.2μs and 0.79V, and ~51.2μs and 0.24V for shale-1, shale-2 and sandstone, respectively. Both the number of laser-excited carriers and electrical-field-polarized charges contribute to the transient photoconductivity process. Sandstone, with larger bandgap width and smaller dielectric constant, has generated the minimum transient voltage. Because of the partially aligned platy clay particles, different polarizations occurred in the shale core with two fields perpendicular to each other, resulting in a variance in the bandgap width and the distribution of polarization charges[22,23]. When the current is parallel to the aligned axis in shale-1, photoexcited carriers are easily generated in grains aligned in the current direction. Nevertheless, shale has a greater dielectric constant in the direction vertical to the beddings, which will accumulate more polarized charges under the electrical field[24]. The competition mechanisms have dwindled the divergence in V_p for shale-1 and shale-2.

Lateral photovoltage effect has been found in different materials and is sensitive to their interior structures[22,25]. Here, laser-spot-position dependent V_p was studied to achieve further insight into the factors affecting the transient voltage response of the core. As depicted in

Figure 3(a), similar variation trends were observed in all the three samples. The signals were stronger when the laser spot was closer to the one measurement electrodes at 0° compared to that at 60°, resulting by the connection between the electrode and the ground wire at the oscilloscope's negative pole. V_p dropped monotonically with the laser spot from 10° to 30°, and became the smallest at the center of the electrodes. As shown in the inset of Figure 3(a), the distance d between the laser spot and the electrodes plane varied with the increasing rotation angle due to the arc core profile. The laser-generated carrier density dropped rapidly with the carrier diffusion depth increase; therefore, the accumulated interface charges reached its minimum when the laser spot located at the center between the electrodes. In Figure 3(b), FWHM remained at ~320μs, ~220μs and ~180μs with the position, indicating that the recombination rates of laser-generated carriers were independent of laser spot positions.

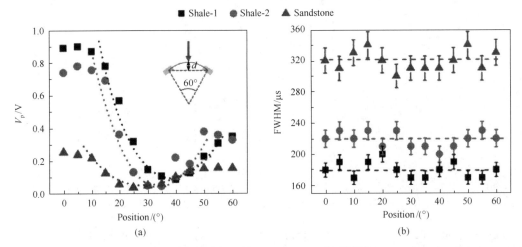

Figure 3   (a) V_p versus laser irradiated position on the core; (b) FWHM versus laser irradiated position on the core. Error bars represent the test accuracy and equal 5% of the data

Shale and sandstone are sedimentary rocks composed of tiny amounts of components with semi-conductor properties and large amount of insulating minerals. For the irradiation of some metallic oxide minerals with the bandgap energy no more than the 248nm laser photon energy (5eV), a single laser photon has sufficient energy to excite an electron from the valence band to the conduction band, which induces interband transition and generation of photoexcited carriers. With the increase of incident laser intensity, simultaneous absorption of multiple photons occur in wide bandgap minerals such as quartz and calcite to excite valence band electrons. If the laser field is intense enough, the excited electrons at the minimum of conduction band will ionize another electrons from the valence band, leading to the so-called electronic avalanche. Because of the high laser pulse energy, the number of photo-induced excitons was much larger than that of the majority carriers in the cores, and the transient voltage was produced by the accumulated and displacement of polarization charges. Once the electronic excitations were generated under the pulsed laser, generation of intrinsic defects can outcome the breakdown of the material[6,26,27]; thus, an investigation about the UV laser ablation was then performed. As shown in Figure 4, features at the core surfaces were examined with an optical microscope, where the surfaces were irradiated by multi-pulses. The action of the laser pulses on rocks for the energy density at 1500mJ/cm^2 results in the formation of craters at several micron thicknesses. In particular, owing to the uneven textures at the lamination plane of the shale, irregularity can be observed in the ablated shale surface because of different strength properties of the stratification layers. Compared to that of the shale, the morphology of the sandstone was irregularly polygon and dispersed well. The present ablation was considered to be associated with rapid buildup of conduction electrons in the shale and sandstone cores.

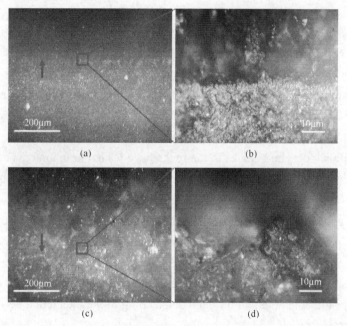

Figure 4 Optical microscope analysis of the surface ablation of (a) shale and (b) sandstone produced by repeated 248nm pulsed laser at the energy 300mJ and irradiation area 2mm^2 for 10 times. The arrows show boundaries between unirradiated and irradiated area

Recently, application of advanced high power laser technology in oil and gas well drilling has been attracting significant research interests[28]. As a source of energy in a highly concentrated form, powerful lasers have become appropriate means for rock perforation[29]. When a high-power laser beam is applied to a rock, different reaction processes can occur, depending on the laser energy power and the way that energy is applied[26]. LIV method, which is sensitive to the electronic excitation and transportation inside the materials' structures, can be a useful tool for real-time monitoring the perforation processes of rocks. The KrF excimer laser has long been used for surface ablation, showing a non-thermal character of the laser-matter interaction because of the electronic excitation leaded by photon absorption[20].

Previously, experimental study of the destruction of carbonate rock was performed using a KrF laser beam monitored by LIV method. The processes of surface spallation, deep spallation, and rock puncture were characterized by analyzing the relationship between voltage and the irradiation time[23]. Here in this study, ablation by the focused UV laser at the surfaces of the dielectric shale and sandstone cores were studied and discussed. The measured LIV signals with partial irradiated at the dielectric rocks' surfaces have indicated that the carriers transferred inside the non-irradiation region could be captured by the silver electrodes, which were possibly leaded by the redistribution of the polarized charges. The results demonstrated that LIV method is effective for dielectrics characterization, and would provide understanding of the interaction between an intense laser and dielectric material.

4 Conclusions

In summary, this study focused on the intense laser-induced voltage in dielectric rock cores. LIV signals in the dielectric core samples were measured with the laser spot partial illuminated at the surfaces under a bias of 210 V, indicating that the carriers transferred inside the non-irradiation region could be captured by the silver electrodes. Heterogeneous distributed minerals, pores and cracks in the rocks have leaded to inhomogeneous polarization charges inside the cores. Accumulation of interface charges have resulted in transient voltage response. The voltage waveforms at the shale and sandstone were observed with different V_p and FWHM in accordance with the components and structures. V_p was also measured as a function of spot position. In addition, because of the rapid buildup of electrons and generation of intrinsic defects in the shale and sandstone cores, ablation occurred on the UV laser irradiated surface. In general, LIV method is a promising means for dielectrics characterization, and there is a considerable potential for an understanding of the interaction between the intense laser and dielectric materials.

Acknowledgements

This work was supported by the National Nature Science Foundation of China (Grant No. 11574401), the Science Foundation of China University of Petroleum, Beijing (Grant Nos. 2462017YJRC029 and yjs2017019), and the Beijing Natural Science Foundation (Grant No. 1184016).

References

[1] Liu R, Törnkvist C, Jeroense M. Space-charge measurement technologies and their potential applications. Sensor. Mater., 2017, 29(8): 1089-1098.

[2] Sessler G M, West J E, Gerhard G. High-resolution laser-pulse method for measuring charge distributions in dielectrics. Phys. Rev. Lett. 1982, 48(8): 563-566.

[3] Liu R S, Takada T, Takasu N. Pulsed electro-acoustic method for measurement of space charge distribution in power cables under both DC and AC electric fields. J. Phys., D: Appl. Phys., 1993, 26(16): 986-993.

[4] Jasapara J, Nampoothiri A V, Rudolph W, et al. Femtosecond laser pulse induced breakdown in dielectric thin films. Phys. Rev. B, 2001, 63(4):045117.

[5] Lednev V N, Pershin S M, Sdvizhenskii P A, et al. Laser induced breakdown spectroscopy with picosecond pulse train. Laser Phys. Lett., 2016, 14(2):026002.

[6] Schultze M. Controlling dielectrics with the electric field of light. Nature, 2013, 493(7430): 75-78.

[7] Costache F, Eckert S, Reif J. Dynamics of laser-induced desorption from dielectric surfaces on a sub-picosecond timescale. Appl. Surf. Sci., 2006, 252(13):4416-4419.

[8] Sommer A, Bothschafter E M, Sato S A, et al. Attosecond nonlinear polarization and light-matter energy transfer in solids. Nature, 2016, 534(7605): 86-90.

[9] Zhan H, Zhao K, Lu H, et al. Oblique-incidence reflectivity difference application for morphology detection. Appl. Optics., 2017, 56(30): 8348-8352.

[10] Miao X Y, Zhan H L, Zhao K, et al. Oil yield characterization by anisotropy in optical parameters of the oil shale. Energ. Fuel., 2016, 30: 10365-10370.

[11] Zhao K, Jin K J, Huang Y H, et al. Ultraviolet fast-response photoelectric effect in tilted orientation $SrTiO_3$ single crystals. Appl. Phys. Lett., 2006, 89: 173507.

[12] Li X M, Zhao K, Ni H, et al. Voltage tunable photodetecting properties of $La_{0.4}Ca_{0.6}MnO_3$ films grown on miscut $LaSrAlO_4$ substrates. Appl. Phys. Lett., 2010, 97: 044104.

[13] Yue Z J, Zhao K, Zhao S Q, et al. Thickness-dependent photovoltaic effects in miscut Nb-doped $SrTiO_3$ single crystals. J. Phys. D Appl. Phys., 2009, 43: 015104.

[14] Stuart B C, Feit M D, Rubenchik A M, et al. Laser-induced ablation in dielectrics with nanosecond to subpicosecond pulses. Phys. Rev. Lett., 1995, 74(12): 2248-2251.

[15] Quéré F, Guizard S, Martin P. Time-resolved study of laser-induced breakdown in dielectrics. Europhys. Lett.,2001, 56 (1): 138-144.

[16] Balling P, Schou J. Femtosecond-laser ablation dynamics of dielectrics: Basics and applications for thin films. Rep. Prog. Phys., 2013, 76(3): 036502.

[17] Schiffrin A, Paasch-Colberg T, Karpowicz N, et al. Optical-field-induced current in dielectrics. Nature, 2013, 493: 70-74.

[18] Xu Z, Reed C B, Kornecki G, et al. Specific energy for pulsed laser rock drilling. J. Laser Appl., 2003, 15(1): 25-30.

[19] Zhubayev A, Houben M E, Smeulders D M J, et al. Ultrasonic velocity and attenuation anisotropy of shales, Whitby, United Kingdom. Geophysics, 2016, 81: D45-D56.

[20] Lazare S, Lopez J, Weisbuch F. High-aspect-ratio microdrilling in polymeric materials with intense KrF laser radiation. Appl. Phys. A, 1999, 69(1): S1-S6.

[21] Miao X, Zhu J, Li Y, et al. Ultraviolet laser-induced voltage in anisotropic shale. J. Phys. D Appl. Phys., 2018, 51: 045503.

[22] Miao X Y, Zhu J, Zhao K, et al. Ultraviolet laser-induced lateral photovoltaic response in anisotropic black shale. Applied Physics. B, 2017, 123(12): 276.

[23] Zhu J, Miao X Y, Zhao K, et al. Characterizing the rock perforation process by laser-induced voltage response. Sci. China: Phys. Mech. Astron., 2018, 61(5): 054221.

[24] López A A, Mousatov A, Markov M, et al. Modeling and inversion of elastic wave velocities and electrical conductivity in clastic formations with structural and dispersed shales. J. Appl. Geophys., 2015, 116: 28-42.

[25] Zhou J, Revil A, Karaoulis M, et al. Image-guided inversion of electrical resistivity data. Geophys. J. Int., 2014, 197: 292-309.

[26] Zhao S, Liu W, Yang L, et al. Lateral photovoltage of B-doped ZnO thin films induced by 10.6μm CO_2 laser. J. Phys. D Appl. Phys., 2009, 42: 185101.

[27] Mao S S, Quéré F, Guizard S, et al. Dynamics of femtosecond laser interactions with dielectrics. Appl. Phys. A., 2004, 79(7): 1695-1709.

[28] Keshavarzi R, Jahanbakhshi R, Bayesteh H, et al. Applying high power lasers in perforating oil and gas wells: Prediction of the laser power loss during laser beam-fluid interaction by using artificial neural networks. Lasers in Engineering. 2011, 21(5): 149-167.

[29] Duan W, Wang K, Dong X, et al. Experimental characterizations of burr deposition in Nd: YAG laser drilling: A parametric study. Int. J. of Adv. Manuf. Tech., 2015, 76(9-12): 1529-1542.

Characterizing the rock perforation process by laser-induced voltage response

Jing Zhu[1,2] Xinyang Miao[2] Kun Zhao[1,2] Honglei Zhan[2]
Qiong Zhou[1] Wenzheng Yue[1]

(1.State Key Laboratory of Petroleum Resources and Prospecting, China University of Petroleum, Beijing 102249, China; 2.Beijing Key Laboratory of Optical Detection Technology for Oil and Gas, China University of Petroleum, Beijing 102249, China)

In the petroleum industry, it is necessary to achieve the exploitation of oil and gas by perforation, and a considerable number of research studies have been conducted with the objective of improving recovery of oil and gas[1-2]. New perforation methods are urgently needed to accelerate development of energy resources. Drilling rock by laser is a non-contact process that can be used to form holes in a wide variety of rocks with a high degree of precision and reproducibility. The feasibility, costs, benefits and environmental impact of adopting high-powered military lasers for a revolutionary application in oil and gas exploration and production were determined by Gas Technology Institute (GTI)[3], proving more efficient and cleaner of laser drilling to perforate wells through hard rock formations at great depths. One of the major advantages of laser drilling is its potential to reduce drilling time (more than 100 times conventional rotary drilling rates)[4]. Seven kinds of lasers were considered promising in gas well drilling and completions[5], including deuterium fluoride (DF)/hydrogen fluoride (HF), free-electron laser (FEL), chemical oxygen-iodine laser (COIL), carbon dioxide (CO_2) laser, carbon monoxide (CO) laser, neodymium: yttrium aluminum garnet (Nd:YAG) and krypton fluoride (KrF) excimer laser, separately[6].

The process of laser rock spallation is complex and depends on many factors, in which the most important variables affecting perforation are laser power and irradiation time[7]. High-power lasers can weaken, spall, melt, and vaporize rocks with a series of complicated dynamic physical and chemical reactions over time. Understanding the damage mechanism of typical rocks has very important value in laser drilling applications. Here, we aim to characterize the dynamic processes of rock perforation by laser-induced voltage response[8-12]. The physical processes of rock perforation and the change of perforation rate are also investigated by analyzing the relationship between peak value of voltage (V_p) and the detecting time (T).

The carbonate rock used here was collected from Sinan county, Guizhou province in Yunnan-Guizhou plateau. The principal component of carbonate rock used in experiment is calcium carbonate (CaCO$_3$). Figure 1(a) shows the surface morphology of carbonate rock by scanning electron microscope (SEM) images. In this work, the sample geometry is 10mm × 2mm with thickness of 2mm [Figure1 (b)]. Four Ag electrodes with the size of 1mm × 2mm separated by 3mm were painted on the surface of the carbonate rock. The schematic of the experimental setup is shown in Figure 1 (c), which consists of a KrF excimer laser (248 nm), a carbonate rock and a digital oscilloscope. The response of carbonate rock was excited by KrF laser at ambient temperature, monitored with a digital oscilloscope with input impedance of 1MΩ. The laser spot was located at the center position between the inner two electrodes. A bias of 210 V was applied to the sample through the outer probes and the inner pair of probes picks up a voltage drop V along the surface. All the electrodes were kept in dark during the measurements. It was reported that the increasing beam repetition rate within the same material removal mechanism would augment the material removal rate because of an increase of maximum temperature, thermal cycling frequency, and intensity of laser-driven shock wave within the rock[13]. The laser pulse repetition rate was 1Hz at a pulse energy of 73.9mJ and an energy density of 36.95mJ/mm^2. The irradiated area was 2mm^2. The measurement was terminated on the condition that V_p remains constant. In oil well rock drilling, rate of penetration (ROP) contributes to determining the efficiency of laser perforating[14]. Here, ROP is quantified as ROP = D/t (mm/s), where D and t are depth of the hole and the laser irradiation time, respectively. In our study, the laser repeat rate is 1Hz, and on-sample energy density is 36.95mJ/mm^2. Thus the average value of ROP is ~0.4 μm/s.

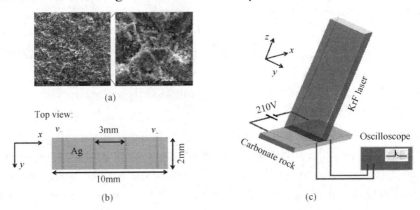

Figure 1 (a) SEM image of carbonate rock; (b)the top view shows the typical device geometry used in our experiments; (c)schematic illustration of the experimental setup

The laser-induced voltage signal of carbonate rock under a 248 nm irradiation is shown in Figure 2. It should be noted that there is a sharp rise of the pulse at the very beginning and a decreasing voltage followed. A rise time of ~25 μs, a fall time of 325 μs and a full width at half maximum (FWHM) of ~114μs are measured when the carbonate rock is irradiated. The selected transient voltages of carbonate rock at T are displayed in Figure 2. The peak values of pulse response at T = 3s, 580s, 1567s and 2930s are equal to 220mV, 80mV, –145mV and –640mV.

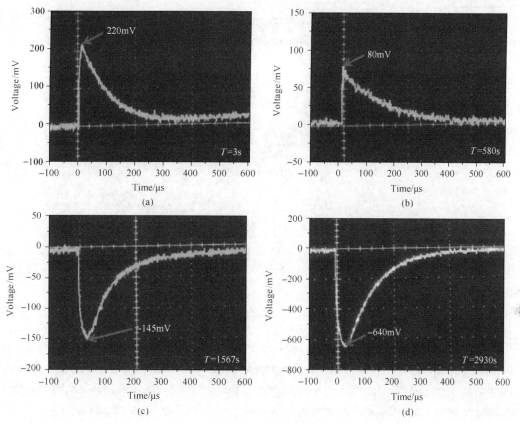

Figure 2　The selected transient voltages of carbonate rock at (a)T = 3s, (b)T = 580s, (c)T = 1567s, (d)T = 2930s

The mechanism of laser-induced voltage response for most single or regular structure materials was studied under a mountain of work[15-19]. For instance, photothermal effect is proved to be the origin of photoconductance and zero-bias photocurrent generation in insulating phase VO_2[17]. Seebeck effect is proposed to elucidate the tilting angle dependence of laser-induced voltage in tilted orientation $SrTiO_3$ single crystal[16]. As the rock sample used in this study is unpurified calcium carbonate with complicated components and structures, a number of factors should be responsible for the laser-induced voltaic response. The voltage response generated in the rock sample was possibly caused by a combinative effect of various mechanisms mentioned above.

In order to analyze the rock perforation process by laser-induced voltage response, T dependent V_P is shown in Figure 3. The whole process could be distinctly divided into five parts: 1s < T < 139s, 139s < T < 1916s, 1916s < T < 2392s, 2392s < T < 4519s, and T > 4519s. The values of V_P at T = 1s, 139s, 1916s, 2392s and 4519s are equal to 210mV, 120mV, −160mV, −720mV and −420mV, respectively. As time increases from 1s to 2392s, V_P continuously decreases with different rates. In this period, there are two sharp transition of V_P at T =139s and 1916s causing three slopes of −0.652mV/s, −0.158mV/s and −1.879mV/s. With

T increasing from 2392s to 4519s, the voltage peak increases linearly with the slope equaling 0.141mV/s. After 4519s, the transient voltage peak becomes unchangeable.

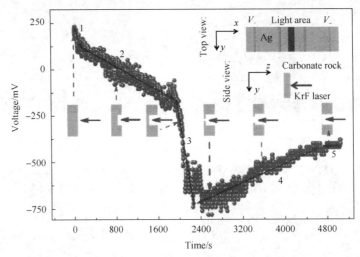

Figure 3　The relationships between the laser irradiation time and voltage response peak

When a high-power laser beam is applied on a rock, different reactions produced such as thermal spallation, melting, or vaporization depending on the exerted laser energy and the way the energy is applied[8]. As is shown in Figure 3, the first three parts are revealed as surface spallation, deep spallation and puncture of rock, respectively. The value of rate is biggest in the third process while smallest in the second process. With the hole depth increasing, the power density of laser decreases resulting in declination of destruction rate[15]. V_P drops rapidly in the third process indicating that the rock is gradually punctured. When T arrives to approximately 2392s, the sample is completely broken into two parts. Then, V_P increases linearly with T increasing from 2392 s to 4519 s. Finally, the value of V_P is invariable that represents that the there are no interaction between the laser and the rock and the width of drilled rock hole arrives into the width of laser spot.

In conclusion, the laser-induced voltage of carbonate rock has been studied under the irradiation of a 248 nm laser. The research focused on reporting the experimental study of destruction carbonate rock performed using KrF laser beam. The process of surface spallation, deep spallation and puncture was markedly characterized by analyzing the relationship between V_P and T. With the rock punctured gradually, V_P drops rapidly. After the sample is completely broken into two parts, V_P increases linearly with T increasing until no interaction between carbonate rock and laser. In addition, the speed variation was discussed in this work according to the power density. This work showed that laser-induced voltage response technique was a useful tool for laser drilling characterization in laser precision machining as well as laser attack-defense field, and would provide great significance in deeply understanding the interaction between ultraviolet laser and rock.

Acknowledgements

This work was supported by the National Nature Science Foundation of China (Grant No. 11574401) Science Foundation of China University of Petroleum, Beijing (No.2462017YJRC029).

References

[1] Jan B M, Rae and Noor M I, et al. Increasing production by maximizing underbalance during perforation using nontraditional lightweight completion fluid. SPE Drilling & Completion, 2009, 24(2): 326-331.

[2] Keshavarzi R, Jahanbakhshi R, Bayesteh H, et al. Applying high power lasers in perforating oil and gas wells: Prediction of the laser power loss during laser beam-fluid interaction by using artificial neural networks. Lasers in Engineering, 2011, 21(5): 329-340.

[3] Graves R, Anibal A, Gahan B, et al. Proceedings of the SPE Annual Technical Conference and Exhibition, 29 September-2 October, San Antonio, USA. Society of Petroleum Engineers, 2007.

[4] Obrien D, Graves R M and Obrien E. Star wars laser technology for gas drilling and completions in the 21st Century. Proceedings of the SPE Annual Technical Conference and Exhibition, 3-6 October, Houston, USA. Society of Petroleum Engineers, 1999.

[5] Sinha P, Gour A. Laser drilling research and application: An update. Proceedings of the Spe/iadc Indian Drilling Technology Conference & Exhibition, 16-18 October, Mumbai, India. Society of Petroleum Engineers, 2006.

[6] Wang F, He H Y, Zhu R X, et al. Laser step-heating $^{40}Ar/^{39}Ar$ dating on young volcanic rocks. Chinese Science Bulletin, 2006, 51(23): 2892-2896.

[7] Yan F, Gu Y, Wang Y, et al. Study on the interaction mechanism between laser and rock during perforation. Optics and Laser Technology, 2013, 54(30): 303-308.

[8] Chang C L, Kleinhammes A, Moulton W G, et al. Symmetry-forbidden laser-induced voltages in $YBa_2Cu_3O_7$. Physical Review B, 1990, 41(16): 11564-11567.

[9] Liu X, Yin X, Chen Q, et al. Effect of annealing oxygen pressure on the enhancement of laser-induced voltage in $La_{2/3}Ca_{1/3}MnO_3$: $Ag_{0.04}$ films. Materials Science and Engineering: B, 2014, 185(1): 105-108.

[10] Liu X, Yan Y Z, Chen Q M, et al. Laser-induced voltage (LIV) enhancement of $La_{2/3}Sr_{1/3}MnO_3$ films with Ag addition. Applied Physics A, 2014, 115(4): 1371-1374.

[11] Liu X, Yin X P, Chen Q M, et al. Improved electrical properties of $La_{2/3}Ba_{1/3}MnO_3$: $Ag_{0.04}$ thin films by thermal annealing. Applied Physics A, 2014, 116(4): 1853-1856.

[12] Zhao S Q, Zhou Y L, Zhao K, et al. Laser-induced thermoelectric voltage in normal state MgB_2 thin films. Applied Surface Science, 2006, 253(5): 2671-2673.

[13] Xu Z, Reed C B and Konercki G. Specific energy for pulsed laser rock drilling. Journal of Applied Physics, 2003, 15(1): 25-30.

[14] Rastegar M, Hareland G, Nygaard R, et al. Optimization of multiple bit runs based on ROP models and cost equation: A new methodology applied for one of the Persian gulf carbonate fields. Medical Physics, 2008, 36(6): 2692-2692.

[15] Zhao K, Jin K J, Huang Y H, et al. Ultraviolet fast-response photoelectric effect in tilted orientation $SrTiO_3$ single crystals. Applied Physics Letters, 2006, 89(17): 173507.

[16] Kasırga T S, Sun D, Park J H, et al. Photoresponse of a strongly correlated material determined by scanning photocurrent microscopy. Nature Nanotechnology, 2012, 7(11): 723-727.

[17] Zhao K, Huang Y, Zhou Q, et al. On microscopic origin of photovoltaic effect in $La_{0.8}Sr_{0.2}MnO_3$ films. Applied Physics Letters, 2005, 86(22): 5178.

[18] Huang Y H, Zhao K, Lu H B, et al. Multifunctional characteristics of $BaNb_{0.3}Ti_{0.7}O_3$/Si p-n junctions. Applied Physics Letters, 2006, 88(6): 1160.

[19] Ahmadi M, Erfan M R, Torkamany M J, et al. The effect of interaction time and saturation of rock on specific energy in Nd: YAG laser perforating. Optics and Laser Technology, 2011, 43(1): 226-231.

页岩各向异性的全光学检测

吕志清[1,2]　杨 肖[2]　魏建新[1]

(1.油气资源与探测国家重点实验室，中国石油大学(北京)，北京 102249；2.油气光学探测技术北京市重点实验室，中国石油大学(北京)，北京 102249)

摘要：本文利用全光学方法对不同地区的页岩样本进行各向异性检测。利用激光在页岩等岩石中激发产生超声波，基于激光干涉原理对超声波速进行检测，从而获得页岩层理结构的各向异性分布特点。检测结果表明超声纵波波速与页岩层理结构吻合得很好。超声纵波沿页岩层理面传播时速度最大，垂直页岩层理面传播时速度最小，超声波速关于页岩层理方向以正弦规律对称分布。不同地区的页岩超声波速也不同。全光学检测方法可以实现对页岩层理结构的快速判断，根据在页岩中传播的超声波速可了解页岩的致密度及各向异性强度，对页岩储层的开发具有一定指导意义。

关键词：页岩；各向异性；层理结构；激光；超声波

Non-contact measurement of the anisotropy in shale with all-optical detection

Zhiqing Lü[1,2]　Xiao Yang[2]　Jianxin Wei[1]

(1.State Key Laboratory of Petroleum Resources and Prospecting, China University of Petroleum, Beijing 102249, China; 2.Beijing Key Laboratory of Optical Detection Technology for Oil and Gas, China University of Petroleum, Beijing 102249, China)

Abstract: The method of all-optical detection was used to investigate the anisotropy in shale samples from different regions. The ultrasonic wave was excited in shales by pulsed laser and detected by optical interferometry. The anisotropic distribution characteristics of the shales were studied by gaining the ultrasonic wave velocity. The results show that ultrasonic P-wave velocity periodically changed with the angle between the detection direction and the bedding in shale. The P-wave velocity reached its maximum when the angle was zero, that the detection direction parallel to the bedding. And the P-wave velocity reached its minimum when the angle was right, that the detection direction perpendicular to the bedding. The all-optical detection method provide a way to rapidly judge the bedding direction in shale or in other rocks, which gives guidance for the development on the shale reservoir.

Keywords: shale; anisotropy; bedding structure; laser; ultrasonic wave

1 引言

随着油气资源勘探开发的不断深入发展以及常规油气的开发难度不断增大，非常规油气资源已经引起全世界的高度关注，并且在世界油气工业中所占的比重日益增大。在非常规油气资源的开采过程中，页岩由于储集了丰富的油气资源而受到极大关注。页岩种类丰富，在石油领域具有重要地位，例如，油母页岩可以提炼石油，黑色页岩可以作为石油的指示地层。随着全球非常规油气资源的勘探开发，页岩气以其巨大的资源量已经成为当今非常规油气藏领域研究的热点。页岩气主要存在于富含有机质页岩的储集岩系中。我国的页岩气资源丰富，具有良好发展前景，对页岩油气的勘探开发也正在全面展开[1-6]。

在对页岩储层开发过程中，由于其层理构造具有明显的各向异性，开展页岩各向异性的研究对钻井和储层改造十分重要。页岩独特的岩石结构特性对勘探开发有很大的影响。在钻井过程中，由于页岩在不同层理结构上表现出不同的强度性质，会加大钻进的工作量，影响钻进效率，还有可能使钻杆在钻进过程中受力不平衡而发生偏斜甚至弯曲，这些都是由地层岩石的各向异性引起的。岩石各向异性是指岩石的某些物理性质会随着空间方向的不同发生改变。影响岩石各向异性的因素有很多，如孔隙、裂纹、颗粒排列、有机物有向发育等，其中岩石的孔隙结构是一个重要因素[7-12]。由于页岩孔隙存有大量的油气资源，对页岩各向异性的测量为研究油气储集结构提供了新方法，对页岩各向异性的研究具有重要意义。

对岩石各向异性的研究主要集中于理论模型和实验室检测。岩石在成岩过程中，矿物颗粒的排列、裂隙的分布方式等随方向而变化，引起力学参数的宏观各向异性。岩石力学性质参数之间的关系，尤其是超声波速度与单轴压缩强度、杨氏模量、巴西劈裂强度的关系，是实验室检测的重要内容[13-15]。利用换能器激发、接收超声波是实验室常用的检测方法，该方法具有快速、无损、准确等优点，对理解超声波速度与岩体强度和变形参数之间的关系具有重要的价值[16]。但换能器检测必须为接触式测量，且需要较大的接触面，同时需要耦合剂的作用，这为检测带来了一定的不便，也不适宜在高温高压等恶劣的环境下进行检测。

利用高能激光脉冲在介质中产生超声波的技术始于1962年，经过几十年的发展，该技术已经逐渐成为材料无损检测的一种重要手段和发展方向。当高能脉冲激光入射到材料表面时，由于热弹效应或烧蚀效应会在材料内部产生超声波。与传统方法相比，这一技术可以重复产生很窄的超声脉冲，具有极高的时间及空间分辨率，同时具有非接触、远距离、快速检测等优点，尤其适用于高温、腐蚀、辐射等恶劣环境及快速在线检测[17-19]。

本文采用全光学方法对岩石的各向异性进行研究，即利用脉冲激光在页岩中激发超声波，并基于光学干涉原理对超声波进行检测。文中分别对理想层状结构的模拟样品、实际采样的不同地区页岩岩心及砂岩的各向异性进行了全光学检测，检测结果表明岩石的超声纵波波速随其层理结构呈周期性变化，利用这一规律可实现对页岩各向异性的检测。

2 实验方法

对页岩各向异性的检测采取全光学方法，即超声波的激发及检测均采用光学方法实现。如图1(a)所示，脉冲功率激光器产生一个一定脉宽和能量的光脉冲，经反射镜M1、透镜及反射镜M2入射至试样一侧表面，其中透镜把激光束汇聚成点状光源。在试样的另一侧用激光干涉仪检测试样内被激发产生的超声波，然后用数字示波器记录其位移波形输入计算机，以便进一步处理。图1(b)是该系统的实物装置图，整套实验系统由激光干涉仪、脉冲功率激光、三维扫描平台和系统控制集成计算机、示波器4部分组成。用于激发超声波的激光器为Tolar-1053，能以1kHz以上的速率产生10~20mJ的单脉冲能量。超声波的接收与检测采用激光干涉仪Tempo 2D，具有高灵敏度及高信噪比等优点。用于记录超声波信号的示波器带宽为100MHz，输入阻抗为1MΩ。这一全光学检测系统的优点在于可以实现激光输出的全自动控制以及物理模型试样超声信号的非接触全自动快速采集。

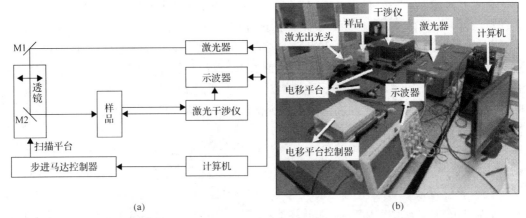

图1 (a)全光学超声波激发与检测原理图；(b)全光学检测系统实验装置图。高能量脉冲激光在样品一侧入射并在样品中激发产生超声波，样品的另一侧用激光干涉仪接收检测携有样品结构信息的超声波信号

Figure 1 (a) The schematic diagram of the all-optical ultrasonic excitation and testing; (b) picture of the experimental setup of the all-optical detection system. High-energy pulsed laser irradiated on one side of the sample to excite ultrasonic waves, which were received and measured by a laser receiver

待测样品均加工成直径为25mm的圆柱体，置于激光光源与干涉仪之间直径为25mm的圆形样品池中，如图2所示，整个样品池固定在一个可进行360°旋转的旋转台上。测试时，激发光与探测光调整在同一水平线上，聚焦于圆柱形样品表面且均通过圆柱的中心轴线。由于样品是圆柱形且安置在旋转台上，测试时可以根据需要进行0°~360°的范围内任意角度的测量。相较于传统的换能器检测技术，全光学方法由于是非接触测量，对样品的形状、大小没有特别的要求，对页岩各向异性进行检测时，对检测方向的控制更加灵活，在0°~360°可以对任意方向的超声波速进行检测。

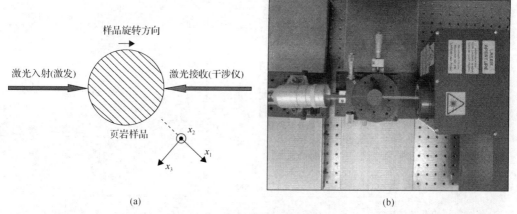

图 2 (a)页岩各向异性全光学检测原理图；(b)实验装置俯视图。检测过程中，页岩样品绕其自身轴线旋转。x_1x_2 平面与页岩层理面平行，x_1 方向与光束方向的夹角为 0°时为测量起点

Figure 2 (a) The schematic diagram of shale anisotropy detection; (b) picture of the experimental setup from the top. The shale sample is rotated around its axis during the measurements. The plane x_1x_2 parallel to the shale bedding. The starting point for measuring is 0° of the angle between the laser beam and x_1 axis

3 实验结果与讨论

3.1 模拟样品的各向异性检测

模拟样品是具有层状结构的粗布胶木和细布胶木，采用上述全光学方法对其各向异性进行检测。样品 A 是直径为 70mm 的圆柱形粗布胶木，将其直接放置在旋转台上进行 180°旋转测量。测量时由平行层理的方向[即图 2(a)中 x_1 方向与光束方向的夹角为 0°]开始每隔 5°进行一次超声纵波的测量，逐步旋转到 x_1 方向与光束方向的夹角为 90°，再旋转到另一侧 x_1 方向与光束方向的夹角为 180°的位置。图 3(a)是相应的实验结果瀑布图，该图显示出样品在 0°~180°的超声波速分布特点。图 3(b)描绘了纵波波速随样品旋转角度的变化关系。由图 3(b)可以看出超声波速在样品中是按正弦规律分布的。超声纵波在平行层理的方向上传播速度最快，在垂直层理方向上传播速度最慢。由于传播路径相等，都是圆柱体的直径，根据 $v=D/t$ 可计算出纵波波速在样品中的传播速率在 2700~3510m/s。样品 B 是直径为 25mm 的层状结构细布胶木。为了进一步验证超声波速在层理结构样品中的正弦分布，对样品 B 进行 360°旋转测量。瀑布图 4(a)显示出一个完整周期的正弦图形，波速随入射角度的变化关系(图 4(b))很清晰地反映了这一特点。模拟样品 B 的实验结果表明，超声波速在平行层理的方向上穿过的介质单一且阻碍最小，传播速度最快(2800m/s)；在垂直层理方向上穿过的介质层理面的次数最多，传播速度最慢(2300m/s)。其他方向的波速随着层理面的旋转呈正弦规律变化，波速从平行层理方向到垂直层理方向慢慢减小，从垂直层理方向到平行层理方向又慢慢增加，周而复始。

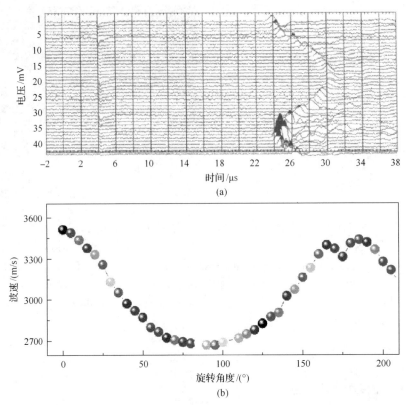

图 3 模拟样品 A(直径为 70mm 的粗布胶木)的激光超声数据(a)瀑布图;
(b)纵波速度与旋转角度的关系曲线

Figure 3 Laser ultrasonic data of the simulation sample A (coarse bakelite with a diameter of 70mm). (a) The waterfall plot; (b) relation curve between the P-wave velocity and the rotation angle. The P-wave velocity reached its maximum when the detection direction parallel to the bedding, and reached its minimum when the detection direction perpendicular to the bedding

3.2 页岩岩心的各向异性检测

页岩岩心取自 3 个不同的地区,共 7 个样品,分别标记为 C、D、E、F、G、H、I。样品 C、D 来自区域一,采样深度分别为 3325.40m 及 3313.09m,均为横向取心;样品 E、F 来自区域二,采样深度分别为 1611.35m 及 3220.15m,均为横向取心;样品 G、H 来自区域三,采样深度为 2317m,均为横向取心;样品 I 来自区域一,采样深度为 3332.43m,为纵向取心。所有页岩岩心样品均加工为直径 25mm、高度 50mm 的圆柱形。从外貌来看,页岩样品可分为两类,一类是通过肉眼可分辨出页岩的层理方向,这种页岩多数是由不同黏土经过长时间的沉积而形成的,纹理比较明显,如图 5(a)所示;另一类如图 5(b)所示,无法通过肉眼分辨其层理方向,主要原因是受其内部有机物的影响而变为黑色,这种页岩较为多见。

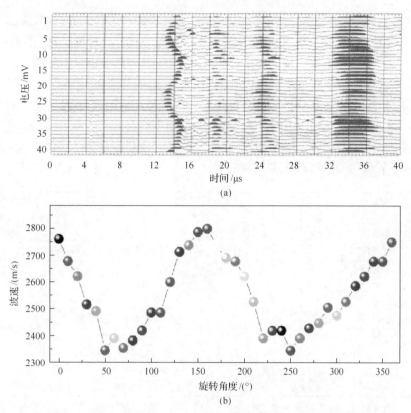

图 4 模拟样品 B(直径为 25mm 的细布胶木)的实验结果瀑布图(a)与纵波波速随样品旋转角度的变化关系(b)。超声纵波波速随样品的层理结构呈周期性变化

Figure 4 Laser ultrasonic data of the simulation sample B (muslin bakelite with a diameter of 25mm). (a) The waterfall plot; (b) relation curve between the P-wave velocity and the rotation angle. The ultrasonic P-wave velocity periodically changed with the rotation angle of the bedding in sample

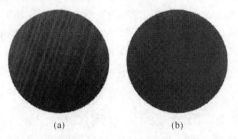

图 5 页岩样品外貌图可见层理(a)与不可见层理(b)(页岩生长过程中受其内部有机物影响而变为黑色,这种页岩较为常见)

Figure 5 Top view of shale samples visible bedding (a) and invisible bedding (b) (which was influenced by its internal organic matter in the process of growth. This kind of shale is common in nature)

7 个样品的检测结果如图 6 所示。随着样品由 0°到 180°旋转,通过样品 C 的超声波速由 3800m/s 减小到 2500m/s,再回升到 3800m/s。根据模拟页岩样品的检测结果(层隙方向和检测方向相同时波速最大,层隙方向和检测方向垂直时波速最小)可推断,波速为

3800m/s 时对应的检测方向是页岩的层理方向，而波速极小值(2500m/s)对应的方向是与页岩层理垂直的方向。实验结果并没有完全关于垂直层理方向对称，这是由于实际的页岩层理并不是严格平行排列的，其内部缺陷及有机物的分布会引起一定的波速变化。样品 D 和 E 的 180°旋转测试也具有相似的变化趋势。随着样品由 0°到 180°旋转，通过样品 D 的超声波速由 3650m/s 减小到 2380m/s，再回升到 3700m/s；通过样品 E 的超声波速由 3400m/s 减小到 2100m/s，再回升到 3400m/s。由于样品 C、D 来自同一区域，且采样深度差异不大，二者的超声波速变化范围相近。样品 E 来自另一区域，其超声波速与样品 C、D 相差较大，反映了样本的区域差异。

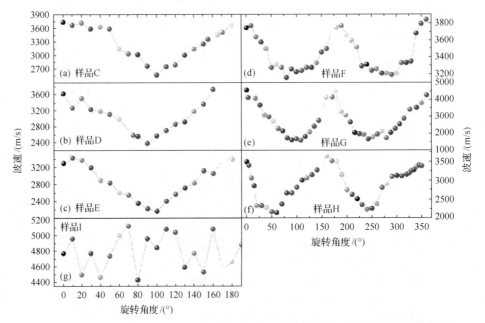

图 6　页岩样品 C、D、E、F、G、H、I 的超声纵波波速随样品旋转角度的变化关系。
超声波速随页岩层理方向以正弦规律对称分布

Figure 6　Relation curves between the P-wave velocity and the rotation angle of the shale samples.
(a) Sample C; (b) sample D; (c) sample E; (d) sample F; (e) sample G; (f) sample H; (g) sample I.
The P-wave velocity symmetrically varied in sine law with the bedding in shale

为了更全面地反映圆柱形样品的各向异性分布，样品 F、G、H 进行了旋转 360°的各向异性检测。随着样品角度 0°、90°、180°、270°、360°旋转，通过样品 F、G、H 的超声波速变化规律分别为 3750m/s、3150m/s、3750m/s、3150m/s、3750m/s，4500m/s、1600m/s、4500m/s、1600m/s、4500m/s，以及 3550m/s、2150m/s、3550m/s、2150m/s、3550m/s。超声波速近似以正弦规律呈周期性变化，360°的旋转测试结果可以看作 180°旋转测试结果的周期性重复，反映了岩心样本致密规则的层状结构。由图 6 可以看出同样条件下通过样品 C、D、E、F、G、H 的超声波速是不相同的，这与页岩本身的材质差异有关。对于同样材质的页岩，疏松材质内部会有更多的空气，延长了超声波的透过时间，因此致密材质的波速要大于疏松材质的波速。相同条件下，通过不同样品的超声波速差异是不同的，说明不同页岩的各向异性强度是不同的。其中样品 G 的超声波速差异最大，平行层理方向的极大波速与垂直层理方向的极小波速相差了 2900m/s，表明其各向异性强度最大。因此通过比较纵波的差异可以判断页岩各向异性强度，对页岩的开采具有一定的指导意义。对于各向异性强度较大的页岩储层，垂直层理方向和平行层理方向的开采难度会有很大区别，这也是页岩开采中很重视其各向异性的原因之一。

样品 I 是纵向取心，即页岩层理面法线与圆柱轴线平行，因此其 360°旋转超声检测结果并未显示出各向异性的特点，如图 6 所示，超声波速在同一层理面内可认为是各向同性的，波速在 4800m/s 附近波动，波动范围小于 200m/s。为了进一步检验这一结论，选择各向异性强度最大的样品 G，在其 50mm 的高度上每隔 10mm 选择一个平面，共选择 4 个平面进行 360°旋转测试，如图 7(a)所示。由图 7(b)可以看出 4 个平面上超声波速随层理角度的正弦变化关系曲线几乎是重合的。这一结果不仅验证了岩心样本致密规则的层状排列，也反映了样本的纵向各向同性。

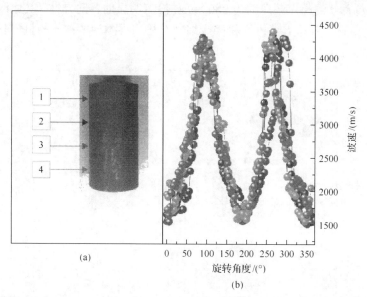

图 7 页岩样品 G 的纵向各向同性示意图(a)实验侧视图；(b)页岩样品的纵波波速与旋转角度的关系曲线；4 个平面的测试结果基本重合，证明了岩心样本致密规则的层状排列

Figure 7 The schematic diagram of longitudinal isotropy of shale sample G(a) Side view schematic of the experimental test and (b) the curves between the P-wave velocity and the rotation angle of the shale sample. The coincident results show the dense and regular bedding structure in the core shale

利用全光学方法不仅可以快速了解页岩的纵波速度分布，结合其他方法测出页岩的横波速度，还可进一步测定页岩的相关力学参数[20]。例如，根据以下公式，利用纵波速度与横波速度计算页岩的动态弹性模量 E_d 与泊松比 v_d。

$$E_d = \frac{\rho V_s^2 (3V_p^2 - 4V_s^2)}{V_p^2 - V_s^2} \tag{1}$$

$$v_d = \frac{V_p^2 - 2V_s^2}{2(V_p^2 - V_s^2)} \qquad (2)$$

式中，ρ 为岩石的密度；V_p 为纵波速度；V_s 为横波速度。

3.3 砂岩的各向异性检测

砂岩由石英颗粒形成，是石英、长石等碎屑成分占 50%以上的沉积碎屑岩，是构成石油、天然气和地下水的主要储集层[21]。由于各向异性的研究多集中于页岩等层理结构较明显的岩石，对砂岩的各向异性研究较少。本文仅对人工砂岩和实地采样砂岩进行对比研究。

砂岩样品 J 为人工砂岩，样品 K 是实地采样砂岩。两种样品均加工成直径 25mm、高度 50mm 的圆柱形。采用同样的全光学方法对砂岩样品进行各向异性检测，检测结果如图 8 所示。样品 J 的超声波速约为 2800m/s，没有明显的各向异性，其原因与样品的制作过程有关。样品 J 是将天然砂岩粉碎成细砂石粉，再添加多种胶凝材料复合制成，因此它在各个方向上表现为各向同性。

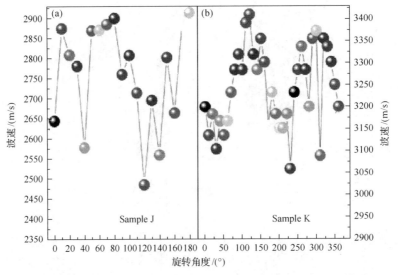

图 8 人工砂岩样品 J(a)与天然砂岩样品 K(b)的超声纵波波速随样品旋转角度的变化关系。和页岩相比，砂岩的各向异性强度较弱

Figure 8 Relation curves between the P-wave velocity and the rotation angle of the sandstone samples. (a) Artificial sample J; (b) natural sample K. The anisotropy in the sandstone sample is weaker than that in shale

样品 K 的纵波波速呈现出与页岩相似的特性，波速呈周期性变化，极大值为 3400m/s，极小值为 3100m/s，波速差异为 300m/s。相较于页岩的各向异性，砂岩的各向异性强度较弱，这与砂岩的形成材质和形成过程有关。研究表明[22]，石英颗粒和裂隙的定向排列是砂岩各向异性的主要原因。干燥砂岩表现出一定的各向异性，随着压力的增大，砂岩的各向异性强度逐渐降低，因此在压力较高的储层条件下可以不考虑砂岩的各向异性。

4 结论

综上所述,利用全光学方法对页岩各向异性的检测结果与页岩的层理结构吻合得很好。激光激发的超声纵波沿着页岩层理面传播时速度最大,垂直页岩层理面传播时速度最小,超声波速的分布关于岩石的层理方向是以正弦规律对称分布的,根据这一结果可以对页岩的各向异性方向进行判定,并借助在页岩中传播的超声波速了解页岩的致密度及各向异性强度,对页岩储层的开采具有直接指导意义。该方法同样适用于其他岩石的各向异性检测。全光学方法在岩石中的应用具有其独特优势。由于采用非接触的激发与接收装置,全光学方法对岩石本身没有破坏作用,属于无损检测;对样品的形状和大小没有特殊要求;采样密度高,样品测试间距可控制在1mm;对测试环境没有特殊要求,可应用于高温高压等恶劣环境下的实时检测。本文应用全光学方法仅对页岩岩心中超声纵波与各向异性的关系进行了研究,随着研究的深入进行,光学方法必将以其独特的优势在油气领域大放光彩。

参 考 文 献

[1] Zou C N, Yang Z, Zhang G S, et al. Conventional and unconventional petroleum "orderly accumulation": Concept and practical significance. Petrol. Explor. Dev., 2014, 41: 14-30.

[2] Zou C N, Zhang G S, Yang Z, et al. Concepts, characteristics, potential and technology of unconventional hydrocarbons: On unconventional petroleum geology. Petrol. Explor. Dev., 2013, 40: 413-428.

[3] 邹才能, 陶士振, 侯连华, 等. 非常规油气地质. 北京: 地质出版社, 2011.

[4] Han X, Kulaots I, Jiang X, et al. Review of oil shale semicoke and its combustion utilization. Fuel, 2014, 126: 143-161.

[5] Lu S F, Huang W B, Chen F W, et al. Classification and evaluation criteria of shale oil and gas resources: Discussion and application. Petrol. Explor. Dev., 2012, 39: 268-276.

[6] Na J G, Im C H, Chung S H, et al. Effect of oil shale retorting temperature on shale oil yield and properties. Fuel, 2012, 95: 131-135.

[7] Wan Y, Pan Z J, Tang S H, et al. An experimental investigation of diffusivity and porosity anisotropy of a Chinese gas shale. J. Nat. Gas. Sci. Eng., 2015, 23: 70-79.

[8] Zeng L, Zhao J, Zhu S, et al. Impact of rock anisotropy on fracture development. Prog. Nat. Sci., 2008, 18: 1403-1408.

[9] 宛新林, 席道瑛, 叶青, 等. 岩石各向异性实验研究及对工程勘探的影响. 中国煤炭地质, 2008, 20: 39-41.

[10] Carter A J, Kendall J M. Attenuation anisotropy and the relative frequency content of split shear waves. Geophys. J. Int., 2006, 165: 865-874.

[11] Best A I, Sothcott J, McCann C. A laboratory study of seismic velocity and attenuation anisotropyinnear-surfacesedimentaryrocks. Geophys. Prospect., 2007, 55: 609-625.

[12] Deng J, Wang S, Han D. The velocity and attenuation anisotropy of shale at ultrasonic frequency. J. Geophys. Eng., 2009, 6: 269-278.

[13] Kuila U, Dewhurst D N, Siggins A F, et al. Stress anisotropy and velocity anisotropy in low porosity shale. Tectonophysics, 2011, 503: 34-44.

[14] 张坤勇, 殷宗泽, 梅国雄. 土体各向异性研究进展. 岩土力学, 2004, 25: 1503-1509.

[15] 孙炜, 李玉凤, 何巍巍, 等. P波各向异性裂缝预测技术在ZY区碳酸盐岩储层中的应用. 石油与天然气地质, 2013, 34: 137-144.

[16] Carcione J M. A model for seismic velocity and attenuation in petroleum source rocks. Geophysics, 2000, 65: 1080-1092.
[17] Matsuda O, Larciprete M C, Voti R L, et al. Fundamentals of picosecond laser ultrasonics. Ultrasonics, 2015, 56: 3-20.
[18] Hernandez-Valle F, Dutton B, Edwards R S. Laser ultrasonic characterisation of branched surface-breaking defects. NDT E Int, 2014, 68: 113-119.
[19] Blum T E, Adam L, van Wijk K, et al. Laboratory measurement of P-wave anisotropy in shales with laser ultrasonics// Proceedings of SEG Technical Program Expanded Abstracts 2012. Las Vegas: Society of Exploration Geophysicists, 2012: 1-6.
[20] 席道瑛, 陈林, 张涛. 砂岩的变形各向异性. 岩石力学与工程学报, 1995, 14: 49-58.
[21] Koesoemadinata A P, McMechan G A. Effects of diagenetic processes on seismic velocity anisotropy in near-surface sand- stone and carbonate rocks. J. Appl. Geophys., 2004, 56: 165-176.
[22] 李阿伟, 孙东生, 王红才. 致密砂岩波速各向异性及弹性参数随围压变化规律的实验研究. 地球物理学进展, 2014, 29: 754-760.

Characterizing of oil shale pyrolysis process with laser ultrasonic detection

Zhiqing Lu[1,2] Xiaoguan Hai[1,2] Jianxu Wei[1] Rima Bao[2]

(1.State Key Laboratory of Petroleum Resources and Prospecting, China University of Petroleum, Beijing 102249, China; 2.Beijing Key Laboratory of Optical Detection Technology for Oil and Gas, China University of Petroleum, Beijing 102249, China)

Abstract: The laser ultrasonic technique was proposed to characterize the pyrolysis process of the oil shale. The ultrasonic velocity was produced by a noncontact all-optical method in the three regions of Barkol, Yaojie, and Longkou of China. It was found that the ultrasonic velocity was related with the pyrolysis temperature of the oil shale. The pyrolysis process was divided into three stages due to the ultrasonic propagation speed in the oil shale. The ultrasonic velocity had small changes from 20℃ to 320℃ in the first stage, a sharp decline between 320℃ and 470℃ in the second stage, and another small change above 470℃ in the third stage. The variation of the velocity was qualitatively explained, which was considered to be closely related with the characteristics of pyrolysis process in oil shale. An empirical equation of the velocity attenuation was proposed to estimate the beginning and the end of the decomposition of the kerogen. It is a new way to characterize the process of the pyrolysis of oil shale by using the laser ultrasonic technique.

1 Introduction

Oil shale, a sedimentary rock considered as a predominant alternative source for alleviating the pressure of petroleum supplies, contains minerals, kerogen, and bitumen[1]. Oil shale has been used for retorting to yield shale oil and burning directly as a fuel to generate electricity or heat for many years. Kerogen is considered as the main organic matter in oil shale yielding a significant amount of oil through pyrolysis, which is one kind of oil shale thermochemical conversion technologies, involving the thermal degradation of virgin material resulting in the production of gas, liquid, and solid[2]. For exploring the optimal operating conditions, many researchers have embarked on the process of pyrolysis of oil shale, and many works have been done about the influences of major parameters including the pyrolysis temperature, heating rate, residence time, pyrolysis atmosphere, and operating pressure as well as the particle size, density, and inorganic matter content of oil shale[3-17]. Thermal analysis is commonly employed in the process of pyrolysis of oil shale, which focuses on the thermodynamic parameters or physical parameters change with temperature[18]. For example, thermogravimetry is used to assess the mass loss of a sample as a function of temperature or time[19]. Differential thermal analysis and differential scanning calorimetry are used to characterize fossil fuels during pyrolysis, so the thermal dynamic parameters based on pyrolysis experiments are vital in the evaluation of oil shale. In addition, some indirect methods have been developed, such as well logging, X-ray diffraction, and nuclear magnetic resonance spectroscopy, among others[20-25].

Laser ultrasonic technology, with ultrasonic excitation by pulsed laser and data received by an optical method, is currently one of hot topics in the field of the nondestructive testing and evaluation of materials[26-28]. This technology has been applied in many fields, such as aerospace, microelectronics, biology, and medicine, among others, with its advantages of noncontact, long distance, high spatial and temporal resolution, and no environmental requirements. In recent years, laser ultrasonic technology in oil field applications has also drawn researchers' attention. The longitudinal wave (P-wave) anisotropy in core shale was measured with the laser ultrasonic technique. The measured velocities showed the significant relevance with the anisotropy of the core shale and provided detailed information on the rock's elastic constants, showing the advantage of laser ultrasonic technology versus traditional transducer technology[29]. Laser ultrasonic technology can be also used in liquid oil-water mixture to detect the water content. The ultrasonic P-wave velocity acquired by this method was found to be accelerated with the increase of water content. Different P-wave velocities corresponding to the different water contents can be used for detecting the crude moisture from 0% to 100%[30].

The present work focused on the P-wave velocity measurements in the pyrolysis process of oil shales by using laser ultrasonic technology as an all-optical method. The laser ultrasonic testing system was set up, and the P-wave velocities corresponding to the pyrolysis of the oil shale were acquired and analyzed. The good correlation of the ultrasonic velocity with the temperature of oil shale pyrolysis indicates the possibility of the laser ultrasonic technique to be a new elevation in oil field.

2 Experimental section

The sketch of the laser ultrasonic testing system was shown in Figure 1. When the short-pulse laser of 1053nm was irradiated on the sample surface, part of that energy was absorbed, and the ultrasonic wave was produced in the sample due to thermoelastic mechanism. The ultrasonic wave was received by the interferometer on the other side of the sample based on photorefractive two-wave mixing. The resulted waveform was recorded by the digital oscilloscope and input into the computer for further processing. In the experiment, the ultrasonic P-wave was recorded, which was the first significant peak that appeared on the interferometer. The laser ultrasonic velocity data of the samples was determined by the time that the ultrasound waves took to pass through the sample.

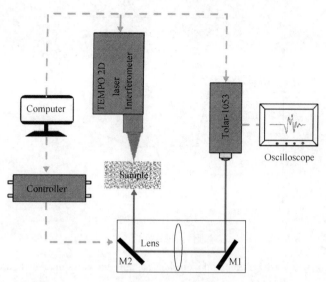

Figure 1　Sketch of the laser ultrasonic testing system

The oil shale used in this work were taken from three regions of China: Kazakhstan Barkol county in Xinjiang, Yaojie street in Lanzhou, and Longkou city in Shandong. The oil shale samples of Barkol, Yaojie, and Longkou in the experiment were 25 × 25mm² after cutting and grinding, with the thicknesses of 3.30mm, 2.22mm, and 1.84mm, respectively. Considering the anisotropy of the rock and the consistency of the experimental conditions, all the samples were cut parallel to the bedding plane of the oil shale. The first laser ultrasonic test of the samples was taken at room temperature before heated. Then they were heated from room temperature (20℃) to 670℃ under a vacuum condition with the temperature interval of 30℃ and the heating rate of 10℃/min, keeping 30min at each temperature point. After each heating, laser ultrasonic tests were performed, and the weight of the samples were recorded.

3　Results and discussion

The corresponding travel time of the oil shale samples under room temperature was shown in Figure 2. A peak picking program has been designed that locates the time of the first peak. The takeoff line was marked in Figure 2. The baseline indicates the delay time of 4.43 μs, so the travel times Δt of the ultrasonic wave through the samples of Barkol, Yaojie, and Longkou are 0.69μs, 0.89μs, and 1.62μs, respectively. Then, the propagation speeds v of the ultrasonic wave are 2667m/s, 2494m/s, and 2037m/s, respectively, which can be calculated by the travel time Δt as a result of

$$v = d/\Delta t \tag{1}$$

where d is the thickness of oil shale sample.

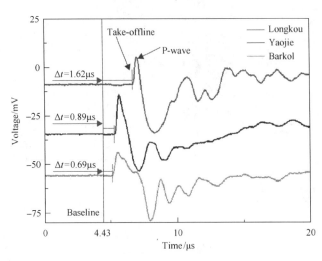

Figure 2 Travel-time of the oil shale samples under room temperature

The curves of the propagation velocity of ultrasonic wave in three samples with the pyrolysis temperature were represented in Figure 3 where the velocity decreased and appeared three stages with the rise of pyrolysis temperature. The raw ultrasonic data of the three samples were listed in Figures 3(a)~(c) to show the velocity change with several pyrolysis temperatures of 20℃, 120℃, 300℃, 360℃, 390℃, 470℃, 520℃, and 570℃, respectively. Taking the oil shale sample in Barkol as an example, the travel time Δt read in Figure 3(a) was 0.69μs, 0.68μs, 0.69μs, 1.21μs, 1.29μs, 1.61μs, 2.21μs, and 2.25μs, respectively. The corresponding velocity can be calculated by Equation (1) to be 2667m/s, 2705m/s, 2667m/s, 1521m/s, 1426m/s, 1143m/s, 832m/s, and 818m/s. As shown in Figure 3 (d), the propagation velocity of the oil shale sample in Barkol changed slightly from the beginning of the pyrolysis to ~320℃. Then, the propagation velocity had a sharp change from 2745.86m/s at 320℃ to 1054.95m/s at 470℃; the velocity subsequently maintained small changes about 200m/s. The change was well verified by the other two samples from Yaojie and Longkou in Figures 3 (b) and (c), which showed that the three oil shale samples present the same change in the process of pyrolysis with their different beginning velocity. The mass variation of the samples in temperature ranges from 20℃ to 600℃ was plotted in Figure 4. The mass loss is approximately 16.5%, 40.9%, and 26.7% for the three samples of Barkol, Yaojie, and Longkou, respectively. The differences of the mass variation related to their difference composition, origin, and their oil yields of 5.66%, 9.05%, and 14.16%, respectively.

The variation of the ultrasonic velocity and the mass of the samples in Figures 3 and 4 were considered to be closely related with the oil shale pyrolysis process. Below 300℃, the mass variation is mainly caused by the release of the adsorbed gas and the water loss, including the gasification or volatilization of the pore water, on-site water, and layer structure water in the oil shale. A series of physical reactions occurred at this stage such as the softness of hard shale particles and recombination between molecules, which also caused some mass loss[31]. The ultrasonic velocities maintain small variations because the homogeneity of the sample was not destroyed below 300℃.

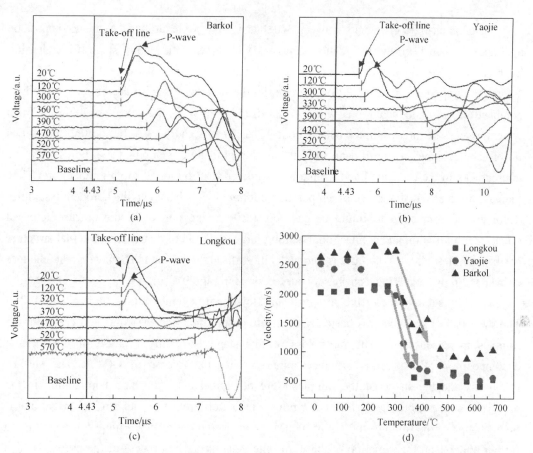

Figure 3 Raw ultrasonic data of the several selected pyrolysis temperature in the sample of (a) Barkol, (b) Yaojie, and (c) Longkou. (d) The ultrasonic propagation speed with the pyrolysis temperature of the three samples

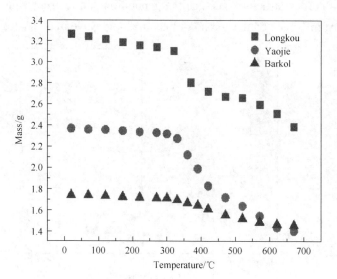

Figure 4 Mass variation of the different pyrolysis temperature of three different areas

The great changes in velocity between 320℃ and 470℃ were attributed to the decomposition of organic matter in the oil shale samples, which was the main stage for releasing hydrocarbons. The decomposition of the kerogen mainly occurred between 300℃ and 500℃. When the temperature reached 320℃, kerogen in the samples began to decompose to produce combustible gas and shale oil, which volatilized at high temperature, resulting in most weightlessness of three samples being about 9.3%, 25.3%, and 13.3%. In addition, porosities in oil shales have a significant impact on the ultrasonic velocity. The pores and the microcracks in the oil shale have a random disordered distribution at room temperature. They increase gradually with the rise of temperature but not obviously. When the kerogen in samples decomposed from solid to liquid or gas above 300℃, the pores and the cracks increased quickly with inflation and connection, resulting large variation of density and internal structure of the samples[32,33], so the ultrasonic velocities through the samples have great changes correspondingly, appearing as a sharply curved slope in Figure 3 (d).

The ultrasonic velocity has a significant slowdown at temperatures higher than 470℃. The pyrolysis of kerogen was basically complete at the time, and the pores and cracks in the samples increased little. The mass loss at this stage is mainly because of the thermal decomposition of clay and carbonate minerals with the release of CO_2[18]. The surface roughness and the shape of the samples were not affected by the high temperature in the pyrolysis progress, but the surface color now is taupe, different from its original bright black. Such change of the samples did not influence the measurement results. Figure 5 was plotted to further understand the variation of the ultrasonic velocity in the process of the pyrolysis. The velocity did not show a linear change with the mass percentage of the sample, which indicated that the organic matter in the oil shale samples was not the only cause of the velocity change. The variation of the rock structure, such as the generation and restructuring of the pores and fractures, has great influence on the spread of the ultrasonic wave in oil shale.

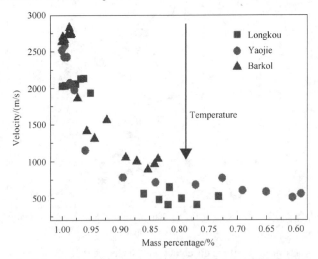

Figure 5 Ultrasonic velocity with the mass percentage of the three samples

Combining the experimental results of the velocity variation with the temperature and the mass of the samples, a preliminary exploration of the velocity attenuation (VD) equation was proposed to estimate the pyrolysis process.

$$VD = \left|1 - \frac{V_P^2}{V_{P0}^2}\right| \times 100\% \qquad (2)$$

where V_P is the ultrasonic velocity in the sample at any time during the pyrolysis and V_{p0} is the velocity before the pyrolysis. The velocity attenuation with the pyrolysis temperature in Figure 6 shows clear staged change in the pyrolysis process. Two parameters were chosen to indicate the beginning and the end of the pyrolysis, as shown in Figure 6. When the VD is ~10%, it is considered the beginning of the decomposition of the kerogen, and when the VD is ~90%, it is considered the end of the decomposition.

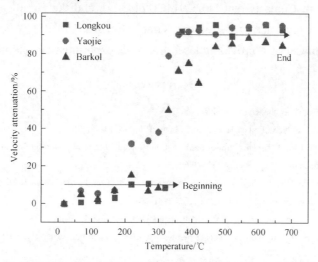

Figure 6   Velocity attenuation VD with the pyrolysis temperature of the three samples

The pyrolysis of oil shale is a very complicated process, accompanied by a variety of complex chemical and physical reactions. The factors that influence the ultrasonic velocities are many and intricate. Here the ultrasonic velocities were only measured and analyzed qualitatively. Its mechanism is not very clear and needs further detailed research. The empirical equation of velocity attenuation from the three samples also needs verified and corrected in further research, but this noncontacting optical method of laser ultrasonic technology provides us a new way to characterize and evaluate the pyrolysis process of oil shale.

4 Conclusions

In this paper, a noncontacting optical method was proposed for characterizing the process of the pyrolysis of the oil shale. The ultrasonic P-wave velocity acquired by this method was found to be interrelated with the pyrolysis temperate of the oil shale. The pyrolysis process was divided into three stages based on the propagation speed of ultrasound in the oil shale. Each stage was analyzed qualitatively combining with the characteristics of the pyrolysis and the variation of the ultrasonic wave velocity in oil shale. An empirical equation based on the pyrolysis characteristics of the three samples was proposed to estimate the beginning and the end of the decomposition of the kerogen. The present results show a new way to characterize the process of the pyrolysis of oil shale by using the laser ultrasonic technique as a noncontact optical method.

Acknowledgements

This work was supported by the Specially Funded Program on National Key Scientific Instruments and Equipment Development (Grant No. 2012YQ140005), the National Key Basic Research Program of China (Grant No. 2014CB744302), and the Excellent Young Teachers Program of China University of Petroleum (Grant No.ZX20150108). We thank CNPC Key Laboratory of Geophysics for providing laser ultrasonic experimental conditions.

References

[1] Schmidt J. Oil shale technology challenge. Oil Shale, 2011, 28: 1-3.
[2] DiRicco L, Barrick P L. Pyrolysis of oil shale. Ind. Eng. Chem., 1956, 48: 1316-1319.
[3] Dung N V. Yields and chemical characteristics of products from fluidized bed steam retorting of Condor and Stuart oil shales: Effect of pyrolysis temperature. Fuel, 1990, 69: 368-376.
[4] Han X X, Jiang X M, Cui Z G. Studies of the effects of retorting factors on the yield of shale oil for a new comprehensive utilization technology of oil shale. Appl. Energy, 2009, 86: 2381-2385.
[5] Wen C S, Kobylinski T P. Low-temperature oil shale conversion. Fuel, 1983, 62: 1269-1273.
[6] Williams P T, Ahmad N. Investigation of oil-shale pyrolysis processing conditions using thermogravimetric analysis. Appl. Energy, 2000, 66: 113-133.
[7] Nazzal J M. Influence of heating rate on the pyrolysis of Jordan oil shale. J. Anal. Appl. Pyrolysis, 2002, 62: 225-238.
[8] Williams P T, Ahmad N. Influence of process conditions on the pyrolysis of Pakistani oil shales. Fuel, 1999, 78: 653-662.
[9] Al-Ayed O S, Suliman M R, Rahman N A. Kinetic modeling of liquid generation from oil shale in fixed bed retort. Appl. Energy, 2010, 87: 2273-2277.
[10] Tong J. Effect of residence time on products yield and characteristics of shale oil and gases produced by low-temperature retorting of Dachengzi oil shale. Oil Shale, 2013, 30: 501-516.
[11] Özbay N, Uzun B B, Varol E A, et al. Comparative analysis of pyrolysis oils and its subfractions under different atmospheric conditions. Fuel Process. Technol., 2006, 87: 1013-1019.
[12] Citiroglu M, Snape C E, Lafferty C J, et al. Oil yield for a Turkish oil-shale-influence of pyrolysis conditions. Erdol, Kohle, Erdgas, Petrochem., 1990, 43: 442-443.
[13] Ahmad N, Williams P T. Influence of particle grain size on the yield and composition of products from the pyrolysis of oil shales. J. Anal. Appl. Pyrolysis, 1998, 46: 31-49.
[14] Zhang J, Kong L W, Bai J R, et al. Pyrolysis characteristics of oil shale of six density sections. Energy Procedia, 2012, 17: 196-201.

[15] Wang Q, Kong L W, Bai J R, et al. The pyrolysis characteristics and pore structure of oil shale of different densities. Energy Procedia, 2012, 17: 876-883.

[16] Gai R, Jin L, Zhang J, et al. Effect of inherent and additional pyrite on the pyrolysis behavior of oil shale. J. Anal. Appl. Pyrolysis, 2014, 105: 342-347.

[17] Wang Z, Deng S, Gu Q, et al. Pyrolysis kinetic study of Huadian oil shale, spent oil shale and their mixtures by thermogravimetric analysis. Fuel Process. Technol., 2013, 110: 103-108.

[18] Kök M V, Pokol G, Keskin C, et al. Combustion characteristics of lignite and oil shale samples by thermal analysis techniques. J. Therm. Anal. Calorim., 2004, 76: 247-254.

[19] Kok M V. Thermal investigation of Seyitomer oil shale. Thermochim. Acta, 2001, 369: 149-155.

[20] Fertl W H. Evaluation of oil shales using geophysical well-logging techniques. Dev. Pet. Sci., 1976, 5: 199-213.

[21] Bardsley S R, Algermissen S T. Evaluating oil shale by log analysis. J. Pet. Technol., 1963, 15: 81-84.

[22] Kumar R, Bansal V, Badhe R M, et al. Characterization of Indian origin oil shale using advanced analytical techniques. Fuel, 2013, 113: 610-616.

[23] Muhammad A F, El Salmawy M S, Abdelaala A M, et al. El-Nakheil oil shale: Material characterization and effect of acid leaching. Oil Shale, 2011, 28: 528-547.

[24] Feng Y, Le Doan T V, Pomerantz A E. The chemical composition of bitumen in pyrolyzed Green River oil shale: Characterization by ^{13}C NMR spectroscopy. Energy & Fuels, 2013, 27: 7314-7323.

[25] Solum M S, Mayne C L, Orendt A M, et al. Characterization of macromolecular structure elements from a Green River oil shale, I. Extracts. Energy & Fuels, 2014, 28: 453-465.

[26] Karabutov A A, Podymova N B. Nondestructive porosity assessment of CFRP composites with spectral analysis of backscattered laser-induced ultrasonic pulses. J. Nondestr. Eval., 2013, 32: 315-324.

[27] Gao W, Glorieux C, Thoen J. Laser ultrasonic study of Lamb waves: Determination of the thickness and velocities of a thin plate. Int. J. Eng. Sci., 2003, 41: 219-228.

[28] Devos A, Cote R. Strong oscillations detected by picosecond ultrasonics in silicon: Evidence for an electronic-structure effect. Phys. Rev. B: Condens. Matter Mater. Phys., 2004, 70: 125208.

[29] Blum T E, Adam L, van Wijk K, et al. Laboratory measurement of P-wave anisotropy in shales with laser ultrasonics. SEG Technical Program Expanded Abstracts, 2012: 4609.

[30] Lu Z Q, Yang X, Zhao K, et al. Non-contact measurement of the water content in crude oil with all-optical detection. Energy & Fuels, 2015, 29: 2919-2922.

[31] Burnham A K, Huss E B, Singleton M F. Pyrolysis kinetics for Green River oil shale from the saline zone. Fuel, 1983, 62: 1199-1204.

[32] Doğan Ö M, Uysal B Z. Non-isothermal pyrolysis kinetics of three Turkish oil shales. Fuel, 1996, 75: 1424-1428.

[33] Yan J, Jiang X, Han X. Study on the characteristics of the oil shale and shale char mixture pyrolysis. Energy & Fuels, 2009, 23: 5792-5797.

Evaluation of simulated reservoirs by using the oblique-incidence reflectivity difference technique

Jin Wang[1]　Honglei Zhan[1]　Liping He[2]　Kun Zhao[1]
Huibin Lü[2]　Kuijuan Jin[2]　Guozhen Yang[2]

(1.Beijing Key Laboratory of Optical Detection Technology for Oil and Gas, China University of Petroleum, Beijing 102249, China; 2.Institute of Physics, Chinese Academy of Sciences, Beijing 100190, China)

　　The evaluation of the oil-gas resources' potential depends on the characterization of reservoirs. The geometry of the pore-crack structures and the compositions of the pore walls have influence on hydrocarbon production about capillarity and permeability[1]. Besides, it has been known that the adsorption phenomenon in micro-structure plays a significant role in oil-gas reservoirs. Some techniques have been applied to characterize conventional and unconventional reservoirs, such as mercury intrusion capillary pressure, low-pressure gas adsorption, scanning electron microscopy (SEM), atomic force microscope (AFM), micro computed tomography (CT)and so on[2]. Some of the mentioned techniques can characterize the surface morphology with different resolution easily, but are not good at the assessment of dynamics in reservoir. The accurate evaluation of the potential of reservoir still needs more approaches[3].

　　In order to realize the evaluation of simulated reservoir, an oblique-incidence reflectivity difference (OIRD) technique was employed to characterize pores, cracks and adsorption of water in reservoir. In recent years, various optical techniques have been used as an important tool for evaluating the potential of reservoir[4-6]. OIRD technique was a sensitive form of detection tool whose signal had much close relationship to the dielectric properties due to the difference in reflectivity between the s- and p-polarized lights[7,8]. According to the OIRD analysis formula: $\varDelta_p - \varDelta_s = \frac{(-i)4\pi d\sqrt{\varepsilon_0}\varepsilon_s cos\varphi_{inc}sin^2\varphi_{inc}}{\lambda(\varepsilon_s - \varepsilon_0)(\varepsilon_s cos^2\varphi_{inc} - \varepsilon_0 sin^2\varphi_{inc})} \frac{(\varepsilon_d - \varepsilon_0)(\varepsilon_d - \varepsilon_s)}{\varepsilon_d}$, OIRD intensity is decided by dielectric constant ε_d and effective thickness d, thus the technique can evaluate the depth of the slits, the amount of absorbed water and the oil-gas-air composition/ratio in real reservoirs. In this research, micron-hole shapes onto silicon wafer, filminess rock slices and water on active carbon were measured to probe the rock micro-structure and the adsorption process, indicating the advantage that OIRD technique is an appropriate method for micor-structure and process monitoring because of the nm-grade-wavelength optical source and the simple measurement conditions when detecting in unconventional petroleum industry.

　　Figure 1(a) shows an OIRD system for the detection of a rock slice with cracks, a micron-hole silicon and active carbon to adsorb water. A two-dimensional motorized stage was

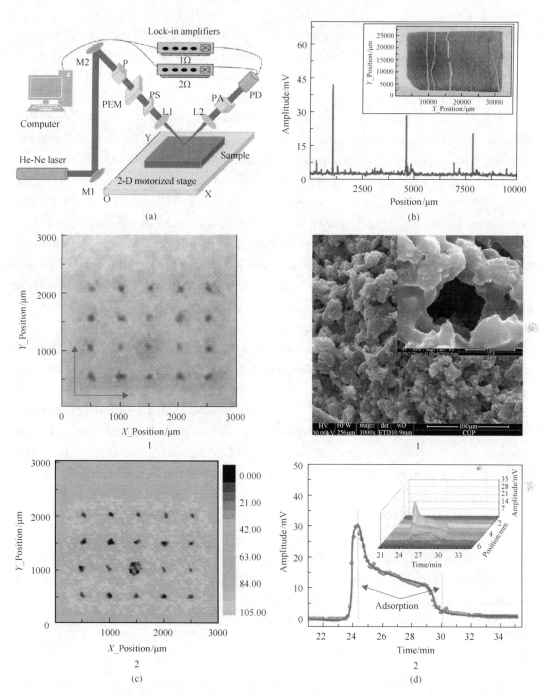

Figure 1 (a) Schematic diagram of optical path of OIRD; (b) OIRD signal amplitude at different position of a rock with cracks; the inset shows the real rock; (c) optical microscope picture (up) and OIRD map (down) of micron-hole silicon wafer; (d) the SEM pictures of active carbon with different enlargement factor (up) and time dependence of OIRD signal amplitude (down)

employed to realize the movement of sample in one or two directions according to the actual need[9]. The scan intervals were set as 4μm. Herein, the measured region of silicon was a square plane with 2mm×2mm sides with the 20 holes possessing the diameter of 20μm; real rock slices with micron slits were irregular pieces with 0.5mm thick laid on glass slide. In our case, the active carbon was applied for adsorption measurement with the void radius from ~100 nm to ~50μm, shown in the SEM image [Figure 1(d)].

Initially, a real rock slice with micron slits were scanned with OIRD. As shown in Figure 1(b), the peaks in OIRD signals were corresponded to rock slits, indicating the rocks and slits had different relative dielectric properties. Secondly, the projection drawing of OIRD signals at different positions on silica wafer were shown in Figure 1(c). The scanning area displayed twenty holes with black color. On one hand, the black micron-hole sections in OIRD image represented the micrograph of silicon wafer. On the other hand, OIRD scanning helped identify the positions and outlines of all pores more clearly. Different from optical microscope, OIRD plot reveals the dielectric image rather than the structural image. In terms of the real samples where holes and cracks were in the middle of the strata, the samples could be measured on a motorized with different directions and angles, or the samples can be cut into slices with directions prior to the OIRD measurement.

In addition, OIRD was used to measure the active carbon adhered with water drops at sequential time. Figure 1(d) indicated the time dependent OIRD signal amplitude of water adhered onto active carbon. The whole process can be divided into three stages: 0~24.3min, 24.3~30min and 30~35min. According to our previous report, the three stages are corresponded to diffusion, adsorption and stable state, respectively[10]. In the adsorption stage, with the increasing molecules adsorbed into active carbon, the water molecules were adsorbed in different holes, thus the OIRD signal was weaker. When the time length equaled 30min, the signal began to keep unchanged. Consequently, the collected tendency of OIRD in different time lengths recorded the dynamics as well as the critical points of different stages.

In this letter, the images of OIRD signal intensities reflected the dielectric properties of the samples mentioned. The results showed that OIRD could not only characterize the defects in simulated rocks, but detect the adsorption dynamics of liquid molecules. Therefore, OIRD provides a new technique together with normal methods to determine reservoir natures.

Acknowledgements

This work was supported by the National Key Basic Research Program of China (Grant No. 2014CB744302), the Specially Funded Program on National Key Scientific Instruments and Equipment Development (Grant No. 2012YQ140005), the China Petroleum and Chemical Industry Association Science and Technology Guidance Program (Grant No. 2016-01-07) and the National Natural Science Foundation of China (Grant No. 11574401).

References

[1] Bao R M, Miao X Y, Feng C J, et al. Characterizing the oil and water distribution in low permeability core by reconstruction of terahertz images. Sci. China: Phys. Mech. Astron., 2016, 59: 664201.

[2] Ma T, Chen P. Study of meso-damage characteristics of shale hydration based on CT scanning technology. Petrol. Explor. Develop., 2014, 41: 249-256.

[3] Bao R M, Li Y Z, Zhan H L, et al. Probing the oil content in oil shale with terahertz spectroscopy. Sci. China: Phys. Mech. Astron., 2015, 58: 114211.

[4] Zhan H L, Wu S X, Bao R M, et al. Qualitative identification of crude oils from different oil fields using terahertz time-domain spectroscopy. Fuel, 2015, 143: 189-193.

[5] Wang D C, Huang Q, Qiu C W, et al. Selective excitation of resonances in gammadion metamaterials for terahertz wave manipulation. Sci. China: Phys. Mech. Astron., 2015, 58: 084201.

[6] Bao R, Wu S, Zhao K, et al. Applying terahertz time-domain spectroscopy to probe the evolution of kerogen in close pyrolysis systems. Sci. China: Phys. Mech. Astron., 2013, 56: 1603-1605.

[7] Liu S, Zhang H Y, Dai J, et al. Characterization of monoclonal antibody's binding kinetics using oblique-incidence reflectivity difference approach. mAbs, 2015, 7: 110-119.

[8] He L P, Liu S, Dai J, et al. Label-free high-throughput and real-time detections of protein interactions by oblique-incidence reflectivity difference method. Sci. China: Phys. Mech. Astron., 2014, 57 (4): 615-618.

[9] Dai J, Li L, He L P, et al. Parallel detection and quantitative analysis of specific binding of proteins by oblique-incidence reflectivity difference technique in label-free format Sci. China: Phys. Mech. Astron., 2014, 57 (11): 2039-2042.

[10] Zhan H L, Wu S X, Bao R M, et al. Water adsorption dynamics into active carbon probed by terahertz spectroscopy. RSC Adv., 2015, 5: 14389-14392.

Real-time detection of dielectric anisotropy or isotropy in unconventional oil-gas reservoir rocks supported by the oblique-incidence reflectivity difference technique

Honglei Zhan[1,2] Jin Wang[2] Kun Zhao[1,2] Huibin Lü[3] Kuijuan Jin[3]
Liping He[3] Guozhen Yang[3] Lizhi Xiao[1]

(1.State Key Laboratory of Petroleum Resources and Prospecting, China University of Petroleum, Beijing 102249, China; 2.Beijing Key Laboratory of Optical Detection Technology for Oil and Gas, China University of Petroleum, Beijing 102249, China; 3. Institute of Physics, Chinese Academy of Sciences, Beijing 100190, China)

Abstract: Current geological extraction theory and techniques are very limited to adequately characterize the unconventional oil-gas reservoirs because of the considerable complexity of the geological structures. Optical measurement has the advantages of non-interference with the earth magnetic fields, and is often useful in detecting various physical properties. One key parameter that can be detected using optical methods is the dielectric permittivity, which reflects the mineral and organic properties. Here we reported an oblique-incidence reflectivity difference (OIRD) technique that was sensitive to the dielectric and surface properties and can be applied to characterization of reservoir rocks, such as shale and sandstone core samples extracted from subsurface. The layered distribution of the dielectric properties in shales and the uniform distribution in sandstones are clearly identified using the OIRD signals. In shales, the micro-cracks and particle orientation result in directional changes of the dielectric and surface properties, and thus, the isotropy and anisotropy of the rock can be characterized by OIRD. As the dielectric and surface properties are closely related to the hydrocarbon-bearing features in oil-gas reservoirs, we believe that the precise measurement carried with OIRD can help in improving the recovery efficiency in well-drilling process.

1 Introduction

Oil-gas industry has been shifting exploration and development activities from conventional hydrocarbon accumulations to unconventional ones. The research of unconventional oil-gas resources focuses on the reservoir space and whether the trap contains petroleum. The physical and structural properties of dense reservoir rocks are related to the recovery efficiency of oil-gas distribution in well-drilling process. Black- or dark-shale deposits in the foreland reflect distinct peculiarities of the deformational activity in any one Paleozoic orogeny because of rapid, loading-related, lithospheric subsidence and sediment starvation in the resulting basins[1,2]. The dielectric property of the rock is an important reference for geoscientists and petrophysicists to more precisely and effectively evaluate oil-gas reservoirs before and during exploration. Because of the quiet and deep anaerobic depositional environment that developed over thousands of years, shale often exhibited well-developed layered structures. The laminations in the layered structures are nearly parallel to each other. The significance of these laminations can be revealed by the geo-mechanics research and extraction of unconventional resources[3]. Generally, the mechanical properties are weakest at the laminations, which can withstand relatively small shear stresses. As a result, the stability of the well wall is degraded[4]. Moreover, porosity and fractures exist near the bedding planes. These unique reservoir structures indicate the existence characteristics of the oil-gas molecules, with significant variations in terms of the permeability in different directions. In sonic logging, the propagation of acoustic waves reflects the obvious anisotropy in the geological strata[5]. After obtaining and precisely analyzing these logging data, hydraulic fracturing technology can be employed. In this technique, fracturing fluid is poured into the reservoir at a pressure that exceeds the absorbing capacity of the oil reservoir. Regarding the location at which the fracturing fluid is poured, the micro-cracks perpendicular to the laminations can be selected to obtain more and larger cracks, thereby creating more fractured channels for hydrocarbon molecules to be extracted. Shale's layered structure is highly significant for traditional mining technologies[6]. Sandstone represents another significant component of tight gas reservoirs, with one-fourth of the world's sedimentary rocks estimated to be sandstone, which mainly consists of sand-sized particles. Sandstone also has a low absolute permeability (between ~ $10\mu D$ and $0.5\mu D$) and a connected porosity less than 10% because of its tight structure. In contrast to shale, sandstone rarely possesses an obvious layered structure. Regarding the research on sandstone reservoirs, the development and preservation of the pores have been the central topic relating to oil-gas reservoirs and are practical problems that must be solved in sedimentary basins[7-9].

The characterization of rock anisotropy is significant for hydrocarbon exploration. Electron microscopy techniques are commonly used for microstructure determination. The planes perpendicular and parallel to the bedding can be analyzed by scanning electron microscopy (SEM), and SEM images have shown some patchy kaolinite and acicular illite with preferred orientations parallel to the bedding[10,11]. High-resolution fractures can also be identified by SEM imaging analysis. Micro-computed tomography (micro-CT) is also an effective method to characterize pore and throat structures in a three-dimensional space. By combining CT scanning and digital image processing technology, the meso-damage characteristics of rock can be investigated during the process of shale hydration[12]. However, such CT-based approaches require sample pretreatment.

We employed an optical technique, referred as oblique-incidence reflectivity difference (OIRD), of imaging the shale and sandstone surfaces for improving characterization of the anisotropy of shale and the isotropy of sandstone. In recent years, various optical techniques have emerged as important tools for evaluating the potential of reservoirs and for characterizing oil-gas and pollutant properties[13,14]. Representative newly developed techniques include terahertz (THz) spectroscopy[15,16], all-optical detection[17], and remote sensing[18]. Similar to the optical methods mentioned above, OIRD is insensitive to the interference of subsurface electromagnetic fields and to high subsurface temperature and pressure. Compared with these methods, the OIRD technique is a more sensitive detection tool. OIRD signal depends on the dielectric and surface properties of the rock surface because this technique measures the difference in reflectivity between s- and p-polarized lights. At present, OIRD is mainly employed in biological molecules identification and imaging. OIRD scan was employed to detection a microarray of 60-base oligonucleotides before and after hybridization. OIRD image can reveal the hybridization and determine the thickness of oligonucleotides in the microarray, simultaneously[19]. Based on a widely used OIRD analysis formula expressed as

$$\Delta_p - \Delta_s = \frac{(-i)4\pi d\sqrt{\varepsilon_0}\varepsilon_s\cos\varphi_{inc}\sin^2\varphi_{inc}}{\lambda(\varepsilon_s - \varepsilon_0)(\varepsilon_s\cos^2\varphi_{inc} - \varepsilon_0\sin^2\varphi_{inc})} \frac{(\varepsilon_d - \varepsilon_0)(\varepsilon_d - \varepsilon_s)}{\varepsilon_d}$$

where ε_0, ε_s, φ_{inc} and λ represent the relative permittivity of environment, relative permittivity of basement, incidence angle and wavelength, thus OIRD signal intensity is a function of dielectric constant ε_d and effective thickness d[20]. OIRD has a time resolution of 20ms, a spatial resolution of 0.4nm, and a detection sensitivity of 14fg of protein per spot, according to previous reports[21,22]. OIRD can be employed for the *in situ* and real-time detection of oxide film epitaxial growth[23]. Recently, more promising applications were reported for the label-free, high-throughput detection of the interactions and dynamic processes of biological molecules and for the identification of the defects as well as the adsorption in simulated reservoirs[24,25]. In this research, we applied the OIRD method to quantify the surface properties of subsurface shale and sandstone cores. Our results indicated that the anisotropy in shale and isotropy in sandstone could be clearly identified by OIRD. The experimental results suggested that the OIRD technique could be used for core analysis required during oil-gas exploration and development, especially in the unconventional petroleum resource industry.

2 Results

Lamination, or a layered structure, is one of the most significant features of shale. The shale rocks from most areas have obvious layered structures. Because of the existence of the bedding surface, the mechanical properties of shale are markedly different from those of rocks with homogeneous structures. Generally, the weakness associated with cementation results in the damage to the lamination surface to precede the damage to the body. Moreover, weak planes cause instability of the wellbore during exploration. A rock core is typically used to analyze the underground geology and mineral resources. In this research, a shale core with a layered structure was measured and analyzed using OIRD. Initially, the sample was located on a two-dimensional (2-D) stage, and we set the direction of sample movement parallel to the bedding plane and adjusted the phase shifter and polarization analyzer to make the $I(2\Omega)$ and $I(1\Omega)$ signals equal to zero. Next, the sample was 2-D scanned in a randomly selected region of 2mm×1mm. During the scanning measurement, the real and imaginary signals, $\text{Im}\{\Delta_p-\Delta_s\}$ and $\text{Re}\{\Delta_p-\Delta_s\}$, and their relative positions were recorded by a computer[23] (Extended Data Figure 1). Figures 1(a) and (b) show the signal intensity images of $\text{Re}\{\Delta_p-\Delta_s\}$ and $\text{Im}\{\Delta_p-\Delta_s\}$, respectively, obtained from a shale core. The signal intensity is different at different positions, revealing the position-dependent variance of the dielectric properties on the shale core surface. A number of striations were parallel to the 1mm side and perpendicular to the 2mm side. Close agreement can be observed between the real map and the imaginary images. The width of the adjacent main beddings was 100~300μm in both images. The secondary laminations were found to have a width ranging from several to dozens of microns. These beddings appeared to be nearly parallel to each other; in fact, some of them were discontinuous. To discuss the lithology more systematically, three dimensional (3-D) cross-sections of the OIRD $\text{Re}\{\Delta_p-\Delta_s\}$ of a selected area were extracted, and the sections with high intensities were observed to be perpendicular and parallel to the bedding directions (Extended Data Figure 2). By connecting the peaks, we can obtain the main bedding plane of shale anisotropy that is perpendicular to the X-axis with an interval of ~300μm. Figure 1(c) shows the 3-D images of six randomly selected regions in the real and imaginary images of the shale sample. The area of all six selected regions equaled 100μm×100μm, and two or three beddings can be clearly observed in these figures. In all six areas, the widths of the peak areas and the distance between the consecutive peaks were 5~30μm, in agreement with the actual layered structures of shale.

Similarly, sandstone, another dense rock that has been widely studied by many geoscientists, is investigated in this paper. Sandstone is a sedimentary rock and is mainly composed of sand grains. Through a series of processes, such as weathering, erosion and transportation, sandstone gradually formed in basins. In contrast to shale, sand-stone rarely has a clear layered structure. Because the composition of sandstone is relatively homogeneous, it is generally considered to be isotropic. Similar to the process for shale, we obtained the real and imaginary OIRD signals in selected areas of the sandstone (Extended Data Figure 3). The signal intensities exhibited several peaks on the surfaces in both the real and imaginary images, showing the position-dependent uniformity of the physical properties. Figures 2(a) and (b) depict the signal intensity images of Re$\{\Delta_p-\Delta_s\}$ and Im$\{\Delta_p-\Delta_s\}$ obtained from the sandstone core. In contrast to the shale images presented in Figure 1, sandstone core does not exhibit any lamination, confirming that the measured striations in Figure 1 are entirely caused by shale's layered structure. It is apparent that sand grains with different sizes are randomly distributed throughout the surface of sandstone. Large granules can have widths as large as ~400μm size. In addition, the locations of the grains in the real image are reflected in the imaginary image. We also plotted the enlarged OIRD signals of six randomly selected regions with areas equaling 100μm×100μm. In the selected areas, the signals continuously changed at near positions. Therefore, OIRD shows great consistency and stability when used to characterize rock surfaces and dielectric properties.

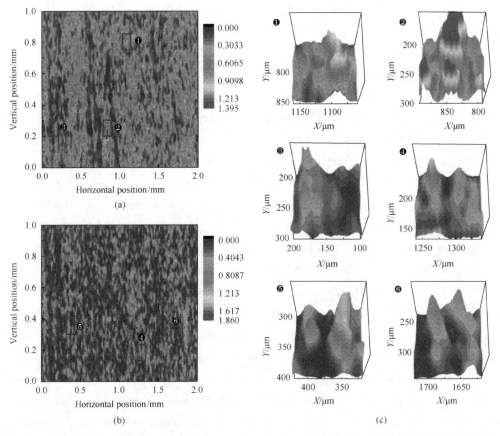

Figure 1 OIRD images of shale core when the scan direction is parallel to the bedding of shale. (a) 2-D image of Re$\{\Delta_p-\Delta_s\}$. A number of striations are obtained parallel to the 1mm side and perpendicular to the 2mm side. (b) 2-D image of Im$\{\Delta_p-\Delta_s\}$. A similar number of striations is obtained parallel to the 1mm side and perpendicular to the 2mm side but show different intensities compared with (a). (c) 3-D images of Re$\{\Delta_p-\Delta_s\}$ of six selected regions with areas of 100μm×100μm. Within the six regions, three are from (a), and the others are from (b). The agreement between the OIRD signals and the bedding properties of the shale is validated

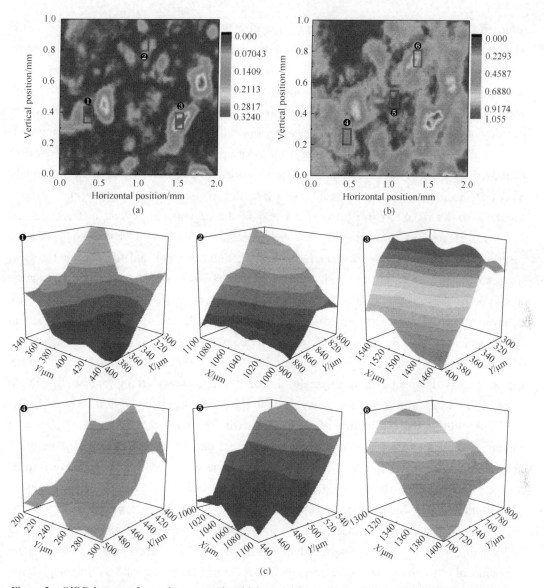

Figure 2 OIRD images of a sandstone core in which a random region was scanned by OIRD. (a) 2-D image of Re$\{\Delta_p-\Delta_s\}$ of sandstone. No striation was observed. In most areas, the signal intensities were uniform; in addition, several peak points existed, but they were not connected. (b) 2-D image of Im$\{\Delta_p-\Delta_s\}$ of sandstone. No striation was observed. In most areas, the signal intensities were uniform; in addition, several peak points existed, but they were not connected. (c) 3-D images of Re$\{\Delta_p-\Delta_s\}$ and Im$\{\Delta_p-\Delta_s\}$ of six selected regions with areas of 100μm×100μm. Areas Nos. 1, 2 and 3 are shown in (a), and Nos. 4, 5 and 6 are shown in (b). The information gathered by this technique indicated the surface properties of the sandstone

3 Discussion

The layered structures of shale and the approximate isotropy of sandstone are clearly reflected in Figures 1 and 2. The rocks exhibit directionality in the OIRD signal intensities. The anisotropy of the rock can be divided into two types: One type is based on the existence of micro-cracks and their arrangements in different directions and is referred to as stress anisotropy; the other type relates to the orientation of the rock particles and does not change with stress. Herein, the phenomena of the layered structures of shale and the approximate isotropy of sandstone were directly validated by optical microscopy (Extended Data Figure 4). To detect the componential and structural information of the rock surfaces, SEM was then employed to analyze the shale and sandstone cores investigated with OIRD. Secondary electron (SE) and back-scattered electron (BSE) modes were used to obtain new set of images. Figure 3 shows the SEM images of shale and sandstone cores at several different scales: Figures 3(a) and (c) show the SE morphology contrast images of shale and sandstone, respectively, with relatively large and small sizes; Figures 3 (b) and (d) present the BSE images of shale and sandstone with different scales. SE was very sensitive to the surface morphology, thus making it an effective method for observing the surface morphology. The contrast of BSE imaging is caused by the different atomic masses. In the first image in Extended Data Figure 5, a number of stripes can be observed and are parallel to each other. The other images represent magnifications of the stripes. In Figure 3, a series of micro-cracks can be clearly distinguished in grayscale; the right image shows an approximately straight crack with a width of 5~10μm that was related to the OIRD peaks shown in the Figure 1. The BSE map reflects the distribution of components with the same atomic mass. As shown in Figure 3(b), oriented lines shown in the same color are parallel to each other; thus, the black areas confirm the arrangement of the orientation of one component with a relatively small atomic mass. The orientation of a higher-atomic-mass component is shown in white (Extended Data Figure 6). The SE and BSE images validated the micro-cracks and orientations of the rock particles in the shale, as clearly shown in the OIRD images with the corresponding real and imaginary signals. We also performed SEM on the surface of sandstone and enlarged the area gradually, as shown in Extended Data Figure 7. In Figure 3(c), the SE imaging shows some differently sized particles with clear annular cracks between them. The shapes of cracks in sandstone were different from those in shale. As a whole, sandstone can be considered as isotropic because of the absence of striations and lack of orientation in the particle arrangement. It can be observed in Figure 3(d) that the components with the same or similar atomic masses are uniformly distributed throughout sandstone. The white areas correspond to particles with larger atomic masses, which are randomly arranged and not oriented. SEM imaging based on SE and BSE validated the distribution characteristics of the real and imaginary signals of OIRD. Thus, this technique is a very promising and practical technology to detect isotropy and anisotropy in rock and is a convenient supplementary method that can be combined with conventional methods.

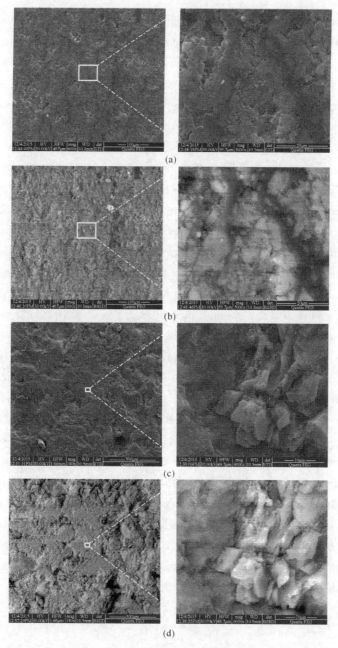

Figure 3　SEM images of shale and sandstone samples that were measured by OIRD. (a) SE morphology contrast images of shale: The right image shows an enlarged view of the selected area in the left image. A directed crevice can be observed in the right image. (b) BSE images of shale: the right image is an enlarged view of the selected area in the left image. An orientational arrangement of the same particles (black) can be observed parallel to the Y-axis. (c) SE morphology contrast images of sandstone: The right image shows an enlarged view of the selected area in the left image. A directed crevice is observed in the right image. An annular crack that is different from the directed crevice in shale is observed. (d) BSE images of sandstone: the right image is an enlarged view of the selected area in the left image. A symmetrical arrangement of the same particles (white) can be seen in this image

This study focused on applying OIRD to shale and sandstone rocks. We clearly observed the anisotropy of shale and isotropy of sandstone in the images of the Re$\{\Delta_p-\Delta_s\}$ and Im$\{\Delta_p-\Delta_s\}$ signal intensities. In shale, the width of the adjacent main beddings was 100~300μm; in addition, the secondary laminations had widths ranging from ~30μm to ~50μm. The detection results in the OIRD images are in agreement with those of the optical microscopy images (Extended Data Figure 4) and SEM maps (Figure 3 and Extended Data Figures 5~7). Accordingly, this study suggests that the methods used to characterize the isotropy or anisotropy of reservoir rock should combine previous techniques, such as sonic wave, and newly developed optical methods, such as OIRD, which reflects the dielectric properties and distributions of rock. Undoubtedly, as exploration continues to deepen, the production of conventional oil and natural gas resources will become increasingly difficult; thus, characterization of rock is becoming important. There is no reason to keep unconventional oil and gas resources "sleeping". The anisotropy in shale and isotropy in sandstone strongly influence various aspects of exploration, such as drilling efficiency and borehole deviation. For example, because of the different strength properties of the stratification layers, the workload of the rock layer increases during drilling; thus, the anisotropy of the rock will influence the drilling efficiency. In addition, because of the anisotropy of the rock, the drill pipe will become mechanically unbalanced, and as a result, the drill rod may deflect at a certain angle and even bend, thereby affecting the deviation of the drilling hole. During oil-gas extraction, the distribution of the strata or rocks' mechanics is the primary feature that should be characterized. Near cracks, the effective density and thickness are clearly different from those at other locations. In the micro-structure, from rock to cracks and from cracks to rock, the dielectric and surface properties undergo repeated variations. As reported in previous articles, the OIRD signal, which reflects the difference between the reflectivity of s- and p-polarized light, is closely related to the sample's dielectric and surface properties, and as a result, OIRD is a very sensitive surface detection technique. OIRD is a very promising and practical technology for detecting the isotropy and anisotropy in rock and is a convenient supplementary technique for conventional methods.

4 Methods

Similar to some other technologies, there also exist some limitations of the new technique. For example, the baseline of signal should be regulated in each measurement; more mathematical methods should be developed to analyze the data of signal. Overall, the research proves that OIRD is a very promising and practical technology for detecting the isotropy and anisotropy in rock so that it can be used for core analysis during oil-gas exploration. Extended Data Figure 8 shows a schematic of the OIRD experimental setup for the detection of rock, including shale and sandstone. The probe beam from a He-Ne laser with wavelength of $\lambda =$

632.8nm is initially p polarized. After reflection by two mirrors (M1 and M2), the laser passes through a polarizer (P) to ensure p-polarized incidence. A photoelastic modulator (PEM) is then employed to oscillate the polarization of the laser beam between p- and s-polarization at a frequency of 50kHz. The laser beam output from the PEM enters a phase shifter (PS), which is utilized to create a fixed phase difference Φ_{ps} between the p- and s-polarization components. After focusing by an optical lens (L1), the laser reaches the surface of the rock to be measured at a incidence angle of $\theta = 60°$. The reflected beam passes through a second lens (L2) that collects the diffuse light into a parallel beam. The light is then introduced into a polarization analyzer (PA) with the transmission axis set at θ_A from s-polarization. The intensity of the reflected beam is detected by a silicon photodiode (PD). Two lock-in amplifiers are used to measure the first harmonic $I(1\Omega)$ and second harmonic $I(2\Omega)$ signals. Before the scanning measurement, we adjust the phase shifter and polarization analyzer so that the $I(2\Omega)$ and $I(1\Omega)$ signals are equal to zero to improve the sensitivity of the measurement. During the scanning measurement, the $I(2\Omega)$ and $I(1\Omega)$ signals of $\text{Im}\{\Delta_p-\Delta_s\}$ and $\text{Re}\{\Delta_p-\Delta_s\}$ are recorded by a computer. The $I(2\Omega)$ and $I(1\Omega)$ signals include information about the surface and the dielectric properties of the sample[24,26].

Generally, there are no special requirements on samples preparation for OIRD detection. To obtain the OIRD images of the rock surface, a 2-D motorized stage is employed. The rock is fastened on the upper surface of the stage, which is controlled by special software. The interval between two adjacent scans can be 1 μm. In the experiment, the scanning step length was 4μm. As shown in Extended Data Figure 9, the scan directions of the laser or the movement directions of the rocks were specifically chosen. In this research, shale and sandstone cores were used. Each core is a cylinder of 25mm diameter and 50mm long. Shale was cored at a depth of 3325.4m in the horizontal direction; thus, the petrophysical layers are perpendicular to the upper and lower surfaces of the core. Sandstone is the main reservoir of oil, gas and groundwater. The shale and sandstone cores were cut and then polished to confirm the parallelism between the upper and lower surfaces of the cores before OIRD measurement. Shale has a vectored stratification plane over the whole space. The direction was confirmed to be parallel relative to the bedding planes during the scan. For the sandstone, however, obvious vectored bedding planes do not exist in 3-D space, and thus, the area to be measured was randomly selected on the surface. All measured regions were rectangular planes of 2mm×1mm. The measurements were performed under ambient conditions.

Acknowledgements

We thank the National Basic Research Program of China (Grant No. 2014CB744302), the Specially Funded Program on National Key Scientific Instruments and Equipment Development (Grant No. 2012YQ140005), the China Petroleum and Chemical Industry Association Science and Technology Guidance Program (Grant No. 2016-01-07), and the National Nature Science

Foundation of China (Grant No. 11574401) for the financial support.

References

[1] Hughes J D. Energy: A reality check on the shale revolution. Nature, 2013, 494: 307-308.

[2] Su W, Huff W D, Ettensohn F R, et al. K-bentonite, black-shale and flysch successions at the Ordovician-Silurian transition, South China: Possible sedimentary responses to the accretion of Cathaysia to the Yangtze Block and its implications for the evolution of Gondwana. Gondwana Res., 2009, 15: 111-130.

[3] Mokhtari M, Tutuncu A N. Characterization of anisotropy in the permeability of organic-rich shales. J. Petrol.Sci. Eng., 2015, 133: 496-506.

[4] Yigit A S, Christoforou A P, Christoforou A P. Stick-slip and bit-bounce interaction in oil-well drillstrings. J. Energ. Resour. Techn., 2006, 128: 268-274.

[5] Pan Z, Ma Y, Connell L D, et al. Measuring anisotropic permeability using a cubic shale sample in a triaxial cell. J. Nat. Gas Sci. Eng., 2015, 26: 336-344.

[6] Roche V, Homberg C, David C, et al. Normal faults, layering and elastic properties of rocks. Tectonophysics, 2014, 622: 96-109.

[7] Bruthans J, Soukup J, Vaculikova J, et al. Sandstone landforms shaped by negative feedback between stress and erosion. Nat. Geosci., 2014, 7: 597-601.

[8] Fu X, Agostini F, Skoczylas F, et al. Experimental study of the stress dependence of the absolute and relative permeabilities of some tight gas sandstones. Int. J. Rock Mech. Min. Sci., 2015, 77: 36-43.

[9] Ahmadi M A, Shadizadeh S R. Experimental investigation of a natural surfactant adsorption on shale-sandstone reservoir rocks: Static and dynamic conditions. Fuel, 2015, 159: 15-26.

[10] Lin B T, Cerato A B. Application of SEM and ESEM in microstructural investigation of shale-weathered expansive soils along swelling-shrinkage cycles. Eng. Geol., 2014, 177: 66-74.

[11] Wells R K, Newman J, Wojtal S. Microstructures and rheology of a calcite-shale thrust fault. J. Struct. Geol., 2014, 65: 69-81.

[12] Ma T, Chen P. Study of meso-damage characteristics of shale hydration based on CT scanning technology. Petrol. Explor. Develop., 2014, 41: 249-256.

[13] Greeney N S, Scales J A. Non-contacting characterization of the electrical and mechanical properties of rocks at submillimeter scales. Appl. Phys. Lett., 2012, 100: 124105.

[14] Zhan H L, Qian L, Zhao K, et al. Evaluating $PM_{2.5}$ at a construction site using terahertz radiation. IEEE T. THz Sci. Techn., 2015, 5: 1028-1034.

[15] Horiuchi N, Zhang X C. Searching for terahertz waves. Nature Photonics, 2010, 4: 662.

[16] Zhan H L, Zhao K, Xiao L Z. Spectral characterization of the key parameters and elements in coal using terahertz spectroscopy. Energy, 2015, 93: 1140-1145.

[17] Lu Z Q, Xiao Y, Zhao K, et al. Non-contact measurement of the water content in crude oil with all-optical detection. Energy & Fuels, 2015, 29: 2919-2922.

[18] Wei L, Hu Z W, Dong L, et al. A damage assessment model of oil spill accident combining historical data and satellite remote sensing information: A case study in Penglai 19-3 oil spill accident of China. Mar. Pollut. Bull., 2015, 91: 258-271.

[19] Landry J P, Zhu X D, Gregg J P. Label-free detection of microarrays of biomolecules by oblique-incidence reflectivity difference microscopy. Opt. Lett., 2004, 29: 581-583.

[20] Landry J P, Gray J, Toole M K, et al. Incidence-angle dependence of optical reflectivity difference from an ultrathin film on a solid surface. Opt. Lett., 2006, 31: 531-533.

[21] Chen F, Lu H B, Chen Z H, et al. Optical real-time monitoring of the laser molecular-beam epitaxial growth of perovskite oxide thin films by an oblique-incidence reflectance-difference technique. J. Opt. Soc. Am. B, 2001, 18: 1031-1035.

[22] Zhu X D, Nabighian E. *In situ* monitoring of ion sputtering and thermal annealing of crystalline surfaces using an oblique-incidence optical reflectance difference method. Appl. Phys. Lett., 1998, 73: 2736-2738.

[23] Zhu X D, Fei Y Y, Wang X, et al. General theory of optical reflection from a thin film on a solid and its application to heteroepitaxy. Phys. Rev. B, 2007, 75: 24.

[24] Liu S, Zhu J H, He L P, et al. Label-free, real-time detection of the dynamic processes of protein degradation using oblique-incidence reflectivity difference method. Appl. Phys. Lett., 2014, 104: 163701.

[25] Wang J, Zhan H L, He L P, et al. Evaluation of simulated reservoirs by using the oblique-incidence reflectivity difference technique. Sci. China: Phys. Mech. Astron., 2016, 59: 114221.

[26] Liu S, Zhang H, Dai J, et al. Characterization of monoclonal antibody's binding kinetics using oblique-incidence reflectivity difference approach. mAbs, 2015, 7(1): 110-119.

Supporting information

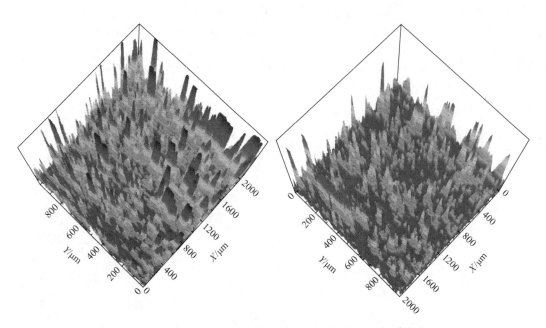

Extended Data Figure 1. Position-dependent OIRD signals of shale.

Real intensities at different scanning positions (up), where the Re $\{\Delta_p-\Delta_s\}$ values reflect the relative variance. Shale's imaginary intensities at different scanning positions (down), where the Im $\{\Delta_p-\Delta_s\}$ values reflect the relative variance. In the 3-D systems of both the real and imaginary parts, the same bedding information can be observed. These beddings appear to be approximately, but not absolutely, parallel to each other; in fact, some are discontinuous. Generally, the beddings are parallel to the 1-mm sides. Moreover, the bedding lines are successive in some areas but not discontinuous in others, reflecting the complex composition and structure on the shale surface.

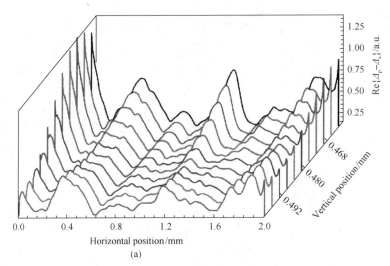

(a)

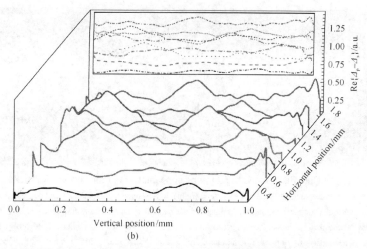

(b)

Extended Data Figure 2. Cross-sectional analysis of the OIRD Re $\{\Delta_p-\Delta_s\}$ signals of shale.
(a) Cross section of the horizontal plane of the OIRD real signals. (b) Cross section of the vertical plane of the OIRD real signals of shale. Shale exhibited obvious anisotropy between the horizontal (X-axis) and vertical (Y-axis) directions. This anisotropy occurs perpendicular and parallel to the bedding directions. To more intuitively observe the layered structure and achieve a higher signal-to-noise ratio, the OIRD data were pre-processed using the Adjacent-Averaging filter. This data pretreatment reduces the noise and smoothes the curves but does not distort the waveforms and features. One of the most important features in Extended Data Figure 2(a) is the Re $\{\Delta_p-\Delta_s\}$ peaks, which are located at the same vertical positions. The unique horizontal positions of these peaks were ~ 0.01, 0.41, 0.76, 1.16, 1.48, and 1.8mm, respectively. Intervals of ~300μm are obtained. Connecting the peaks, we can determine the main bedding plane of the shale anisotropy, which is perpendicular to the X-axis. However, the OIRD signals of the cross section parallel to bedding planes in Extended Data Figure 2(b) indicated small difference relative to that in Figure 5(a) over the vertical position range. The largest value was ~ 0.5 mm, and the average value over the whole range was approximately 0.25 mm. These results indicated that the approximate isotropic properties of the cross sections were parallel to shale's laminations.

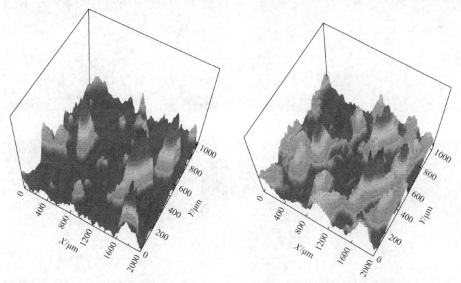

Extended Data Figure 3. Position dependent OIRD signals of sandstone.
Real part intensities at different scanning positions (up), where the Re$\{\Delta_p-\Delta_s\}$ values revealed several peaks. Shale's imaginary part intensities at different scanning positions (down). In the 3-D systems of both the real and imaginary parts, no bedding information can be observed, and instead, the signals indicated overall uniformity. Differently sized sand grains are randomly distributed on the sandstone surface. Large granules can have widths as large as ~ 400 μm. The isotropy properties of the sandstone and some single particles can be clearly observed.

Extended Data Figure 4. Optical microscope analysis of the surfaces of rocks.
Microscope pictures of the surfaces of shale (left) and sandstone (right). In the shale, the laminations can be clearly observed. The main beddings are depicted with black dotted lines, and the laminations are parallel to each other, with an interval of ~ 300μm between the adjacent bedding lines. These features are in good agreement with those in Extended Data Figure 2. Meanwhile, among the main laminations, a number of secondary striations are discontinuous and parallel to each other. High homogeneity can be observed between the features. The sandstone surface includes with some random sand grains and can be considered to be isotropic. The OIRD signal was strongly sensitive to the features shown in Figure 2. Herein, the optical microscope image shows these features and reveals that the sandstone surface is indeed isotropic. The OIRD method thus constitutes a simple and direct approach for the detection of surface properties, including isotropy and anisotropy caused by the chemical components and structures of rocks, such as shale and sandstone.

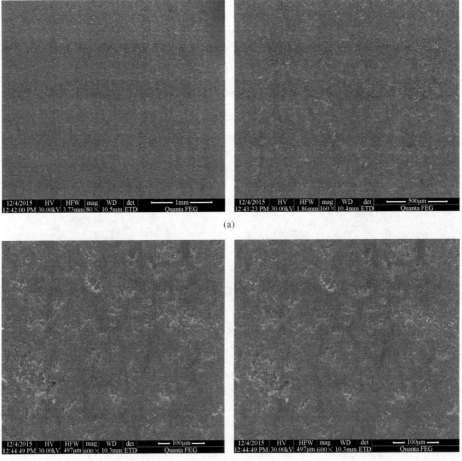

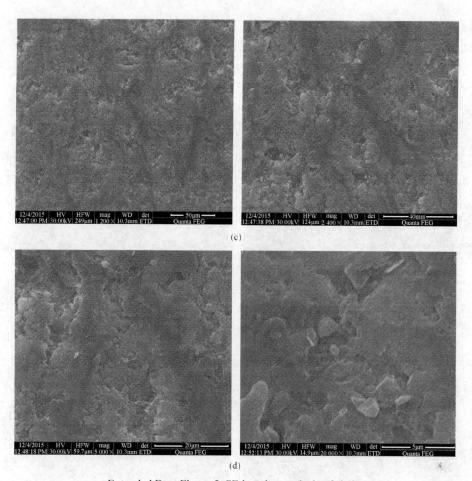

(c)

(d)

Extended Data Figure 5. SE imaging analysis of shale.
The first images shows the shale surface on the millimeter scale. A number of parallel stripes can be observed. This area is gradually enlarged, from left to right and from top to bottom. As the amplification factor increases, some micro-cracks with micron widths are gradually observed. In SE imaging, micro-cracks are clearly distinguished in grayscale; in the bottom-right image, the approximately straight crack in the middle of picture is related to that shown by the OIRD peaks in Figure 1.

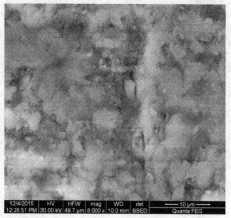

Extended Data Figure 6. BSE imaging analysis of shale surface.
The white areas validated the orientated arrangement of a material with a relatively large atomic mass that has the same orientation direction as the cracks shown in Extended Data Figure 5.

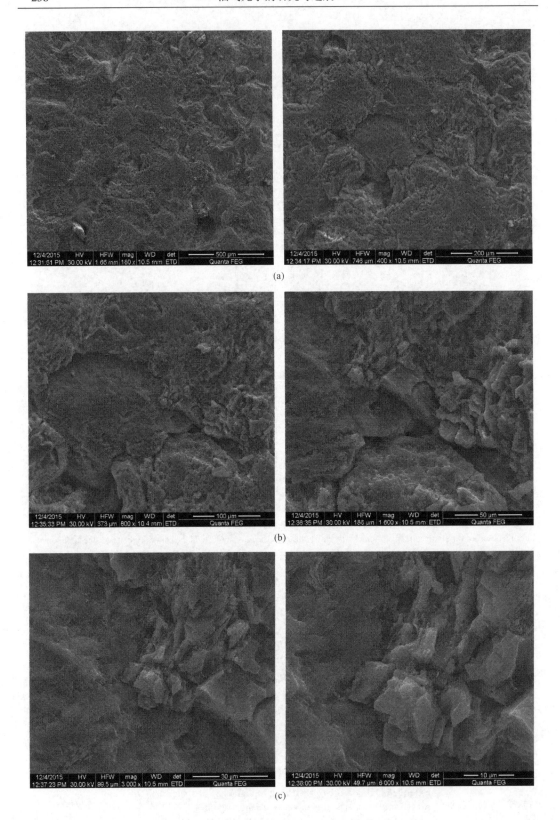

(d)

Extended Data Figure 7. SE imaging analysis of sandstone.

The first image shows the sandstone surface on the millimeter scale. This area is gradually enlarged from left to right and from top to bottom. As the amplification factor increases, some micro-cracks between particles can be gradually observed. The SE imaging shows that differently sized particles are separated by annular cracks. The crack shape in sandstone is different from that in shale. Overall, sandstone can be considered as an isotropic material because of the absence of orientation, in agreement with the information shown in Figure 2.

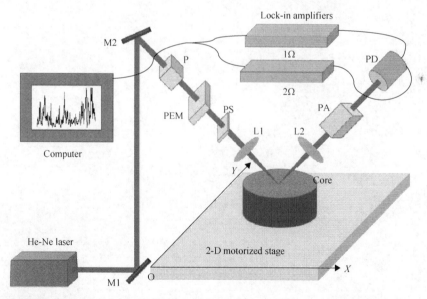

Extended Data Figure 8. Schematic diagram of OIRD system used to analyze shale and sandstone.

The rocks is fixed on a pair of translation stages that can be moved in the X and Y directions: PEM, photoelastic modulator; PS, Phase shifter; and PD, photodiode. The PEM oscillates the polarization of the laser beam between p-and s-polarization at a frequency of 50kHz. The PS adjusts the phase difference Φ_{ps} between the p-and s-polarization components. The PD detects the intensity of the reflected laser beam.

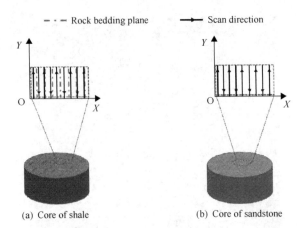

Extended Data Figure 9. Schematic diagram of the OIRD scan.
Laser scanning of the surfaces of shale (a) and sandstone (b). The rock is fastened to the upper surface of the stage, which is controlled by special software. The interval between two adjacent scans was 4μm. The scan directions of the laser or the movement directions of the rocks were specifically chosen.

Oblique-incidence reflectivity difference application for morphology detection

Honglei Zhan[1,2,3]　Kun Zhao[1,2]　Huibin Lü[4]　Kuijuan Jin[4]
Guozhen Yang[4]　Xiaohong Chen[1]

(1.State Key Laboratory of Petroleum Resources and Prospecting, China University of Petroleum, Beijing 102249, China; 2.Beijing Key Laboratory of Optical Detection Technology for Oil and Gas, China University of Petroleum, Beijing 102249, China; 3.Department of Material Science and Engineering, China University of Petroleum, Beijing 102249, China; 4.Institute of Physics, Chinese Academy of Sciences, Beijing 100190, China)

Abstract: Analogy with scanning electron microscopy, we use an oblique-incidence reflectivity difference (OIRD) approach for morphology detection. By scanning the active carbon clusters with one-dimensional way and the reservoir rocks in a two-dimension way, the morphology of samples' surface can be revealed in OIRD signal images. High OIRD signals of active carbon samples refer to the centralized distribution areas of carbon, and the fluctuations are caused by the uneven distribution carbon pellets. OIRD intensity is proportional to the thickness of materials. In terms of rocks, the trough areas with smaller values refer to the low-lying fields. The areas with relatively large OIRD intensities correspond to the protuberance areas of rocks. Consequently, OIRD is a sensitive yet rapid measure of surface detection in material and petrogeology science.

Keywords: OIRD; morphology; reservoir rocks; SEM

1 Introduction

Measuring the surface dielectric properties of materials directly involves a number of theoretical and technical challenges, including realizing nondestructive detection, understanding the signal information of the sample, and measuring the sample under the common condition. An optical microscope is a common optical instrument to recognize the magnification imaging that cannot be distinguished with eyes. Scanning electron microscopy (SEM) and atomic force microscopy (AFM) are appropriate ways to describe the surface structures and morphology information[1]. However, some materials cannot be measured by SEM, because a vacuum environment is necessary when measuring, and liquid water is not allowed because it may volatilize and harm the SEM setup. Although AFM has a very large resolution of atoms, its scanning velocity is very small and time length is very long.

In recent years, various optical techniques have emerged as important tools for detecting the morphology of materials and rocks[2,3]. Representative and newly developed techniques include terahertz (THz) spectroscopy and oblique-incidence reflectivity difference (OIRD)[4-7]. A THz wave can be applied to rocks mapping the mineralogy and used to obtain high-spatial resolution maps of the spatially varying dielectric permittivity of heterogeneous materials[8,9]. The anisotropy in THz parameters of the oil shale can be extracted and employed to characterize the oil yield precisely[10]. The THz spectrum and its imaging are gradually and successfully employed in the petroleum field[11-14]. Meanwhile, THz chemical imaging is appropriate to reveal hydrogen bond distributions[15]. Consequently, they both can measure the physical information of materials, but relative to THz spectroscopy, OIRD has a larger spatial resolution, which is of significance for some substances such as reservoir rocks.

OIRD measures the difference in reflectivity between s- and p-polarized lights; thus, it is very sensitive to the dielectric and surface properties of the material surface. An OIRD image can reveal the hybridization and determine the thickness of oligo-nucleotides in the microarray, simultaneously. According to previous reports, OIRD has a time resolution of 20 μs, a spatial resolution of 0.4nm, and a detection sensitivity of 14fg of protein per spot[16-18]. OIRD signal intensities are related to the physical properties of materials. According to the classical three-layer model,

$$\Delta_p - \Delta_s = \frac{(-i)4\pi d\sqrt{\varepsilon_0}\varepsilon_s\cos\varphi_{inc}\sin^2\varphi_{inc}}{\lambda(\varepsilon_s - \varepsilon_0)(\varepsilon_s\cos^2\varphi_{inc} - \varepsilon_0\sin^2\varphi_{inc})} \frac{(\varepsilon_d - \varepsilon_0)(\varepsilon_d - \varepsilon_s)}{\varepsilon_d}$$

where ε_0, ε_s, φ_{inc}, and λ represent the relative permittivity of the environment, relative permittivity of the basement, incidence angle, and wavelength, the OIRD signal intensity is a function of dielectric constant ε_d and effective thickness d. For the real-time detection of morphology, d would change as the scanning position is moving. In the OIRD system, a two-dimensional motorized stage is used to fix a sample and obtain the OIRD signals of samples. The stage can be motioned in a two-dimensional range to obtain images and in a one-dimensional range to obtain a line information.

Recently, more promising applications were reported for the label-free, high throughput detection of interactions of biological molecules[19]. The OIRD technique can not only characterize the defects but also detect the adsorption dynamics of liquid molecules[20]. The layered distribution of the dielectric properties in shales and the uniform distribution in sandstones can be clearly identified using OIRD signals. The dielectric and surface properties are closely related to the hydrocarbon-bearing features in oil gas reservoirs so that the precise measurement carried with OIRD can help improve the recovery efficiency[21]. In this research, OIRD was used to detect the morphology information with one-dimensional and two-dimensional scanning. Active carbon and reservoir rocks were measured and their OIRD images were validated by SEM analysis. The work allows us to achieve more approaches for the detection of morphology information in materials science and the petrogeology industry.

2 Experimental methods

As depicted in Figure 1, we show the scheme of the OIRD system for the label-free and real-time detection of reservoir rocks. A p-polarized He-Ne laser with a wavelength of $\lambda = 632.8$nm is selected as the detection light beam. The laser passes through a polarizer (P) to ensure p-polarized incidence. The light beam then passes through a photoelastic modulator (PEM) so that the polarization of the light beam oscillates from p- to s-polarization at the frequency of 50kHz. The laser beam output from the PEM enters a phase shifter that would introduce a variable phase difference Φ_{ps} between the p- and s-polarization components. The light beam is focused by an optical lens into the sample surface near the Brewster angle $\theta = 57.6°$. The reflected beam passes through a second lens and enters a polarization analyzer (PA) with the transmission axis from s-polarization. At last the intensity of the reflected beam is detected by a silicon photodiode (PD). Two lock-in amplifiers are employed to measure the first harmonic $I(1\Omega)$ and the second harmonic $I(2\Omega)$ signals. Before the scanning measurement, we adjust the phase shifter and polarization analyzer so that the $I(2\Omega)$ and $I(1\Omega)$ signals are equal to zero to improve the sensitivity of the measurement[22,23].

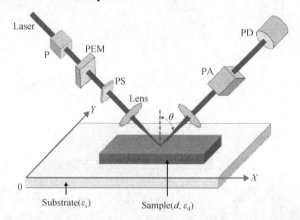

Figure 1 Sketch of OIRD system for the morphology detection of the active carbon and reservoir rocks

In the experiment, an active carbon and two different rocks were used to obtain one-dimensional and two-dimensional morphology information. Generally, there are no special requirements on sample preparation for OIRD detection. The active carbon pellets adhered tightly onto the fiber cloth to form the active carbon fiber cloth (ACFC), which was taken from the active carbon breathing mask. The rocks were cored at a depth of 3000m in the horizontal direction from different oil fields. The rocks were cut and then polished to confirm the parallel-ism between the upper and lower surfaces of the cores before OIRD measurement. After the samples were prepared, the active carbon or reservoir rocks were fixed on the two-dimensional stage. The scanning range was a 1mm×1mm rectangle area. The scanning step length was set as 4μm. The measurements were performed under ambient conditions. Finally, we detected the real part Re$\{\Delta_p-\Delta_s\}$ and the imaginary part Im$\{\Delta_p-\Delta_s\}$ signal of the samples. In this research, Im$\{\Delta_p-\Delta_s\}$ was used as the parameter to detect the morphology.

3 Results and discussion

A basic investigation was initially performed about the OIRD reflectivity response of ACFC by two-dimensional scanning. Figure 2(a) is the intensity profile taken along a line in the active carbon sample. Im$\{\Delta_p-\Delta_s\}$ alters with the change of scanning distance. Several maximum signals can be observed and respective peak wave are located at different positions; meanwhile, fluctuations are also observed in some special areas, indicating that the morphology information of the sample sur-faces are reflected in the OIRD signals. To determine the structural information of the ACFC sample surfaces, SEM was then employed to analyze active carbon investigated with OIRD. Secondary electron (SE) was used to obtain a

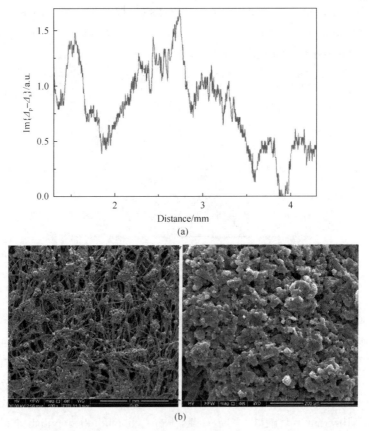

Figure 2 Morphology analysis of active carbon samples. (a) Distance dependent Im$\{\Delta_p-\Delta_s\}$ intensities of active carbon sample; (b) SEM images of ACFC: The below picture refers to the enlarged image of the upper one

new set of images shown in Figure 2(b). SE was very sensitive to the surface morphology, thus making it an effective method for observing the surface morphology. According to Figure 2(b), the carbon reflected a centralized distribution in some areas other than distributed evenly. Besides, the enlarged image shows that the pellets were also uneven in the active carbon areas. Active carbon was a kind of porous material whose void radius approximately varied from ~100nm to ~50μm. Comparing the OIRD signals and the SEM images, it can be revealed that the areas with high OIRD signals refer to the centralized distribution areas of active carbon, and the fluctuations of OIRD signals are caused by the uneven distribution carbon pellets. In terms of the areas with smaller OIRD signal intensities, the ACFC sample has little carbon pellets. The fiber cloth wires are criss-crossed between the areas of carbon pellets. The OIRD signal is related to the distribution of active carbons, which can be revealed by the comparison of OIRD response and SEM images. Consequently, OIRD measurement characterized the surface morphology of ACFC samples with one-dimensional scanning.

To further explore the practicality of the OIRD approach being applied in morphology detection of the petroleum reservoir, we selected two pieces of rocks 3km underground and performed the OIRD measurement. An optical microscope analysis was initially performed before the rocks were measured by OIRD. Figure 3 shows a set of optical microscope images of the mentioned two reservoir rocks. The spatial resolution in horizontal and vertical directions were about 200μm. The gray level distribution mainly reflects the particle distribution that was caused by the rocks' components. Consequently, optical microscope images indicated that the morphology of rocks were irregularly polygon and dispersed well.

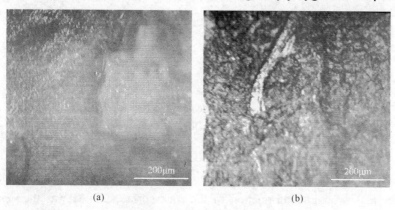

Figure 3 Optical microscope analysis of the surfaces of reservoir rocks

Figure 4 shows three-dimensional OIRD signal mapping of two reservoir rocks cored at a depth of 3000m in the horizontal direction. The square of each image is 300μm×300μm, and the Im{$\Delta_p - \Delta_s$} values reflect the relative variance. The signal intensities exhibited several peaks and troughs on the surfaces in imaginary images, showing the position-dependent uniformity of the physical properties. It is apparent that the trough areas with different sizes are randomly distributed throughout the surface of reservoirs rocks. Large granules can have a width range from 20μm to 130μm shown in Figure 4(a). In the three-dimensional systems of the imaginary parts, the trough areas appear to be approximately, but not absolutely, equal to each other. In contrast to Figure 4(a), Figure 4(b) plots a lot of trough areas with different sizes smaller than 100μm. Similarly, the areas are randomly distributed throughout the surfaces and are nondirectional relative to being over the whole areas. Compared to Figure 3 obtained with an optical microscope, Figure 4 shows the more obvious contrast at different points. Meanwhile, OIRD signal images reflect the dielectric properties' distribution, which can help reveal more information of physical properties in the concentrated areas of high or low intensities. The signals continuously change at near positions. According to the OIRD images, it is revealed that OIRD images of rock surfaces contribute variously at different positions, indicating that OIRD signals are related to the physical properties of materials.

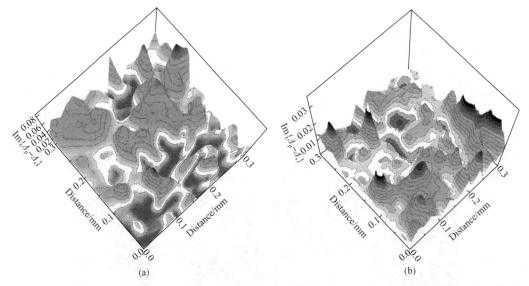

Figure 4 OIRD images of two reservoir rocks in which random regions were scanned by OIRD system. Red and cyan represent small and large OIRD signal intensities, respectively

The approximately irregular structures and the uneven morphology features of rocks are clearly reflected in Figure 4. The type relates to the orientation of rock particles or a crevice. To detect the real structural information of the rock surfaces, SEM was then employed to analyze the rocks measured by OIRD. SE modes were used to obtain the morphology images of relative reservoir rocks. Figure 5 shows the SEM images of two rocks: (a) shows the SE morphology image of a reservoir rock with a scale line of 200μm; (b) shows the SE morphology image of another reservoir rock with a scale line of 100μm. The SE images revealed the microdefects such as cracks and orientations of rock particles in the rocks, as clearly shown in the OIRD images with the corresponding imaginary signals. In Figure 5(a), some differently sized particles were observed with clear annular cracks between them. These annular cracks surround some low-lying areas. In Figure 5(b), the rock is denser, and the defects have relatively smaller size.

According to comparison between Figures 4 and 5, high similarity can be observed in terms of the structures in two reservoir rocks. OIRD signal intensity is a function of effective thickness d and dielectric constant ε_d on the basis of the widely used formula in the experimental section. OIRD intensity is proportional to the thickness of materials. Consequently, the trough areas with smaller values in Figure 4 refer to the low-lying fields. They possess similar-size ranges and contour shapes in both reservoir rocks. In terms of cyan areas, which have relatively large OIRD intensities, they correspond to the protuberances areas in Figure 5. The differences of OIRD intensities in these areas may be caused by the dielectric constant variances. The elemental components with different atomic masses are uniformly distributed through the cyan areas. Therefore, the OIRD technique is a very promising and practical technology to detect the morphology information and is a convenient supplementary method that can be combined with conventional methods.

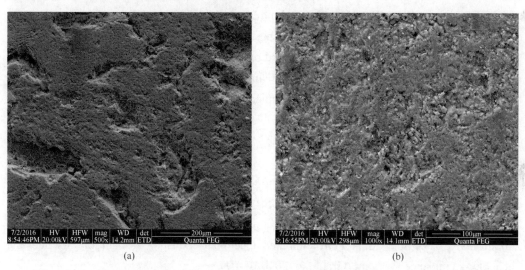

Figure 5　SEM images of reservoir rocks which were measured by OIRD

Similar to some other technologies, there are some limitations of the new technique. For instance, the baseline of the signal should be regulated in each measurement. Besides, more mathematical methods should be developed to analyze the data of the signal. Overall, this study focuses on applying the OIRD approach to active carbon and reservoir rocks, and proves that the OIRD is a very practical technique for detecting morphology. The morphology features can be clearly observed in the images of Im$\{\varDelta_p-\varDelta_s\}$ by both one-dimensional and two-dimensional scanning. The detection results in the OIRD images are in agreement with those of the SEM mappings. Accordingly, the research suggests that this approach used to characterize the morphology of the reservoir rock should combine previous techniques, such as optical microscopy, the SEM technique, and newly developed optical methods, such as terahertz (THz) spectroscopy[3,8,9]. Compared to the THz technique, OIRD has a higher spatial resolution due to the light source possessing a shorter wavelength. Relative to microscope and SEM techniques, OIRD can detect the dielectric properties of the materials that are useful in revealing the physical or chemical information. As reported in previous articles, the OIRD signal, which reflects the differences between the reflectivity of s- and p-polarized light, is related to the material's dielectric and surface properties so that it is a sensitive surface detection technique[21,24]. OIRD is a very promising and practical technology for detecting the morphology of general materials and reservoir rocks.

4 Conclusions

In summary, by using a newly developed OIRD technique, we were able to detect the morphology features of samples with both one-dimensional and two-dimensional scanning. This optical approach has extremely high sensitivity to the distribution of structural and componential properties. We tested this technique on active carbons and reservoir rocks, and compared the OIRD results to SEM mappings, indicating that the OIRD analysis can be clearly validated. The technique is fast (limited only by the speed of the mechanical scanning) and is easily applied to any space due to the simple system composition. It can be expected that the OIRD technique is a convenient supplementary morphology detection technique for conventional methods.

Acknowledgements

This work was supported by the National Natural Science Foundation of China (NSFC) (Grant No. 11574401); the National Basic Research Program of China (Grant No. 2014CB744302); the Science Foundation of China University of Petroleum, Beijing (Grant No. 2462017YJRC029); the China Petroleum and Chemical Industry Association Science and Technology Guidance Program (Grant No. 2016-01-07).

References

[1] Sujka M, Jamroz J. α-Amylolysis of native potato and corn starches - SEM, AFM, nitrogen and iodine sorption investigations. LWT - Food Science and Technology, 2009, 42(7): 1219-1224.

[2] Tianshou M A, Chen P. Study of meso-damage characteristics of shale hydration based on CT scanning technology. Petroleum Exploration & Development, 2014, 41(2): 249-256.

[3] Greeney N S, Scales J A. Non-contacting characterization of the electrical and mechanical properties of rocks at submillimeter scales. Applied Physics Letters, 2012, 100(12): 222909.

[4] Zhan H L, Li Q, Zhao K, et al. Evaluating $PM_{2.5}$ at a construction site using terahertz radiation. IEEE Transactions on Terahertz Science & Technology, 2015, 5(6): 1028-1034.

[5] Kar A, Saha A, Goswami N. Long-range surface plasmon-induced tunable ultralow threshold optical bistability using graphene sheets at terahertz frequency[J]. Applied Optics, 2017, 56(8): 2321-2329.

[6] Nasari H, Abrishamian M S. Terahertz bistability and multistability in graphene/dielectric Fibonacci multilayer. Applied Optics, 2017, 56(19): 5313-5322.

[7] Landry J P, Gray J, Toole M K, et al. Incidence-angle dependence of optical reflectivity difference from an ultrathin film on a solid surface. Opt. Lett., 2006, 31: 531-533.

[8] Greeney N S, Scales J A. Dielectric microscopy with submillimeter resolution. Applied Physics Letters, 2007, 91: 124105.

[9] Scales J A, Batzle M. Millimeter wave spectroscopy of rocks and fluids. Applied Physics Letters, 2006, 88(6): 221-282.

[10] Miao X Y, Zhan H L, Zhao K, et al. Oil yield characterization by anisotropy in optical parameters of the oil shale. Energy & Fuels, 2016, 30(12): 10365-10370.

[11] Zhan H L, Zhao K, Xiao L Z. Spectral characterization of the key parameters and elements in coal using terahertz spectroscopy. Energy, 2015, 93: 1140-1145.

[12] Zhan H L, Wu S X, Bao R M, et al. Qualitative identification of crude oils from different oil fields using terahertz time-domain spectroscopy. Fuel, 2015, 143: 189-193.

[13] Ge L N, Zhan H L, Leng W X, et al. Optical characterization of the principal hydrocarbon components in natural gas using terahertz spectroscopy. Energy & Fuels, 2015, 29: 1622-1627.

[14] Zhan H L, Wu S X, Bao R M, et al. Water adsorption dynamics into active carbon probed by terahertz spectroscopy. RSC Adv., 2015, 5: 14389-14392.

[15] Ajito K, Ueno Y. THz chemical imaging for biological applications. IEEE Transactions on Terahertz Science & Technology, 2011, 1(1): 293-300.

[16] Landry J P, Zhu X D, Gregg J P. Label-free detection of microarrays of biomolecules by oblique-incidence reflectivity difference microscopy. Opt. Lett., 2004, 29: 581-583.

[17] Zhu X D, Nabighian E. *In situ* monitoring of ion sputtering and thermal annealing of crystalline surfaces using an oblique-incidence optical reflectance difference method. Appl. Phys. Lett., 1998, 73: 2736-2738.

[18] Liu S, Zhu J H, He L P, et al. Label-free, real-time detection of the dynamic processes of protein degradation using oblique-incidence reflectivity difference method. Appl. Phys. Lett., 2014, 104: 163701.

[19] Chen F, Lu H B, Chen Z H, et al. Optical real-time monitoring of the laser molecular-beam epitaxial growth of perovskite oxide thin films by an oblique-incidence reflectance-difference technique. J. Opt. Soc. Am. B, 2001, 18: 1031-1035.

[20] Wang J, Zhan H L, He L P, et al. Evaluation of simulated reservoirs by using the oblique-incidence reflectivity difference technique. Sci. China: Phys. Mech. Astron., 2006, 59: 114221.

[21] Zhan H L, Wang J, Zhao K, et al. Real-time detection of dielectric anisotropy or isotropy in unconventional oil-gas reservoir rocks supported by the oblique-incidence reflectivity difference technique. Sci. Rep., 2016, 6: 39306.

[22] Wang J Y, Dai J, He L P, et al. Label-free and real-time detections of the interactions of swine IgG with goat anti-swine IgG by oblique-incidence reflectivity difference technique. Journal of Applied Physics, 2012, 112(6): 064702.

[23] He L P, Liu S, Dai J, et al. Label-free and real-time monitor of binding and dissociation processes between protein A and swine IgG by oblique-incidence reflectivity difference method. Chinese Physics Letters, 2015, 32(2): 020703.

[24] Liu S, Zhang H Y, Dai J, et al. Characterization of monoclonal antibody's binding kinetics using oblique-incidence reflectivity difference approach. mAbs, 2015, 7: 110-119.

In situ monitoring of water adsorption in active carbon using an oblique-incidence optical reflectance difference method

Honglei Zhan[1,2,3] Kun Zhao[1,2] Huibin Lü[4] Jing Zhu[2,3] Kuijuan Jin[4]
Guozhen Yang[4] Xiaohong Chen[1]

(1.State Key Laboratory of Petroleum Resources and Prospecting, China University of Petroleum, Beijing 102249, China; 2.Beijing Key Laboratory of Optical Detection Technology for Oil and Gas, China University of Petroleum, Beijing 102249, China; 3.Department of Material Science and Engineering, China University of Petroleum, Beijing 102249, China; 4.Institute of Physics, Chinese Academy of Sciences, Beijing 100190, China)

The adsorption of water molecules in active carbon in normal condition can be monitored *in situ* with an oblique-incidence polarization-modulated optical reflectance difference technique. The optical response cannot only characterize the time length of adsorption, but reveal the tendency of dielectric properties of active carbon during the adsorption process. Therefore, the whole adsorption can be described by the permittivity obtained by OIRD measurement. Such a technique is also effective under high ambient pressure and temperature because of the optical detection, indicating that the precise measurement carried with this technique can help in building adsorption theory system in some different conditions.

One of the main issues remaining to be solved in unconventional petrogeology exploration is the adsorption behavior and its dynamics. Shale gas is a common unconventional natural gas hidden in the strata or mudstone layers in a free or an adsorbed state[1]. Adsorption should be physical or chemistry behavior[2,3]. Different reservoirs able to adsorb oil and gas have been studied. However, most articles are related to the adsorption and absorption of organic molecules or ions from water or solutions. In terms of petroleum resources, the adsorption dynamics is a rather more significant issue due to the migration characteristics. Active carbon is a typical adsorbate and is used to adsorb water molecules to simulate the adsorption dynamics[4,5].

At present, the study about adsorption dynamics focuses on the detection methods and mathematical models[6-8]. Scanning electron microscopy(SEM) and atomic force microscopy (AFM) are appropriate to describe the surface and structure of the holes; however, they can hardly observe the dynamics process and need an extreme condition. Several adsorption kinetic models have been established to understand the adsorption kinetics and the rate-limiting step[9]. Pseudo-first and second-order rate models are often used to quantify the extent of uptake in sorption kinetics. Pseudo-second-order kinetic model investigates the relationship between adsorption and diffusion. These models describe the adsorption dynamics theoretically[10-12].

Originally published in *AIP Advance*, 2017, 7(9): 095219.

Therefore, the physical parameters of adsorbates and experimental observation are relatively insufficient and possibly more useful to characterize the adsorption theory.

Permittivity is a basic physical parameter and can be used to determine the polar properties of materials. By monitoring the permittivity change of active carbon, the adsorption and its process can be characterized. Oblique-incidence reflectivity difference(OIRD) is sensitive to the interference of subsurface electromagnetic fields. It depends on the dielectric and surface properties because this technique measures the difference in reflectivity between s-and p-polarized lights[13-18]. In this study, we applied the OIRD method to describe the adsorption information of water molecules in active carbon.

We chose active carbon for this proof-of-principle study because it is a typical porous material whose void radius approximately varied from 100nm to 50μm. The active carbon pellets adhered tightly onto the fiber cloth to form the active carbon fiber cloth[19]. A single drop of water whose volume is 40μL was dropped onto the middle of the active carbon. For the oblique-incidence polarization-modulated optical reflectance difference measurement, we used the OIRD setup in Figure 1 whose light beam and detection method were described previously by Liu and Zhan et al[20,21]. We fabricated the water adsorption into active carbon for the OIRD measurements using conventional procedure which was used for protein detection as reported previously. This measurement was carried out over the duration of the experiments at room temperature and relative humidity. In order to measure the adsorption process of water molecules in active carbon, the sample was fixed on a chamber in the two dimensional scanning stage of the OIRD system. We detected the real part Re$\{\Delta_p-\Delta_s\}$ and the imaginary part Im$\{\Delta_p-\Delta_s\}$ signal and took an advantage of the scanning stage with one dimensional shown in Figure 1. Consequently, we obtained the OIRD signal data of water adsorption with different position and different time length.

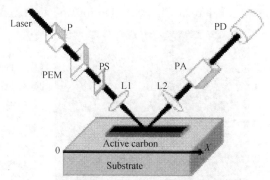

Figure 1 Sketch of OIRD system for *in situ* monitoring of adsorption process

Figure 2 illustrates the results obtained from above adsorption performed on the active carbon. We displayed the optical reflectance difference signals over 9mm of adsorption with the sample at 0min (No adsorption) and 35min respectively. As illustrated in the plot, Re$\{\Delta_p-\Delta_s\}$ signal intensities are different from samples without water to those with water 35min, even though these active carbons are fixed at the same positions. This indicates that after adsorption the active carbons had differences in optical dielectric constant and the water were adsorbed into the holes of most active carbons. As expected, the $\text{Re}\{\Delta_p-\Delta_s\}$ signal intensities of adsorbed sample are different from that without adsorption in the position range from 0.3mm to 6.6mm. The closer the point is to the middle, the larger difference the phenomenon can be found. In the range of 6.6~9.0mm, the signal intensities of two measurements are same, representing that the water molecules were not adsorbed by active carbons at these positions which are far from the initial water location.

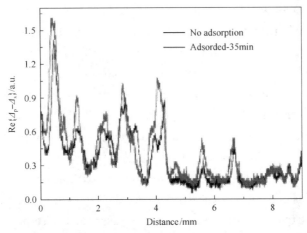

Figure 2　The analysis of OIRD signal about whether the water is adsorbed in active carbon. The real part $\text{Re}\{\Delta_p-\Delta_s\}$ signal of active carbon without and with water at 35 minute

To further explore the kinetics of the adsorption, we have monitored the process continuously by a constant linear scanning. When the water drop has been adhered onto the superficies of active carbon, the water cannot decentralize immediately because of the large tension surface and the OIRD signals have little fluctuation in Figure 3. During the time frame of > 23.7min, the

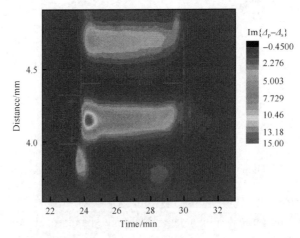

Figure 3　The measurement results of adsorption dynamics using the OIRD. The color bar represent the $\text{Im}\{\Delta_p-\Delta_s\}$ signal intensities of samples at different scanning positions and at different time frames

imaginary Im{$\Delta_p-\Delta_s$} signals rapidly increase and keep a high level. This stage is homologous with the adherence time of the water drop. When the adsorption ends, the samples remains unchanged, thus the OIRD signal intensities at different time frames keep unchanged, indicating that the adsorption process has ended and the water molecules water has been adsorbed in a stable state.

According to a widely used formula, OIRD signal intensity is a function of dielectric constant ε_d and effective thickness d[22]. OIRD signal can be analyzed by

$$\begin{cases} \mathrm{Im}\{\Delta_p - \Delta_s\} \propto d \\ \mathrm{Im}\{\Delta_p - \Delta_s\} \propto \varepsilon_d - (\varepsilon_0 + \varepsilon_s) + \dfrac{\varepsilon_0 \varepsilon_s}{\varepsilon_d} \end{cases}$$

where ε_0 and ε_s represent the relative permittivity of the environment and the basement. According to the OIRD response in Figure 3, when the time length goes to 23.7min, the water drop has been adhered onto the superficies of active carbon and diffused to the holes due to the concentration gradient. This diffusion mainly refers to the depth diffusion and experiences a short time frame (0.6min), thus the depth of water-carbon layer increases rapidly. Then the active carbon with a very large specific surface area adsorbed the water molecules into the voids. In this stage, with the increasing molecules adsorbed into active carbon, the water molecules were scattered in different holes. The dielectric constants of carbon adsorbing water molecules augment with the time increasing. However, the OIRD signal intensities dwindle after the time exceed amplitude point 24.3min, indicating the permittivity of carbon-water $\varepsilon_d < \sqrt{\varepsilon_0 \varepsilon_s}$. Consequently, OIRD traced the adsorption process depending on the sensitivity of OIRD to the dielectric properties of sample.

As shown in Figure 3, the bright scales represent the adsorption dynamics of water molecules in active carbons and have the width of 350m. To determine whether the OIRD results are resulted by the structure and properties of the sample, scanning electron microscopy (SEM) was then employed to detect the componential and structural information of active carbon. Spray-gold pretreatment was used for the electric conduction of the sample in the SEM and EDS measurement. Secondary electron (SE) mode was used to obtain new set of images. Figure 4 (a) shows the SEM images of active carbon adhered on fiber. A single active carbon has a width of 345m, which is close to the width of OIRD signals. Figure 4 (b)~(d) illustrate the morphology of the holes and the chemical composition of the active carbon. According to the EDS spectrum, the mass fraction and atom fraction of the chemical element C are 77.52% and 91.81%, while those of O are 8.03% and 7.14%. Au was introduced by the spray-gold treatment.

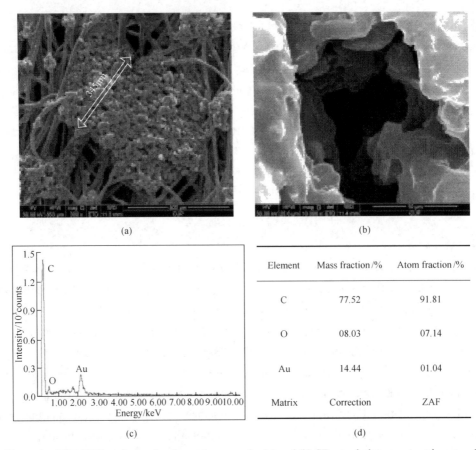

Figure 4　SEM-EDS analysis of active carbon sample. (a) and (b) SE morphology contrast images of active carbon: (b) shows an enlarged view of a selected area in (a); (c) and (d) elemental component of active carbon measured by EDS

Active carbon is composed of graphene layers and the surface of the carbon is rather large. A hydrogen bond O—H···O is formed between the element of the hydroxyl group at the surface of carbon and water molecules. The participation of hydrogen bonding can promote the adsorption behavior and improve the adsorption rate[23]. Besides, the molecules with smaller size are easier to scatter into inner holes and to be adsorbed[24,25]. Due to the adsorption of water molecules, the dielectric properties of active carbon altered gradually. Consequently, OIRD, which reflects the difference between the reflectivity of s- and p-polarized light, is closely related to the sample's dielectric properties. It is a very promising and practical technology for explore the adsorption theories in the unconventional oil-gas recovery[26].

In summary, we have monitored the adsorption processes of water into active carbon using OIRD method in a label-free and real-time manner. Based on the OIRD signal intensities, the adsorption time and scale can be obtained. The SEM-EDS analysis about active carbon also validate the OIRD results. The research demonstrated that OIRD offers an approach to monitor adsorption and its process by a non-contact and simple way.

Acknowledgements

This work was supported by the National Basic Research Program of China (Grant No. 2014CB744302), Science Foundation of China University of Petroleum, Beijing (No. 2462017YJRC029), the Specially Funded Program on National Key Scientific Instruments and Equipment Development (Grant No. 2012YQ140005), the China Petroleum and Chemical Industry Association Science and Technology Guidance Program (Grant No. 2016-01-07), and the National Nature Science Foundation of China (Grant No. 11574401).

References

[1] Castillo J, Vargas V, Piscitelli V, et al. Study of asphaltene adsorption onto raw surfaces and iron nanoparticles by AFM force spectroscopy. Journal of Petroleum Science & Engineering, 2017, 151: 248-253.

[2] Izquierdo M T, Yuso A M D, Valenciano R, et al. Influence of activated carbon characteristics on toluene and hexane adsorption: Application of surface response methodology. Applied Surface Science, 2013, 264: 335-343.

[3] Kim K J, Kang C S, You Y J, et al. Adsorption-desorption characteristics of VOCs over impregnated activated carbons. Catalysis Today, 2006, 111(3-4): 223-228.

[4] Franz M, Arafat H A, Pinto N G. Effect of chemical surface heterogeneity on the adsorption mechanism of dissolved aromatics on activated carbon. Carbon, 2000, 38(13): 1807-1819.

[5] Roostaei N, Tezel F H. Removal of phenol from aqueous solutions by adsorption. Journal of Environmental Management, 2004, 70(2): 157-164.

[6] Kim K J, Kang C S, You Y J, et al. Adsorption-desorption characteristics of VOCs over impregnated activated carbons. Catalysis Today, 2006, 111(3): 223-228.

[7] Daniel T, Georg H, Martin G H, et al. Theoretical study of adsorption sites on the (001) surfaces of 1:1 clay minerals. Langmuir, 2002, 18(1): 139-147.

[8] Sun M S, Shah D B, Xu H H, et al. Adsorption equilibria of C_1 to C_4 alkanes, CO_2, and SF_6 on silicalite. The Journal of Physical Chemistry B, 1998, 102(8): 1466-1473.

[9] Febrianto J, Kosasih A N, Sunarso J, et al. Equilibrium and kinetic studies in adsorption of heavy metals using biosorbent: A summary of recent studies. Journal of Hazardous Materials, 2009, 162(2-3): 616-645.

[10] Pavasant P, Apiratikul R, Sungkhum V, et al. Biosorption of Cu^{2+}, Cd^{2+}, Pb^{2+}, and Zn^{2+} using dried marine green macroalga caulerpa lentillifera. Bioresource Technology, 2006, 97(18): 2321-2329.

[11] Rabiaa D, Hamdaoui O. Sorption of copper (II) from aqueous solutions by cedar sawdust and crushed brick. Desalination, 2008, 225(1-3): 95-112.

[12] Apiratikul R, Pavasant P. Batch and column studies of biosorption of heavy metals by Caulerpa lentillifera. Bioresource Technology, 2008, 99(8): 2766-2777.

[13] Zhu X D, Nabighian E. In situ monitoring of ion sputtering and thermal annealing of crystalline surfaces using an oblique-incidence optical reflectance difference method. Applied Physics Letters, 1998, 73(19): 2736-2738.

[14] Zhu X D, Lu H B, Yang G Z, et al. Epitaxial growth of $SrTiO_3$ on $SrTiO_3$ (001) using an oblique-incidence reflectance-difference technique. Physical Review B, 1998, 57(4): 2514-2519.

[15] Liu S, Zhu J H, He L P, et al. Label-free, real-time detection of the dynamic processes of protein degradation using oblique-incidence reflectivity difference method. Applied Physics Letters, 2014, 104(16): 163701.

[16] Wang J Y, Dai J, He L P, et al. Label-free and real-time detections of the interactions of swine IgG with goat anti-swine IgG by oblique-incidence reflectivity difference technique. Journal of Applied Physics, 2012, 112(6): 064702.

[17] Landry J P, Zhu X D, Gregg J P. Label-free detection of microarrays of biomolecules by oblique-incidence reflectivity difference microscopy. Optics Letters, 2004, 29(6): 581-583.

[18] Chen F, Lu H B, Chen Z H, et al. Optical real-time monitoring of the laser molecular-beam epitaxial growth of perovskite oxide thin films by an oblique-incidence reflectance-difference technique: erratum. Journal of the Optical Society of America B: Optical Physics, 2001, 18(7): 1031-1035.

[19] Zhan H L, Wu S X, Bao R M, et al. Water adsorption dynamics in active carbon probed by terahertz spectroscopy. RSC Advances, 2015, 5: 14389.

[20] Wang J, Zhan H L, He L P, et al. Evaluation of simulated reservoirs by using the oblique-incidence reflectivity difference technique. Science China: Physics, Mechanics & Astronomy, 2016, 59(11): 114221.

[21] Liu S, Zhang H Y, Dai J, et al. Characterization of monoclonal antibody's binding kinetics using oblique-incidence reflectivity difference approach. mAbs, 2015, 7(1): 110-119.

[22] Landry J P, Gray J, O'Toole M K, et al. Incidence-angle dependence of optical reflectivity difference from an ultrathin film on a solid surface. Optics Letters, 2006, 31(4): 531-533.

[23] Bittner E W, Smith M R, Bockrath B C. Characterization of the surfaces of single-walled carbon nanotubes using alcohols and hydrocarbons: A pulse adsorption technique. Carbon, 2003, 41(6): 1231-1239.

[24] Zhu J, Zhan H L, Miao X Y, et al. Terahertz double-exponential model for adsorption of volatile organic compounds in active carbon. Journal of Physics D: Applied Physics, 2017, 50(23): 235103.

[25] Zhu J, Zhan H L, Miao X Y, et al. Adsorption dynamics and rate assessment of volatile organic compounds in active carbon. Physical Chemistry Chemical Physics, 2016, 18: 27175-27178.

[26] Zhan H L, Wang J, Zhao K, et al. Real-time detection of dielectric anisotropy or isotropy in unconventional oil-gas reservoir rocks supported by the oblique-incidence reflectivity difference technique. Scientific Reports, 2016, 6: 39306.

Real-time monitoring the formation and decomposition processes of methane hydrate with THz spectroscopy

Xinyang Miao Shining Sun Yizhang Li Wei wang Rima Bao Kun Zhao

(Beijing Key Laboratory of Optical Detection Technology for Oil and Gas,
China University of Petroleum, Beijing 102249, China)

Gas hydrates are crystalline clathrate compounds comprised of hydrogen-bonded water cavities, consisting of guest molecules trapped in a lattice of polyhedral water cages under elevated pressures and low temperatures[1]. The structures of the gas hydrates are closely related to the types and sizes of the guest species, and three distinct structures, including cubic structure-I (sI), cubic structure-II (sII), and hexagonal structure-H (sH), are known to form[2]. Being expected as a kind of future energy resource, natural gas hydrates are found worldwide in locations such as the permafrost regions and beneath the sea. Besides, since hydrates store large quantities of natural gas (e.g., 180 volumes per each volume of hydrate at standard condition), hydrates have drawn much attention as a novel means of natural gas storage and transportation in recent years[3,4]. Thus, it is vital to evaluate the formation process of gas hydrates. Properties such as stability, structure and cage occupancy have been widely investigated, and various spectroscopic methods (e.g., solid-state ^{13}C, ^{1}H NMR, Raman, XRD, and ESR) were performed[5,6]. In this letter, the formation and dissociation processes of gas hydrate, ice as well as mixture of ice and methane were studied by THz spectroscopy.

The experimental equipment used in this study consisted of a reaction kettle, real-time monitors for temperature and pressure as well as a THz-time domain spectroscopy (TDS) system. A detailed description and schematic drawing of the THz system based on photo conductive antenna and electro-optical sampling has been stated in our previous study[7-9]. As a well-known surfactant, solution of sodium dodecyl sulfate (SDS) a 0.1wt % was used to shorten the time of nucleation and promote the hydrate formation on condition that methane was pumped in. The experiment was divided into 4 steps: (i) The reactor was purged with methane, then it was pressurized with methane to ~5MPa at 275.2K; (ii) after reaching equilibrium at the initial temperature and pressure, the system was cooled to the hydrate formation temperature at 272.2K; (iii) the pressure and temperature were maintained in the experimental apparatus for 48h to provide sufficient time for the hydrate formation; (iv) after hydrate formation was complete, THz measurements were carried out to evaluate the influence of temperature and pressure to the hydrates. The temperature was lowered to 266.2 K at first; the vent valve was then opened and the remaining gas was purged; after that, the temperature was raised to 272.2 K, and the system was allowed to remain at atmospheric pressure for hydrate decomposition. The whole process was recorded in real-time by THz-TDS with the setup covered with dry nitrogen at room temperature (294.1K ± 0.3K) and the signal-to-noise ratio equaled ~1500. In addition, variation of the SDS solution with the temperature was observed by THz-TDS in the range of 272.2~266.2K, and the gas-liquid mixture (methane and the solution) was also obtained and tested by repeating step (i), (ii) and (iv) for comparison.

Figure 1(a) shows the formation and decomposition processes of gas hydrate with respect to the variation of THz amplitude values (A_{mp}). As shown in Figure 1, the whole process can be divided into 3 sequential stages where temperature and pressure change. During the initial freezing stage, A_{mp} increases linearly, reaching 0.014 at 266.2K and 5.17MPa. Methane fills the cage-structure built by water molecules. In the second stage, temperature is maintained at 266.2K but the pressure is reduced. Thus hydrates formation process moderates, and accordingly, the increasing of A_{mp} slows down. Besides, A_{mp} reaches a nearly constant value after ~32min. Most of methane has escaped from the reaction kettle at the second stage, while a fraction of gas enters the cage-structure and props it up with the changing temperature. van der Waals forces between CH_4 and H_2O molecules play an important role in stabilizing cage-structure of gas hydrates against changing temperature. Therefore, decreasing pressure and rising temperature followed in the third stage will not change A_{mp}.

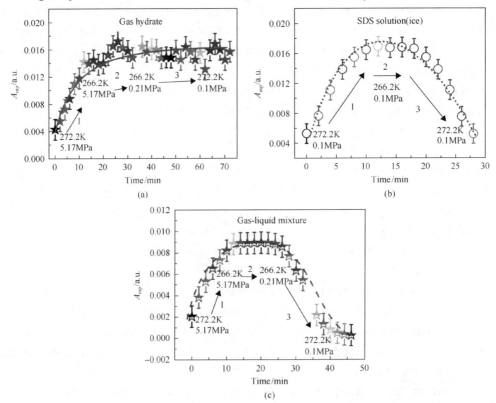

Figure 1 Time dependent THz A_{mp} of (a) gas hydrate; THz observed variation of (b) SDS solution and (c) gas-liquid mixture with different conditions. Error bars represent 0.5 fluctuation of standard deviation of A_{mp}

Despite similar appearance, ice with hexagonal structure distinguishes itself from hydrates[10]. Figure 1 (b) shows the A_{mp} of ice as a function of time. Changes in A_{mp} appear to mirror the changes in temperature during experiment, which coincides with the effect of temperature on hydrogen bond. To be specific, A_{mp} is 0.005 around in the beginning with temperature of 272.2 K and rises to 0.017 with temperature of 266.2 K after 10 min. However, A_{mp} starts to fall when the temperature begins to rise and turns back to 0.005 around finally. The curve of the gas-liquid mixture resembles that of ice (Figure 1 (c)), indicating that ice within the mixture undergoes the same process indicated by Figure 1 (b). Unlike gas hydrate, the cage-structure hardly forms and the mixture is not stable when the temperature goes up. Moreover, owing to the transparency of nonpolar molecules in THz range, the concentration of methane has little effect on A_{mp}.

In this letter, THz-TDS was utilized to evaluate the formation and decomposition processes of gas hydrate. Alteration of the ice as well as gas-liquid mixture against temperature and pressure was also observed by THz-TDS for comparison. Our results demonstrate the unique capability of THz-TDS to monitor the synthesis of gas hydrate. Therefore, THz technology is supposed to be a new means of studying gas hydrate. Future applications involving THz-TDS will be available in related research.

Acknowledgements

This work was supported by the National Nature Science Foundation of China (Grant No. 61405259 and 11574401), National Basic Research Program of China (Grant No. 2014CB744302), the Specially Founded Program on National Key Scientific Instruments and Equipment Development, China (Grant No. 2012YQ140005) and the China Petroleum and Chemical Industry Association Science and Technology Guidance Program (Grant No. 2016-01-07).

References

[1] Zhao J F, Zhu Z H, Song Y C, et al. Analyzing the process of gas production for natural gas hydrate using depressurization. Applied Energy, 2015, 142: 125-134.

[2] Zhuang C, Chen F, Cheng S H, et al. Light carbon isotope events of foraminifera attributed to methane release from gas hydrates on the continental slope, northeastern South China Sea. Science China: Earth Sciences, 2016, 59(10): 1981-1995.

[3] Kim N J, Lee J H, Cho Y S, et al. Formation enhancement of methane hydrate for natural gas transport and storage. Energy, 2010, 35(6): 2717-2722.

[4] Mimachi H, Takeya S, Yoneyama A, et al. Natural gas storage and transportation within gas hydrate of smaller particle: Size dependence of self-preservation phenomenon of natural gas hydrate. Chemical Engineering Science, 2014, 118(18):208-213.

[5] Kim D Y, Lee H. Spectroscopic identification of the mixed hydrogen and carbon dioxide clathrate hydrate. Journal of the American Chemical Society, 2005, 127(28): 9996-9997.

[6] Eslamimanesh A, Mohammadi A H, Richon D, et al. Application of gas hydrate formation in separation processes: A review of experimental studies. The Journal of Chemical Thermodynamics, 2012, 46: 62-71.

[7] Wang D D, Miao X Y, Zhan H L, et al. Non-contacting characterization of oil-gas interface with terahertz wave. Science China:Physics, Mechanics & Astronomy, 2016, 59(7):674221.

[8] Zhan H L, Sun S N, Zhao K, et al. Less than 6 GHz resolution THz spectroscopy of water vapor. Science China: Technological Sciences, 2015, 58(12):2104-2109.

[9] Bao R M, Miao X Y, Feng C J, et al. Characterizing the oil and water distribution in low permeability core by reconstruction of terahertz images. Science China: Physics, Mechanics & Astronomy, 2016, 59(6):664201.

[10] Tajima H, Kiyono F, Yamasaki A. Direct observation of the effect of sodium dodecyl sulfate (SDS) on the gas hydrate formation process in a static mixer. Energy & Fuels, 2010, 24(1): 432-438.

第三篇　光学技术在油气储运中的应用

Simultaneous characterization of water content and distribution in high-water-cut crude oil

Yan Song[1,2]　Honglei Zhan[1,2]　Kun Zhao[1,2]　Xinyang Miao[2]　Zhiqing Lu[2]
Rima Bao[2]　Jing Zhu[2]　Lizhi Xiao[1]

(1.State Key Laboratory of Petroleum Resources and Prospecting, China University of Petroleum, Beijing 102249, China; 2.Beijing Key Laboratory of Optical Detection Technology for Oil and Gas, China University of Petroleum, Beijing 102249, China)

Abstract: High-water-cut crude oil is currently the main form of underground oil in China. The high water content and the distribution of oil and water have an important impact on the rheological property and exploration efficiency. Terahertz (THz) waves are extremely sensitive to hydrogen bonds and intermolecular forces, and consequently, the responses of oil, water and air exhibit obvious differences. In this work, a series of crude oils with high water content were investigated by using THz time-domain spectroscopy (THz-TDS). Linear models were built between amplitudes of THz-TDS and the absorbance and water content from 50.05% to 100.00% with correlation coefficients (R) of 0.99 and >0.93, respectively. In addition, oil and water were determined to be distributed at different positions, based on THz signals in a series of oriented scans. Moreover, the transparent parts for visible light are identified by scanning the local points. The positions with low and high THz intensities refer to water and air regions, respectively. Therefore, THz technology is expected to act as a supplementary method to characterize high-water-cut crude oil, which will promote the efficiency and safe operation of pumping units and oil pipelines.

1 Introduction

Oil, which is also known as the "blood" of industrial production, has become one of the most important energy resources for the past 100 years. However, the high price of crude oil and future worldwide energy demands make it necessary to enhance oil recovery processes as well as the oil production capacity[1]. In order to meet this demand, secondary recovery strategies have been implemented, which rely on maintaining or increasing the reservoir pressure by injecting water into the reservoir. The water injection schemes are quite successful in enhancing hydrocarbon production; these strategies lead to the generation of water content in crude oil[2,3]. The measurement of water content in crude oil is of great significance. The existence of large quantities of water in crude oil can introduce serious problems in the petroleum refining process, transportation, and other related chemical or petroleum engineering processes. The increase of oily sewage in the petroleum production process and damage to the lubrication films on mechanical equipment can cause the equipment to corrode as well as cause instability of the distillation operation for the petroleum refining process. In addition to these problems, high-water-content oil causes issues regarding the distribution of oil, water, and the remaining oil. The distribution of oil and water becomes more complex in the ground, and the remaining oil is highly dispersed but locally enriched in space. Thus, the high water content has an influence on the characteristics of oil, which has generated widespread consideration of oil-water mixtures. It has been previously difficult to accurately measure the water content of crude oil as a result of several factors including the time-variant mixture state of crude oil and water, the multi flow pattern, the complex ingredients, the poor oil field environment, etc.

Many measurement technologie, have been developed for determining water content in crude oil, such as the electrical desorption, the dehydration, the distillation, the Karl Fischer titration, the capacitance, and the ray method. These methods are employed in different situations according to specific needs. The distillation, the Karl Fischer titration and the ray methods are preferred for the sake of high precision. In order to operate easily, the electrical desorption and the ray methods are competitive. Besides, the capacitance method has advantage over others in economy and maintainability. Nevertheless, the aforementioned methods have their respective limitations under different conditions[4-8]. Moreover, some new techniques have been proposed in recent years. For example, a noncontact optical method, called all-optical detection, has been proposed to measure the water content ranging from 0 to 100% in crude oil by a noncontact laser source and receiver (adaptive laser interferometer)[9]. A system with capacitance to detect phase angle conversion is designed to measure ultralow water content in crude oil, where the resolution can reach ±50ppm (1ppm=10^{-6})[10]. THz spectroscopy is a developing spectral technique bridging the gap between microwave and infrared spectroscopy and is becoming a hot research topic, because of its unique advantages[11-13]. It has been shown that some low-energy events such as molecular torsions or vibrations, as well as intramolecular or intermolecular hydrogen bonding in part intrinsically correspond to THz frequencies[14-17]. In addition, differences in molecular configuration and the polarity of oil, water, and gas make the THz-TDS a technique that can distinguish between them readily. Therefore, THz-TDS technology has attracted significant attention for applications in the energy field such as in the petroleum industry[18-20]. The THz-TDS technology, combined with multivariate statistical methods, has been shown to be an effective method to identify crude oils from different oil fields and pattern transitions of oil-water two-phase flow, which are the principal components of natural gas[21-25].

Because of the different absorption properties of oil, water, and gas in the THz range, their respective distributions in crude oil can be determined qualitatively and quantitatively. Recently, THz-TDS was used to measure the water content for crude oil with low water content ranging from 0.01% to 25.00% (w/w)[26]. However, because of the strong absorption of the THz pulse in water, the method has trouble measuring the water content (as a percentage) for high-water-content oils. In this research, the oil-water mixtures with different water contents (50.05%~100.00%) were studied by THz-TDS. Initially, the water content was quantitatively determined by linear models based on THz peak intensities and absorbance values. By comparing THz intensities at different locations, the distribution of oil and water, especially water or air, were clearly distinguished. This investigation shows that THz spectroscopy is an effective technique for the comprehensive detection of underground high water-cut-crude oil.

2 Experimental methods

A conventional transmission THz-TDS system was used in this research. The principal measurement of this setup was described in a previous report[23]. As shown in Figure 1, the collimated THz pulse was focused and reflected by a lens (L1) and mirror (M1), respectively, and transmitted through the oil-water mixture. The sample-encoded THz beam then reached the ZnTe crystal detector. The THz signal was transformed to an electrical signal by the electro-optic effect and was detected by a lock-in amplifier. The experiment was operated at a constant temperature of 294K±0.3K.

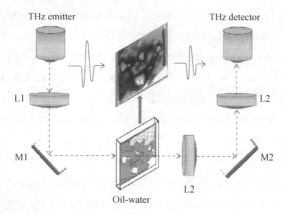

Figure 1 Experimental setup for the detection of oil-water distributions with transmission THz-TDS

The water-oil mixtures with varying water content used in this experiment were obtained by mixing different proportions of crude oil to water. The crude oil samples used in this work were Brazilian crude oil with a water content of 0.1%. The sample cells consisted of two plastic plates with a thickness of ~40μm, to ensure a high enough signal amplitude, since the water has a strong absorption in the target THz frequency range. The sample cells had dimensions of 5cm × 5cm. The oil and water was added dropwise into the sample cell using syringes. The total number of droplets was eight, with the number of oil droplets decreased from four to zero while water droplets increased from four to eight. Therefore, oil-water mixtures were obtained with water content from 50.05% to 100.00%. Since oil and water are not miscible, the distribution of oil and water was not uniform. As a result, measurements were taken at several locations on the sample with near-symmetrical distributions. In each measurement, the mixture is located at the focus of the lens and the THz wave is vertically incident to the surface of the plastic plates.

3 Results and discussion

The water content of crude oil in oil exploitation, transportation and storage is a vital parameter to assess crude oil production accurately. Because of the water flooding technique currently used in oil exploitation, the water content in the majority of mined oil is high. Accurately measuring the water content of high-water content crude oil has historically been difficult because of the oil medium conditions, working conditions, measurement technologies and instruments, and other factors[27-29]. Several points were measured on the oil-water mixture samples with a distribution that was almost symmetric because of the uneven distribution of oil and water. The peak intensities E_p value of each point were extracted from the THz time-domain spectra. As shown in Figure 2, the smaller black points represent the peak intensities of the time-domain spectra measured at different sampling locations on the oil-water samples. The bigger colored stars are a function of the water content acquired by averaging all the values of the measured peak intensities. The average peak value decreased from 140mV to 74mV with increasing water content from 50.05% to 100.00%. The peak intensities were linearly fitted to the equation $y = -0.00139x+0.21328$ and had a correlation coefficient (R) of ~0.99, which is shown as the line in Figure 2. This indicates that the linear fit represents the data well. This is explained by noting that water has a significantly stronger absorption in the THz range compared with crude oil. The THz time-domain spectra of the mixtures with varying water content were obtained using the spectra of the points whose peak intensities were closest to the average values. As shown in the inset in Figure 2, the peak intensities of the THz waveform decreased with increasing water content in the mixture. It can be seen that there was a phase shift as the water content increased. Therefore, THz-TDS can provide information about water content, and it can be used to measure the water content for oil exploitation, transport, or storage.

Both phase and amplitude information of the THz pulses after propagating through the reference (empty sample cell) and oil-water mixtures were measured using THz-TDS[30]. By comparing the pulse of the reference with the mixtures and applying a numerical fast Fourier transform (FFT), the frequency domain spectra were obtained. The absorbance spectra were calculated using the definition of absorbance, $A=\lg(I_0/I)$, where I_0 and I are the values of the THz frequency domain spectrum (FDS) of empty sample cell and sample, respectively.

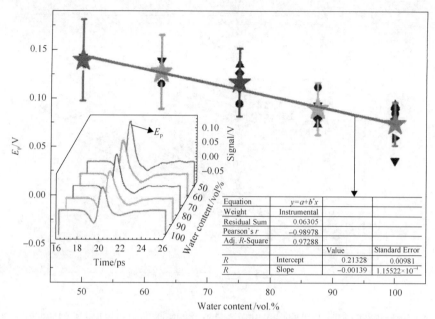

Figure 2　THz peak intensities (E_p) of the oil-water mixtures with water content ranging from 50.05% to 100%. The inset shows the time-domain waveforms corresponding to different oil content. The table lists the parameters of linear fitting

The absorbances of the mixtures were obtained using the method described above. The principle behavior of the absorbances of the different mixtures investigated was very similar to the THz peaks, which increased with increasing frequency. The absorbance spectra of the mixtures are shown in the inset of Figure 3. The slopes of the absorbance spectra increased with higher water content in the mixtures. The available dynamic range of the system was approximately limited to the values from 0.5THz to 1.2THz. The sphere symbols in Figure 3 represent the absorbance of the mixtures obtained from the absorbance spectra at select frequencies (0.47THz, 0.55THz, 0.70THz, 0.87THz, and 1.00THz). The results show that the absorbance increased as the water content increased, and the corresponding liner fitting lines are shown with correlation of $R > 0.93$. This liner relationship strongly indicates that the water content in oil-water mixtures can be estimated using the relationship between THz absorbance and water content.

It is important to accurately measure the concentration of oil, water and gas to monitor the production status of oil wells, predict oil reserves, optimize the method of oil exploitation, and manage the production process of crude oil[31-34]. Because of the inhomogeneous distribution of oil and water, and the obvious difference between the absorption of water, gas, and oil in the THz range, the 50.05% water content oil-water mixture was used to investigate the distribution of oil, water, and gas for the mixture.

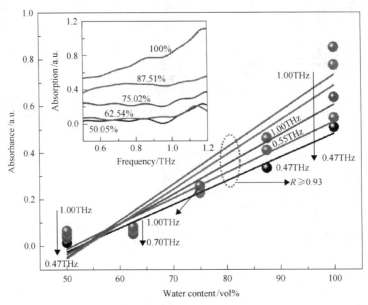

Figure 3　Absorbance of the oil-water mixtures with mentioned water content at five randomly selected THz frequencies. The insert graph shows the absorbance spectra for 0.5~1.2THz

The distribution of oil was studied first. A total of 33 points were measured from the bottom of the sample to the top along three vertical lines, denoted Left (11 points), Middle (11 points) and Right (11 points), as shown in Figure 4. The Left sampling line only transmits through one point at 1.4 cm from the bottom of the sample, which is shown in Figure 4 as corresponding to the one strong decrease of E_p. The Middle sampling line has two transparent points at 1cm and 3.2cm with an obvious decrease in the E_p. The Right sampling line was a single transparent point at 3.2cm, and a deep decrease in E_p occurred. These phenomena indicate that THz-TDS technology effectively realize the identification of water content via comparing the THz time-domain spectra of the mixtures, and it is capable of distinguishing between water and oil distributions.

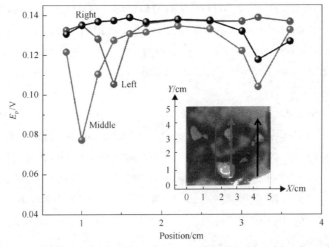

Figure 4　THz peak intensities (E_p), as a function of the sampling point's position from the bottom to the top along the y-axis. The arrows in the inset indicate the trend of sampling

In addition, another 28 points dispersed in three horizontal directions were measured. As shown in Figure 5, these points were measured from left to right and were named Top (9 points), Middle (9 points), and Bottom (10 points). It can be seen that the extracted peak intensities (E_p) values of the THz time-domain spectra for these points show a distinction between the water and oil regions. The Top sampling line transmitted through three transparent points at the left, middle and right of parts of the line. Transmitted light corresponded to the two deep decreases of E_p at 2.1cm and 3.7cm in Figure 5 but only a minor decrease at the left point. It was determined that the transmission from the left transparent region was gas by later experimentation. The Middle sampling line had no transparent points, and correspondingly, no obvious decrease in the E_p had obvious. The Bottom sampling line had one transparent point in the middle of the line at 2.6cm, and a corresponding deep decrease in E_p was observed. With regard to the measurement in horizontal and vertical directions, the E_p values of the total of 9 intersections show the distribution of the oil or moisture phase. Taking the intersection of the vertical Left and the horizontal Middle direction as an example, E_p = 0.13592V in Figure 4 and E_p = 0.13942V in Figure 5. By comparing the E_p values of the crossing points, the THz absorption of oil-water mixture is similar to each other in same area.

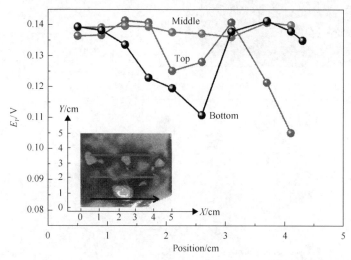

Figure 5 THz peak intensities (E_p), as a function of the sampling point locations along the *x*-axis. The inset shows the sampling direction from left to right

As discussed above, Figures 4 and 5 clearly display the primary distributions of water and oil compared to gas. The positions with large and small THz intensities corresponded to oil and water regions, which was in agreement with the physical inspection of color differences in the sample. However, a new phenomenon was observed from the relatively large THz intensity decreases at the first two measurement points on the up-horizon line, which were also observed to be transparent in the photograph. It was determined that both water and air molecules existed in the transparent regions. To determine the contents of these transparent regions, the five transparent regions were designated as Labels 1~5 and were carefully measured. Three positions were chosen for Label 1, located approximately at the center, the left edge, and the right edge of this region. Because of the size of Labe 2, only one position at the center was measured. The points measured for Label 3 were located about the tail of this region, the center, and the border of the inner circle. Two points were measured for Labels 4 and 5 located at the core of the embedded circle, as well as the bottom of the region for Label 4 and the upper middle for Label 5, respectively. All chosen points for each region are illustrated in Figure 6(a). The raw experimental transmission time-domain spectra are plotted in Figure 6(b) with time values from 17 ps to 23 ps and correspond to the five transparent regions. According to THz-TDS, a difference in the spectra from the sampling points in one transparent region (Label 3) can be clearly observed.

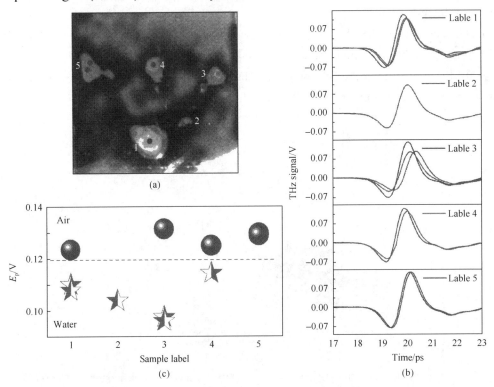

Figure 6 Qualitative analysis of water and gas in transparent regions of the sample: (a) The photograph identifying specific locations of each sample point in the transparent regions, which are denoted as Labels 1~5; (b) the THz time-domain spectra of the sampling points at the five distinct transparent regions; (c) the THz peak intensities of each transparent region for each sampling point with the color corresponding to the spectra in panel (b)

Different responses in the THz range were observed, depending on the position of the sample point in each transparent region. It was apparent that the THz signal intensities deviated in Labels 1, 3 and 4 more so than those in Labels 2 and 5. Figure 6(c) illustrates the E_p value of the THz time-domain spectra shown in Figure 6(b). For Labels 1 and 4, the E_p

values of the points measured in the marginal annular section were lower than that of the middle section. In contrast, the E_p values of the points located on the edge were relatively greater than those in the middle section of Label 3. For Label 5, there were no obvious differences between E_p values in the spectra. Since the size of Label 2 was small, only one point was measured, and the E_p = 104mV. The decrease of the peak intensities is due to the existence of water. These features are in good agreement with this fact and show that water and gas can be differentiated using THz spectroscopy. As for Labels 1~5, the sampling points are measured randomly and some of them are close to the measurement regions in Figures 4 and 5. The left location of Label 1 is taken as an example in this part, and the E_p = 0.11014V and 0.10544V when measured vertically in Figure 4. The results suggest that this technology has high repeatability.

The measurement of water content in high-water-content crude oil has been problematic for a long time. The ability of THz-TDS to characterize the water content for crude oil emulsions with low water content (0.01%~25.00%) has been shown previously. A linear relationship was found between the absorption coefficient or the amplitude of the THz pulse and the water content of crude oil emulsions in the observed range. This relationship was used to evaluate the water content in oil emulsions[26]. In this work, the potential of THz-TDS to accurately determine the water content in high-water-content crude oil and the distribution of oil, water, and gas has been shown. The hydrogen-bond collective network formed by water molecules changes on a picosecond (ps) timescale, thereby causing the THz spectrum to be sensitive to the fluctuations of the water dipole moments[35,36]. Because the main component of petroleum is hydrocarbons, which are mostly nonpolar molecules, the absorption of oil from THz radiation is relatively small. Thus, in the crude oil production process, because of the strong attenuation of THz radiation by water, THz radiation becomes a highly sensitive, noncontact probe of water content in high-water-content oil-water mixtures, as well as in the distinction between oil, water, and gas. The repeatability has been proved in this paper, thus the THz technique allows us to follow the online examination of water content for oil exploitation and should be regarded as a complementary method for qualitative and quantitative characterization, especially for the online determination of water content of crude oil in the oil industry.

4 Conclusions

In summary, the THz time-domain spectra of crude oil mixtures with water contents from 50.05% to 100.00% were measured. The experimental results indicate that both the amplitude and absorbance of the THz pulse can be used to characterize the water content in water-oil mixtures. The mixture with a water content of 50.05% was also measured to study the distribution of oil, water, and gas. The results demonstrated that water content in high-water-cut crude oil can be precisely characterized using linear fits. The distribution of oil, water, and air is directly predicted with THz signal intensities. The THz technique can not only provide an accurate method for quantitative analysis of water content in crude oil but also proves to be a promising technique for determining the distribution of oil, water, and gas.

Acknowledgements

This work was supported by the National Basic Research Program of China (Grant No. 2014CB744302), the Specially Funded Program on National Key Scientific Instruments and Equipment Development (Grant No. 2012YQ140005), and the National Nature Science Foundation of China (Grant Nos. 11574401 and 61405259).

References

[1] Seyyedi M, Sohrabi M, et al. Enhancing water imbibition rate and oil recovery by carbonated water in carbonate and sandstone rocks. Energy & Fuels, 2006, 30: 285-293.

[2] Olajire A A. Review of ASP EOR (alkaline surfactant polymer enhanced oil recovery) technology in the petroleum industry: Prospects and challenges. Energy, 2014, 77: 963-982.

[3] Borges G R, Farias G B, Braz T M, et al. Use of near infrared for evaluation of droplet size distribution and water content in water-in-crude oil emulsions in pressurized pipeline. Fuel, 2015, 147: 43-52.

[4] Kestens V, Conneely P, Berneuther A. Vaporisation coulometric Karl Fischer titration: A perfect tool for water content determination of difficult matrix reference materials. Food Chem., 2008, 106: 1554-1559.

[5] Fortuny M, Oliveira C B Z, Mole R L F V, et al. Effect of salinity, temperature, water content, and pH on the microwave demulsification of crude oil emulsions. Energy & Fuels, 2007, 21: 1358-1364.

[6] MacLeod S K. Moisture determination using Karl Fischer titrations. Anal. Chem., 1991, 63: 557A-566A.

[7] Tang T B, Lim Y L, Aslam M Z. Detecting trace amount of water in crude oil with capacitance sensors. IEEE Sensors, 2014: 2030-2033.

[8] Roshani G H, Feghhi S A H, Mahmoudi-Aznaveh A, et al. Precise volume fraction prediction in oil-water-gas multiphase flows by means of gamma-ray attenuation and artificial neural networks using one detector. Measurement, 2014, 51: 34-41.

[9] Lu Z Q, Yang X, Zhao K. Non-contact measurement of the water content in crude oil with all-optical detection. Energy & Fuels, 2015, 29: 2919-2922.

[10] Aslam M Z, Tong B T. A high resolution capacitive sensing system for the measurement of water content in crude oil. Sensors, 2014, 14: 11351-11361.

[11] Horiuchi N. Searching for terahertz waves. Nat. Photonics, 2010, 4: 662-662.

[12] Zhang Y S, Han Z. Spoof surface plasmon based planar antennas for the realization of Terahertz hotspots. Fuel, 2015, 5: 18606.

[13] Lourembam J, Srivastava A, Laovorakiat C, et al. New insights into the diverse electronic phases of a novel vanadium dioxide polymorph: A terahertz spectroscopy study. Sci. Rep., 2014, 5: 9182.

[14] Liu L, Pathak R, Cheng L J, et al. Real-time frequency-domain terahertz sensing and imaging of isopropyl alcohol-water mixtures on a microfluidic chip. Sens. Actuators B Chem., 2013, 184: 228-234.

[15] Ng B, Hanham S M, Giannini V, et al. Lattice resonances in antenna arrays for liquid sensing in the terahertz regime. Opt. Express, 2011, 19: 14653-14661.

[16] Smith A E, Hawkins B G, Kirby B J, et al. Microfluidic devices for terahertz spectroscopy of biomolecules. Opt. Express, 2008, 16: 1577-1582.

[17] Zhan H L, Wu S, Bao R, et al. Water adsorption dynamics in active carbon probed by terahertz spectroscopy. RSC. Adv., 2015, 5: 14389-14392.

[18] Al-Douseri F M, Chen Y, Zhang X C, et al. THz wave sensing for petroleum industrial applications. Int. J. Infrared Milli., 2006, 27: 481-503.

[19] Jin Y S, Kim G J, Shon C H, et al. Analysis of petroleum products and their mixtures by using terahertz time domain spectrocopy. J. Korean. Phys. Soc., 2008, 53: 1879-1885.

[20] Gorenflo S, Tauer U, Hinkov I, et al. Dielectric properties of oil-water complexes using terahertz transmission spectroscopy. Chem. Phys. Lett., 2006, 421: 494-498.

[21] Zhan H L, Wu S X, Bao R M, et al. Qualitative identification of crude oils from different oil fields using terahertz time-domain spectroscopy. Fuel, 2015, 143: 189-193.

[22] Feng X, Wu S X, Zhao K, et al. Pattern transitions of oil-water two-phase flow with low water content in rectangular horizontal pipes probed by Terahertz Spectrum. Opt Express, 2015, 23: 1693-1699.

[23] Zhan H L, Zhao K, Xiao L Z. Spectral characterization of the key parameters and elements in coal using terahertz spectroscopy. Energy, 2015, 93: 1140-1145.

[24] Ge L N, Zhan H L, Leng W X, et al. Optical characterization of the principal hydrocarbon components in natural gas using terahertz spectroscopy. Energy & Fuels, 2015, 29: 1622-1627.

[25] Leng W X, Zhan H L, Ge L N, et al. Rapidly determinating the principal components of natural gas distilled from shale with terahertz spectroscopy. Fuel, 2015, 159: 84-88.

[26] Jin W J, Zhao K, Yang C, et al. Experimental measurements of water content in crude oil emulsions by terahertz time-domain spectroscopy. Apply Geophys, 2013, 10: 506-509.

[27] Zheng X Y, Jin Y Q, Chi Y, et al. Simultaneous determination of water and oil in oil sludge by low-field ^{1}H NMR relaxometry and chemometrics. Energy & Fuels, 2013, 27: 5787-5792.

[28] Yu Z C, Li L, Liu K, et al. Petrological characterization and reactive transport simulation of a high-water-cut oil reservoir in the Southern Songliao Basin, Eastern China for CO_2 sequestration. Energy, 2015, 37: 191-212.

[29] Fingas M, Fieldhouse B. Studies on water-in-oil products from crude oils and petroleum products. Mar. Pollut. Bull., 2012, 64: 272-283.

[30] Kubacka T, Johnson J A, Hoffmann M C, et al. Large-amplitude spin dynamics driven by a THz pulse in resonance with an electromagnon. Science, 2014, 343: 1333-1336.

[31] Fakhru'l-Razi A, Pendashteh A, Abdullah L C, et al. Review of technologies for oil and gas produced water treatment. J. Hazard. Mater., 2009, 170: 530-551.

[32] Khoshnevis G N, Mailybaev A A, Marchesin D, et al. Effects of water on light oil recovery by air injection. Fuel, 2014, 137: 200-210.

[33] Zhong D L, Ding K, Lu Y Y, et al. Methane recovery from coal mine gas using hydrate formation in water-in-oil emulsions. Appl. Energ., 2016, 162: 1619-1626.

[34] Souza V B, Mansur C R E. Oil/Water nanoemulsions: Activity at the water-oil interface and evaluation on asphaltene aggregates. Energy & Fuels, 2015, 29: 7855-7865.

[35] Castro-Camus E, Palomar M, Covarrubias A A, et al. Leaf water dynamics of Arabidopsis thaliana monitored in-vivo using terahertz time-domain spectroscopy. Sci Rep, 2013, 3: 2910.

[36] Choi D H, Son H, Jung S, et al. Dielectric relaxation change of water upon phase transition of a lipid bilayer probed by terahertz time domain spectroscopy. J. Chem. Phys., 2012, 137: 175101.

Non-contacting characterization of oil-gas interface with terahertz wave

Dandan Wang   Xinyang Miao   Honglei Zhan   Jin Wang   Kun Zhao

(Beijing Key Laboratory of Optical Detection Technology for Oil and Gas,
China University of Petroleum, Beijing 102249, China)

The interface of oil and gas has been a key issue in the process of oil and gas storage and transportation, which is a preparatory work to improve the transport efficiency and prevent oil and gas leaks. There are many methods to characterize the interface of oil and gas, such as the ultrasonic transducers and multi-electrode capacitance level sensors. Recently, the application of terahertz (THz) technique has attracted much attention in the characterization and evaluation of oil and gas[1-7], such as a magnetic-field-induced disaggregation of the size of suspended colloidal particles in crude oil and pattern transitions of oil-water two-phase flow[6,7]. THz time-domain spectroscopy (THz-TDS) was expected to detect the interface of oil and gas since oil and gas can be characterized with THz wave. In this letter, a drop of crude oil with its volatile gas from an oilfield, which is inserted in the middle of a resin craft, was selected to probe the interface between oil and gas. The test size was 26mm×24mm×28mm where the sample was illuminated with THz wave through the opaque area.

As shown in Figure 1(a) a typical THz-TDS setup with transmission geometry was used to scan the craft from top to bottom and left to right with a step interval of 2mm[8]. Each spectrum was measured three times to improve the repeatability and stability. In this research, about 160 groups of data were got through scanning point by point. The THz field amplitude as a function of time was shown in Figure 1(b) for several representative data above the oil-gas interface, at the interface and below the interface, respectively. In our case, the resin, crude oil and volatile gas have different refractive indexes n_r, n_o and n_g. Since $n_r > n_o > n_g$ the propagation velocity of THz wave is the fastest in volatile gas while the slowest in resin. Thus three peaks E_r, E_o and E_g in THz-TDS were founded at 75.7ps, 63.3ps and 36.7ps corresponding to resin, crude oil and volatile gas, respectively. When the THz beam was closed to the interface two peaks of E_o and E_g appeared together due to a 2mm diameter of THz beam in our setup. With THz beam scanning through the interface into crude oil, E_g decreased gradually and disappeared at last. However, E_o was not gradually increased, but eventually disappeared. There are a variety of factors responsible for this phenomenon, such as high water content of the crude oil since water has strong absorption in THz range. Furthermore, the THz frequency domain spectra (THz-FDS) were calculated in Figure 1(c) after the application of fast Fourier transform (FFT), which strongly depended on the illuminated place. Due to the strong absorption of oil, the effective frequency range was reduced to 0.02~0.5THz. Specially the oscillation of THz power was observed at the vicinity of oil and gas interface which was caused by multiple peaks in the THz-TDS in Figure 1(b). Thus the interface of oil and gas can be probed with oscillated THz-FDS.

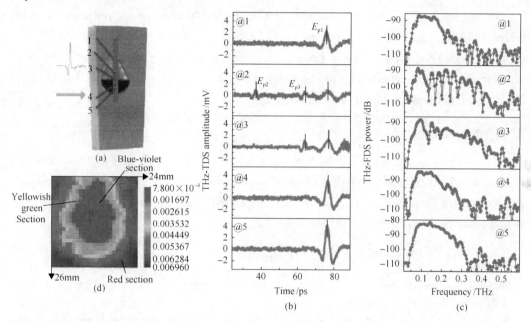

Figure 1 (a) Schematic diagram of the experimental setup. The test size was 26mm×24mm×28mm where the sample was illuminated with THz wave through the opaque area. Different positions of the samples were numbered 1, 2, 3, 4 and 5 which represent volatile gas, oil and gas interface, and oil, respectively; (b) THz-TDS of samples for different positions. Three peaks of E_r, E_o and E_g correspond to resin, crude oil and volatile gas, respectively; (c) THz-FDS of samples for different positions; (d) THz peak amplitude images of the sample by recording the resin peak amplitude of each point. Red represents the maximum of peak amplitude, and purple represents the minimum of peak amplitude

In addition, the THz image was obtained by recording the amplitude of the peak E_r as shown in Figure 1(d), and the level of the intensity was represented by various colors gradually from dark red to yellowish green then blue-violet with the decrease of THz peak amplitude. Since the E_r amplitude covered the information of crude oil and volatile gas, it can be concluded that the red, yellowish green and blue-violet sections represent the resin, transition layer and oil or gas, respectively. Here E_r was selected for THz imaging, so the contour of the resin can be identified clearly, which was coincided with the test sample.

The non-contact measurement has its unique advantages in oil and gas optics engineering. Some technologies are evolving quickly for oil and gas characterization, such as laser ultrasonic detection and laser induced voltage[9,10]. In addition to the above methods, the interface of oil and gas can be now characterized by the THz wave due to its better penetration, partial resolution and less scattering characteristic. Considering its feasibility, there are lots of issues that need to be studied further, such as the precision of real-time detection, the stability of the device, etc. Although the current study is preliminary and is not widely used now, the non-contact THz technology is believed to have great potential for application in oil and gas characterization.

In this letter, THz amplitude images were performed to characterize the contour of the sample, and the interface of oil and gas could be clearly indicated by the THz spectra. The result proved THz wave to be a promising means in detecting oil and gas properties. Furthermore, the non-contact THz testing method may provide new ideas for the key issues in the storage and transportation process of oil and gas.

Acknowledgements

This work was supported by the National Key Basic Research Program of China (Grant No. 2014CB744302), the Specially Funded Program on National Key Scientific Instruments and Equipment Development (Grant No. 2012YQ140005) and the National Natural Science Foundation of China (Grant No. 11574401).

References

[1] Al-Douseri F M, Chen Y, Zhang X C. THz wave sensing for petroleum industrial applications. International Journal of Infrared & Millimeter Waves, 2006, 27(4): 481-503.

[2] Bao R M, Li Y Z, Zhan H L, et al. Probing the oil content in oil shale with terahertz spectroscopy. Science China: Physics, Mechanics & Astronomy, 2015, 58(11): 114211.

[3] Zhan H, Wu S, Bao R, et al. Qualitative identification of crude oils from different oil fields using terahertz time-domain spectroscopy. Fuel, 2015, 143: 189-193.

[4] Qin F L, Qian L, Zhan H L, et al. Probing the sulfur content in gasoline quantitatively with terahertz time-domain spectroscopy. Science China: Physics, Mechanics & Astronomy, 2014, 57(7): 1404-1406.

[5] Ge L N, Zhan H L, Leng W X, et al. Optical characterization of the principal hydrocarbon components in natural gas using terahertz spectroscopy.Energy & Fuels, 2015, 29(3): 1622-1627.

[6] Feng Xin, Wu S X, Zhao K, et al. Pattern transitions of oil-water two-phase flow with low water content in rectangular horizontal pipes probed by terahertz spectrum. Optics Express, 2015, 23(24): A1693-A1699.

[7] Jiang C, Zhao K, Zhao L J, et al. Probing disaggregation of crude oil in a magnetic field with terahertz time-domain spectroscopy. Energy & Fuels, 2014, 28(1): 483-487.

[8] Zhan H L, Sun S N, Zhao K, et al. Less than 6 GHz resolution THz spectroscopy of water vapor. Science China: Technological Sciences, 2015, 58(12): 2104-2109.

[9] Lu Z Q, Yang X, Zhao K, et al. Non-contact measurement of the water content in crude oil with all-optical detection. Energy & Fuels, 2015, 29(5): 2919-2922.

[10] Lu Z Q, Sun Q, Zhao K, et al. Laser-induced voltage of oil shale for characterizing the oil yield. Energy & Fuels, 2015, 29(8): 4936-4940.

Terahertz dependent evaluation of water content in high-water-cut crude oil using additive-manufactured samplers

Limei Guan Honglei Zhan Xinyang Miao Jing Zhu Kun Zhao

(Beijing Key Laboratory of Optical Detection Technology for Oil and Gas,
China University of Petroleum, Beijing 102249, China)

Abstract: The evaluation of water content in crude oil is of significance to petroleum exploration and transportation. Terahertz (THz) waves are sensitive to fluctuations in the dipole moment of water. However, due to the strong absorption of water in the THz range, it is difficult for the THz spectrum to determine high water content with the common sampler. In this research, micron-grade samplers for THz detection were designed and manufactured using additive manufacturing (AM) technology. Oil-water mixtures with water content from 1.8% to 90.6% were measured with the terahertz time-domain spectroscopy (THz-TDS) system using sample cells. In addition, a detailed analysis was performed of the relationships among THz parameters such as signal peak, time delay, and refractive index as well as absorption coefficient and high water content (>60%). Results suggest that the combination of THz spectroscopy and AM technique is effective for water content evaluation in crude oil and can be further applied in the petroleum industry.

Keywords: water content; crude oil; additive manufacturing; terahertz spectroscopy

1 Introduction

Water content in crude oil is closely related to the prediction of the water level in an oil well, the evaluation of oil reservoir exploitation, and the recovery and development of mining schemes[1]. Therefore, the accurate measurement of water content is significant not only for forecasting the exploiting capacity of an oil field before mining and estimation of working status, but also for grading and evaluating oil products. In addition, the existence of water in crude oil may lead to serious security problems such as pipeline corrosion during the petroleum refining process. There are many methods to detect the water content in crude oil. At present, the most common laboratory methods are centrifugation, distillation, and electrical dewatering[2]. Other approaches such as ray casting, and shortwave, microwave, and capacitance methods, are commonly used for online measurements of water content in crude oil[3,4].

In order to optimize production and confirm integrity, simple and secure methods have been proposed for detecting the water content in crude oil. THz radiation is a newly developed, non-contact, safe method that can be absorbed by organics and water molecules to varying degrees[5-8]. Because water molecules form labile hydrogen-bond collective networks which are constantly changed on a picosecond time scale[9], the THz spectrum is sensitive to the fluctuations of water dipole moments due to the constant breaking and forming of hydrogen bonds[10]. However, main components of crude oil, such as alkane, have very weak polarity; as a result, THz waves are more weakly absorbed than water. Accordingly, THz radiation is very sensitive to the change of water content in crude oil. THz spectroscopy technique has been used to detect leaf water content due to the sensitivity of THz to detect slight changes in leaf-tissue water content produced by the closing and opening of stomata[11]. THz spectroscopy is a nondestructive and non-contact method that allows continuous monitoring [9]; thus it is a highly sensitive probe of water content.

Until recently, the use of THz spectroscopy for water content detection has been limited to measuring samples with relatively low water content, which sometimes can be made into thin slices (as with coal and paper) to reduce the amount of water[12-15]. Our previous report demonstrated a linear relation between absorption coefficient α and water content (0.01%~25%) in emulsions by using samplers with the thickness of 1mm[16]. We used THz-TDS to characterize 5 oil-water mixtures with water content ranging from 50.05% to 100%; the samples were produced by dropping the mixtures between two pieces of plastic plate. That research was mainly focused on the distribution of oil, gas, and water in high-water-content crude oil mixtures. Measurement accuracy can be improved with smaller measurement intervals and more samples contained by uniform thickness samplers[17].

Precise evaluation of crude oil with water content >60%, known as high-water-cut oil, is affected by the strong absorption of water in the THz range. Significant signal attenuation leads to a decrease in signal-to-noise ratio (SNR), which makes it difficult to detect the THz signal of oil that contains a large amount of water. The effective solution to this problem is to reduce the sample-pool size to micron level. Rapid prototyping technology is a promising choice for the fabrication of sample cells with high precision and small size. Additive manufacturing (AM), generally known as 3D printing, a rapid prototyping technology of direct digital manufacturing, possesses the advantages of nanometer accuracy and cost efficiency[18-21]. AM is based on the data of a 3D model to rapidly solidify material and fabricatea physical object. AM has been used in the energy industry to print test specimens of rock crevices and joint structures to investigate the stress changes of rocks during mining[22,23]. AM was employed to manufacture models of underground gas channels to guide the development of natural gas fields and to render a print porosity model to characterize pore connectivity before hydraulic exploitation[24]. Because samplers with mm-grade thickness have difficulty measuring samples with water proportions >25% using THz spectroscopy, in the current study we measured the water content in oil-water mixtures ranging from 1.8%~90.6% with specially designed samplers for high-water-content oil that were manufactured with a 3D-printing system. Thus the water content of oil-water mixtures with high water content was quantitatively characterized by the combination of THz spectroscopy and 3D-printing technology.

2 Sample preparation and experiment

The AM processes start from a software model to fully describe the external geometry (Figure 1). In order to obtain the THz spectra of crude oil with high water content, a 3D-printed cell was used. In our previous work[16], we tested some samples with water content >25% using 1mm cells, but did not obtain an effective THz signal because of the strong absorption due to the excessive amount of water. Therefore, to improve the SNR, in the current work we designated the thickness of the measured region as 500μm and the wall as 1000μm thick. Two holes were placed at the top of the digital cell model for injecting liquid sample and for releasing exhaust gases. The digital models of the cells was designed with Pro-Engineering and then transported into STL (stereolithography) file format (most AM machines accept the STL file format, which has become the standard[25]). The STL file describes the external closed surfaces of the original 3D digital model and forms the basis for calculation of the slices, which can then be successfully read by a 3D printer[26].

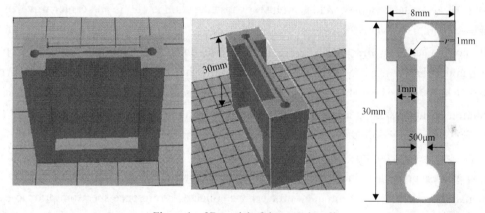

Figure 1 3D model of the sample cell

Next, the digital model was processed by 3D printer control software into sectional images containing information about the section outline and the filling track. Appropriate printing speed and layer thickness are necessary to the AM system for the following reasons: (i) Using higher speed may lead to problems in material solidification and be harmful to the quality of printed product; and (ii) applying a thinner layer can reduce physical roughness and reduce system error. Synthetically considering these factors, building parameters such as material constraints, printing speed, and layer thickness were properly set up prior to the printing process; these were acronitrile-butadiene-styrene (ABS), low speed, and 0.01cm, respectively. The sample-pool material was heated, melted, and then extruded from the 3D printer's two nozzles for direct printing. The printer we used (XYZ printing da Vinci 2.0, resolution of 100 μm) uses fused deposition modeling technology (FDM), a technique for curing polymer during the 3D printing process. ABS wire rod, a kind of thermoplastic material commonly employed in FDM, was used as the raw material for the sample cells. All of the 27 samples we employed were 3D-printed using the same design model, the same 3D-printing device, and the same printing parameters, in order to shorten the product development cycle and reduce production costs. Analysis of the test area thickness showed that the average relative error was less than 5% and the R_a of the surface of the sample cells was 0.1mm.

For the present research, crude oil with a water content of 1.8% was obtained from Changqing oilfield, China. The oil-water samples were prepared by blending the crude oil with additional water. Water concentrations of various percentages were calculated w/w (ratio of water mass to mixture mass). To confirm the uniform distribution of oil and water, the mixtures were stirred for 5m, immediately injected into the sample cells through the two circular holes, and tested using THz-TDS system. The time period from completed sample to test completion was <1m.

The THz-TDS system contains a diode-pump mode-locked laser with a repetition of 80MHz; it generates femtosecond laser pulses whose duration is 100 fs and center wavelength equals 800nm. The average power of the input laser is <150mW. The laser pulse is initially split into two beams by the polarization beam splitter to obtain the pump pulse and probe beam. The pump beam is used to generate THz radiation using photoconductive antennas. After being focalized by a lens, the collimated THz pulse transmits into the detection system. The resolution of this system is less than 6GHz, as reported in one of our previous articles[27]. The THz-TDS of each mixture was obtained by this THz transmission system.

In fact, systematic error derived from the instrument and random errors originating in the testing process are unavoidable in the process of an experiment such as this. To minimize the systematic errors in this measurement process, we collected a reference spectrum prior to each sample filling of the cell. The oil-water mixtures were investigated using the corresponding numbered cells and every sample was tested twice at different locations. Moreover, the experiment was conducted under a condition of (21±0.2)℃ in dry nitrogen, with humidity <2% to avoid interference by vapor in the air.

3 Results and discussion

Oil-water samples were measured with a THz system. The THz-TDS of the mixtures with varying water cuts show obvious differences (results for selected mixtures are shown in Figure 2, and selected signals with high-water-content samples are shown in the insert of Figure 2). The effective THz waveform of THz signal appears between 18.5ps and 25.5ps. The signal amplitude (E_P) of sample with 1.8% water content reached 88.23mV, whereas the E_P of crude oil with 90.6% water content is only 23.50mV. As a whole, the E_P dwindles with the increasing of water content, mainly because water molecules absorb significantly more than crude oil in the THz range. Meanwhile, a phase shift occurs with the augmentation of water content.

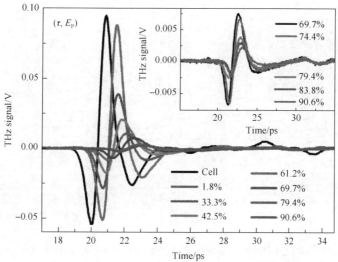

Figure 2　THz-TDS of oil-water mixtures with different water contents

The THz-FDS of different water contents were calculated from THz-TDS by fast Fourier transform (FFT) on the basis of

$$\omega = 2\pi f \tag{1}$$

The ratio of the sample signal to the empty sample cell signal can be written as

$$\frac{A_s(\omega)}{A_c(\omega)} = \rho(\omega) e^{-j\varphi(\omega)} \tag{2}$$

where f represents frequency; $\rho(\omega)$ is the module of the ratio; and $\varphi(\omega)$ is the phase of the ratio, which can be calculated by THz-FDS. The refractive index n and absorption coefficient α can be calculated by the following formula:

$$n_s(\omega) = \varphi(\omega) \frac{c_0}{\omega d} + 1 \tag{3}$$

$$\alpha_s(\omega) = \frac{d}{2} \ln \frac{4n_s(\omega)}{\rho(\omega)[n_s(\omega)+1]^2} \tag{4}$$

where c_0 is the velocity of light in vacuum, and d represents the optical path of the sample cell[28]. The frequency-dependent refractive index and absorption coefficient of the oil-water mixtures are shown schematically in Figures 3(a) and (b), respectively. Due to the strong absorption of oil-water mixture, we reduced the effective frequency range to 0.65THz. We observed that $α$ changed little with the frequency and there is no obvious absorption peak, which is in agreement with the previous report[29]. Here, it is shown that samples with higher water content have greater refractive indices and absorption coefficients.

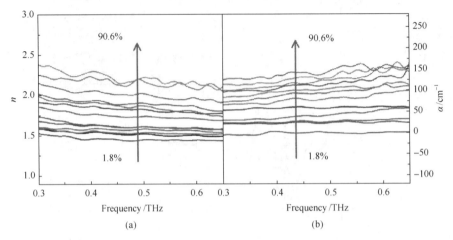

Figure 3 (a) Refractive index (n) of different oil-water mixtures; (b) absorption coefficient ($α$) of different oil-water mixtures

According to the results above, we can see that there is a relationship between the THz parameter and the water content. To accurately analyze the relationship between them, we separately analyzed the E_P and time delay ($τ$, the respective transverse coordinate of E_P). Figure 4(a) shows the scatter distribution of $τ$ as well as the water content and Figure 4(b) shows the relationship between E_P and water content. According to the tendency shown in these collected samples, an exponential relationship can be built between $τ$ and water content as well as E_P and water content. E_P decreases with the increase of water content; $τ$ increases with the addition of water. To ensure that the calculations in these error bars did come from multiple measurements, the signal peak and time delay shown in the Figure 4(a) and (b) were the average of the two measurements.

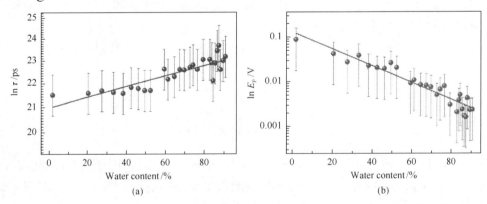

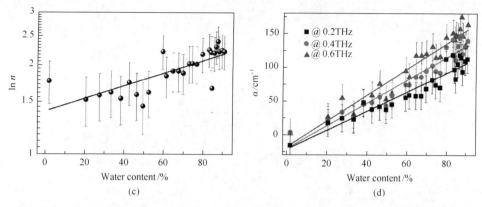

Figure 4 (a) lnτ of the oil-water mixtures with water content ranging from 1.8% to 90.6%; (b) lnE_P of the oil-water mixtures with water content ranging from 1.8% to 90.6%; (c) ln$n$ of different water-content oil-water mixtures; (d) α of different water-content of oil-water mixtures at 0.2THz, 0.4THz, and 0.6THz

We further analyzed the changes of refractive index and absorption as water content according to the refractive index and absorption coefficient of the oil-water mixtures. The results, in Figure 4(c), exhibit exponential damping as the water content in the emulsion increases. We also extracted the absorbance data at selected frequencies. Figure 4(d) shows the water-content-dependent absorption coefficient α of oil-water mixtures at 0.2~0.6THz with the interval of 0.2THz. Positive correlation relationships can be built in terms of the absorbance at the selected frequencies. The slopes increase with the augmenting of frequency.

In actual production, the water cut of crude oil in most oilfields, especially in China, has exceeded 60%. For example, the Daqing oilfield has entered the late stage of high-water-cut and the water content is more than 80%[30]. The water cut of crude oil in the Changqing oilfield has already gone over 60%, as it has in the Shengli and Liaohe oilfields[31]. We investigated the relationship of THz response and water content in crude oil with >60% water concentration to simulate the actual conditions in present-day oilfields. A total of 18 samples with water content >60% were analyzed. As shown in Figure 5(a), the E_P of the 3D-printed cells (E_P-cell) were extracted and equal ~96mV. The relative error is <5% calculated by the equation $e = (E_{MAX} - E_{AVE})/E_{AVE}$, where E_{MAX} and E_{AVE} represent the maximum value and the average of E_P, respectively, whose errors fall in an acceptable range. In order to improve the accuracy of the analysis and reduce system error, a special THz parameter, the ratio of signal peak, was used as the characterization parameter Figure 5(b). Water content was the abscissa; the ordinate was the ratio of E_P of the empty sample cells (E_P-cell) and the oil-water mixtures (E_P-sam). The stars with error bars represent the average E_P ratio of the two measurements. Results show that the E_P ratio decreases with the increasing of water content.

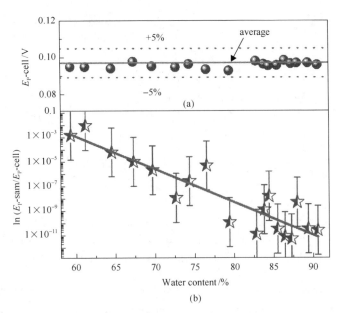

Figure 5 Analysis of the mixtures with high water content. (a) E_P of the signal of 3D-printed sample cells; (b) the ratio that takes into account E_P of the empty sample cells (E_P-cell) and the mixtures (E_P-sam)

In the process of exploitation and management of oilfields, water content in high-water-cut crude oil is an important technical indicator. THz technique is an effective way to estimate the water content. THz technology can be used to detect high water content due to the vastly different transmission properties of water and crude oil in the THz range. In fact, THz-TDS has been evaluated for discriminating among water contamination levels (0%, 0.1%, and 0.2%) in diesel engine oil. In addition, pure water has been studied using different techniques to observe its properties at THz frequencies. The refractive index values equaled >2 and absorption coefficients were found to be >150cm^{-1} at 1THz[32,33]. In addition, THz ATR spectroscopy was used to monitor highly absorbent liquids with water content <40%[34].

The precise measurement of high-water-cut oil, however, is affected by the strong absorption of water in the THz range. In order to solve this problem, all of the standard sample pools in this work were designed and customized with 3D printing technology. The results indicate our initial estimation that the combination of THz spectroscopy and AM technique can be applied in oil characterization, even when the oilfield enters the high-water-bearing period. The added value of our research is the combination of terahertz spectroscopy and AM technology, which has not previously been reported. This combination could be a promising selection for measuring water content of fluid with good accuracy, which in turn could open up a new area of research.

4 Conclusions

We investigated the water content in crude oil by using THz spectroscopy combined with an AM technique, and were able to produce an improved sample cell manufactured by 3D printing that resolved the problem of water content in high-water-cut samples of a common thickness for centimeter level samplers, which had been generally hard to distinguish in the THz region. Results suggest that THz parameters, including E_P, τ, n, and α, can be used to evaluate water content and that this method is accurate for quantitative analysis of water content in crude oil, even high-water-cut samples. This research represents a new attempt to apply AM technology in the oil exploration and development field. The combination of THz spectroscopy and AM technology seems to be a promising choice for the determination of water content, and the combination of the two technologies will promote innovative applications in oil and gas engineering.

Acknowledgements

This work was partially funded by the National Basic Research Program of China (Grant No. 2014CB744302), the Specially Funded Program on National Key Scientific Instruments and Equipment Development (Grant No. 2012YQ140005), the China Petroleum and Chemical Industry Association Science and Technology Guidance Program (Grant No. 2016-01-07) and the National Nature Science Foundation of China (Grant No. 11574401).

References

[1] Zhao Z, Wang Y. Application of neural network in the measuring system on-line for water content of crude oil. Advance Journal of Food Science & Technology, 2013, 5(3): 276-279.

[2] Fingas M, Fieldhouse B. Studies on water-in-oil products from crude oils and petroleum products. Marine Pollution Bulletin, 2012, 64(2): 272-283.

[3] Bo O L, Nyfors E. Application of microwave spectroscopy for the detection of water fraction and water salinity in water/oil/gas pipe flow. Journal of Non-Crystalline Solids, 2002, 305(1-3): 345-353.

[4] Zubair M, Tang T. A high resolution capacitive sensing system for the measurement of water content in crude oil. Sensors, 2014, 14(7): 11351-11361.

[5] Mittleman D M. Frontiers in terahertz sources and plasmonics. Nature Photonics, 2013, 7(9): 666-669.

[6] Zhan H L, Zhao K, Xiao L Z. Non-contacting characterization of $PM_{2.5}$ in dusty environment with THz-TDS. Science China: Physics, Mechanics & Astronomy, 2016, 59(4): 644201.

[7] Burnett A D, Fan W, Upadhya P C, et al. Broadband terahertz time-domain spectroscopy of drugs-of-abuse and the use of principal component analysis. Analyst, 2009, 134(8): 1658-1668.

[8] Fecko C J. Ultrafast hydrogen-bond dynamics in the infrared spectroscopy of water. Science, 2003, 301(5640): 1698-1702.

[9] Yu B L, Yang Y, Zeng F, et al. Reorientation of the H_2O cage studied by terahertz time-domain spectroscopy. Applied Physics Letters, 2005, 86(6): 72.

[10] Leitner D M, Gruebele M, Havenith M. Solvation dynamics of biomolecules: Modeling and terahertz experiments. HFSP Journal, 2008, 2(6): 314-323.

[11] Castrocamus E, Palomar M, Covarrubias A A. Leaf water dynamics of Arabidopsis thaliana monitored in-vivo using terahertz time-domain spectroscopy. Scientific Reports, 2013, 3(10): 2910.

[12] Hor Y L, Federici J F, Wample R L. Nondestructive evaluation of cork enclosures using terahertz/millimeter wave spectroscopy and imaging. Applied Optics, 2008, 47(1): 72-78.

[13] Gente R, Born N, Voß N, et al. Determination of leaf water content from terahertz time-domain spectroscopic data. Journal of Infrared Millimeter & Terahertz Waves, 2013, 34(3-4): 316-323.

[14] Vassilev V, Stoew B, Blomgren J, et al. A mm-wave sensor for remote measurement of moisture in thin paper layers. IEEE Transactions on Terahertz Science and Technology, 2015, 5(5): 770-778.

[15] Banerjee D, Spiegel W V, Thomson M D, et al. Diagnosing water content in paper by terahertz radiation. Optics Express, 2008, 16(12): 9060-9066.

[16] Jin W J, Zhao K, Yang C, et al. Experimental measurements of water content in crude oil emulsions by terahertz time-domain spectroscopy. Applied Geophysics, 2014, 10(4): 506-509.

[17] Song Y, Bao R M, Lu Z Q, et al. Simultaneous characterization of water content and distribution in high-water-cut crude oil. Energy & Fuels, 2016, 30(5): 3929-3933.

[18] Pires D, Hedrick J L, De Silva A, et al. Nanoscale three-dimensional patterning of molecular resists by scanning probes. Science, 2010, 328(5979): 732-735.

[19] Onses M S, Song C, Williamson L, et al. Hierarchical patterns of three-dimensional block-copolymer films formed by electrohydrodynamic jet printing and self-assembly. Nature Nanotechnology, 2013, 8(9):667-675.

[20] Sun K, Wei T S, Ahn B Y, et al. 3D printing of interdigitated Li-ion microbattery architectures. Advanced Materials, 2013, 25(33):4539-4543.

[21] Fischer A C, Belova L M, Rikers Y G M, et al. 3D free-form patterning of silicon by ion implantation, silicon deposition, and selective silicon etching. Advanced Functional Materials, 2012, 22: 4004-4008.

[22] Ju Y, Xie H, Zheng Z, et al. Visualization of the complex structure and stress field inside rock by means of 3D printing technology. Chinese Science Bulletin, 2014, 59(36): 5354-5365.

[23] Jiang Q, Feng X T Feng, Gong Y H, et al. Reverse modelling of natural rock joints using 3D scanning and 3D printing. Computers and Geotechnics, 2016, 73: 210-220.

[24] Ishutov S, Hasiuk F J, Harding C, et al. 3D printing sandstone porosity models. Interpretation, 2015,3(3):SX49-SX61.

[25] Brown A C, De Beer D, Conradie P. Development of a stereolithography (STL) input and computer numerical control (CNC) output algorithm for an entry-level 3-D printer. South African Journal of Industrial Engineering, 2014, 25(2): 39-47.

[26] Wu M, Tinschert J, Augthun M, et al.Application of laser measuring, numerical simulation and rapid prototyping to titanium dental castings. Dental Materials Official Publication of the Academy of Dental Materials, 2001, 17(2): 102-108.

[27] Zhan H L, Wu S X, Bao R M, et al. Qualitative identification of crude oils from different oil fields using terahertz time-domain spectroscopy. Fuel, 2015, 143: 189-193.

[28] Leng W X, Ge L N, Xu S S, et al. Pressure-dependent terahertz optical characterization of heptafluoropropane. Chinese Physics B, 2014, 23(10): 107804.

[29] Luong T Q, Verma P K, Mitra R K, et al. Do hydration dynamics follow the structural perturbation during thermal denaturation of a protein: A terahertz absorption study. Biophysical Journal, 2011, 101(4): 925-933.

[30] Gao D, Ye J, Liu Y, et al. Coupled numerical simulation of multi-layer reservoir developed by lean-stratified water injection. Journal of Petroleum Exploration and Production Technology, 2016, 6(4): 719-727.

[31] Shangguan Y, Zhang Y, Xiong W. The effect of physical property change on the water flooding development in Changqing oilfield Jurassic low permeability reservoir. Petroleum, 2015, 1(4): 300-306.

[32] Abdul-Munaim A M, Reuter M, Abdulmunem O M, et al. Using terahertz time-domain spectroscopy to discriminate among water contamination levels in diesel engine oil. Transactions of the ASABE. 2016, 59(3): 795-801.

[33] Wang T, Klarskov P, Jepsen P U. Ultrabroadband THz time-domain spectroscopy of a free-flowing water film. IEEE Transactions on Terahertz Science and Technology, 2014, 4(4): 425-431.

[34] Soltani A, Busch S, Plew P, et al. THz ATR spectroscopy for inline monitoring of highly absorbing liquids. Journal of Infrared Millimeter & Terahertz Waves, 2016, 37(10): 1001-1006.

Non-contact measurement of the water content in crude oil with all-optical detection

Zhiqing Lu[1,3] Xiao Yang[1,2] Kun Zhao[1,2] Jianxin Wei[1]
Wujun Jin[3] Chen Jiang[3] Lijun Zhao[3]

(1.State Key Laboratory of Petroleum Resources and Prospecting, China University of Petroleum, Beijing 102249, China; 2.Beijing Key Laboratory of Optical Detection Technology for Oil and Gas, China University of Petroleum, Beijing 102249, China; 3.Key Laboratory of Oil and Gas Terahertz Spectroscopy and Photoelectric Detection, China Petroleum and Chemical Industry, Beijing 100723, China)

Abstract: Water content in crude oil is very important in an oilfield production logging system. A non-contact optical method, called all-optical detection, was proposed for measuring the water content in the full range of 0~100% in crude oil by a non-contact laser source and receiver (adaptive laser interferometer). The laser-induced ultrasonic P-wave velocities of three series of an oil-water mixture accelerated significantly from 1289m/s, 1342m/s, and 1343m/s, respectively, to 1458m/s with the increase of the water content from 0% to 100%. The theoretical values of the effective medium model were agreement with the experimental results of three series of an oil-water mixture.

1 Introduction

The water content ratio of the crude oil in the oil tank, oil well, and oil pipeline is a vital parameter to accurately measure the crude oil production, which has a direct impact on the mining, dehydration, transportation, measurement, marketing, and refining of the crude oil[1,2]. During the process of exploitation, storage, and transportation of an oilfield, highly precise measuring of the water content in crude oil can optimize production parameters and improve the oil recovery rate[3,4]. In addition, crude oil is always treated using an electrostatic extraction technique to remove the water[5,6], and it is desired to monitor continuously the relative concentrations of the oil and water of the treated mixture to ensure that the water content remains below a certain redetermined level.

It is very difficult to measure accurately the water content of the crude oil because of many factors, including the complex ingredient as alkanes, hydrocarbons, and sulfur in crude oil, multiphase mixed fluid pattern in pipeline transport, and poor oil field environment with strong corrosion, scaling, and wax[7-10]. A variety of measurement methods have been put forward to probe the water content in crude oil[11-17]. Now, the measurements of the water content of crude oil can be divided as offline and online measuring methods. The basic principle of offline analysis is to separate the moisture from crude oil and calculate the water mass fraction by volume ratio or density ratio, such as the electric dehydration method, distillation, and Karl Fischer titration method. The traditional electric dehydration method can measure the water content generally lower than 30% with simple operation, while its accuracy is affected by the temperature, demulsifier, and properties of crude oil, such as viscosity and volatile and wax contents. Distillation measuring needs about 2h to complete a sample analysis, including sampling, heating, and analysis, which is not suitable for online measurement. A Karl Fischer titration method is fast and precise for the trace water cut oil between 1 ppm and 5% according to Mettler Toledo.

Online measurement becomes the main development trend with the increase of automation in the process of oilfield production, including conductivity, density method, electromagnetic method, capacitance method, ray method, etc. The capacitance method is one of the most mature measurement methods, which uses the quite difference in the dielectric constant of oil and water to determine the water content. However, the capacitance probe is generally suitable for a constant temperature and low water content less than 10%. The short wave method based on the index law between the incident intensity of the electromagnetic wave and water cut is not affected by the temperature and pressure but quite complicated for its several different test parameters.

Here, we propose a way to measure the full-range water content up to 100% by acquiring laser ultrasonic velocity data. The method uses a non-contacting laser source and receiver (adaptive laser interferometer), which is called all-optical detection[18,19]. When a short-pulse laser irradiated on an optically absorbing surface, part of that energy is absorbed and converted into heat, resulting in localized thermal expansion, which, in turn, generates elastic waves in the ultrasonic range. Then, a second laser is coupled to an optical interferometer for the remote detection of the ultrasonic surface motion. The output from this interferometer is sent to the oscilloscope and is also triggered by the short-pulse laser[20-22]. Thus, we can then determine the time that the ultrasound waves take to pass through the sample by measuring when the first significant peak appears on the interferometer[23-26].

2 Experimental apparatus

Longitudinal wave (P-wave), one important branch in bulk waves, is the first significant peak and has often been used for characterizing the thickness, density, and elasticity of the material[27-31]. In our case, as shown in Figure 1(a), a high-energy pulsed Tolar-1053 laser (1053nm wavelength and 20mJ/pulse) is to excite ultrasonic waves via thermoelastic expansion and a He-Ne laser is used as an indicator light. Here, we partially focus the laser source beam of ~6mm in radius on the crude oil sample. The resulting waves were measured with a laser ultrasonic receiver. The adaptive laser interferometer is a specifically interference receiver, and a constant wave (CW) beam at a wavelength of 532nm is generated. The receiver uses two-wave mixing in a photorefractive crystal to deliver the displacement of the sample surface. This receiver measures the in-plane and out-of-plane displacement simultaneously and has a high sensitivity with a bandwidth from 10kHz to 10MHz.

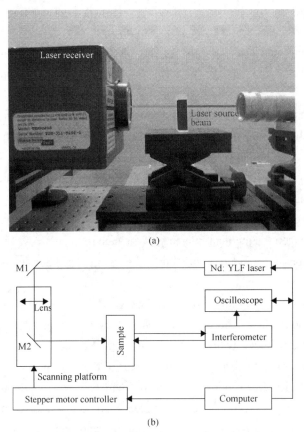

Figure 1 (a) Picture of the experimental setup and (b) schematic diagram of the ultrasonic detection

Figure 1(b) is the schematic diagram of the ultrasonic detection. Laser pulses irradiate to the surface of the sample on one side through the mirrors M1, lens, and mirrors M2. On the other side of the sample, a laser interferometer detects the ultrasonic signal, which is recorded by a digital oscilloscope and input into a computer for further processing. The merits of the all-optical detection system are the automatic control of laser output and the non-contact automatic rapid acquisition of the ultrasonic signal.

The experimental samples were prepared with anhydrous oils. Sample A was from the Liaohe oil field, and samples B and C were from two different areas in the Dagang oil field. Different water contents of crude oils A, B, and C from 0% to 100% in weight was prepared and put inside a 1mm thick quartz vessel of 10mm×20mm×40mm, which could keep for 30min without separation. All of the water contents of samples were verified by distillation measuring. Finally, the sample was mounted in a stage for laser ultrasonic detection, and the source and receiver beams are aligned on antipodes for a transmission experiment. Because the bottle is transparent quartz and the crude oil samples are dark materials, an aluminum foil has been applied to the surface as the reflective tape to enhance the amount of light reflected back to the laser receiver.

3 Results and discussion

The corresponding travel-time map of the crude oil with different water contents from 0 to 100% of samples A, B, and C was measured by an all-optical method. The resulting consistent time-voltage plots of sample A, as an example of the raw ultrasound data with several water contents, are shown in Figure 2. Each line in Figure 2 is the output voltage from the oscilloscope corresponding to the water content. A peak picking program has been designed that locates the time of the first peak. The red short vertical line in Figure 2 marks the first peak of each measurement. It was obvious that the resulting travel time of the P-wave was strongly dependent upon the water content for the arrival of the first peak delayed in turn with the decrease of the water content. The travel time τ_1 was 8.11μs, 8.05μs, 7.64μs, 7.50μs, 7.34μs, and 7.21μs, and the propagation speed v was 1289m/s, 1299m/s, 1372m/s, 1399m/s, 1431m/s, and 1458m/s, for the selected water contents of 0%, 20.19%, 40.12%, 59.34%, 89.99%, and 100% in Figure 2, respectively. The propagation speed v of ultrasound can be calculated by the travel time τ_1 as a result of

$$v = d/\tau = d/(\tau_1 - \tau_2) \tag{1}$$

where d is 10mm, which is the thickness of aqueous crude oil, and τ_2 is the travel time through the 2mm quartz vessel and is 0.35μs according to the P-wave speed of 5639m/s in quartz.

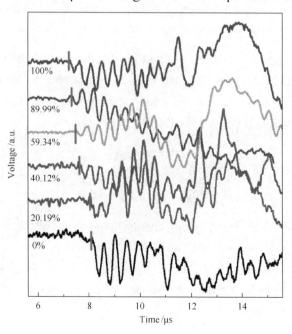

Figure 2   Time-voltage dependence of the selected aqueous crude oil A

The same measurements have been performed for all samples A, B, and C with different water contents from 0% to 100%, and the results were plotted as solid points in Figure 3. The ultrasonic propagation speed was different in the three anhydrous crude oils ($c_m = 0$), which was 1289m/s, 1342m/s, and 1343m/s, respectively. The P-wave travel time τ with ±0.2μs standard deviations, decreased with the increase of the water content in crude oil and reached its minimum in water ($c_m = 0$), as shown in Figure 3(a). Correspondingly, the propagation speed of ultrasound v, with ±20m/s standard deviations, accelerated with the increase of the water content and reached its maximum in water, as shown in Figure 3(b).

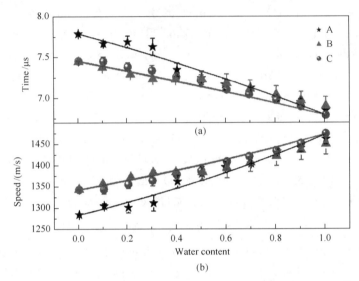

Figure 3 (a) Time delay τ and (b) ultrasound velocity v of different water contents. The solid points and lines represent the experiment and simulation results, respectively

The theoretical expression of the relationship by the formula of the effective medium model was consistent with the experimental results. The effective medium model hypothesizes that the mixture fluid, whose components were not dissolved, keeping its volume unchanged before and after mixing[32]. Therefore, the P-wave travel time

$$\tau = d_1/v_1 + d_2/v_2 \tag{2}$$

and

$$d_1/d_2 = \rho_2 c_m / \rho_1 (1-c_m) \tag{3}$$

where ρ_1 and ρ_2 (0.8455g/cm^3, 0.8987g/cm^3, and 0.8742g/cm^3 for oils A, B, and C, respectively) are the densities of water ($c_m = 100\%$) and oil ($c_m = 0$) and v_1 (1458m/s) and v_2 (1289m/s, 1342m/s, and 1343m/s for oils A, B, and C from Figure 3) are the P-wave speed of water ($c_m = 100\%$) and oil ($c_m = 0$). Because the volume of aqueous crude oil is equivalent to the volume of water and oil, thus

$$d = d_1 + d_2 \tag{4}$$

where d_1 and d_2 are the thicknesses of water and oil. The effective medium model is described as shown in Figure 4. The white part indicates the water, and the black part indicates the crude oil. The volume of aqueous crude oil can be considered as the sum of the water volume and oil volume. The theoretical results were described as solid lines in Figure 3, which are in good agreement with the experimental results. The maximum and minimum relative errors between the predicted and experimental results are 1.39% and 0.04%, respectively. According to the effective medium model, the travel time τ in samples has a nonlinear relationship with the water content c_m.

$$\tau = \frac{c_m \rho_2 d_1}{\rho_1 - \rho_1 c_m + \rho_2 c_m} \left(\frac{v_2 - v_1}{v_1 v_2} \right) + \frac{d}{v_2} \tag{5}$$

Therefore, Figure 3(a) shows a nonlinear relationship between the lines of C and water content, as well as in Figure 3(b).

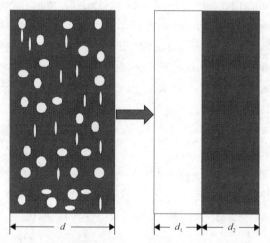

Figure 4 Schematic diagram of the effective medium model. d_1 and d_2 are the thickness of water and oil

The non-contact measurement has its unique advantages in online detection. The microwave method can better solve the problems, such as pressure and temperature changes and adverse conditions, using the different permittivity in the oil-water mixture. However, this method is less in the actual application for its more complex operation and high cost. The ray method has more field applications with it being rapid, accurate, and easy to measure, while its radiation to the human body cannot be ignored. A non-contact method of terahertz time-domain spectroscopy was researched to measure the water content in crude oil[33], suggesting the linear relation for the terahertz absorption coefficient and the water content of crude oil. In addition to the above methods, the water content of crude oil can be characterized by the ultrasonic longitudinal wave velocity acquired by the non-contact optical method. As a way for detecting the water content of crude oil, the noncontacted all-optical method ignored the impact of conductivity and had no more desire for samples, being a simple, fast, and low-cost detection. The results in this paper show that the all-optical method identified the water content effectively by the ultrasonic P-wave velocity. Considering its feasibility, there are lots of details that need to be studied further, such as the precision of real-time detection, the portability of the detection device, etc. Although the current study is preliminary and is not widely used now, the non-contact optical method is believed to have great potential for application in an oil field.

4　Conclusions

In this paper, a non-contacting optical method was proposed for measuring the water content in crude oil. The ultrasonic P-wave velocity acquired by this method was found to be accelerated with the increase of the water content. Different P-wave velocities correspond to the different water contents that can be used for detecting the water content from 0% to 100%. An effective medium model was used to explain the relationship between the P-wave velocity and the water content, and the theoretical values were in agreement with the experimental results of three different series of crude oil samples. The noncontact optical method provides us a new way to detect the water content in crude oil and is expected to have further application in the petroleum industry.

Acknowledgements

The authors acknowledge the National Key Basic Research Program of China (Grant No. 2014CB744302) and the Specially Funded Program on National Key Scientific Instruments and Equipment Development (Grant No. 2012YQ140005) for the financial support of this work.

References

[1] Filho D C M, Ramalho J B V S, Spinelli L S, et al. Aging of water-in-crude oil emulsions: Effect on water content, droplet size distribution, dynamic viscosity and stability. Colloid. Surf. A: Physicochem. Eng. Asp., 2012, 396: 208-212.

[2] Abdurahman N H, Rosli Y M, Azhari N H, et al. Pipeline transportation of viscous crudes as concentrated oil-in-water emulsions. J. Petr. Sci. Eng., 2012, 90: 139-144.

[3] Merv F, Ben F. Studies on crude oil and petroleum product emulsions: Water resolution and rheology. Colloid. Surf. A: Physicochem. Eng. Asp., 2009, 333: 67-81.

[4] Ye G X, Lü X P, Shen X, et al. Pretreatment of crude oil by ultrasonic-electric united desalting and dewatering. Chin. J. Chem. Eng., 2008, 16: 564-569.

[5] Lee C M, Sams G W, Wagner J P. Power consumption measurements for AC and pulsed DC for electrostatic coalescence of water-in-oil emulsions. J. Electrostat., 2001, 53: 1-24.

[6] Less S, Hannisdal A, Bjørklund E, et al. Electrostatic destabilization of water-in-crude oil emulsions: Application to a real case and evaluation of the Aibel VIEC technology. Fuel, 2008, 87: 2572-2581.

[7] Martinez-Palou R, Mosqueira M D, Zapata-Rendon B, et al. Transportation of heavy and extra-heavy crude oil by pipeline: A review. J. Petr. Sci. Eng., 2011, 75: 274-282.

[8] Creek J L, Lund H J, Brill J P, et al. Wax deposition in single phase flow. Fluid Phase Equilibra, 1999, 158: 801-808.

[9] Frigaard I, Vinay G, Wachs A. Compressible displacement of waxy crude oils in long pipeline startup flows. J. Non-Newton. Fluid Mech., 2007, 147: 45-64.

[10] Mohammed I Z. Shear driven crude oil wax deposition evaluation. J. Petr. Sci. Eng., 2010, 70: 28-34.

[11] Mohameda A M O, Elgama I M, Said R A. Determination of water content and salinity from a producing oil well using CPW probe and eigendecomposition. Sensor. Actuat. A, 2006, 125: 133-142.

[12] Gholam R C. Two-stage ultrasonic irradiation for dehydration and desalting of crude oil: A novel method. Chem. Eng. Proc., 2014, 81: 72-78.

[13] Tsouris C, Tavlarides L L. Volume fraction measurements of water in oil by an ultrasonic technique. Ind. Eng. Chem. Res., 1993, 32: 998-1002.

[14] Huynh Q, Phan T D, Thieu V Q Q. Research on distillation technology to extract essential oil from Melaleuca alterfornia (TTO). Int. Proc. Chem., Biol. Environ. Eng., Phuket, 2012, 43: 125-130.

[15] Kestens V, Conneely P, Bernreuther A. Vaporisation coulometric Karl Fischer titration: A perfect tool for water content determination of difficult matrix reference materials. Food Chem., 2008, 106: 1454-1459.

[16] Fortuny M, Oliveira C B Z, Melo R L F V, et al. Effect of salinity, temperature, water content, and pH on the microwave demulsification of crude oil emulsions. Energy & Fuels, 2007, 21: 1358-1364.

[17] MacLeod S K. Moisture determination using Karl Fischer titrations. Anal. Chem., 1991, 63: 557A-566A.

[18] Lee J R, Chia C C, Shin H J, et al. Laser ultrasonic propagation imaging method in the frequency domain based on wavelet transformation. Opt. Laser Eng., 2011, 49: 167-175.

[19] Jhang K Y, Kim H M, Ha Y. Laser generation of focused ultrasonic wave. AIP Conf. Proc., 2003: 305-310.

[20] Karabutov A A, Podymova N B. Nondestructive porosity assessment of CFRP composites with spectral analysis of backscattered laser-induced ultrasonic pulses. J. Nondestruct. Eval., 2013, 32: 315-324.

[21] Nelson K A, Lutz D R, Fayer M D, et al. Laser-induced phonon spectroscopy. Optical generation of ultrasonic waves and investigation of electronic excited-state interactions in solids. Phys. Rev. B, 1981, 24: 3261-3275.

[22] Devos A, Côte R. Strong oscillations detected by picosecond ultrasonics in silicon: Evidence for an electronic-structure effect. Phys. Rev. B, 2004, 70: 125208.

[23] Gao W, Glorieux C, Thoen J. Laser ultrasonic study of Lamb waves: determination of the thickness and velocities of a thin plate. Int. J. Eng. Sci., 2003, 41: 219-228.

[24] Hess P, Lomonosov A M, Mayer A P. Laser-based linear and nonlinear guided elastic waves at surfaces (2D) and wedges (1D). Ultrasonics, 2013, 54: 39-55.

[25] Dubois M, Enguehard F, Bertrand L. Analytical one-dimensional model to study the ultrasonic precursor generated by a laser. Phys. Rev. E, 1994, 50: 1548-1551.

[26] Jia H, Lu M H, Wang Q C, et al. Subwavelength imaging through spoof surface acoustic waves on a two-dimensional structured rigid surface. Appl. Phys. Lett., 2013, 103: 103505.

[27] Hayashi Y, Ogawa S, Cho H, et al. Non-contact estimation of thickness and elastic properties of metallic foils by laser-generated Lamb waves. NDT & E International, 1999, 32: 21-27.

[28] Shen D, Qi K, Ping Z, et al. Frequency response to liquid density of a piezoelectric quartz crystal sensor with longitudinal wave. Anal. Chim. Acta, 2004, 525: 205-211.

[29] Murayama R, Kobayashi M, Jen C K. Study of material evaluation probe using a longitudinal wave and a transverse wave. J. Sensor Tech., 2013, 3: 25-29.

[30] Amziane A, Amari M, Mounier D, et al. Laser ultrasonics detection of an embedded crack in a composite spherical particle. Ultrasonics, 2012, 52: 39-46.

[31] Moreau A, Levesque D, Lord M, et al. On-line measurement of texture, thickness and plastic strain ratio using laser-ultrasound resonance spectroscopy. Ultrasonics, 2002, 40: 1047-1056.

[32] Gong L L, Wang Y H, Cheng X D. A novel effective medium theory for modelling the thermal conductivity of porous materials. Int. J. Heat Mass Transfer, 2014, 68: 295-298.

[33] Jin W J, Zhao K, Yang C, et al. Experimental measurements of water content in crude oil emulsions by terahertz time-domain spectroscopy. Appl. Geophys., 2013, 10: 506-509.

The detection of water flow in rectangular microchannels by terahertz time-domain spectroscopy

Yan Song[1,2]   Kun Zhao[1,2]   Jian Zuo[3]   Cuicui Wang[3]
Yizhang Li[1,2]   Xinyang Miao[1,2]   Xiaojing Zhao[3]

(1.State Key Laboratory of Petroleum Resources and Prospecting, China University of Petroleum, Beijing 102249, China; 2.Beijing Key Laboratory of Optical Detection Technology for Oil and Gas, China University of Petroleum, Beijing 102249, China; 3.Department of Physics, Key Laboratory of Terahertz Optoelectronics, Ministry of Education, Capital Normal University, Beijing 100048, China)

Abstract: Flow characteristics of water were tested in a rectangular microchannel for Reynolds number (Re) between 0 and 446 by terahertz time-domain spectroscopy (THz-TDS). Output THz peak trough intensities and the calculated absorbances of the flow were analyzed theoretically. The results show a rapid change for $Re<250$ and a slow change as Re increases, which is caused by the early transition from laminar to transition flow beginning nearly at $Re = 250$. Then this finding is confirmed in the plot of the flow resistant. Our results demonstrate that the THz-TDS could be a valuable tool to monitor and character the flow performance in microscale structures.

Keywords: water flow; THz-TDS; rectangular microchannel; flow-rate; pressure drop; flow resistance

1 Introduction

The emergence of micro-electro-mechanical systems has attracted significant interest in the field of microscale devices. During the last few years, because of the immense potential of micro systems such as higher accuracy, lower power and lower cost, micromachined fluidic systems have had an important impact on a wide variety of areas such as medicine, chemical analysis, bioengineering, molecular separation and other industries. However, due to the higher surface to volume ratio caused by the smaller typical length, the surface force as well as microchannel wall wettability has great influence on the fluid movement; thus, there will be significant departure of flow characteristics from the conventional flow as fluid is induced to flow through the microchannel[1,2]. As such, in order to predict the flow performance in such micro devices accurately, it is necessary to understand the fluid flow characteristics on the microscale.

In fact, the design and control of microfluidic devices require the fundamental understanding of flow characteristics such as flow pattern transition and pressure loss. As reported by early investigators, the flow characteristics in microchannels are different from that in the normal situation described by the Navier-Stokes equations[3]. It has been found that the transition of the flow pattern from laminar to turbulent flow might take place much earlier than that for flow going through conventional larger-sized channels[4,5]. Other investigations have supported the earlier findings and have served to illustrate that the flow characteristics are strongly affected by the surface roughness[6,7].

Vicente et al.[8] employed water and ethylene glycol as working fluids in dimpled tubes to measure the laminar and transitional flow characteristics. They reported a relatively low critical Re down to 1400 owing to the surface roughness and then proposed the correlation for the prediction of critical Re in dimpled tubes. They declared that dimple height of dimpled tubes was the main factor affecting the hydraulic behavior of the flow. Similarly, Kandlikar et al.[9] investigated the flow of distilled water through small circular tubes with various hydraulic diameters and surface roughness. They stated that the surface roughness had a dramatic effect on the pressure drop for the smaller diameter tube and the critical Re for smaller rough tube was much lower than 2300. Li et al.[10] applied microscopic particle image velocimetry (microPIV) to measure instantaneous velocity fields in a polydimethylsiloxane (PDMS) microchannel at various Re. By analyzing the great variation caused by velocity fluctuations in the individual velocity fields, the calculated velocity fluctuations could be used to predict the critical Re. They found laminar flow ceased at Re of 1535, and fully turbulent flow was achieved at $2630 < Re < 2853$, both of which were lower than classical results.

Optical measurements of the microfluidic flow have been proposed recently. Lauri et al.[11] measured the flow velocity profile in a capillary with two Doppler optical coherence tomography (DOCT) systems, and depth scanning was also achieved by moving the whole measurement system with the reference mirror fixed. Lucchetta et al.[12] utilized an optofluidic device consisting of a Bragg grating written on a soft wall to measure flow rate in a microfluidic channel. They used diffraction of a white-light probe beam as detecting wave and established a simple theoretical model for the response time of the diffracted signal to determine the flow rates. However, there is still little research on understanding the characteristics and affecting factors of the flow in microchannels by THz-TDS. Preliminary studies on the flow pattern and the slip phenomenon for oil-water two phase flow in macroscale pipes based on THz-TDS have been made previously[13,14]. Owing to the sensitivity of THz waves to the fluctuations of water dipole moments occurring on the picosecond (ps) timescale, THz technology can detect the subtle changes of water[15,16], so attempts have been made to predict the flow pattern transition of single phase flow in a rectangular microchannel using a method based on THz-TDS in this work. The flow-rates, the pressure drops and the THz signals of water in microchannel were measured and analyzed and the interaction between THz wave and flow pattern was analyzed. We then calculated the flow resistances of the flow, which verified the experimental analysis further. The results of this study demonstrate the potential of THz-TDS for microfluidic studies.

2 Materials and methods

A schematic diagram of the experimental apparatus is shown in Figure 1. The experiments were carried out with a single channel plastic microfluidic chip which was fabricated on a 29mm×11mm×5.6mm PDMS plate. The microdevice was fabricated by the molding method which includes two steps, namely the production of a SU-8 glue mold and casting molding of the PDMS chip. There are four sub-steps to get the SU-8 glue mold. The first step is to design the microchip shape with CAD and transfer it to the chrome plate. The second step is to cast a thin layer of SU-8 photoresist on silicon wafer and then carry out high temperature treatment. The third step is exposure and cure, in which the chrome plate is used as the mask and the cured SU-8 negative photoresist is exposed with the lithography machine. The fourth step is to get the SU-8 glue mold by dipping the silicon wafer in the developer. The casting molding of PDMS chip is described in brief. First, PDMS prepolymer should be prepared. The ratio of PDMS and curing agent is 10:1, and the prepolymer needs to be degassed for 10min. Then the degassed PDMS prepolymer is poured onto the SU-8 mold, and is heated in a 75℃ drying chamber for about 1h. After that, the cured PDMS is removed from the SU-8 glue mold. When the two PDMS plates are put in the plasma cleaning apparatus, it is worthwhile to note that the two sides being bonded should remain upwards, and the plates should be taken out 2min after the appearance of violet light and then the two plates should be fitted within 30s. The fourth step is to place the fitted chip in the oven at 80℃ for 1h. The obtained microdevice is shown in Figure 1. It consists of a capillary and two stainless steel needles whose dimensions are 0.5mm×0.8mm×15mm. The cross-section of the microchannel is rectangular, with dimensions of 200μm×50μm×20mm($W×H×L$), resulting in the hydraulic diameter $D_H = 80$μm:

$$D_H = 4S / (2H + 2W) \tag{1}$$

where S is the cross-sectional area and the draw ratio L/D_H is equal to 250, thus the entrance effects can be ignored[17]. The digital pressure gauges were connected to the microdevice by tubes in the experiments. All the connection tubes are 1.6mm outer diameter and 0.6mm inner diameter polytetrafluoroethylene (PTFE) tubes. The working fluid was deionized water and it was introduced to the horizontal rectangular microchannel by digital injection pump. The flow velocity of the water was increased gradually at the range of 0~5.58m/s. When setting a new flow-rate, the system ran for at least 5min until the flow remained stable with very few fluctuations in the pressure drop readings. The density and viscosity of the water are 998kg/m³ and 1.01mPa·s respectively.

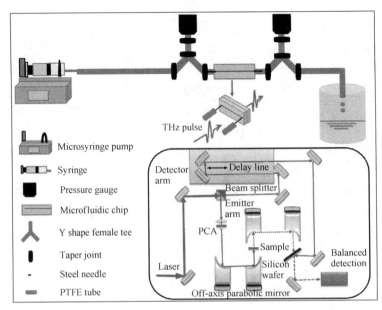

Figure 1 Sketch of the experiment arrangement for detection of deionized water flowing in microchannel with transmission THz-TDS and setup of transmission THz-TDS systems (inset)

The behavior of deionized water in the microchannel was followed by a conventional transmission THz-TDS whose bandwidth is beyond 2.5THz. The detailed transmission principal of THz-TDS system is depicted in Figure 1. The femtosecond pulses are provided by a mode-locked Ti-sapphire laser (MaiTai) whose center wavelength and average power are 790nm and 200mW respectively. The amplitude and phase information of the sample are obtained by THz-TDS in a coherent way. The femtosecond laser pulse is divided into two beams, namely pump beam and the probe beam. The average power of pump beam is maintained at 40mW while that of probe beam is 8mW. The pump beam is used to generate THz pulse at the emitter which is composed of a photoconductive antenna (PCA) while the probe beam transmits and is controlled by a delay stage. In the system, the probe beam acts as a gated detector to monitor the temporal waveform of THz field. After being focalized and reflected by a lens and a mirror, the collimated THz pulse transmits the sample and then the sample-information-carried THz pulse reaches the silicon wafer, meeting the delayed probe beam. The optical signal is detected by a balance detector, amplified by a lock-in amplifier and finally processed by the LabVIEW software [18-20].

The interaction between water molecules leads to the formation of a complex multi-body system, which has strong absorption of terahertz waves. Therefore, attention must be paid while measuring water with THz-TDS. The first is that the thickness of water layer must be thin enough for the transmission measurement. In addition, the sample needs to be placed in a nitrogen atmosphere to eliminate the influence of water vapor. Besides, to obtain optimal signal it is important to place the sample at the focus.

The degree of attenuation of the THz radiation in the chip can be related to the flow pattern. The loop-pipe wall was pre-wetted by water and the flow monitoring experiment consisted of three types of measurements: the water volume flow-rate (Q), the pressure drops and the THz-TDS. The flow-rate was controlled accurately by the pre-set digital injection pump while the pressure drop was measured using two digital pressure gauges across the stream wise length of the channel, and the THz-TDS was recorded when the flow remained a stable state. In the experiment, to simplify the data processing, we ensured the THz wave transmitted the microchannel perpendicularly. The temperature and humidity were controlled at the range of 21.2~21.4℃ and 1.0%~1.3%, respectively, in the experiments.

3 Results and discussion

The flow monitoring experiment consisted of a recorded THz pulse waveform in the time domain and pressure drop at varying velocity intervals. By adjusting the parameters of the microsyringe pump, the flow-rate was controlled to range from 0 to 3.35mL/min. Each of the waveforms represents a "snapshot" of flowing performance at the flow-rate recorded. The empty PDMS chip was used as the reference measurement.

Figure 2(a) shows the output THz-TDS of flows at several flow rates and the difference between peak trough intensities can be observed from the partial enlarged detail. The absorbances at these flow rates [Figure 2(b)] are calculated by a numerical fast Fourier transform (FFT). The FFT of the time-domain waveforms of reference and sample enables determination of the sample absorbance and permittivity et al., which are the basis of THz-TDS[21-23]. The absorbances (A) are calculated as a function of THz frequency (ν) using the following equation:

$$A(\nu) = -\ln[E_{sample}(\nu) / E_{reference}(\nu)] \tag{2}$$

where E_{sample} and $E_{reference}$ refer to the THz amplitude of sample and reference respectively[24].

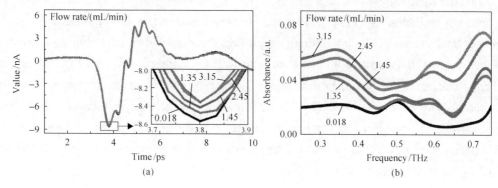

(a)   (b)

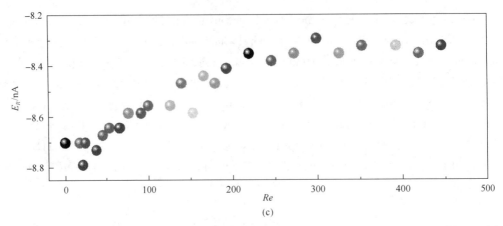

Figure 2 (a) Time domain trace of the THz pulse through the flow at different flow-rates and the partial enlarged detail (inset); (b) the absorbance curves against frequency for several flow rates; (c) output THz peak trough intensities versus Re for single water flow

Figure 2(c) shows the derived peak trough intensities at 27 Reynolds numbers. Re is a dimensionless parameter to determine if a flow is laminar or turbulent, and can be calculated as

$$Re = \frac{\rho v D_H}{\mu}, v = \frac{Q}{S} \tag{3}$$

where ρ is the fluid density, v is the average flow velocity, and μ is the dynamic viscosity. The signals transmitting through the chip show the presence of peak trough intensity features at different Re. As expected, the signal loss increases with the increase of Re, but it seems to occur in two-phases which is consistent with the results in our previous research[14]: a first fast phase until Re reaches about 250, and next a second phase showing a slower signal loss. Then the question arises why, although the flow seems to be in the laminar range, there is such an inversion point in the E_P changing along with Re. Therefore, the experimental results were further analyzed in the next section.

In order to investigate the dependence of THz signal on Re, we extracted the absorbances at certain frequencies, as shown in Figure 3. It clearly shows that the higher the Re is, the stronger the absorbance is. Interestingly, regarding the dependence of Re on absorption in Figure 3, it is self-evident to approximate both linear relationships for the two stages with different slopes. In the first stage (the small Re region) there is an obvious increase of the absorbance. However, with an increase in Re (the second stage), the absorbance grows much more slowly. As per our understanding, it is believed that the slope change is caused by the early transition of flow pattern from laminar to transition flow in the rectangular microchannel which has been observed by some scholars[25].

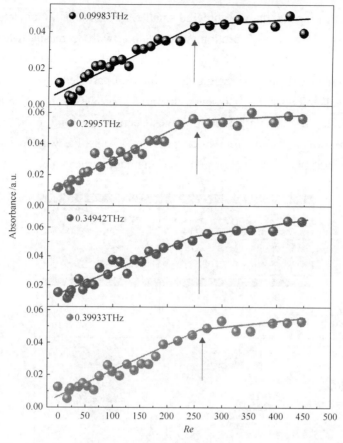

Figure 3   The THz absorbance as a function of Re at various frequencies

From the traditional theory, it is known that the internal flow undergoes a remarkable transition from laminar to transition and then a turbulent regime when the Re raises to certain values. The origin of turbulence and the accompanying transition from laminar to transition and to turbulent flow is of great importance to the studies in fluid mechanics. The possible flow patterns here are shown in Figure 4(a). In the flow through the uniform straight microchannel at low Re, every fluid particle moves with a uniform velocity with velocity grades along vertical direction of fluids flow despite the existence of slippage caused by the hydrophobicity of channel surface. Due to the effect of viscous forces on the flow particles, the velocity of the particles near the wall is smaller than those in the center core. The flow is well ordered and the particles travel along neighboring layers in the straight microchannel (laminar flow). As the Re value increases, the number of particles increases and the maximum particle size reduces. However, for further increase of Re, the momentum exchange from different layers takes place, making strong mixing of the particles from the layers and the orderly pattern of flow cease to exist (transition flow) [26].

Figure 4(b) depicts the theoretical analysis for the experimental results. In laminar region, along with the increasing *Re*, the well-ordered small water particles in the detecting area will linearly increase the THz wave absorption. What's more, with decreasing water particle size, the scattering effect may be enhanced as well, which would be another factor contributing to the THz absorption[13]. In transition region, the number of particles tend to be more while the maximum size tends to be smaller. However, the mixed particles from different layers reduce the augment of water to a certain degree in the test section, resulting in the slower growth in the THz wave absorbance. Our observation provides a sensitive way to monitor the flow characteristic of single phase water flow in the rectangular microchannel.

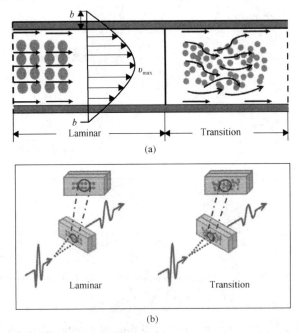

Figure 4 The schematic diagram for the interaction between THz wave and the flow. (a) The sketch map of laminar and transition; (b) interaction between THz wave and the laminar and transition flow

The most fundamental feature to discriminate the flow pattern is a noticeable change in the pattern of flow resistance (Darcy friction factor *f*). Therefore, the experimental pressure gradients ($\Delta P/\Delta L$ in Figure 5(a)) were compared with the theoretical ones calculated by the Darcy-Weisbach equation:

$$\Delta P / \Delta L = \frac{f\rho v^2}{2D_H} \quad (4)$$

where ΔL is the channel length and the theoretical flow resistances can be obtained by

$$f = \frac{Po}{Re} \quad (5)$$

for fully developed laminar flow in rectangular microchannel, the theoretical Poiseuille number Po can be computed numerically as the following equation given by Hartnett and Kostic[27]:

$$Po = 96(1 - 1.3553\beta + 1.9467\beta^2 - 1.7012\beta^3 + 0.9546\beta^4 - 0.2537\beta^5) \quad (6)$$

Where β is the channel aspect ratio which must be less than 1, and the inverse should be taken if it is greater than 1.

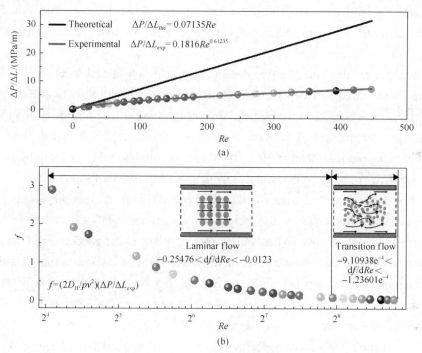

Figure 5 (a) The curve for the pressure gradient as a function of Re; (b) the flow resistance versus Re for the single flow in microchannel

It is found that the experimental pressure gradients indicate a significant departure from the theoretical predictions as the black line shows in Figure 5(a). The departure is caused by the hydrophobicity of the microchannel surface. As channel size decreases, the hydrophobicity of microchannel surface strongly affects the lubrication and sliding friction of the flow[28,29] and plays a critical role in pressure drop across the channel[30-32]. The PDMS plate used in this experiment has not been surface modified further and thus the microfluidic chip remains hydrophobic in the experiment. Because of the wall hydrophobicity, apparent water slip was observed, resulting from the obvious decrease of pressure gradients. Different from the linear relationship between the pressure gradient required to force liquid through the microchannel and Re in conventional laminar flow theory, a nonlinear relationship ($\Delta P/\Delta L_{exp} = 0.1816 Re^{0.16235}$) is observed here, which is caused by the turbulent flow.

Then some valuable information can be gathered by analyzing the variation tendency of experimental f. as depicted in Figure 5(b), f is calculated by the above-mentioned Darcy-Weisbach equation. For smaller Re, f decreases rapidly along with the rising Re, which is the characteristic of laminar flow resistance. By contrast, for larger Re, f almost keeps unchanged, which is the characteristic of transition flow resistance[33,34]. To make sure the value of critical Re, the differential values of f (df/dRe) were derived. $-0.25476 < df/dRe < -0.0123$ when Re is less than 250 while df/dRe approaches to zero, when Re is greater than 250, which means f decreases sharply until Re reaches to about 250. Overall, from the above analysis, it can be concluded that for the microchannel, there is an early transition from laminar to transition flow at Re about 250.

The need for reliable and effective design and fabrication of such microdevices has been the driving force behind an extensive effort to understand the fluid flow fundamentals on the microscale[35]. It has been reported that the smaller the typical length scale of the microfluid flow is, the smaller the critical Re from laminar to fully developed turbulent flow becomes compared with the ordinary channel flow[36,37]. We have employed THz-TDS to investigate the flow characteristics of single phase water flow in a rectangular PDMS microchannel under the condition of varying water flow-rates in the experiment. It is found that the THz peak trough intensities and the absorbances of the flowing water can accurately determine the transition from laminar to transition flow. In the laminar region, owing to the good order of the abundant small water particles, the THz parameters increase linearly with the increasing Re while in the transition region, the THz parameters increase much more slowly because of the mixed particles from different layers in the test section. By analyzing the pressure gradient and the flow resistance, the results are further demonstrated in theory. This approach can serve as a novel and practicable technique to determine the flow characteristics of single water flow. Further studies should be conducted to comprehend more about the potential flow mechanisms in microfluidic flow based on THz-TDS.

4 Conclusions

The flow characteristics of water with different flow-rates in a rectangular microchannel whose hydraulic diameter was 80μm were investigated experimentally and theoretically. It is observed that for smaller Re(less than 250), the THz peak trough intensities as well as the THz wave absorbances of the flow increases were sharply compared with larger Re, which indicates that the transition from laminar to transition flow takes place at an early Re compared with the conventional theory, and that the THz-TDS is indeed an effective tool for the study of flow characteristics. Based on the measured pressure drop, the flow resistances were derived as well. It is observed that for smaller Re, f decreases much rapidly than that for larger Re, which further proves the early transition of the flow pattern. As an initial effort, this study used pure water as an idealized case to show the potential of THz-TDS in studying the flow in microchannels, and the mixture of water, oil and gas should be considered in further studies to produce results closer to practical situations.

Acknowledgements

This Work is surpported by the National Natural Science Foundation of China (Grant No. 11574401).

References

[1] Squires T M, Quake S R. Microfluidics: Fluid physics at the nanoliter scale. Rev. Mod. Phys., 2015, 77: 977-1026.
[2] Asadi M, Xie G, Sunden B. A review of heat transfer and pressure drop characteristics of single and two-phase microchannels. Int. J. Heat Mass Tranf., 2014, 79: 34-53.
[3] Morini G L. Single-phase convective heat transfer in microchannels: A review of experimental results. Int. J. Therm. Sci., 2004, 43: 631-651.
[4] Qu W L, Mala G M, Li D Q. Heat transfer for water flow in trapezoidal silicon microchannels. Int. J. Heat Mass Tranf., 2000, 43: 3925-3936.
[5] Yang C Y. Friction characteristics of water, R-134a, and air in small tubes. Nnosc. Microsc. Therm., 2003, 7: 335-348.
[6] Yang W H, Zhang J Z, Cheng H E. The study of flow characteristics of curved microchannel. Appl. Therm. Eng., 2005, 25: 1894-1907.
[7] Zhou G, Yao S C. Effect of surface roughness on laminar liquid flow in micro-channels. Appl. Therm. Eng., 2011, 31: 228-234.
[8] Vicente P G, García A, Viedma A. Experimental study of mixed convection and pressure drop in helically dimpled tubes for laminar and transition flow. Int. J. Heat Mass Tranf., 2002, 45: 5091-5105.
[9] Kandlikar S G, Joshi S, Tian S. Effect of surface roughness on heat transfer and fluid flow characteristics at low reynolds numbers in small diameter tubes. Heat Transf. Eng., 2003, 24: 4-16.
[10] Li H, Ewoldt R, Olsen M G. Turbulent and transitional velocity measurements in a rectangular microchannel using microscopic particle image velocimetry. Exp. Therm. Fluid Sci., 2005, 29: 435-446.
[11] Lauri J, Wang M, Kinnunen M, et al. Measurement of microfluidic flow velocity profile with two Doppler optical coherence tomography systems. //Proceedings of the SPIE-The International Society for Optical Engineering, San Jose, CA, USA, 2008.
[12] Lucchetta D E, Vita F, Francescangeli D, et al. Optical measurement of flow rate in a microfluidic channel. Microfluid. Nanofluid., 2016, 20: 1-5.
[13] Feng X, Wu S X, Zhao K, et al. Pattern transitions of oil-water two-phase flow with low water content in rectangular horizontal pipes probed by Terahertz Spectrum. Opt. Express, 2015, 23: 1693-1699.
[14] Song Y, Miao X Y, Zhao K, et al. Reliable evaluation of oil-water two phase flow using a method based on terahertz time-domain spectroscopy. Energy & Fuels, 2017, 31: 2765-2770.
[15] Castro-Camus E, Palomar M, Covarrubias A A. Leaf water dynamics of Arabidopsis thaliana monitored in vivo using terahertz time-domain spectroscopy. Sci. Rep., 2013, 3: 2910.
[16] Song Y, Zhan H L, Zhao K, et al. Simultaneous characterization of water content and distribution in high water cut crude oil. Energy & Fuels, 2016, 30: 3929-3933.
[17] Veltzke T, Baune M, Thöming J. The contribution of diffusion to gas microflow: An experimental study. Phys. Fluids, 2012, 24: 082004.
[18] Li Y Z, Wu S X, Yu X L, et al. Optimization of pyrolysis efficiency based on optical property of semicoke in terahertz region. Energy, 2017, 126: 202-207.

[19] Jang C, Zhao K, Zhao L J, et al. Probing disaggregation of crude oil in a magnetic field with terahertz time-domain spectroscopy. Energy & Fuels, 2014, 28: 483-487.

[20] Zhan H L, Wu S X, Bao R M, et al. Qualitative identification of crude oils from different oil fields using terahertz time-domain spectroscopy. Fuel, 2015, 143: 189-193.

[21] Baxter J B, Guglietta G W. Terahertz spectroscopy. Anal. Chem., 2011, 83: 4342-4368.

[22] Li R Y, D'Agostino C, McGregor J, et al. Mesoscopic structuring and dynamics of alcohol/water solutions probed by terahertz time-domain spectroscopy and pulsed field gradient nuclear magnetic resonance. J. Phys. Chem. B, 2014, 118: 10156-10166.

[23] Jiang C, Zhao K, Fu C, et al. Characterization of morphology and structure of wax crystals in waxy crude oils by terahertz time-domain spectroscopy. Energy & Fuels, 2017, 31: 1416-1421.

[24] Naftaly M, Miles R E. Terahertz time-domain spectroscopy for material characterization. IEEE Proc., 2007, 95: 1658-1665.

[25] Morini G L. Laminar-to-turbulent flow transition. Nanosc. Microsc. Therm., 2004, 8: 15-30.

[26] Mala G M, Li D Q. Flow characteristics of water in microtubes. Int. J. Heat Fluid Flow, 1999, 20: 142-148.

[27] Hartnett J P, Kostic M. Heat transfer to Newtonian and non-Newtonian fluids in rectangular ducts. Adv. Heat Transf., 1989, 19: 247-356.

[28] Kawakatsu T, Trägårdh G, Trägårdh C, et al. The effect of the hydrophobicity of microchannels and components in water and oil phases on droplet formation in microchannel water-in-oil emulsification. Colloid Surf. A Physicochem. Eng. Asp., 2001, 179: 29-37.

[29] Sikarwar B S, Khandekar S, Muralidhar K. Simulation of flow and heat transfer in a liquid drop sliding underneath; a hydrophobic surface. Int. J. Heat Mass Transf., 2013, 57: 786-811.

[30] Dilip D, Bobji M S, Govardhan R N. Effect of absolute pressure on flow through a textured hydrophobic microchannel. Microfluid. Nanofluid., 2015, 19: 1409-1427.

[31] Chakraborty S, Anand K D. Implications of hydrophobic interactions and consequent apparent slip phenomenon on the entrance region transport of liquids through microchannels. Phys. Fluids, 2008, 20: 043602.

[32] Papageorgiou D P, Tsougeni K, Tserepi A, et al. Superhydrophobic, hierarchical, plasma-nanotextured polymeric microchannels sustaining high-pressure flows. Microfluid. Nanofluid., 2013, 14: 247-255.

[33] Hrnjak P, Xiao T. Single phase pressure drop in microchannels. Int. J. Heat Fluid Flow, 2007, 28: 2-14.

[34] Qi S L, Zhang P, Wang R Z, et al. Single-phase pressure drop and heat transfer characteristics of turbulent liquid nitrogen flow in micro-tubes. Int. J. Heat Fluid Flow, 2007, 50: 1993-2001.

[35] Young E W K, Simmons C A. Macro-and microscale fluid flow systems for endothelial cell biology. Lab Chip, 2010, 10: 143-160.

[36] Harms T M, Kazmierczak M J, Gerner F M. Developing convective heat transfer in deep rectangular microchannels. Int. J. Heat Fluid Flow, 1999, 20: 149-157.

[37] Peng X F, Peterson G P. Convective heat transfer and flow friction for water flow in microchannel structure. Int. J. Heat Mass Transf., 1996, 39: 2599-2608.

Pattern transitions of oil-water two-phase flow with low water content in rectangular horizontal pipes probed by terahertz spectrum

Xin Feng[1,2]　Shixiang Wu[3]　Kun Zhao[1,2]　Wei Wang[1,2]
Honglei Zhan[1,2]　Chen Jiang[1,2]　Lizhi Xiao[1]　Shaohua Chen[1,2]

(1.State Key Laboratory of Petroleum Resources and Prospecting, China University of Petroleum, Beijing 102249, China; 2.Beijing Key Laboratory of Optical Detection Technology for Oil and Gas, China University of Petroleum, Beijing 102249, China; 3.Petroleum Exploration and Production Research Institute, China Petroleum and Chemical Corporation, 100083, China)

Abstract: The flow-pattern transition has been a challenging problem in two-phase flow system. We propose the terahertz time-domain spectroscopy (THz-TDS) to investigate the behavior underlying oil-water flow in rectangular horizontal pipes. The low water content (0.03%~2.3%) in oil-water flow can be measured accurately and reliably from the relationship between THz peak amplitude and water volume fraction. In addition, we obtain the flow pattern transition boundaries in terms of flow rates. The critical flow rate Q_c of the flow pattern transitions decreases from $0.32 m^3/h$ to $0.18 m^3/h$ when the corresponding water content increases from 0.03% to 2.3%. These properties render THz-TDS particularly powerful technology for investigating a horizontal oil-water two-phase flow system.

1 Introduction

Oil and water two-phase flow in horizontal pipes is a common occurrence in the petroleum industry for long-distance transportation. The investigation of pattern transitions is very important in horizontal oil-water two-phase flow system due to the existence of various flow patterns and different mechanisms governing them[1-5], especially for the prediction of rheological behavior in the horizontal oil wells, measurement of flow parameters and optimization of industrial production process. Usually, the flow pattern of oil-water in horizontal pipe switches from stratified structure to dispersed structure with increasing flow rate and the oil-water interaction is also presented during this change.

On account of its significant importance in physical and chemical research fields, the horizontal oil-water two-phase flow has attracted a considerable research effort. The hydrodynamic characteristic of each flow pattern has been addressed and a flow pattern contains the shape and spatial distribution of the two-phase flow in the pipe[6-23]. Despite extensive efforts, the dynamics of various patterns in horizontal oil-water two-phase flow is still controversial. In particular, an ability to distinguish liquid-liquid flow behavior accurately is of fundamental importance. How to reveal the complex oil-water flow structure from experimental signals still represents a significant challenge.

The amplitude and phase composition of THz light transmitted or reflected from various materials can be obtained in THz-TDS. THz-TDS has been used to study the dielectric properties of oil-water complexes[24]. THz radiation was also used to study the moisture content and distribution in various materials, such as paper, biomolecules, food wafers, plant leaves and crude oil[25-28]. THz-TDS has undergone a remarkable development in the last decade and provided us a powerful tool for investigating complex systems from different disciplines in fast and *situ* phase transitions.

In this work we investigated the flow structures of oil-water two phase fluid systems with the low water contents (0.03%~2.3%) by THz-TDS. A correlation is presented for the prediction of the inversion point of an oil-water dispersion system. The formation and evolution of different oil-water flow structures are also discussed.

2 Experiments

Before the measurement, the conventional 0# diesel and tap water were mixed using a homogenizer for more than 30min in the storage tanks(Figure 1) at room temperature and atmospheric outlet pressure. Then the oil-water mixtures were pumped into pipe and the flow rate from $0.1m^3/h$ to $0.36m^3/h$ was accurately measured by using fluid flowmeter. The input water volume fractions increased from 0.03% to 2.3%. The oil-water mixtures flowed along a 3m long horizontal pipe from the entry point to the test section. A long flow distance provided sufficient entrance length to stabilize the flow. The THz-TDS measurement for different water content oil-water mixtures was carried out when the steady state reached in a certain flow condition.

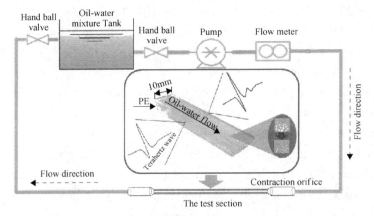

Figure 1　Experimental flow loop facility. The insert shows a schematic of the THz-TDS measurement process

The amplitude and phase composition of THz-TDS can be used to investigate various materials by coherence measurements of THz light[28-34]. Previously, a typical THz-TDS system was used to measure the flow states[28,29]. Here, the focus diameter of the THz beam was about 1mm. In order to reduce optical absorption by the curved (transparent) tube wall, a high-purity polyethylene pipe with a square cross section (10mm×10mm) was chosen as the flow pipe and the pipe thickness was 1mm. To minimize the influence of moisture, the measurements were performed at 20℃ under dry nitrogen. The pipe was located in the focus of the two Si lens, as shown in Figure 1, and perpendicular to the incident THz beam. The polyethylene pipe is transparent for visible light and has a low refractive index and absorption in THz range. Both the time-domain sample and reference spectra were obtained by testing the polyethylene pipe holding the mixtures and empty pipe, respectively.

3 Results and discussion

THz-TDS of diesel oil-water mixtures with different water content (0.03%, 0.7%, 1%, 1.3%, 1.5%, 1.7%, 2% and 2.3%) was measured and the waveforms at flow rate 0.28m³/h are shown in Figure 2. The signal of the THz waveform decreases as the water content of the mixture increases and the peak signal of the transmitted THz pulse dropped from 67mV to 19mV. Due to the absorption characteristics of water in THz range[25,28], the water content in oil-water mixtures has great influence on the THz signal. Hence, THz-TDS can provide information on the water content of mixtures by using the amplitude and phase change. Generally, a phase shift relative to the reference pulse occurs for THz pulses transmitted through the empty cell and mixture which is from the time delay between their THz amplitude signals. The phase changes of THz-TDS can be used to obtain the optical parameters of samples, such as the refractive index and the extinction coefficient.

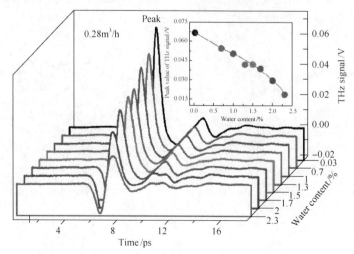

Figure 2 THz-TDS for various water content of diesel oil-water mixtures at flow rate 0.28m³/h. The insert shows the peak value of THz signal for different water contents in oil-water mixtures

For the various flow structures (flow patterns), the horizontal oil-water flow patterns can be classified into segregated flow and dispersed flow. The segregated flow includes a stratified flow pattern (ST) and a stratified flow with mixing at interface pattern (ST and MI), while the dispersed flow includes a dispersion of oil in water and water flow pattern (D O-in-W and W), a dispersion of water in oil and oil in water flow pattern (D W-in-O and D O-in-W), a dispersion of oil in water flow pattern (D O-in-W) and a dispersion of water in oil flow pattern (D W-in-O)[2]. Here 0# diesel fuel belongs to light oil ($\rho_o/\rho_w = 0.85$, $\mu_o/\mu_w = 2.7$ and $\varepsilon_w = 0.03\%$, where ρ_o and ρ_w are oil and water densities, μ_o and μ_w are oil and water viscosities as well as ε_w is water volume fraction.). The flow pattern of the diesel oil-water flow is a dispersion of water in oil flow pattern (D W-in-O). The water phase is dispersed as droplets in oil continuous phase for water content at a low level ($\varepsilon_w<3\%$)[2,6,17]. For different water content, the THz-TDS peak intensity of diesel-water mixtures shows the transition point at flow rates from $0.1m^3/h$ to $0.36m^3/h$, as shown in Figures 3(a)~(h). These flow rates of THz-TDS amplitude transition points are named critical flow rates Q_c. Water content dependence of Q_c is shown in Figure 4 where Q_c decreases from $0.32m^3/h$ to $0.18m^3/h$ with the water content ε_w increasing from 0.03% to 2.3%.

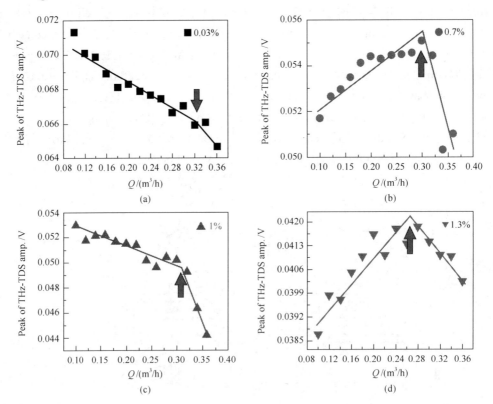

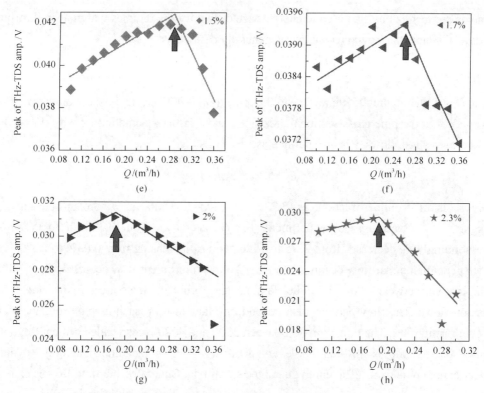

Figure 3 (a)~(h) Peak of THz-TDS amplitude for different water content diesel oil-water mixtures with increasing flow rates from 0.1m³/h to 0.36m³/h

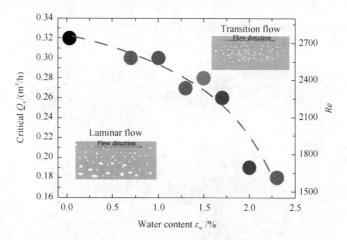

Figure 4 Critical flow rate Q_c and Reynolds number Re for different water content diesel oil-water mixtures. The inserts show schematics of the flow patterns: laminar flow pattern (left image) and transition flow pattern (right image)

In fluid mechanics, the Reynolds number (Re) is a dimensionless quantity that is used to predict similar flow patterns in different fluid flow situations. The Reynolds number can be used to determine dynamic similitude between two different cases of fluid flow. Moreover, the

Reynolds number can be also used to characterize different flow regimes within a similar fluid. For flow in a pipe, the Reynolds number is generally defined as

$$Re = QD_H / (vA) \quad (1)$$

where D_H is the hydraulic diameter of the pipe ($D_H = 0.01$m), Q is the volumetric flow rate(m^3/s), A is the pipe cross-sectional area (m^2), and v is the kinematic viscosity (m^2/s). For oil-water two phase flows, v is also defined as[35]

$$v = (\varepsilon_o\mu_o + \varepsilon_w\mu_w)/(\rho_o\varepsilon_o + \rho_w\varepsilon_w) \quad (2)$$

where ε_o is oil volume fraction, $\mu_o = 2.73$mPa·s, $\mu_w = 1.01$mPa·s, $\rho_o = 852$kg/m^3, and $\rho_w = 998$kg/m^3 [35-37]. Based on our experiment, the Q_c decreases from 0.32m^3/h to 0.18m^3/h, correspondingly Re decreases from 2775 to 1586, with ε_w increasing from 0.03% to 2.3%.

In general, laminar flow occurs when $Re < 2300$ and turbulent flow occurs when $Re > 4000$. In the interval between 2000 and 4000, laminar and turbulent flows are possible and called transition flows. The flow patterns also depend on other factors, such as pipe roughness and flow uniformity[36,37]. The transition Re between 2300 and 4000 is also called critical Reynolds number Re_c and Re_c is different for every geometry[37,38]. In fact, the matching Reynolds number is not on its own sufficient to guarantee similitude. Generally, the fluid flow is chaotic, thus the very small changes on shape and surface roughness can result in very different flows. As shown in Figure 4, Re is strongly dependent on the water content, indicating that the viscosity is not the only factor to determine pattern transition. Distribution of the dispersed phase in the continuous phase, the shape and particle size of the dispersed phase and the contact form between dispersed phase and the tube wall may also affect flow pattern transition.

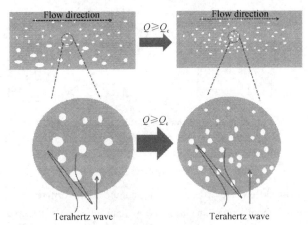

Figure 5   The schematic for the flow patterns transitions at $Q \geq Q_c$

For $Q < Q_c$, the flow pattern of oil-water mixtures in the pipe is a stratified flow pattern with oil continuous phase, thus the energy of shear stress is not strong enough to break water droplets and the movements of water phase comes along with oil continuous phase regularly. The energy of shear stress increases along with the increase of flow rate Q, but the increase of shear stress is limited. Whereas for $Q \geqslant Q_c$, the flow patterns become transition flows, the turbulence energy of the mixtures flow increases further and the interaction between the oil and water becomes enhanced. Under this condition, the turbulence energy is high enough to break water phase into smaller droplets above Q_c and there exist more water drops in the pipe center than that near pipe wall. The flow of abundant water droplets may increase the water distribution of the detecting areas(Figure 5), which denotes the increase of THz wave absorption. In addition, the scattering effect may also be enhanced with decreasing size of water drops, which would be another factor contributing to the THz absorption. Therefore, the transition of the oil-water flow pattern may increase the absorption of THz radiation. It also indicates the flow pattern changes with different water contents in critical flow rates.

4 Conclusions

In summary, we use THz-TDS system to investigate the flow states in diesel oil-water two-phase flow. The experimental results suggest that THz-TDS can be used to detect the water content and the pattern transitions in the oil-water two-phase flow. The amplitude of the THz pulse is strongly correlated to the water content. Moreover, the change of oil-water two-phase flow pattern at low water content can be also measured by THz-TDS system. The critical flow rates of the flow pattern transitions point show a non-linear trend with the water content in the mixtures. Such model can be used to qualitatively and quantitatively characterize the molecular state of oil-water flow. Accordingly, this work provides a fast and *in situ* way to investigate the oil-water flow system in the process of crude oil storage and transportation.

Acknowledgements

This work was supported by the Specially Funded Program on National Key Scientific Instruments and Equipment Development (Grant No. 2012YQ140005) and the National Basic Research Program of China (Grant No. 2014CB744302).

References

[1] Russell T W F, Hodgson G W, Govier G W. Horizontal pipeline flow mixtures of oil and water. Can. J. Chem. Eng., 1959, 37(1): 9-17.

[2] Trallero J L, Sarica C, Brill J P. A study of oil-water flow patterns in horizontal pipes. SPE Production and Facilities, 1997, 12(3): 165-172.

[3] Nadler M, Mewes D. Flow induced emulsification in the flow of two immiscible liquids in horizontal pipes. Int. J. Multiphase Flow, 1997, 23(1): 55-68.

[4] Angeli P, Hewitt G F. Flow structure in horizontal oil-water flow. Int. J. Multiphase Flow, 2000, 26(7): 1117-1140.

[5] McKibben M J, Gillies R G, Shook C A. A laboratory investigation of horizontal well heavy oil-water flows. Can. J. Chem. Eng., 2000, 78(4): 743-751.

[6] Brauner N, Maron M. Flow pattern transitions in two-phase liquid-liquid flow in horizontal tubes. Int. J. Multiphase Flow, 1992, 18(1): 123-140.

[7] Ng T S, Lawrence C J, Hewitt G F. Interface shapes for two-phase laminar stratified flow in a circular pipe. Int. J. Multiphase Flow, 2001, 27(7): 1301-1311.

[8] Ng T S, Lawrence C J, Hewitt G F. Laminar stratified pipe flow. Int. J. Multiphase Flow, 2002, 28: 963-996.

[9] Brauner N, Ullmann A A. Modeling of phase inversion phenomenon in the two-phase flows. Int. J. Multiphase Flow, 2002, 28(6): 1177-1204.

[10] Lovick J, Angeli P. Experimental studies on the dual continuous flow pattern in oil-water flows. Int. J. Multiphase Flow, 2004, 30(2): 139-157.

[11] Rodriguez O, Oliemans R. Experimental study on oil-water flow in horizontal and slightly inclined pipes. Int. J. Multiphase Flow, 2006, 32(3): 323-343.

[12] Al-Wahaibi T, Angeli P. Transition between stratified and non-stratified horizontal oil-water flows. Chem. Eng. Sci., 2007, 62(11): 2915-2928.

[13] Piela K, Delfos R, Ooms G, et al. On the phase inversion process in an oil-water pipe flow. Int. J. Multiphase Flow, 2008, 34(7): 665-677.

[14] Kumara W A S, Halvorsen B M, Melaaen M C. Particle image velocimetry for characterizing the flow structure of oil-water flow in horizontal and slightly inclined pipes. Chem. Eng. Sci., 2010, 65(15): 4332-4349.

[15] Al-Wahaibi T, Angeli P. Experimental study on interfacial waves in stratified horizontal oil-water flow. Int. J. Multiphase Flow, 2011, 37(8): 930-940.

[16] Al-Wahaibi T, Yusuf N, Al-Wahaibi Y. Experimental study on the transition between stratified and non-stratified horizontal oil-water flow. Int. J. Multiphase Flow, 2012, 38(1): 126-135.

[17] Morgan R G, Markides C N, Hale C P, et al. Horizontal liquid-liquid flow characteristics at low superficial velocities using laser-induced fluorescence. Int. J. Multiphase Flow, 2012, 43: 101-117.

[18] Cem S, Ekarit P. Review of paraffin deposition research under multiphase flow conditions. Energy & Fuels, 2012, 26(7): 3968-3978.

[19] Zhang H Q, Cem S, Eduardo P. Review of high-viscosity oil multiphase pipe flow. Energy & Fuels, 2012, 26(7): 3979-3985.

[20] Gao Z K, Zhang X W, Jin N D, et al. Multivariate recurrence network analysis for characterizing horizontal oil-water two-phase flow. Phys. Rev. E, 2013, 88(3): 032910.

[21] Ekarit P, Cem S. Experimental study of single-phase and two-phase water-in-crude-oil dispersed flow wax deposition in a mini pilot-scale flow loop. Energy & Fuels, 2013, 27: 5036-5053.

[22] Zhai L S, Jin N D, Gao Z K. Cross-correlation velocity measurement of horizontal oil-water two-phase flow by using parallel-wire capacitance probe. Exp. Therm. Fluid Sci., 2014, 53: 277-289.

[23] Gao Z K, Yang Y X, Fang P C, et al. Multi-frequency complex network from time series for uncovering oil-water flow structure. Sci. Rep., 2015, 5: 08222.

[24] Gorenflo S, Tauer U, Hinkov I, et al. Dielectric properties of oil-water complexes using terahertz transmission spectroscopy. Chem. Phys. Lett., 2006, 421(4-6): 494-498.

[25] Banerjee D, von Spiegel W, Thomson M D, et al. Diagnosing water content in paper by terahertz radiation. Opt. Express, 2008, 16(12): 9060-9066.

[26] George P A, Hui W, Rana F, et al. Microfluidic devices for terahertz spectroscopy of biomolecules. Opt. Express, 2008, 16(3): 1577-1582.

[27] Hadjiloucas S, Karatzas L S, Bowen J W. Measurements of leaf water content using terahertz radiation. IEEE T. Microw. Theory, 1999, 47(2): 142-149.

[28] Jin W J, Zhao K, Yang C, et al. Experimental measurements of water content in crude oil emulsions by terahertz time-domain spectroscopy. Appl. Geophys., 2013, 10(4): 506-509.

[29] Jiang C, Zhao K, Zhao L J, et al. Probing disaggregation of crude oil in magnetic field with terahertz time-domain spectroscopy. Energy & Fuels, 2014, 28(1): 483-487.

[30] Ge L N, Zhan H L, Leng W X, et al. Optical characterization of the principal hydrocarbon components in natural gas using terahertz spectroscopy. Energy & Fuels, 2015, 29(3): 1622-1627.

[31] Qin F L, Li Q, Zhan H L, et al. Probing the sulfur content in gasoline quantitatively with terahertz time-domain spectroscopy. Sci. China: Phys. Mech. Astron., 2014, 57(7): 1404-1406.

[32] Li Q, Zhao K, Zhang L W, et al. Probing $PM_{2.5}$ with terahertz wave. Sci. China: Phys. Mech. Astron., 2014, 57(12): 2354-2356.

[33] Bao R M, Wu S X, Zhao K, et al. Applying terahertz time-domain spectroscopy to probe the evolution of kerogen in close pyrolysis systems. Sci. China: Phys. Mech. Astron., 2013, 56(8): 1603-1605.

[34] Zhao H, Zhao K, Tian L, et al. Spectrum features of commercial derv fuel oils in terahertz region. Sci. China: Phys. Mech. Astron., 2012, 55(2): 195-198.

[35] Dukler A E, Wicks M, Cleveland R G. Frictional pressure drop in two-phase flow. AIChE J., 1964, 10(1): 44-51.

[36] Rott N. Note on the history of the Reynolds number. Annu.Rev. Fluid Mech., 1990, 22(1): 1-11.

[37] Holman J P. Heat Tansfer. 10th ed. New York: McGraw-Hill, 2002.

[38] Potter M C, Wiggert D C, Ramadan B H, et al. Mechanics of Fluids. 4th ed. Boston: Cengage Learning Inc. 2012.

Reliable evaluation of oil-water two-phase flow using a method based on terahertz time-domain spectroscopy

Yan Song[1,2] Xinyang Miao[1] Kun Zhao[1,2] Honglei Zhan[1,2]
Chen Jiang[2] Dandan Wang[2] Lizhi Xiao[1]

(1.State Key Laboratory of Petroleum Resources and Prospecting, China University of Petroleum, Beijing 102249, China; 2.Beijing Key Laboratory of Optical Detection Technology for Oil and Gas, China University of Petroleum, Beijing 102249, China)

Abstract: Oil-water two-phase flow widely exists in the petroleum industry, such as crude oil production and transportation. The flow pattern and slip phenomenon of oil-water two-phase flow have an important impact on the rational design and management of oil wells and pipelines. Because of the extremely high sensitivity of the terahertz(THz) wave to hydrogen bonding, terahertz time-domain spectroscopy (THz-TDS) is used to study oil-water two-phase flows with the input water fraction ranging from 0.5% to 5.0%. The flow rate is controlled in the range of 0~0.6m³/h, and the flow pattern transition and slip phenomenon in horizontal circular pipes have been studied experimentally and theoretically. In a word, the flow pattern and critical flow rate for the flow pattern transition can be distinguished by amplitudes of THz-TDS of the oil-water two-phase flows. The work indicates that, at low Q value, the flow pattern is an oil layer over a dispersion of water-in-oil (O & DW/O) flow and the ratio of *in situ* oil to water velocity is smaller than those at a high Q value, where the flow is a water-in-oil (W/O) flow.

Keywords: oil-water two phase flow; THz-TDS; flow pattern; horizontal pipes; slip phenomenon

1 Introduction

Oil-water two-phase flow is frequently encountered in a diverse range of processes and equipment[1]. With the increase of the water content in the production fluid, oil-water two-phase flow will be observed in the rock voids, the oil wells in the strata, and the long-distance oil pipelines on the ground. Recently, the interest in oil-water two-phase flow has greatly increased as a result of the outstanding interdisciplinary problems to the crude oil exploitation, transportation, reservoir, and production performance monitoring[2]. Understanding the dynamic behaviors of oil-water two-phase flow is crucial to important problems, such as pressure drop or water holdup predicting in the oil wells and pipelines[3-5]. The flow pattern, pressure drop, and water holdup are the primary concerns in the previous studies, and their measurement of two-phase flow is related to not only the flow stability but also the phase inversion phenomena[6-8]. Therefore, to realize the reasonable design and safe operation of pipelines, it is necessary to clarify the flow pattern as well as slip degree of oil-water two-phase flow in the pipeline. However, as a result of the interplay among many complex factors, such as the pipe diameter parameter, fluid property, experimental environment condition, and way of introduction of each phase, the study of oil-water two-phase flow exhibits a characteristic of diversity.

The study of horizontal oil-water two-phase flow can date back to the 1950s and can be divided into three stages. Early investigations on oil-water two-phase flow were mainly focused on experimental observations in small size glass or organic glass tubes based on the naked eye. Then, the related studies were gradually developed; however, the definition or discrimination of flow patterns still mainly depended upon the unaided eye, with the shortage of subjectivity[9]. In the flowing stage, with the development of the science and oil production process, scholars raised the research of oil-water two-phase flow into the experimental stage combined with high-speed photography to eliminate the subjectivity of flow pattern identification[10]. In recent years, the research entered the third stage; scholars studied the horizontal oil-water two-phase flow with analytical and numerical simulation analyses. On the basis of the Young-Laplace equation, scholars predicted the boundary shape of stratified flow when the flow of water and oil were both laminar flow. Moreover, the global and local flow characteristics of this fluid were estimated using the boundary element method[11,12].

A new oil-water test facility suitable for optical measurement is designed, constructed, and operated in this work to investigate the flow pattern transition and slip phenomenon in a horizontal circular pipe. The THz wave is used in the experiment because of its different responses of oil, water, and gas as a result of divergences in molecular configuration of them[13-15]. THz-TDS technology has drawn more and more attention on applications in the energy field, especially in the petroleum industry[16-19]. Owing to the sensitivity of THz to the fluctuations of water dipole moments that occur on the picoseconds (ps) time scale, the THz technology can detect the subtle change of water molecules[20]. A new means has been proposed to monitor the dynamic process of oil-water two-phase flow using THz-TDS. By comparison of the amplitudes of the maximum peak in THz-TDS, the ability of THz-TDS has been shown previously to discriminate the flow pattern for oil-water two-phase flow with a low water content (0.03%~2.30%) in a rectangular pipe[21]. For further study on oil-water two-phase flow, here, our emphasis is placed on the flow pattern and slip phenomenon changing along with the varied mixture flow rate (Q) as well as input water volume fraction (ε_w) in horizontal circular pipe.

2 Experimental section

A schematic diagram of the test facility is shown in Figure 1. The flow loop installation is built to simulate the flow of two immiscible liquids within a horizontal circular pipe. All experiments are conducted using diesel and tap water ($\mu_o = 3.30 \text{mPa} \cdot \text{s}$ and $\mu_w = 1.01 \text{mPa} \cdot \text{s}$) at a constant temperature 294K±0.3K and atmospheric outlet pressure. The fluids used are initially stored in a tank without mixing evenly. The mixing fluid is then sent into the test section using a pump from the storage tank. Before entry into the test section, the flow rate of the oil-water mixture is measured by a turbine flowmeter. The system consisted of transparent polytetrafluoroethylene horizontal circular pipes with 8mm internal diameter (D) and 1mm wall thickness, leading to a total 2.5m long (L) pipe (the aspect ratio L/D is equal to ~312) and some metal parts. The transparent test pipe allows for the observation of the flow pattern. Oil and water flow through the pipe and are then collected in the tank, where they can then be pumped back to the pipe. Conventional transmission THz-TDS is used in the experiment. The detailed transmission principal measurement of the THz-TDS system for analysis has been described in the previous report[22,23]. As shown in Figure 1, the focused THz pulse whose spot size and spatial resolution are 1.4mm and 150μm, respectively, transmits through the oil-water two-phase flow. The sample-encoded THz beam is then received by the detector, and the THz signal is transformed into an electrical signal, which will be sent to the computer.

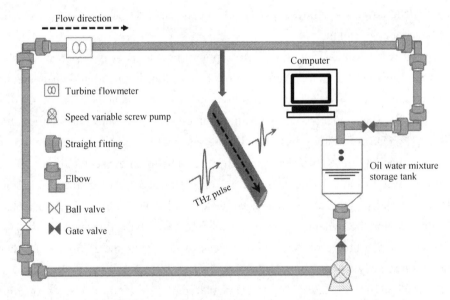

Figure 1 Schematic of the experimental facility for the detection of oil-water two-phase flow with transmission THz-TDS

The input water volume fraction of samples ranges from 0.5% to 5.0% in the measurements. As we all know, the Reynolds number is a dimensionless number that can be used to characterize fluid flow, and the critical Reynolds number is the Reynolds number at the transition point where flow structure changes; the critical Reynolds number includes upper and lower critical Reynolds numbers. In comparison to the upper critical Reynolds number, the lower critical Reynolds number is relatively stable; therefore, it is usually used as the criterion to judge laminar flow and turbulent flow in the hydromechanics[24]. To ensure the evenly mixing of oil and water at the same time, the mixture flow rate is controlled to decrease from $0.6m^3$/h to 0 in the measurements. The density and viscosity at 20℃ of the oil are $830.6kg/m^3$ and $3.30mPa \cdot s$, respectively. Furthermore, the water used in this study is tap water, whose density and viscosity are $998kg/m^3$ and $1.01mPa \cdot s$, respectively, at the same condition. The basic physical properties of the oil and water samples are listed in Table 1. In the work presented here, each experimental point consists of two types of measurements: the mixture flow rate and the THz signal. When the steady state is reached under a flow condition, the mixture flow rate as well as the raw data of the THz measure-ment are measured. The time that it takes to obtain a stable flow state is about 4min, and the principle to judge the steady state is keeping the data fluctuation of the flow rate under $0.005m^3$/h in the experiment.

Table 1 Liquid-phase properties measured at 20℃ and atmospheric outlet pressure

Liquid phase	Density ρ/(kg/m^3)	Viscosity μ /(mPa · s)
water	998	1.01
diesel	830.6	3.30

3 Results and discussion

THz-TDS of diesel oil-water two-phase flows with different input water volume fractions (0.5%, 1.0%, 1.5%, 2.0%, 2.5%, 3.0%, 3.5%, and 5.0%) has been measured, and the transmission waveforms (raw data) whose input water volume fraction is 2.5% are illustrated in Figure 2. It is apparent that the THz signal is changing with the varied flow rate of mixing fluid from two aspects (amplitude and time delay), which is consistent with the results in our previous research[21]. For amplitude, the THz signal has a trend of descent first and then stationary, while the time delay increases with the augment of the flow rate. The differences in the transmitted signal are believed to result from the different flow structures in the two-phase flows. To build a relationship between the transmitted signal and the mixture flow rate, the peak amplitudes for the two-phase flows with a diverse flow rate are derived from the measured THz wave forms. It can be seen that there is a transition point named the critical flow rate (Q_c) for the peak intensity from the inset picture of Figure 2. When Q is lower than Q_c, it's clear that the peak outputs decrease sharply with increasing Q; nevertheless, they almost remain unchanged when Q exceeds Q_c.

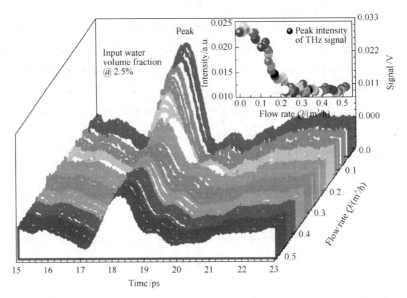

Figure 2 THz-TDS of the oil-water two-phase flow with input water volume fraction of 2.5% for different flow rates. The inset shows the peak intensities of THz signals for the two-phase flows with the changing flow rate

For different input water volume fractions, all of the THz-TDS peak intensities of diesel oil-water two-phase flows exist at the transition points when the flow rate ranges from $0.6 m^3/h$ to 0, as presented in Figure 3. In general, under the condition of input water volume fraction ranging from 0 to 100%, there are six basic flow patterns in horizontal circular pipes, which can be divided into two categories (segregated flow and dispersed flow) according to the classification of Trallero et al[25]. Stratified flow (ST) and stratified flow with some mixing at the interface (ST&MI) are segregated flow patterns, while the dispersed flow patterns include water dominated (a dispersion of oil-in-water over a water layer (DO/W & W) flow and an emulsion of oil-in-water (O/W) flow) or oil dominated (an emulsion of water-in-oil (W/O) flow and a dual dispersion (DW/O & DO/W) flow). However, limited by the low input water fraction in the experiment, the flow patterns mentioned above are not all observed here. In combination with the results of experimental observation and analysis, the flow structures of oil-water two-phase flow are distinguished into two types, including an oil layer over a dispersion of water-in-oil (O & DW/O) flow and an emulsion of W/O flow. For high oil-water mixture flow rates, the water phase is intensively impacted by the oil phase and dispersed almost evenly in the continuous oil phase with a small water drop concentration gradient and W/O flow is found when the water content in the test part is almost kept unchanged and the THz signal remains unchanged as well. With the decrease of the mixture flow rate, the droplets are mainly affected by gravity and the forces associated with the motion reduces. The mentioned forces are not big enough to overcome the effects of gravity to maintain all of the droplets suspended. For a further decrease in the mixture flow velocities, the number of droplets reduces, the maximum drop size increases, the flow is gravity dominated, and then the

O & DW/O flow pattern is formed. The decreasing amount of droplets and increasing drop size make the decrease of the water content in the test section and cause smaller absorbance for the THz pulse. Consequently, the flow pattern transition can be judged by the THz technique.

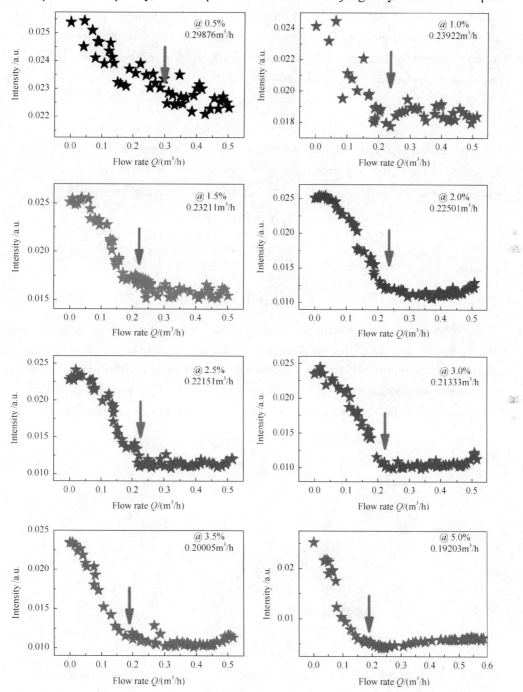

Figure 3 THz-TDS amplitudes for different diesel oil-water two-phase flows with input water volume fractions ranging from 0.5% to 5.0% as the measuring flow rate decreases from 0.6m³/h to 0m³/h

The flow-pattern map and water fraction dependence Q_c are depicted in Figure 4, where Q_c decreases from 0.28976m³/h to 0.19203m³/h and the Reynolds number (Re) ranges from 2238 to 3342 when the water fraction increases from 0.5% to 5%. The Reynolds number is the similarity criterion of viscosity in fluid mechanics and a dimensionless number that can be used to characterize fluid flow. The smaller the Re, the more significant influence of viscous force on flow, and the larger the Re, the more significant effect of inertia force on flow[26,27]. The Reynolds number can be expressed as

$$Re = \frac{\rho u D_H}{\mu} \quad (1)$$

where ρ, u, and μ are the density, velocity, and dynamic viscosity of the fluid, respectively. D_H represents the hydraulic diameter, and for a circular tube, it takes D as D_H. According to the homogeneous model, ρ, u, and μ can be calculated by the flowing equations:

$$\rho = \varepsilon_w \rho_w + (1 - \varepsilon_w)\rho_o \quad (2)$$

$$u = \frac{Q}{A} = \frac{4Q}{\pi D^2} \quad (3)$$

$$\mu = \varepsilon_w \mu_w + (1 - \varepsilon_w)\mu_o \quad (4)$$

In combination with the above relations, Re of the two-phase flow can be gained. In general, the Reynolds number when the flow pattern turns from laminar to turbulent is called the critical Reynolds number, denoted as Re_c. Experimental results have shown that the range of Re_c for flow in circular pipes is 2000~2600, which is related to the pipe geometry[28]. As observed in Figure 4, the corresponding Re_c for the critical flow rate is calculated as 3342~2238

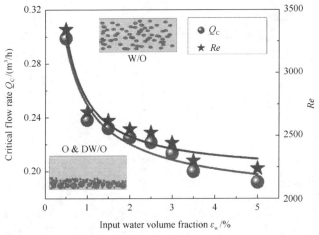

Figure 4   Critical flow rate Q_c and Reynolds number Re for different input water volume fractions of diesel oil-water two-phase flows. The insets present sketches of the flow patterns: an O & DW/O flow and an emulsion of W/O flow

along the increasing water fraction, which is in agreement with the transmission point calculated from the THz measurement.

In addition, the peak intensities of the THz signal are kept unchanged when the flow pattern remains as W/O flow, and under these circumstances, Q is bigger than Q_c in Figure 3. To confirm the relationship between the THz response of the emulsion of water in oil and the water fraction, the stable THz signals are acquired by averaging each intensity whose corresponding Q is greater than Q_c. The obtained stable signals are given in Figure 5 as the bigger stars, and the smaller black points represent the peak intensities of THz-TDS measured at different flow rates that are bigger than Q_c. A function of the water fraction is obtained by linearly fitting with the equation $y = -0.00421x + 0.0235$, whose correlation coefficient R is approximately 0.952, which is in agreement with the results in our previous articles[22,29]. THz-TDS of the flows with varying input water volume fractions at a flow rate of $0.50 m^3/h$ are plotted in the inset in Figure 5; it is obvious that there is a decrease in the peak intensities of the THz waveform with an increasing water fraction in the flows, whereas the phase shifts in an increased direction as the water content raises. THz-TDS can provide information about the water content by both amplitude and phase.

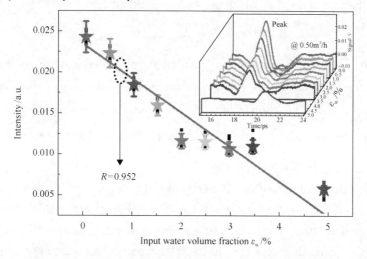

Figure 5　THz peak intensities of the oil-water two-phase flows with varying input water volume fraction ranging from 0.5% to 5.0%. The inset describes the time-domain waveforms corresponding to different water fractions at the flow rate of $0.50 m^3/h$

In terms of oil-water two-phase flow, as a result of the difference in physical properties of the two phases, there is a velocity difference between the dispersed phase and the continuous phase, causing the *in situ* volume fraction of oil and water to differ from the input volume fractions of them, which is generally called the slip phenomenon[30,31]. The slip can be expressed by the velocity ratio, S, which can be expressed as follows:

$$S = \frac{\varepsilon_o / \varepsilon_w}{\beta_o / \beta_w} \tag{5}$$

where ε_o and ε_w are the input volume fractions of oil and water, while β_o and β_w represent the *in situ* volume fractions of oil and water, respectively.

To investigate the effect of the input mixture flow rate on the slip, simulations have been performed under selected mixture flow rates for the two-phase flows with different input water fractions in this experiment. As shown in Figures 6(a) and (b), the calculation principle of water holdup and the ratio of *in situ* oil to water velocity is based on the energy conservation law to make the circular tube equal to a square tube whose height and width (L) are 8mm and 2π mm; thus, the absorbance spectra of the two-phase flows with different input water fractions are acquired according to the formula

$$A = \lg\left(I_0 / I\right) \tag{6}$$

where I_0 and I indicate the values of the THz frequency domain spectrum (FDS) of empty sample cell and sample, respectively[32,33]. For the mixture with two substances, the mixture absorbance is the sum of the two, which can be expressed as

$$A = \alpha_w L_w + \alpha_o L_o, \quad L = L_w + L_o \tag{7}$$

where α_w and α_o are the absorption coefficients of water and oil to the THz wave, respectively. L, L_w and L_o are the equivalent width of the sample cell, water, and oil. Using Equations 6 and 7, the equivalent width of water is acquired and then the water holdup can be obtained by formula $\beta_w = \frac{S_w}{S}$, as illustrated in Figure 6(c); calculated β_w almost constant in the range of 0.2~0.35THz.

Then, the curves of changing S acquired using Equation5 against the input water fraction at selected mixture flow rates are depicted in Figure 6(d), which shows that S is mainly dominated by the mixing flow rate at low flow rates, whereas the effect decreases gradually with an increasing flow rate. In the test range, S has a trend of increase first and then decrease with increasing ε_w, and it is the same with the result of the experiment of Xu et al[34]. As mentioned above, the flow patterns at low and high Q are O & DW/O and W/O flow, respectively. From the result in Figure 6(d), it appears that, at a given input water fraction, S increases with increasing Q when $Q<0.21 m^3/h$, which is close to most of the critical flow rates for the two-phase flows and is then maintained unchanged. In comparison of the data obtained at the five flow rates, it can be found that the ratio of *in situ* oil to water velocity for O & DW/O flow is smaller than that at W/O flow. This is caused by the discrepancies of the flow characteristic for the two flow types. For O & DW/O flow with low flow rates, the *in situ* water fraction should be less than the input water fraction as a result of the faster speed of water than oil, which makes S less than 1 according to Equation5. However, for W/O flow, oil is the faster flowing phase; thus, S is close to or bigger than 1 based on the results of Xu et al., Hapanowicz, and Rodriguez et al[34-37].

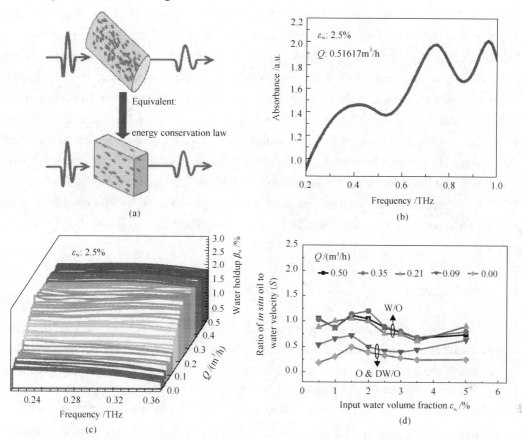

Figure 6 (a) Calculation principle of water holdup and S used in our experiments; (b) Absorbance spectrum of the oil-water two-phase flow with 2.5% ε_W at Q of 0.51617m³/h from 0.2THz to 1.0THz; (c) Obtained water holdup at different Q ranging from 0.51617m³/h to 0 for the two-phase flow, as mentioned above; (d) Curves of changing S against ε_W at selected Q, where the flow pattern is O & DW/O flow for low flow rates, while it is W/O flow for higher flow rates

The need for reliable detecting methods for two-phase flow in horizontal circular pipes has been the driving force behind an extensive research effort in this area over the past few decades[38]. THz-TDS analysis is able to retrieve both the physical properties and dynamic information (flow pattern) of oil-water two-phase flow by studying the changes in amplitude of the maximum peak in the impulse function[19,21,22]. In this experiment, in combination with the analysis of the changes in amplitude of the maximum peak, it is found that two flow patterns can form in the entire test range because of the higher mixture velocity and lower input water fraction. At high mixture flow rates, the water is intensively impacted and disturbed by the flow, the water phase is likely to disperse into the oil phase in the form of "water droplets", and then a homogeneous DW/O flow pattern forms. Under this circumstance, the water volume in the test section changes little, making THz-TDS change barely as well. As the flow rate decreases gradually, the two-phase interfacial tension greatly reduces and the forces associated with the motion become too small to maintain all of the water droplets suspended; thus, some of them eventually settle, but water in oil still exists in the bottom. Meanwhile, the free oil presents in the upper part and occupies a larger cross-section of the pipe, leading to less amount of water transmitted by THz waves. Hence, the THz signal augments with the decrease of the flow rate. Because of the physical property differences in oil and water, the flow speeds of the two vary with the flow pattern of two-phase flow. In this work, the potential of THz-TDS is also revealed to accurately calculate the slip between the two phases. Attention has been paid to the effect of the input mixture flow rate on the slip, and it is found that the flow rate has a dominate effect on the slip at low flow rates, while this influence is negligible at high flow rates. In other words, the THz technique allows us to achieve the characterization and quantitative analyses of oil-water two-phase flow in the oil industry.

4 Conclusions

An experimental study of oil-water two-phase flows has been conducted with the input water fraction ranging from 0.5% to 5.0% through the horizontal circular pipe using THz-TDS. The effects of the mixture flow rate on the flow pattern as well as slip have been analyzed both experimentally and theoretically. The work reveals that, at a low Q value, the flow pattern is O & DW/O flow and the ratio of *in situ* oil to water velocity is smaller than that at a high Q value, where the flow is W/O flow. In summary, the amplitude of THz-TDS of oil-water two-phase flow is corresponding to the flow pattern; the flow pattern and critical flow rate can be distinguished by THz-TDS. The flow pattern map is obtained by THz-TDS in this work, from which the flow pattern under a certain mixture flow rate can be identified. The water holdups at the flow rate ranging from 0.6m^3/h to 0 are also calculated, and then the comparison of S at several Q is made. The results reveal that the mixture flow rate has a prominent influence on S at low Q value, and the influence reduces gradually with Q increasing.

Acknowledgements

This work was supported by the National Basic Research Program of China (Grant No. 2014CB744302), the Specially Funded Program on National Key Scientific Instruments and Equipment Development (Grant No. 2012YQ140005), the China Petroleum and Chemical Industry Association Science and Technology Guidance Program (Grant No. 2016-01-07), and the National Natural Science Foundation of China (Grant No. 11574401).

References

[1] Morgan R G, Markides C N, Zadrazil I, et al. Characteristics of horizontal liquid-liquid flows in a circular pipe using simultaneous high-speed laser-induced fluorescence and particle velocimetry. Int. J. Multiphas. Flow., 2013, 49: 99-118.

[2] Gao Z K, Jin N O, Wang W X, et al. Motif distributions in phase-space networks for characterizing experimental two-phase flow patterns with chaotic features. Phys. Rev. E., 2010, 82: 016210.

[3] Ullmann A, Brauner N. Closure relations for two-fluid models for two-phase stratified smooth and stratified wavy flows. Int. J. Multiphas. Flow, 2006, 32: 82-105.

[4] Cai J, Chong L, Tang X, et al. Experimental study of water wetting in oil-water two phase flow—Horizontal flow of model oil. Chem. Eng. Sci., 2012, 73: 334-344.

[5] Ehizoyanyan O, Appah D, Sylvester O. Estimation of pressure drop, liquid holdup and flow pattern in a two phase vertical flow. IJET, 2015, 4: 241-253.

[6] Rhys G M, Markides C N. Horizontal liquid-liquid flow characteristics at low superficial velocities using laser-induced fluorescence. Int. J. Multiphas. Flow, 2012, 43: 101-117.

[7] Poesio P. Experimental determination of pressure drop and statistical properties of oil-water intermittent flow through horizontal pipe. Exp. Therm. Fluid Sci., 2008, 32: 1523-1529.

[8] Kumara W A S, Halvorsen B M, Melaaen M C. Single-beam gamma densitometry measurements of oil-water flow in horizontal and slightly inclined pipes. Int. J. Multiphas. Flow, 2010, 36: 467-480.

[9] Russell T W F, Hodgson G W, Govier G W, et al. Horizontal pipeline flow of mixtures of oil and water. Chem. Eng. Sci. Can. J. Chem. Eng., 1959, 37: 9-17.

[10] Arirachakaran S, Oglesby K D, Malinowsky M S, et al. An Analysis of Oil/Water Flow Phenomena in Horizontal Pipes. Oklahoma: SPE Production Operations Symposium, 1989.

[11] Ng T S, Lawrence C J, Hewitt G F. Interface shapes for two-phase laminar stratified flow in a circular pipe. Int. J. Multiphas. Flow, 2001, 27: 1301-1311.

[12] Ng T S, Lawrence C J, Hewitt G F. Laminar stratified pipe flow. Int. J. Multiphas. Flow, 2002, 28: 963-996.

[13] Horiuchi N. Searching for terahertz waves. Nat. Photonics, 2010, 4: 662.

[14] Zhang Y S, Han Z. Spoof surface plasmon based planar antennas for the realization of Terahertz hotspots. Fuel, 2015, 5: 18606.

[15] Lourembam J, Srivastava A, Laovorakiat C, et al. New insights into the diverse electronic phases of a novel vanadium dioxide polymorph: a terahertz spectroscopy study. Sci. Rep., 2014, 5: 9182.

[16] Zhan H L, Xiao L, Ge L, et al. Water adsorption dynamics in active carbon probed by terahertz spectroscopy. RSC Adv., 2015, 5: 14389-14392.

[17] Ge L N, Zhan H L, Leng W X, et al. Optical characterization of the principal hydrocarbon components in natural gas using terahertz spectroscopy. Energy & Fuels, 2015, 29: 1622-1627.

[18] Leng W X, Zhan H L, Ge L N, et al. Rapidly determinating the principal components of natural gas distilled from shale with terahertz spectroscopy. Fuel, 2015, 159: 84-88.

[19] Zhan H L, Wu S X, Bao R M, et al. Qualitative identification of crude oils from different oil fields using terahertz time-domain spectroscopy. Fuel, 2015, 143: 189-193.

[20] Choi D H, Son H, Jung S, et al. Non-bulk-like behavior of hydration water on fluid phase lipids revealed by terahertz (THz) spectroscopy. International Conference on Infrared Millimeter, and Terahertz Waves. 2012: 1-3.

[21] Feng X, Wu S X, Zhao K, et al. Pattern transitions of oil-water two-phase flow with low water content in rectangular horizontal pipes probed by Terahertz Spectrum. Opt. Express, 2015, 23: 1693-1699.

[22] Yan S, Zhan H L, Zhao K, et al. Simultaneous characterization of water content and distribution in high water cut crude oil. Energy & Fuels, 2016, 30: 3929-3933.

[23] Zhan H L, Zhao K, Xiao L Z, et al. Spectral characterization of the key parameters and elements in coal using terahertz spectroscopy. Energy, 2015, 93: 1140-1145.

[24] Schlichting H, Gersten K. Boundary-Layer Theory. 8th ed. Berlin: Springer, 2000: 3-26, 799.

[25] Trallero J L. A study of oil-water flow patterns in horizontal pipes. SPE Production & Facilities, 1997, 12: 165-172.

[26] Lebental B, Bourquin F. Visco-acoustic modelling of a vibrating plate interacting with water confined in a domain of micrometric size. J. Sound Vib., 2012, 331: 1870-1886.

[27] Petford N. Which effective viscosity? Mineral Mag., 2016, 73: 167-191.

[28] Trinh K T. On the critical Reynolds number for transition from laminar to turbulent flow. Phys., 2010: 1-39.

[29] Jin W J, Zhao K, Yang C, et al. Experimental measurements of water content in crude oil emulsions by terahertz time-domain spectroscopy. Apply. Geophys., 2013, 10: 506-509.

[30] Lucas G P, Panagioto poulos N. Oil volume fraction and velocity profiles in vertical, bubbly oil-in-water flows. Flow Meas. Instrum., 2009, 20: 127-135.

[31] Abubakar A, Al-Wahaibi T, Al-Hashmi A R, et al. Influence of drag-reducing polymer on flow patterns, drag reduction and slip velocity ratio of oil-water flow in horizontal pipe. Int. J. Multiphas. Flow, 2015, 73: 1-10.

[32] Heyden M, Sun J, Funkner S, et al. Dissecting the THz spectrum of liquid water from first principles via correlations in time and space. P. Natl. Acad. Sci. USA, 2010, 107: 12068-12073.

[33] Vieira F S, Pasquini C. Determination of cellulose crystallinity by terahertz-time domain spectroscopy. Anal. Chem., 2014, 86: 3780-3786.

[34] Xu J Y, Wu Y X, Chang Y, et al. Experimental investigation on the slip between oil and water in horizontal pipes. Exp. Therm. Fluid. Sci., 2008, 33: 178-183.

[35] Hapanowicz J. Slip between the phases in two-phase water-oil flow in a horizontal pipe. Int. J. Multiphas. Flow, 2008, 34: 559-566.

[36] Rodriguez O M H, Oliemans R V A. Experimental study on oil-water flow in horizontal and slightly inclined pipes. Int. J. Multiphas. Flow, 2006, 32: 323-343.

[37] Gao Z K, Shim H, Osher S. Level set based simulations of two-phase oil-water flows in pipes. J. Sci. Comput., 2007, 31: 153-184.

[38] Carneiro J N E, Fonseca J R, Ortega A J, et al. Statistical characterization of two-phase slug flow in a horizontal pipe. J. Braz. Soc. Mech. Sci., 2011, 33: 251-258.

Probing disaggregation of crude oil in a magnetic field with terahertz time-domain Spectroscopy

Chen Jiang[1,2] Kun Zhao[1,2] Lijan Zhao[2] Wujan Jin[2,3]
Yuping Yang[3] Shaohua Chen[2]

(1.State Key Laboratory of Petroleum Resources and Prospecting, China University of Petroleum, Beijing 102249, China; 2.Key Laboratory of Oil and Gas Terahertz Spectroscopy and Photoelectric Detection, China Petroleum and Chemical Industry Federation (CPCIF), Beijing 100723, China; 3.College of Science, Minzu University of China, Beijing 100081, China)

Abstract: Magnetic technology can significantly change the characteristics of crude oil, and the physical mechanism has received much attention for several years. To date, there is not yet an accurate conclusion. Terahertz time-domain spectroscopy(THz-TDS) is employed in this paper to characterize the optical parameters of selected wax-bearing crude oil under the action of a magnetic field. The corresponding absorption coefficient and extinction coefficient of samples are obtained by a fast Fourier transform frequency domain spectrum, and the aggregation characteristics of particles are analyzed in crude oil under a magnetic field. The decrease of the extinction coefficient may be suggested as qualitative proof of a magnetic-field-induced disaggregation of the suspended colloidal particles. THz-TDS may be a powerful tool to characterize the particle disaggregation state in crude oil.

1 Introduction

Petroleum, as the blood of the industry and a kind of nonrenewable resource, is becoming shorter, scarcer, and more and more valuable than before. Hence, enhancing the recovery rate is the key step. Over the past few years, the petroleum industry has paid more attention to heavy oil production. Because of high viscosity, poor fluidity, and high content wax, crude oil is difficult to exploit and gather. Some physical and chemical methods, such as heat treatment and chemical viscosity reduction[1-4], were used to reduce its viscosity. Although these techniques are commonly used, they have some limitations, such as large energy consumption, high cost, poor efficiency, and environmental pollution.

In recent years, magnetic technology has been promoted and applied in some oil fields because of low investment, high efficiency, convenience of installment, and easy maintenance[5-14]. Some authors showed that magnetic fields may improve the rheological properties of crude oil[10-13]. Loskutova et al.[14] found that magnetic treatment led to substantial changes in the paramagnetic, antioxidant, and viscosity characteristics of paraffin-based and high-viscosity oils.

Originally published in *Energy & Fuels*, 2014, 28(1): 438-487.

In the works above, the physical mechanism of magnetic treatment has been searched using many different kinds of technical methods. For example, there are the ultraviolet-visible (UV-vis) extinction experiments and theories on the physical mechanism of magnetic treatment[5]. More recent work has described both experimental and theoretical studies on electron paramagnetic resonance spectroscopy, nuclear magnetic resonance spectroscopy, and X-ray fluorescence[7,13]. The dark color of crude oil prevented us from identifying any aggregated particles with microscopes. To date, the movement characteristics of the particle in crude oil are still not clear, and various mechanisms of magnetic action have been proposed but not proven conclusively[14-22]. The key idea is that magnetic (or electric) field treatment aggregates asphaltene (or paraffin) particles inside crude oil into larger particles, while Evdokimov et al. have shown that magnetic treatment results in disaggregation of colloidal particles in a crude oil[5]. In recent years, terahertz (THz) spectroscopy has been used to study a variety of physical phenomena from atomic transitions to dynamics of biological molecules because it allows for the far-infrared optical properties of a material to be determined as a function of frequency[23]. THz-TDS is a coherent technique, in which both the amplitude and phase of a THz pulse are measured. Coherent detection enables direct calculations of both the imaginary and real parts of the refractive index. One can obtain the absorption coefficient and dielectric constant of the material. In this paper, to acquire more information concerning the magnetic behavior of crude oil, the magnetic field dependence of THz optical property of a selected crude oil was experimentally characterized using transmission THz-TDS. From the measured absorption coefficient and extinction coefficient in the THz region, the movement of particles in crude oil under a magnetic field are analyzed and identified. The present study suggests that a magnetic field can result in disaggregation of colloidal particles in a crude oil.

2 Experimental section

The Zahra crude oil sample is obtained from China National Petroleum Corporation (CNPC). The dehydration processing of the studied virgin crude oil is carried out by the electric method to reduce the water content. The exact water concentration of 0.01% is measured with Karl Fischer titration. The contents of asphaltene, colloid, and paraffin are 0.44%, 11.51%, and 8.29%, respectively. The density, viscosity, and American Petroleum Institute (API) gravity are confirmed to be $0.8968 g/cm^3$ (20℃), 245.22mPa·s (20℃), and 26.99, respectively. A transmission type of the THz-TDS detection method is adopted[24-26]. A polyethylene cuvette is selected as the sample cell because it has little absorption in the THz frequency range that could be deemed to be nearly transparent. The liquid oil sample is sealed in a 10mm sample cell with a thickness of 1mm. Thus, the sample thickness d is 8mm. As shown in Figure 1, a set of NdFeB permanent magnets are fixed on both sides of the sample cell and 14 groups of magnets are selected with different magnetic flux densities B (B =

11.3mT, 13.6mT, 15.6mT, 33.0mT, 33.6mT, 48.8mT, 50.1mT, 51.7mT, 55.2mT, 57.1mT, 69.1mT, 78.4mT, 99.2mT, and 118.5mT). Without removing the magnets in the process of testing, this paper implemented *in situ* detection of crude oil under magnetic treatment. The empty cell with a magnetic field is tested first as the reference signal, and crude oil is injected in the cell and laid aside for 30min. Then, the THz spectroscopy is measured within 2min. The temperature is maintained at 20℃ during the whole experimental process. For accuracy, all THz-TDS measurements are performed 3 times for a single sample and the average value was used to calculate the optical parameters. A fast Fourier transform is applied to the time-domain data to yield the power $P(d,v)$ and phase $\varphi(d,v)$ of each frequency component v. The absorption coefficient (α) and extinction coefficient (κ) are calculated from the ratio and the relative phase difference between the sample and reference power spectra.

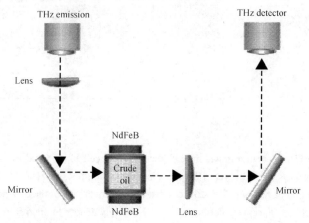

Figure 1 Experiment setup for magnetic treatment of crude oil with transmission THz-TDS

3 Results and discussion

The THz pulse waveforms transmitted through the empty cell are displayed in Figure 2(a) with selected magnetic flux densities ($B = 0$, 13.6mT, 15.6mT, 33.0mT, and 78.4mT), which are called as the reference pulses. It is clear that the maximum amplitude V_0 is about 0.175V and slightly dependent upon B. Figure 2(b) shows the THz time-domain spectra (TDS) of crude oil in the sample cell under different magnetic fields. The corresponding fast Fourier transforms are shown in Figure 2(c). A phase shift relative to the reference pulse occurs for THz pulses transmitted through the empty cell and sample, and a significant decrease in the amplitude is also observed. The distinct temporal shift δt is related to the thickness of the sample (d) and group index (n_g), with $n_g = 1 + c\delta t/d$, where c is the velocity of light in free space, and $n_g = 1.612$ for $B = 0$ is obtained. The maximum amplitude V_s is strongly dependent upon B and increased rapidly from 0.069V without B to 0.082V at $B = 13.6$mT. Figure 2(d) presents the B dependence of ΔV, defined as $V_s - V_0$. $\Delta V = -0.106$V before sample is exposed to the magnetic field. Then, $|\Delta V|$ is decreased to 0.081V at 13.6mT, 0.067V at 33.0mT, and 0.052V at 118.5mT, respectively.

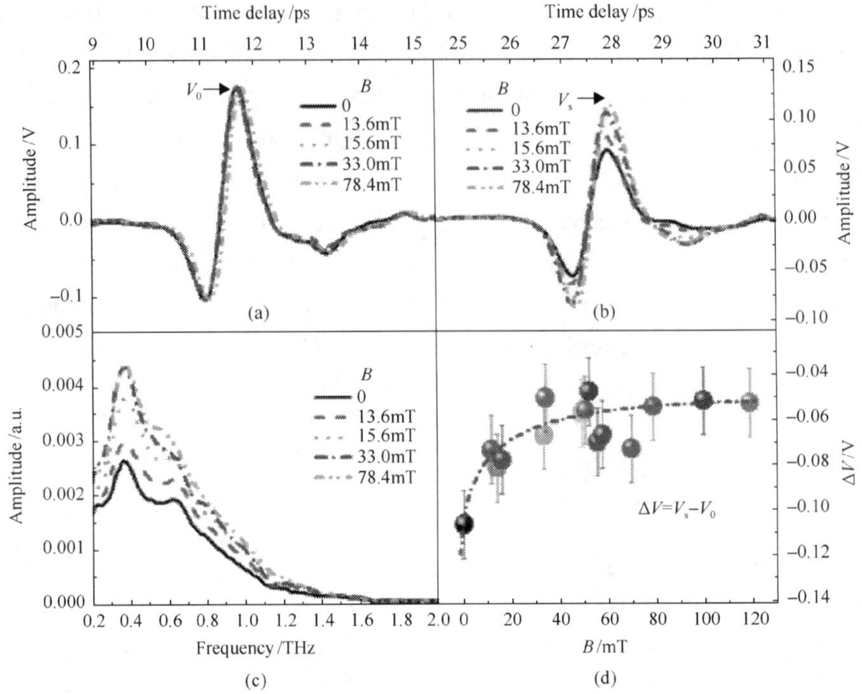

Figure 2 TDS of the THz wave transmitted through the (a) empty cell and (b) crude oil sample under selected magnetic flux density B. V_0 and V_s represent the maximum amplitudes of THz signals. (c) Frequency dependence of amplitude spectra is derived from fast Fourier transform of the data presented in (b). (d) Difference ΔV between V_0 and V_s, defined as $V_s - V_0$, strongly depended upon B from 0 to 118.5mT

The absorption coefficient is calculated as $\alpha(v) = -\ln[\beta P_s(d,v)/P_r(d,v)]/d$, where β is a correcting factor and $P_s(d,v)/P_r(d,v)$ is the ratio of transmission of the sample to the reference cell. An empty cell is used as a reference for the THz beam passing through four N_2/polyethylene interfaces, while in the filled cell with oil, it passes through two N_2/polyethylene and two oil/polyethylene interfaces. Thus, β is presented as $\beta = (n_1/n_3)^2 [(n_3 + n_2)/(n_1 + n_2)]^4$, where n_1, n_2, and n_3 denote the refractive indices of N_2, polyethylene, and oil, respectively. In our case, n_1 and n_2 are confirmed as 1.000298 and 1.513. The refractive index of the present crude oil is not a constant in the THz frequency range, which leads the correcting factor β to depend upon the frequency. To simplify the calculations, the average value of $n_3 = 1.612$ is used and $\beta \approx 0.920$. Figures 3(a) and (b) give the $\alpha(v)$ and $\kappa(v)$ of crude oil in the frequency range from 0.2THz to 2.0THz, where the extinction coefficient $\kappa(v)$ is determined by $\kappa(v) = c\alpha(v)/4\pi v$. According to the frequency dependence of power spectra [Figure 2(c)], a lower signal-to-noise ratio exists above 1.5THz and an effective frequency range is located in 0.2~1.5THz. The insets show the dependences of α and κ upon B at selected frequencies of 0.5THz, 0.7THz, 0.9THz, and 1.1THz. A decrease of α and κ is clearly seen at 0.2~1.5THz; e.g., α and κ at 0.5THz is changed from 1.99cm^{-1} and 0.0095cm^{-1} in the absence of a field to 0.30cm^{-1} and 0.0025cm^{-1} in a field of 78.4mT, respectively.

It is noticed that the original crude oil had no obvious absorption peak at 1.8THz, while the THz response forms a peak at this value with applied magnetic fields. The absorption of electromagnetic radiation at different frequency ranges corresponds to different physical processes and reveals information about those processes. Frequencies in the THz range correspond to motions of the entire molecular structure, involving relatively large masses and relatively shallow potentials. Previously, the absorption peak at 1.8THz has been obtained in the THz-TDS of grease, paraffin, aviation kerosene, and many other substances.[24,27-36] Considering that crude oil is a very complex polymer, it is difficult to determine which vibration of the concrete chemical bonds gives rise to the special phenomena around 1.8THz in Figure 3. In addition, according to the frequency dependence of power spectra in Figure 2(c), the effective frequency range is 0.2~1.5THz. Thus, it is difficult to determine the credibility of the peak at 1.8THz. We are conducting a further in-depth study with the hope to accurately confirm the effect of magnetization on the intermolecular force in crude oil by THz-TDS.

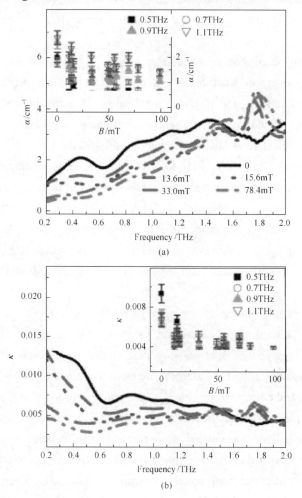

Figure 3　(a) Absorption coefficient α and (b) extinction coefficient κ versus frequency ν at selected magnetic flux densities B. The insets show the dependences of α and κ upon B

In addition to attenuation of the transmitted light by absorbance, light scattering by particles aggregated in crude oil also contributes to the light attenuation. In the Rayleigh limit, the light attenuation results from the absorbance and scattering contributions.[37,38] The extinction cross-section σ_{ext} is considered to be a sum of the absorbance cross-section σ_{abs} and scattering cross-section σ_{scat}. The influence of scattering increases extensively with an increasing diameter of the particles because $\sigma_{scat} = 2^7\pi^5 [(\eta^2-1)/(\eta^2+2)]^2 r^6 v^4/3c^4 \propto r^6$, where r is the radius of colloidal particles/aggregates and η is the ratio of the particle to continuous phase index of refraction.[27] Previous work showed that aggregation effects on absorbance are negligible at asphaltene concentrations characteristic of non-diluted crude oils.[39,40] Accordingly, the decrease of extinction κ may be suggested as qualitative proof of a magnetic-field-induced disaggregation of the size of suspended colloidal particles. Assuming an index ratio $\eta\sim1.2$, one evaluates the particle radius to be $\sim1.01\mu m$ in the absence of a magnetic field and $\sim0.64\mu m$ in a field of 78.4mT using 0.5THz for the frequency. Here, the suggested sizes of the aggregates would be merely speculative. Further research is needed to reveal the true aggregate nature.

Various mechanisms of magnetic action have been proposed but not conclusively proven. There exist controversies on whether an external magnetic field promotes aggregation or disaggregation. For example, the physical mechanism of a magnetically treated crude oil suggests that magnetic field treatment aggregates asphaltene (or paraffin) particles inside crude oil into large particles, and particle aggregation results in the observed viscosity decrease.[11,12,41] On the contrary, indirect evidence of post-treatment disaggregation is obtained by microscopy of deposits and the charged species in petroleum arise Lorentz forces, which can destroy molecular aggregates.[8,10,42] The magnetic treatment effect is also discussed in terms of intrinsic magnetic properties of asphaltene colloids, and disaggregation is suggested by measuring the inhibition centers.[14,15] In particular, the optical experiments reveal that magnetic treatment can result in disaggregation of colloidal particles in a crude oil.[5,39,40]

In addition, a shape change of the aggregated flocks can also be responsible for the change of the aggregation/disaggregation state in crude oil in the presence of a magnetic field. The shape of bulk aggregates, such as micelles, is due to a combination of interactions, including the hydrophobic effect, headgroup interaction (such as charge-charge interactions), packing constraints, and entropy of mixing.[43] For surface aggregates, the forces specific to the interface, such as electrostatic interactions and the hydrogen bond, should also be considered.[44-46] Asphaltene and colloid are surface-active substances in crude oil; therefore, asphaltene could be adsorbed on the wax particles, and then it works together with the colloid to form micelles.[47,48] Micelles in oil could aggregate into flocks.[49] When crude oil is exposed to an adequate magnetic field, the hydrogen bond between asphaltene molecules can be easily broken down, which may lead to the fracture of flocks.[20] In addition, paraffin molecules tend to align their poles along the magnetic field direction, and weak dipoles generate a repulsion force between these molecules, which results in a disturbance in the crystal agglomeration process and a change in the morphological property.[8,13] As a result, under the given magnetic fields, the aggregated flocks can be dispersed with the shape change from more extended to more globular. To date, nevertheless, there is a lack of conclusive data on the effects of magnetic fields on the phase behavior of crude oil.

Crude oil is composed of various organic matters, including alkanes, paraffin, and saturated hydrocarbons. In our case, the present crude oil is wax-bearing oil. The melting point of paraffin is generally higher than the experiment temperature of 20℃. Thus, wax crystal particles will precipitate from crude oil and aggregate flocks with the other components in crude oil. When paraffin with a colloidal state is dissolved in crude oil under a certain temperature, paraffin molecules will gather together and then crystallize. However, a magnetic field can induce the magnetic moment, change the molecular aggregation state, and make the wax crystal broken and dispersed.[8,13] It is suggested that a transient breakup of hydrogen bonds between asphaltene molecules can lead to the disaggregation of petroleum colloids.[20] In the original crude oil, as shown in Figure 4, some particles interlock each other, such as paraffin particles, asphaltene particles, etc. When crude oil is exposed to the magnetic field, the interlocked particles are dispersed and disaggregation occurs; thus, the average size of the suspended particle decreases. In fact, at the moment, the possible molecular mechanism of magnetic treatment in crude oil is still insufficient and under-investigated, which is expected to be proven by further studies.

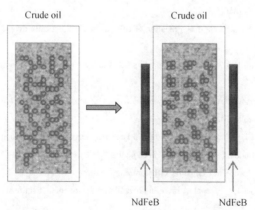

Figure 4 Schematic illustration of the proposed effect of the magnetic field influence on crude oil properties

4 Conclusions

In this study, crude oil under a magnetic field was characterized by THz-TDS, and the absorption coefficient and extinction coefficient were obtained. On the basis of these parameters, the particle movement characteristics were analyzed and identified in crude oil. A magnetic-field-induced disaggregation of the size of suspended colloidal particles is suggested by the decrease of the extinction coefficient with applied magnetic fields. THz-TDS represents a powerful tool for unearthing information about the particle aggregation characteristics for the petroleum industry.

Acknowledgements

The authors acknowledge the National Key Basic Research Program of China (Grant No. 2013CB328706), the Specially Funded Program on National Key Scientific Instruments and Equipment Development (Grant No. 2012YQ140005), the National Natural Science Foundation of China (Grants Nos. 11104360 and 11204191), and the Beijing National Science Foundation (Grant No. 4122064) for the financial support of this work.

References

[1] Ricardo D, Thomas F E. Study of heavy crude oil flows in pipelines with electromagnetic heaters. Energy & Fuels, 2012, 26: 4426-4437.

[2] Abuhesa M B, Hughes R. Comparison of conventional and catalytic *in situ* combustion processes for oil recovery. Energy & Fuels, 2009, 23: 186-192.

[3] Wang H, Yan W, He L, et al. Supporting tungsten oxide on zirconia by hydrothermal and impregnation methods and its use as a catalyst to reduce the viscosity of heavy crude oil. Energy & Fuels, 2012, 26: 6518-6527.

[4] Maity S K, Ancheyta J, Marroquin G. et al. Catalytic aquathermolysis used for viscosity reduction of heavy crude oils: A review. Energy & Fuels, 2010, 24: 2809-2816.

[5] Evdokimov I N, Kornishin K A. Apparent disaggregation of colloids in a magnetically treated crude oil. Energy & Fuels, 2009, 23: 4016-4020.

[6] Tao R, Huang K, Tang H, et al. Response to the comments: Fuel efficiency of internal combustion engines. Energy & Fuels, 2009, 23: 3339-3342.

[7] Goncalves J L, Bombard A, Soares D A W, et al. Study of the factors responsible for the rheology change of a Brazilian crude oil under magnetic fields. Energy & Fuels, 2011, 25: 3537-3543.

[8] Tung N P, Vinh N Q, Phong N T P. Perspective for using Nd-Fe-B magnets as a tool for the improvement of the production and transportation of Vietnamese crude oil with high paraffin content. Phys. B, 2003, 327: 443-447.

[9] von Flatern R. Steering clear of problem well paths. Offshore Eag., 1997: 37-39.

[10] Rocha N, González C, Marques L C D, et al. A preliminary study on the magnetic treatment of fluids. Pet. Sci. Technol., 2000, 18: 33-50.

[11] Tao R, Xu X. Reducing the viscosity of crude oil by pulsed electric or magnetic field. Energy & Fuels, 2006, 20: 2046-2051.

[12] Tao R. The physical mechanism to reduce viscosity of liquid suspensions. Int. J. Mod. Phys. B, 2007, 21: 4767-4773.

[13] Goncalves J L, Bombard A J F, Soares D A W, et al. Reduction of paraffin precipitation and viscosity of Brazilian crude oil exposed to magnetic fields. Energy & Fuels, 2010, 24: 3144-3149.

[14] Loskutova Y V, Yudina N V, Pisareva S I, et al. Effect of magnetic field on the paramagnetic, antioxidant, and viscosity characteristics of some crude oils. Pet. Chem., 2008, 48: 51-55.

[15] Morozov V I, Usatenko S T, Savchuk O V. Influence of a magnetic field on the physical properties of hydrocarbon fluids. Chem. Technol. Fuels Oils, 1977, 13: 743-746.

[16] El-Mohamed S, Achard M F, Hardouin F, et al. Correlations between diamagnetic properties and structural characters of asphaltenes and other heavy petroleum products. Fuel, 1986, 65: 1501-1504.

[17] Jordan T C, Shaw M T. Electrorheology. IEEE Trans. Dielectr. Electr. Insul., 1989, 24: 849-878.

[18] Semple K M, Cyr N, Fedorak P M, et al. Characterization of asphaltenes from Cold Lake heavy oil: Variations in chemical structure and composition with molecular size. Can. J. Chem., 1990, 68: 1092-1099.

[19] Tsouris C, Scott T C. Flocculation of paramagnetic particles in a magnetic field. J. Colloid Interface Sci., 1995, 171: 319-330.

[20] Evdokimov I N, Eliseev N Y. Thermally responsive properties of asphaltene dispersions. Energy & Fuels, 2006, 20: 682-687.

[21] Edamura K, Otsubo Y. Electrorheology of dielectric liquids. Rheol. Acta, 2004, 43: 180-183.

[22] Wen W, Huang X, Sheng P. Electrorheological fluids: Structures and mechanisms. Soft Matter, 2008, 4: 200-210.

[23] Ferguson B, Zhang X C. Materials for terahertz science and technology. Nat. Mater., 2002, 1: 26-33.

[24] Tian L, Zhou Q L, Jin B, et al. Optical property and spectroscopy studies on the selected lubricating oil in the terahertz range. Sci. China, Ser. G: Phys., Mech. Astron., 2009, 39: 1938-1943.

[25] Bao R M, Wu S X, Zhao K, et al. Applying terahertz time-domain spectroscopy to probe the evolution of kerogen in close pyrolysis systems. Sci. China: Phys., Mech. Astron., 2013, 56: 1603-1605.

[26] Hui Z, Zhao K, Zhao S Q, et al. Spectrum features of commercial derv fuel oils in terahertz region. Sci. China: Phys. Mech. Astron., 2012, 55: 195-198.

[27] Tian L, Zhou Q L, Zhao K, et al. Consistency-dependent optical properties of lubricating grease studied by terahertz spectroscopy. Chin. Phys. B, 2011, 20: 010703.

[28] Laman N, Harsha S S, Grischkowsky D, et al. High-resolution waveguide THz spectroscopy of biological molecules. Biophys. J., 2008, 94: 1010-1020.

[29] Yang Y P, Harsha S S, Shutler A J, et al. Identification of genistein and biochanin A by THz (far-infrared) vibrational spectra. J. Pharm. Biomed., 2012, 62: 177-181.

[30] Laman N, Harsha S S, Grischkowsky D. Narrow-line waveguide terahertz time-domain spectroscopy of aspirin and aspirin precursors. Appl. Spectrosc., 2008, 62: 319-326.

[31] Laman N, Harsha S S, Grischkowsky D, et al. 7 GHz resolution waveguide THz spectroscopy of explosives related solids showing new features. Opt. Express, 2008, 16: 4094-4105.

[32] Melinger J S, Laman N, Harsha S S, et al. High-resolution waveguide terahertz spectroscopy of partially oriented organic polycrystalline films. J. Phys. Chem. A, 2007, 111: 10977-10987.

[33] Melinger J S, Harsha S S, Laman N, et al. Temperature dependent characterization of terahertz vibrations of explosives and related threat meterials. Opt. Express, 2010, 18: 27238-27250.

[34] Melinger J S, Harsha S S, Laman N, et al. Guided-wave terahertz spectroscopy of molecular solids. J. Opt. Soc. Am. B, 2009, 26: 79-89.

[35] Harsha S S, Melinger J S, Qadri S B, et al. Substrate independence of THz vibrational modes of polycrystalline thin films of molecular solids in waveguide THz-TDS. J. Appl. Phys., 2012, 111: 023105.

[36] Liu H B, Zhang X C. Dehydration kinetics of D-glucose monohydrate studied using THz time-domain spectroscopy. Chem. Phys. Lett., 2006, 429: 229-233.

[37] Joshi N B, Mullins O C, Jamaluddin A, et al. Asphaltene precipitation from live crude oil. Energy & Fuels, 2001, 15: 979-986.

[38] Aske N, Kallevik H, Johnsen E E, et al. Asphaltene aggregation from crude oil and model system studied by high-pressure NIR spectroscopy. Energy & Fuels, 2002, 16: 1287-1295.

[39] Evdokimov I N, Eliseev N Yu, Akhmetov B R. Assembly of asphaltene molecular aggregates as studied by near-UV/visible spectroscopy. II. Concentration dependencies of absorptivities. J. Pet. Sci. Eng., 2003, 37: 145-152.

[40] Evdokimov I N, Losev A P. On the nature of UV/vis absorption spectra of asphaltenes. Pet. Sci. Technol., 2007, 25: 55-66.

[41] Tao R, Huang K, Tang H, et al. Electrorheology leads to efficient combustion. Energy & Fuels, 2008, 22: 3785-3788.
[42] Marques L C C, Rocha N O, Machado A L C, et al. Study of paraffin crystallization process under the influence of magnetic fields and chemicals. Proceedings of the 5th Latin American and Caribbean Petroleum Engineering Conference and Exhibition, Rio de Janeiro, 1997.
[43] Evans D F, Wennerström H. The Colloidal Domain. New York: VCH Publishers, 1994.
[44] Ducker W A, Wanless E J. Surface-aggregate shape transformation. Langmuir, 1996, 12: 5915-5920.
[45] Wanless E J, Davey T W, Ducker W A. Surface aggregate phase transition. Langmuir, 1997, 13: 4223-4228.
[46] Manne S, Schäffer T E, Huo Q, et al. Gemini surfactants at solid-liquid interfaces: Control of interfacial aggregate geometry. Langmuir, 1997, 13: 6382-6387.
[47] Lian H, Lin J R, Yen T F. Peptization studies of asphaltene and solubility parameter spectra. Fuel, 1994, 73: 423-428.
[48] Ali M F, Alqam M H. The role of asphltenes, resins and other solids in the stabilization of water in oil emulsions and its effects on oil production in Saudi oil fields. Fuel, 2000, 79: 1309-1316.
[49] Yen T F, Chilingarian G V. Asphaltenes and Asphalt. Amsterdam, Netherlands: Elsevier Science, 1994: 95-110.

Characterization of morphology and structure of wax crystals in waxy crude oils by terahertz time-domain spectroscopy

Chen Jiang[1,2] Kun Zhao[2] Cheng Fu[2] Lizhi Xiao[1]

(1.State Key Laboratory of Petroleum Resources and Prospecting, China University of Petroleum, Beijing 102249, China; 2.Beijing Key Laboratory of Optical Detection Technology for Oil and Gas, China University of Petroleum, Beijing 102249, China)

Abstract: The content, morphology, and structure of precipitated wax crystals are major factors affecting crude oil rheology. In this paper, model oils obtained by dissolving a realistic mixture of long-chain *n*-octacosane in diesel fuels were studied using terahertz time-domain spectroscopy (THz-TDS) and microscopy to gain insight into clusters composed of asphaltene and wax with increasing wax content. The fractal dimension was used for quantitative characterization of the morphology and structure of clusters in the model oils. From the measured absorption and extinction coefficients in the THz region, dynamic processes of the clusters in the model oils were analyzed and identified. The extinction coefficient in the THz region strongly depended on the dispersed and aggregated states of the asphaltene and wax crystals. These observations suggest that the aggregation state of the particles in model oils can be monitored with THz-TDS. In the future, THz-TDS technology may be used to effectively analyze particle dispersion or the aggregation state in crude oil and may thus be useful for rapid assessment of the effect of pour-point depressant on wax crystal aggregates.

1 Introduction

The wax content, morphology, and structure of wax crystals are among the most important factors influencing the flow properties of waxy crude oils[1-3]. The macroscopic rheology of crude oil is affected by heat and shear conditions, which affect the shape and structure of the wax crystals. Modification (including heat treatment of waxy crude oil and addition of admixtures) primarily alters the size of the wax crystals or the form and state of aggregation, and in this way achieves the goal of improving the macro rheological properties of crude oil. Therefore, study of the microscopic characteristics of the particles has an important guiding role in improving the low-temperature fluidity of crude oil.

Crude oil can be classified according to wax content. Various methods are applied to measure the wax content, for example, the standard acetone method, gas chromatography, nuclear magnetic resonance, density measurement techniques, and differential scanning calorimetry (DSC)[4-7]. Furthermore, the optical properties of petroleum and processed fuels, such as the refractive index and absorption coefficient, are already regarded as important[8-10].

The morphology and structure of the wax crystals play an important role in the flow properties of waxy crude oils and affect the apparent viscosity, pour point, and temperature of wax crystallization[11-16]. Kok et al.[17] studied the wax appearance temperatures (WAT) of crude oils by DSC, thermomicroscopy, and viscometry. They concluded that DSC and thermomicroscopy should be used together for a more accurate determination of the WAT of crude oils and that viscometry should be used to study the flow properties of crude oils below the WAT. Using thermomicroscopy, Létoffé et al.[18] observed that the crystals in mixtures of pure paraffins and a crude oil matrix are small and that their size depends on the length of the paraffinic chains at the pour point. Variation of the precipitation rate does not markedly affect the crystal size, and the crystals remain small (1~3μm). In addition, Radlinski et al.[19] studied the microstructures of model diesel fuels by dissolving a realistic mixture of n-alkanes in toluene or nitrobenzene. The microstructures of these solutions were comprehensively studied using absolute-calibrated small angle neutron scattering (SANS), small-angle X-ray scattering (SAXS), and dynamic light scattering techniques. The results indicated that flat aggregates of several paraffin-like molecules formed spontaneously when the paraffin concentration was above 10wt %. Furthermore, it was shown that the only solid component of the gelled crude oil was crystalline[20,21]. here have thus been many studies demonstrating that shear alters both the rheological behavior of waxy crude oils and the morphology of the wax crystals.

However, the work described above on the relationship between the oil composition and the morphology and structure of the wax crystals is primarily qualitative. Because of the high complexity and irregularity of wax crystal microstructures, quantitative characterization of their morphology and structure is difficult. The fractal dimension could help to solve this problem. Lorge, Kané, and co-workers assumed that wax clusters had a fractal structure and developed a simple relationship between the viscosity and fractal dimension[20-22]. Then, based on the scaling theory in the framework of colloidal suspensions, da Silva et al.[23] deduced that the fractal dimension was associated with the elastic modulus and determined fractal dimensions of 1.7, 1.9, and 2.2 for three waxy oils. Using the gel storage modulus, Uriev[23] also developed a correlation between the elastic modulus and fractal dimension. Yi et al.[24,25] used the fractal dimension of wax crystal microstructures to characterize the morphology and structure of wax crystals. This has made it easier to develop a relationship between the oil composition and the morphology and structure of the wax crystals.

Wax crystals dispersed in oil play a critical role in structural and dynamic processes. THz-TDS can be used as an experimental probe for wax crystals dispersed in oil. As a coherent technique, the amplitude and phase of a THz pulse can be measured at the same time. Both the imaginary and real parts of the refractive index are available. One can determine the absorption coefficient and dielectric constant of the material. Recently, the operation of spectrometers covering the entire THz range has been considerably simplified, and as a result the development of THz techniques for the generation and detection of THz radiation based on time-domain spectroscopy has received more and more attention[26]. THz-TDS is particularly sensitive to the collective vibration modes of liquids; the amplitude and spectral position of the resonance depend on the size of the water pools at the core of sodium bis-2-ethylhexyl-sulfosuccinate (AOT) micelles, and the signal is absent in bulk water[27,28]. Differences in the absorption intensities and the refractive indices obtained using this technique reflect the correlation of the organic components with the material properties and can be used to distinguish inflammable liquids and quantitatively determine their presence in fuels[29-32]. Research on the dielectric properties of water-oil mixtures, in particular alkanes, using THz-TDS has been reported[33,34].

Because crude oil is a complex mixture of aromatics, paraffins, naphthenes, asphaltenes, resins, and other organic components, we studied a model oil to gain insight into the structure of clusters composed of asphaltene and wax with increasing wax content using THz-TDS together with microscopic imaging techniques. The model oil was prepared by dissolving a mixture of long-chain n-octacosane in diesel fuel. The fractal dimension of the clusters composed of asphaltene and wax was determined on the basis of microscopic images of model oil samples and was used to characterize the morphology and structure of the clusters in the model oil. The same samples were then experimentally characterized using transmission THz-TDS. The measured absorption and extinction coefficients in the THz region were used to analyze and identify the dynamic processes of particles in the model oils.

2 Materials and methods

2.1 Materials and sample preparation

The diesel fuels and analytically pure n-octacosane were supplied by Sinopec (Beijing) and Aladdin Chemistry (Shanghai) Co., respectively. The n-octacosane had a purity of more than 97%, a molecular weight of 394.76g/mol, and a melting point of approximately 62℃. The exact water concentration of the diesel fuels was 71.6ppm. The density, viscosity, and solidifying point of the diesel fuels were determined to be 0.8306g/cm^3 (20℃), 3.2975mPa·s (20℃), and −16℃, respectively. Asphaltene was initially separated from diesel fuel, and the content of asphaltene was determined to be 0.32% according to the Chinese Standard Petroleum Test Method SH/T 0509—92. Using the weighing method, experimental samples with different wax contents, C_W, were prepared by dissolving n-octacosane in diesel fuels. The solutions were not oversaturated, and a consistent one-phase mixture (slightly yellow transparent liquid) was obtained. Each sample was heated at 80℃ for 1 h in a closed container and shaken thoroughly to obtain complete dissolution of the wax. The specimens were then kept at room temperature and cooled statically for at least 24h before use to achieve better reproducibility. All DSC thermal analyses were performed using a TA2000/MDSC2910 DSC apparatus (TA

Instruments, New Castle, DE) according to the Chinese Standard Petroleum Test Method SY/T 0545-2012. Tests were performed from 50℃ to −20℃ at a cooling rate of 5℃/min to determine the WAT and the concentration of precipitated wax at temperatures between the WAT and −20℃. The WAT was 0.63℃, −0.04℃, −0.23℃, 1.39℃, 1.01℃, 1.96℃, 3.97℃, 5.11℃, 6.25℃, 7.88℃, 10.55℃, and 16.27℃ for samples 1~12, respectively. Optical microscopy provided information about particle aggregation kinetics, fractal properties, and floc compactness. A Nikon OPTIPHOT2-POL polarizing microscope was used to obtain microscope images of the samples at room temperature. Taking oil sample 12 as an example, its micrograph is shown in Figure 1. The long gray objects in Figure 1 indicate tiny scratches on the surface of the glass slide.

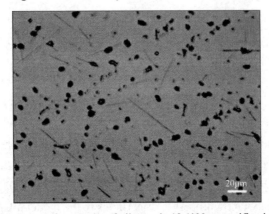

Figure 1   Micrographs of oil sample 12 (400×magnification)

2.2 THz-TDS Spectroscopy

As shown in Figure 2, a conventional transmission THz-TDS system with a mode-locked Ti-sapphire laser (MaiTai, Spectra Physics) was used for this study. The amplitude and phase information on the sample were obtained by THz-TDS utilizing the principle of coherent measurement of light. A Ti-sapphire laser with a center wavelength of 800nm, a repetition rate of 80MHz, a pulse width of 100fs, and an output power of 960mW were used. THz radiation was generated by an emitter composed of photoconductive antenna. An optical lens was used to focus the THz pulses onto a sample, and then a THz beam carrying information from the sample met the probe laser beam at the ZnTe crystal in the THz detector[35]. A lock-in amplifier was used to amplify the signal. The THz beam path was purged with nitrogen (N_2) to minimize the absorption of water vapor. The humidity was kept at less than 3.0%. A 1mm thick polystyrene (PS) vessel with a size of 10mm×10mm×45mm was selected as the sampler to improve the signal-to-noise ratio. PS has little absorption in the THz range, and a PS cell is thus an ideal sampler for THz measurements. Sample preparation was carried out in a glass container. Polystyrene sample vessels were used only for the THz-TDS measurements. The entire test process lasted no more than 1min, during which corrosion did not occur to any notable extent. THz-TDS was measured for the reference and samples by scanning the empty

PS cell and the PS cells containing the crude oils. Spectral information was obtained after applying fast Fourier transform (FFT).

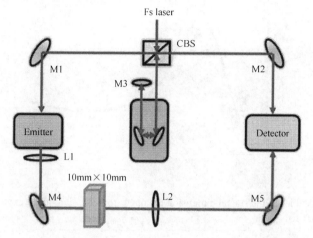

Figure 2　Diagram of the transmission THz-TDS setup and sample cell

3　Results and discussion

All of the microphotographs were processed and analyzed with ImageJ software. The fractal dimension, F, and the average diameter, D, of the particles were also calculated with ImageJ. F was determined from the slope of the regression line for the log-log plot of box size (or scale) and count. Figures 3 and 4 show the dependence of F and D on C_W. Changes in the morphology and structure of the particles were accurately reflected in the value of F. When C_W<0.2wt %, the particles were dispersed uniformly and the fractal dimension was small. In contrast, as the wax content increased, the particles assembled as aggregates or masses with higher fractal dimensions and more intricate structures. When C_W was 0.4027wt %, F increased to a maximum of 1.19767. Thereafter, F declined to 1.16233 at 0.491wt %, 1.13896 at 0.769wt %, and 1.11679 at 0.976wt %. Figure 4 shows that D was between 3μm and 6μm. An increase in D was clearly observed between 0 and 1.19767wt %, and the value of D changed from 3.515μm at 0.1082wt % to 4.630μm at 1.9157wt %. The maximum value of D was 5.493μm at 0.976wt %. The WAT of diesel oil (C_W = 0) was 0.63℃. However, when C_W was increased to 0.1082wt %, the WAT decreased to a minimum of −0.04℃. An increase in the WAT was then observed between 0.1082wt % and 1.9157wt %, with a maximum value of 16.27℃ at 1.9157wt %. This indicates that changes in the wax content of the samples have taken place, and changes in the parameters F and D show that the size and aggregation state of the particles has also changed.

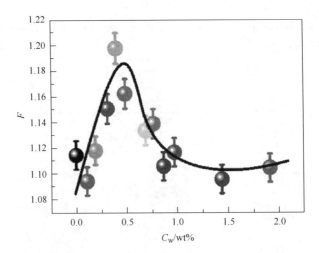

Figure 3 Values of the fractal dimension obtained from the 12 model waxy oil samples

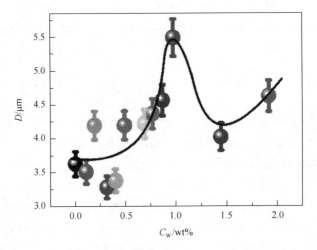

Figure 4 Average diameter of the particles in the model waxy oils

Figure 5(a) shows the THz waveforms, measured in transmission, of samples with several different wax contents; these represent typical experimental results. In Figure 5(a), the abscissa represents the time delay, and the ordinate represents the THz signal strength. For these time-domain data, the sample was held in a cell with a fixed path length of 10mm, and a reference pulse was first obtained by scanning the empty PS cell. The waveforms of all of the spectra were different, not only in amplitude, but also in peak time, which indicates that the THz-TDS technique can simultaneously give amplitude and phase information about the samples. A phase shift of approximately 15ps relative to the reference pulse occurred for THz pulses transmitted through the samples. After application of FFT, the THz frequency-domain spectra (THz-FDS) were calculated; the results were shown in the inset of Figure 5(a). The abscissa and ordinate of the inset in Figure 5(a) represent the frequency and amplitude, respectively. The frequency dependence of the amplitude spectra shows that a lower signal-to-noise ratio exists above 1.5THz and that the effective frequency range is located at 0.2~1.5THz. The maximum amplitude, E_P, is strongly dependent on C_W. As shown in Figure 5(b), EP increased rapidly from 0.126V for the sample without *n*-octacosane to 0.147V at 0.4wt %, then decreased to 0.135V at approximately 0.9wt %, and finally increased again with increasing C_W.

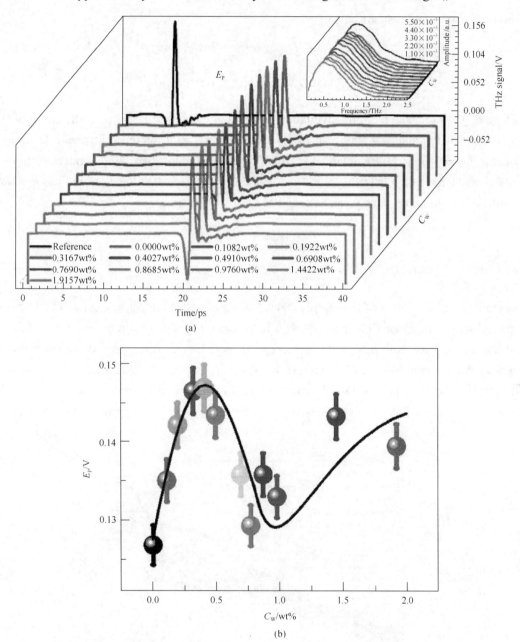

Figure 5 (a) Terahertz time-domain spectroscopy and terahertz frequency-domain spectroscopy of the samples and reference; (b) maximum amplitude, E_P, for oils with different wax contents

THz-TDS can be used to obtain the THz signal, $E_{sam}(t)$, containing the sample information and the reference signal, $E_{ref}(t)$, without the sample information. $\tilde{E}_{sam}(\omega)$ and $\tilde{E}_{ref}(\omega)$ are obtained by Fourier transform. The complex transmission function of the sample, $H_{measure}(\omega)$ can be calculated by $\tilde{E}_{sam}(\omega)/\tilde{E}_{ref}(\omega)$ [36,37]. The frequency dependence of the refractive index and absorption coefficient are calculated as

$$n_1(\omega) = \arg[H_{measure}(\omega)]c/(2\omega d_1) + n_0 \qquad (1)$$

$$\alpha_1(\omega) = d_1^{-1} \ln 16 n_0^2 n_1^2 / \{[|H_{measure}(\omega)|](n_1+n_0)^4\} \qquad (2)$$

where n_0 and n_1 denote the refractive indices of N_2 and PS, respectively; α_1 is the absorption coefficient of PS, and $d_1 = 1$mm. An empty cell was used as a reference for the THz beam passing through four N_2/PS interfaces, while it passes through two N_2/PS and two oil/PS interfaces in the cell filled with oil[38,39]. According to the transfer function, the absorption coefficient, α, of liquid sample can be calculated by

$$n_2(\omega) = \arg[H_{measure}(\omega)]c/\omega d_2 + n_0 \qquad (3)$$

$$\alpha_2(\omega) = 2d_2^{-1} \ln n_2(n_0+n_1)^2 / \{[|H_{measure}(\omega)|](n_1+n_2)^2 n_0\} \qquad (4)$$

where d_2 is the thickness of the sample (8mm). These data are shown in Figure 6(a). In this spectral range, the absorption spectrum does not contain any sharp features, but rather shows an obvious increase with increasing frequency. At room temperature, most of the vibrational peaks exhibit inhomogeneous line broadening (>0.2THz of the full width at half maximum (FWHM)), and the superposition of vibrational peaks results in complicated spectral features that are often difficult to resolve. Consequently, the analysis of THz spectra is often difficult

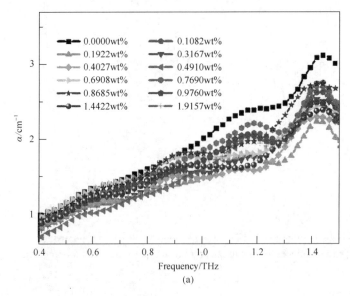

(a)

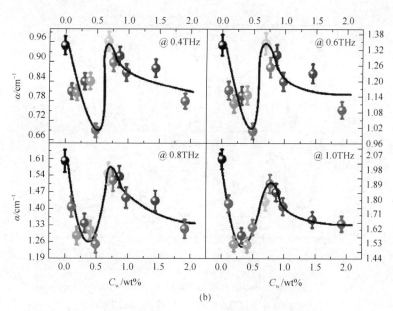

Figure 6 (a) Change in THz absorption coefficients over the entire frequency range and (b) THz absorption coefficients of oils with different wax contents at 0.4THz, 0.6THz, 0.8THz, and 1.0THz

owing to the broadened and overlapped nature of the vibrational peaks. Figure 6(b) illustrates the dependence of α on C_w at selected frequencies of 0.4THz, 0.6THz, 0.8THz, and 1.0THz. From these spectra it can be concluded that α did not change monotonically with C_w, for example, the value of α at 1.0 THz increased from 1.48cm^{-1} to 1.85cm^{-1} when C_w was increased from 0.19wt % to 0.77wt % and then decreased to 1.61cm^{-1} when C_W was 1.91wt %.

Previously, a decrease in the extinction coefficient, κ, has been suggested as qualitative proof of the magnetic-field-induced disaggregation of suspended colloidal particles[40]. In this work, light scattering due to particle aggregation in the sample plays an important role in the light attenuation in addition to attenuation of the transmitted light caused by absorbance[41]. Rayleigh scattering is suitable if $A = 2\pi D/\lambda \ll 1$, where D is the diameter of the particle and λ is the wavelength. Here, the average diameter, D, is 3~6μm and A is 0.06~0.12 at 1.0THz. At the Rayleigh limit, the extinction cross section is the sum of the absorbance cross section and the scattering cross section. The influence of the scattering increases dramatically with an increase in the diameter of the particles because the scattering cross section is proportional to D^6[40]. A decrease in the extinction coefficient, κ, may thus be suggested as a qualitative proof of disaggregation of the particles and aggregates. Because $\kappa = c\alpha/(4\pi v)$, where c is the velocity of light and v is the THz frequency, κ has the same dependence on C_w as α, as shown in Figure 7, which indicates that the particles disaggregated, aggregated, and then eventually disaggregated with increasing C_w.

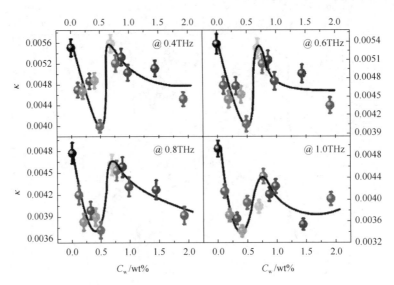

Figure 7 Terahertz extinction coefficients of oils with different wax contents at 0.4THz, 0.6THz, 0.8THz, and 1.0THz

The aggregation state of the particles is mainly influenced by intermolecular forces. These forces include attractive and repulsive intermolecular forces. As nonpolar molecules, n-alkanes have no measurable dipole moment, and the attractive dipole-dipole force between two molecules is exceedingly small. The principal contribution to the intermolecular attraction is thus this very weak induced-dipole interaction[42,43]. Therefore, the intermolecular interaction between wax and oil results mainly from weak van der Waals forces[22]. Owing to competition between attractive and repulsive forces, the particles will tend to be connected if the total effect of the interactions is attractive, whereas they will remain in a state of dispersion if it is repulsive.

4 Conclusions

In this work, the microstructures of model oils were studied using a polarizing microscope. The fractal dimension accurately reflected changes in the morphology and structure of the particles in the model oils. In addition, the THz optical properties of model waxy oils were investigated. From the measured absorption and extinction coefficients in the THz region, dynamic processes of the particles in the model oils were analyzed and identified. The extinction coefficient in the THz range strongly depended on the dispersed and aggregated state of the asphaltene and wax crystals. These observations suggest that the aggregation state of the particles in the model oils can be monitored with THz-TDS. In the future, THz-TDS technology may be used for the effective analysis of particle dispersion or the aggregation state in crude oil. When the aggregation states of particles before and after adding a pourpoint depressant are compared, THz spectroscopy can evaluate beneficial or deleterious effects produced by the addition of a pour-point depressant.

Acknowledgements

This work was supported by the National Basic Research Program of China (Grant No. 2014CB744302), the Specially Funded Program on National Key Scientific Instruments and Equipment Development (Grant No. 2012YQ140005), and the National Nature Science Foundation of China (Grant No. 11574401).

References

[1] Yang X L, Peter Kilpatrick. Asphaltenes and waxs do not interact synergistically and coprecipitate in solid organic deposits. Energy & Fuels, 2005, 19: 1360-1375.

[2] Visintin R F G, Lockhart T P, Lapasin R, et al. Structure of waxy crude oil emulsion gels. J. Non-Newtonian Fluid Mech., 2008, 149: 34-39.

[3] Venkatesana R, Nagarajanb N R, Pasoa K, et al. The stength of paraffin gels formed under static and flow conditions. Chemical Engineering Science, 2005, 60: 3587-3598.

[4] Yang C D, Gu K Y, Wu W H. Petrochemical Analysis. Beijing: Science Press, 1990.

[5] Chinese Standard Petroleum Test Method SY/T 7550-2000. Determination of Wax, Resins, and Asphaltenes Contents in Crude Oil.

[6] Chen J, Zhang J J, Li H Y. Determining the wax content of crude oils by using differential scanning calorimetry. Thermochimica Acta, 2004, 410: 23-26.

[7] Castillo J, Gutierrez H, Ranaudo M, et al. Measurement of the refractive index of crude oil and asphaltene solutions: Onset flocculation determination. Energy & Fuels, 2010, 24: 492-495.

[8] Tian L, Zhou Q L, Zhao K, et al. Consistency denpendent optical properties of lubricationg grease studied by terahertz spectroscopy. Chinese Physics B, 2011, 20(1): 010703.

[9] Tian L, Zhou Q L, Jin B, et al. Optical property and spectroscopy studies on the selected lubricating oil in the terahertz range. Sci. China: Phys. Mech. Astron., 2009, 39: 1938-1943.

[10] Al-Douseri F M, Chen Y Q, Zhang X C. THz wave sensing for petroleum industrial application. Int. J. Infrared Millimeter Waves, 2006, 27: 481-503.

[11] Pedersen K S, Rønningsen H P. Influence of wax inhibitors on wax appearance temperature, pour point, and viscosity of waxy crude oils. Energy & Fuels, 2003, 17: 321-328.

[12] Agarwal K M, Purohit R C, Surianarayanan M, et al. Influence of waxes on the flow properties of bombay high crude. Fuel, 1989, 68: 937-939.

[13] Rønningsen H P, Bjørndal B, Hansen A B, et al. Wax precipitation from North Sea crude oils. 1. Crystallization and dissolution temperatures, and Newtonian and non-Newtonian flow properties. Energy & Fuels, 1991, 5: 895-908.

[14] Lakshmi D S, Purohit R C, Srivastava S P, et al. Low temperature flow characteristics of some waxy crude oils in relation to their composition: part II. Effect of wax composition and concentration on the dew axed crude oils with/without additives. Petrol. Sci. Technol., 1997, 15: 685-697.

[15] Chanda D, Sarmah A, Borthakur A, et al. Combined effect of asphaltenes and flow improvers on the rheological behaviour of Indian waxy crude oil. Fuel, 1998, 77: 1163-1167.

[16] Zhang S F, Sun L L, Xu J B, et al. Aggregate structure in heavy crude oil: Using a dissipative particle dynamics based mesoscale platform. Energy & Fuels, 2010, 24: 4312-4326.

[17] Kok M V, Létoffé J M, Claudy P, et al. Comparison of wax appearance temperatures of crude oils by differential scanning calorimetry, thermomicroscopy and viscometry. Fuel, 1996, 17: 787-790.

[18] Létoffé J M, Claudy P, Kok M V, et al. Crude oils: Characterization of waxes precipitated on cooling by DSC. and thermomicroscopy. Fuel, 1995, 74: 810-817.

[19] Radlinski A P, Barré L, Espinat D. Aggregation of *n*-alkanes in organic solvents. Journal of Molecular Structure, 1996, 383: 51-56.
[20] Kané M, Djabourov M, Volle J L, et al. Correction of biased time domain NMR estimates of the solid content of partially crystallized systems. Appl. Magn. Reson., 2002, 22: 335-346.
[21] Kané M, Djabourov M, Volle J L, et al. Morphology of paraffin crystals in waxy crude oils cooled in quiescent conditions and under flow. Fuel, 2003, 82: 127-135.
[22] Lorge O, Djabourov M, Brucy F. Crystallisation and gelation of waxy crude oils under flowing conditions. Oil. Gas. Sci. Technol., 1997, 52: 235-239.
[23] da Silva J A L, Coutinho J A P. Dynamic rheological analysis of the gelation behaviour of waxy crude oils. Rheol. Acta, 2004, 43: 433-441.
[24] Yi S Z, Zhang J J. Relationship between waxy crude oil composition and change in the morphology and structure of wax crystals induced by pour-point-depressant beneficiation. Energy & Fuels, 2011, 25: 1686-1696.
[25] Yi S Z, Zhang J J. Shear-induced change in morphology of wax crystals and flow properties of waxy crudes modified with the pour-point depressant. Energy & Fuels, 2011, 25: 5660-5671.
[26] Mittleman D. Sensing with Terahertz Radiation. Heidelberg: Spring-Verlag, 2002.
[27] Boyd J E, Briskman A, Colvin V L. Direct observation of terahertz surface modes in nanometer-sized liquid water pools. Phys. Rev. Lett, 2001, 87: 147401.
[28] Boyd J E, Briskman A, Sayes C M, et al. Terahertz vibrational modes of inverse micelles. J. Phys. Chem. B, 2002, 106: 6346-6353.
[29] Ikeda T, Matsushita A, Tatsuno M, et al. Investigation of inflammable liquids by terahertz spectroscopy. Appl. Phys. Lett, 2005, 87: 034105.
[30] Naftaly M, Foulds A P, Miles R E, et al. Terahertz transmission spectroscopy of nonpolar materials and relationship with composition and properties. Int. J. Infrared Millimeter Waves, 2005, 26: 55-64.
[31] Al-Douseri F M, Chen Y Q, Zhang X C. THz wave sensing for petroleum industrial applications. Int. J. Infrared Millimeter Waves, 2006, 27: 481-503.
[32] Zhan H L, Wu S X, Bao R M, et al. Qualitative identification of crude oil from different oil fields using terahertz time-domain spectroscopy. Fuel, 2015, 143: 189-193.
[33] Gorenflo S, Tauer U, Hinkov I, et al. Dielectric properties of oil-water complexes using terahertz transmission spectroscopy. Chem. Phys. Lett, 2006, 421: 494-498.
[34] Laib J P, Mittleman D, Temperature-dependent terahertz spectroscopy of liquid *n*-alkanes. J. Infrared. Milli. Terahz. Waves, 2010, 31: 1015-1021.
[35] Löffler T, Hahn T, Thomson M, et al. Large-area electro-optic ZnTe terahertz emitters. Opt. Express, 2005, 13: 5353-5362.
[36] Dorney T D, Baraniuk B G, Mittleman D M. Material parameter estimation with terahertz time-domain spectroscopy. J. Opt. Soc. Am. A, 2001, 18: 1562-1571.
[37] Duvillaret L, Garet F, Coutaz J L. Highly precise determination of optical constants and sample thickness in terahertz time-domain spectroscopy. Appl. Opt., 1999, 38: 409-415.
[38] Mickan S P, Zhang X C. T-ray sensing and imaging. Int. J. High Speed Electron. Syst., 2003, 13: 601-676.
[39] Dorney T D, Baraniuk B G, Mittleman D M. Material parameter estimation with terahertz time-domain spectroscopy. J. Opt. Soc. Am. A, 2001, 18: 1562-1571.
[40] Jiang C, Zhao K, Zhao L J, et al. Probing disaggregation of crude oil in magnetic field with terahertz time-domain spectroscopy. Energy & Fuels, 2014, 28: 483-487.
[41] Kerker M. The Scattering of Light and Other Electromagnetic Radiation. London: Academic Press, 1969.
[42] Carey F A. Organic Chemistry. New York: McGraw-Hill Inc., 1992.
[43] Wessel R, Ball R C. Fractal aggregates and gels in shear flow. Phys. Rev. A: At., Mol., Opt. Phys., 1992, 46: R3008-R3011.

Optical characterization of the principal hydrocarbon components in natural gas using terahertz spectroscopy

Lin Ge[1,2] Honglei Zhan[2,3] Wenxiu Leng[3] Kun Zhao[3] Lizhi Xiao[1]

(1.State Key Laboratory of Petroleum Resources and Prospecting, China University of Petroleum, Beijing 102249, China; 2.Beijing Key Laboratory of Optical Detection Technology for Oil and Gas, China University of Petroleum, Beijing 102249, China; 3.Key Laboratory of Oil and Gas Terahertz Spectroscopy and Photoelectric Detection, China Petroleum and Chemical Industry Federation (CPCIF), Beijing 100723, China)

Abstract: A rapid technique is necessary to detect the natural gas which is a more and more significant fuel resource in modern industry. Terahertz (THz) technique was employed in this research to detect the principal hydrocarbon components of natural gas including methane, ethane, and propane. Two- and three-component mixtures were measured by THz setup, respectively. The amplitude ratio and time delay deviation of THz peaks between samples and reference were calculated. Phase projection pictures were obtained between the component concentrations and the amplitude ratio as well as time deviation. The phase figures evidently reflected the concentration dependent THz response, and a greatly different distribution was located in the whole phase projection area. In addition, back-propagation artificial neural networks method was utilized for the quantitative determination of components concentration and total pressure, and the correlation coefficient of the prediction set was proved to be 0.9859. Therefore, THz technique can satisfy the increasing need of rapid and efficient detection in the natural gas industry.

1 Introduction

Natural gas is an extremely vital fuel resource and will play a more and more significant role in worldwide industry because of the larger and larger energy consumption. As a fuel resource, natural gas was found to possess large reserves and great exploration values[1-3]. In brief, natural gas is a complicated mixture mainly composed of hydrocarbons. The chief components of the hydrocarbon in natural gas include methane (CH_4) along with ethane (C_2H_6) and propane (C_3H_8). To realize more efficient exploration and detection, a rapid technique is necessary to realize the identification of the principal hydrocarbon components in the natural gas industry[4,5].

THz spectroscopy is a newly developed spectral technique due to the rapid development of ultra-short pulse lasers, semiconductors, and optical detectors, which has received increasing attention in many fields in recent years. THz spectroscopy ranging from 0.1THz to 10THz bridges the gap between microwave and infrared spectroscopy. As a newly developed spectral technique, THz spectroscopy is a very promising method for natural gas detection because of unique properties. THz spectroscopy can provide rich intermolecular and intramolecular vibration modes and give the amplitude as well as phase information on the sample simultaneously. In addition, THz is little sensitive to thermal background radiation and the scattering effect in the gas, and scarcely causes any damage to the tested organic gas. Generally, the high signal-to-noise ratio (>1000) makes it an effective tool for both qualitative and quantitative method. THz time-domain spectroscopy (THz-TDS) is a normal and significant THz method based on the THz electric field with time resolution which is generated by a femtosecond laser pulse. After employing the fast Fourier transform, the frequency dependent spectra can be obtained. Some spectral features can be observed from the THz spectra and can be used as the standard to qualitatively and quantitatively determine the natural gas. THz-TDS can be used to characterize the principal components of natural gas in a simple measurement condition in that the samples do not need any pretreatment[6-12]. Several reports were found to study the THz response of gas and gas mixtures[13-19]. The humid air was mostly investigated and was observed to possess several absorption characteristic peaks in virtue of the water vapor in air[13-15]. In addition, polar molecules reflect great absorption effect in THz range; in brief, nonpolar molecules has a relative small absorption of THz pulse[16-19]. The research about the principal hydrocarbon ingredients of natural gas has been very significant in actual industry and has been found little using THz technique in previous work.

In this research, the principal hydrocarbon components of natural gas, including CH_4, C_2H_6, and C_3H_8, were qualitatively and quantitatively analyzed using THz-TDS. The two-and three-component mixtures were discussed, respectively. The two-component system, such as $CH_4 + C_2H_6$, was compounded as follows: The gas cell was filled by CH_4 in 1.0atm and then C_2H_6 was introduced at the pressure intervals of 0.1atm, so the pressure ratios of CH_4/C_2H_6 in this system equaled 1∶0, 1∶0.1, 1∶0.2, ⋯, 1∶2. Other systems were obtained similarly. The two-component systems, such as $CH_4 + C_2H_6$, $C_2H_6 + CH_4$, $C_3H_8 + CH_4$ and $CH_4 + C_3H_8$, were initially measured; then, the THz-TDS of the three-component systems were also scanned with THz setup. The amplitude and the delay of samples and reference were extracted to calculate the amplitude ratio and the delay time deviation. The ratio and deviation were found to be related to the concentration of different components, and the plotted phase projections between concentrations and ratio as well as deviation obviously reflected the concentration dependent THz response. Moreover, back-propagation artificial neural networks (BPANN) method was used to build a quantitative model between the THz spectra and concentrations as well as the total pressure of three-component system. These results indicate that THz is a promising tool for the qualitative and quantitative detection of natural gas.

2 Experimental methods

As shown in Figure 1, the measurement system was built based on a commercial transmission THz-TDS setup with a multipass cell which has been discussed in our previous report. In brief, a mode-locked Ti-sapphire laser, whose central wavelength was 800nm, was used as the source to generate and detect terahertz waves[20]. The laser beam was split into pump and probe beams. The pump beam (~100mW) was focused onto the surface of a biased GaAs photoconductive antenna for terahertz generation and the probe beam for electrooptic detection. In this study, a 50cm long multipass cell was designed and built so as to fit the default optical path[21]. The cell was attached on the top of a vacuum chamber. By the insertion of two flat mirrors, the THz wave was reflected into the cell so that the total path length of the sample equaled 1m. The cell was made of an airtight cylinder, and two vacuum ports were attached on the sidewall. The pressure in the cell was monitored by a capacitance manometer. The cell was sealed with O-rings and 10mm thick polytetrafluoroethylene windows fixed on the entrance and the exit of the THz beam, which had little reflection at THz frequencies. The sample chamber surrounding the THz path is purged by a continuous flow of nitrogen gas to reduce the effect of water-vapor absorption. In this research, the principal hydrocarbon gases, including CH_4, C_2H_6, and C_3H_8 with the purity of 99.95%, were used for THz measurement. A reference spectrum was first measured using the vacuum cell evacuated below 10^{-4} atm by a turbo molecular pump. The reference spectrum was obtained by the averaging of five scans. The concentration ratio of components in the mixtures system was calculated by controlling the pressure of each ingredient, displayed by the capacitance manometer. In this research, the THz spectra of samples were measured in common room condition. To avoid vapor absorption in air and enhance the signal-noise ratio, the setup, including the THz spectrometer and gas cell, was covered with dry nitrogen. The detailed measurement temperature and relative humidity can be tested by a sensor in the THz setup and displayed in the computer, and were 295.1K±0.4K and 0~0.3% in the experiment, respectively.

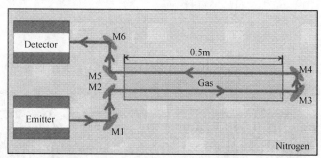

Figure 1　Sketch map of THz-TDS setup

3　Results and discussion

In the actual natural gas industry, both pressure and hydrocarbon components have been significant parameters to be characterized. It is noted that THz is not only sensitive to pressure effects but also to the components. Figure 2 showed the THz-TDS of the CH_4, C_2H_6, and C_3H_8, all pressures of which were 1atm. Therefore, both the pressure and components should be considered and quantitatively characterized in the two-or three- component system. As shown in Figure 3(a), the reference spectra (black line, ref.) was initially measured, which indicates the THz field amplitude as a function of time after the transmission of the THz pulses through a gas cell with vacuum phenomenon (pressure<0.0001atm). The gas cell filled by CH_4 in 1.0atm was scanned as the first sample and then filled out with ethane at the pressure intervals of 0.1atm so that 21 groups of the $CH_4 + C_2H_6$ mixture were obtained and measured by the THz-TDS setup. Similarly, the $C_2H_6 + CH_4$, $C_3H_8 + CH_4$, and $CH_4 + C_3H_8$ mixtures at the intervals of 0.1atm were then scanned by THz-TDS setup one by one. The THz-TDS of the four systems were depicted in Figures 3(a)~(d), respectively. A special tendency is observed with regard to the time delays, which increase gradually with the increasing input of the second component in all mixture systems, such as C_2H_6 in the $CH_4 + C_2H_6$ system, CH_4 in the $C_2H_6 + CH_4$ system, CH_4 in the $C_3H_8 + CH_4$ system, and C_3H_8 in the $CH_4 + C_3H_8$ system. The spectral phenomenon also proves the stability of the setup performance.

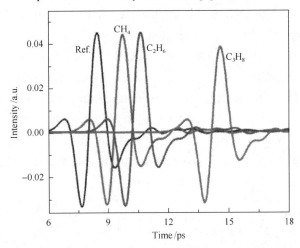

Figure 2 THz-TDS of CH_4, C_2H_6, and C_3H_8 with the pressure of 1atm

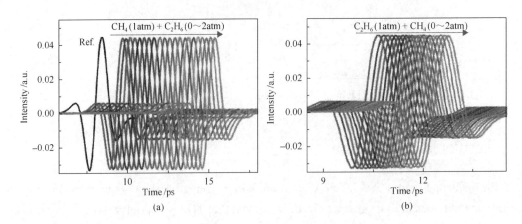

(a) (b)

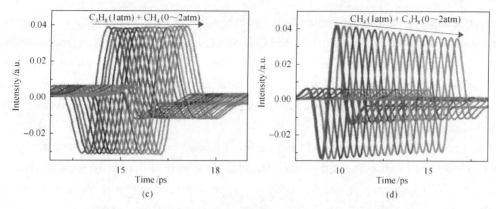

Figure 3　THz-TDS for two-component systems: (a) $CH_4 + C_2H_6$;
(b) $C_2H_6 + CH_4$; (c) $C_3H_8 + CH_4$; (d) $CH_4 + C_3H_8$

Depending on the system stability and measurement conditions, the three-component hydrocarbon system was also analyzed and discussed with THz technique in this research. The gas cell filled by CH_4 in 0.6atm was scanned as the first sample, and then filled out with ethane (0.6atm) so that the volume ratio of CH_4/C_2H_6 equaled. 1∶1. C_2H_6 gas was continuously introduced into the cell at the pressure intervals of 0.1atm. Finally, 20 samples were obtained,

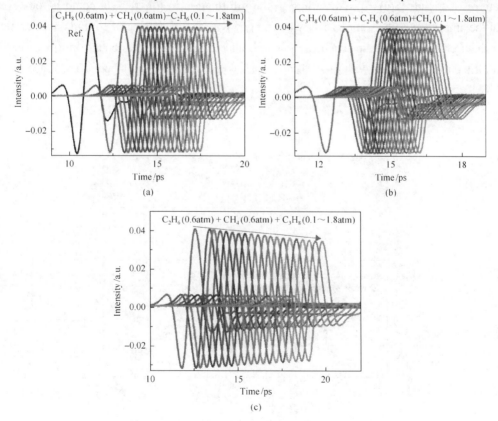

Figure 4　THz-TDS for three-component systems: (a) $C_3H_8 + CH_4 + C_2H_6$;
(b) $C_3H_8 + C_2H_6 + CH_4$; (c) $C_2H_6 + CH_4 + C_3H_8$

and every sample was scanned in the terahertz range. Similarly, another two systems of CH_4, C_2H_6, and C_3H_8 mixtures, including $C_3H_8 + C_2H_6 + CH_4$ and $C_2H_6 + CH_4 + C_3H_8$, were manufactured with the different input order of gases. The THz-TDS of the three three-component systems were plotted in Figures 4(a)~(c), respectively. Similar tendency is found among them that the time delays regularly change with the gradual input of the second and third components with the different pressures.

It is noted that the waveforms of the samples are quite similar in Figures 3 and 4 with the input of new gas in all systems, but the THz responses seem different from each other. To discuss the THz responses of samples with the different components and concentrations, the amplitudes and time delays of the THz pulses are extracted and correlated with certain component concentrations. Here the amplitude ratio ($I_{sample}/I_{reference}$) and delay deviation (Δt, $T_{sample} - T_{reference}$) were used for the sake of removing the small differences of references caused by different measurement. The left pictures of Figures 5(a)~(d) reflected the amplitude ratio as a function of the CH_4 concentration in the two-component system and the CH_4, C_2H_6, and C_3H_8 concentrations in the three-component system. Besides, the right pictures of Figures 5(a)~(d) indicate the Δt as a function of the same component concentration with that in the left pictures of Figures 5(a)~(d). According to the information in Figure 5, it could be observed that the amplitude ratios were basically unchanged with the increasing of certain concentration, but this trend cannot remain when C_3H_8 is introduced as the last component. In this case, the amplitude ratio increases with the CH_4 and C_2H_6 concentrations; on the contrary, it decreases

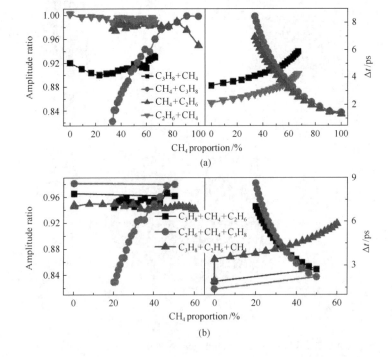

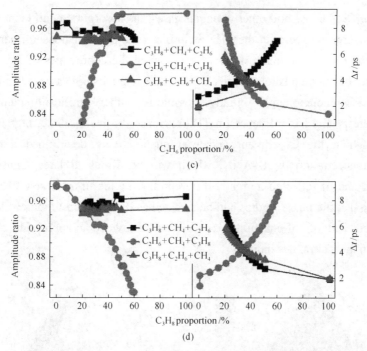

Figure 5　Volume concentration of different gas component dependent amplitude ratio (left) and Δt (right): (a) CH_4 proportion dependence of amplitude ratio (left) and Δt (right) in two-component systems; (b) CH_4 proportion in three-component systems; (c) C_2H_6 proportion in three-component systems; (d) C_3H_8 proportion in three-component systems

with the increasing of the C_3H_8 content. However, an obvious phenomenon can be observed that Δt regularly changes with the input of all components, including CH_4, C_2H_6, and C_3H_8 both in two-and three-component systems. Therefore, results based on Figure 5 indicate the probability of the THz technique for the identification of the principal components and their mixtures in the natural gas industry.

Furthermore, in order to study the THz response of CH_4, C_2H_6, and C_3H_8 systems with different concentrations, a phase diagram was employed to build a relationship between the THz parameter and components' concentrations. As shown in Figure 6, the phase projection pictures of amplitude ratio (a) and delay deviation (b) were graphically described between samples and reference in the $CH_4 + C_2H_6 + C_3H_8$ system, respectively. An interesting tendency was observed that the smallest amplitude ratio and the largest Δt were located in the same field of phase projection in Figures 6(a) and (b). In this field, the C_3H_8 concentration approximated to 60% and both CH_4 and C_2H_6 concentrations equaled about 20%. It could be concluded that with the input of C_3H_8, the THz response of the system gradually changed, which was related to the intrinsic nature of C_3H_8. C_3H_8 is an asymmetric top with C_{2v} symmetry, of which the dipole moment is small but strong enough to make the rotational spectrum be observed obviously. Consequently, C_3H_8 provided the torsional and rotational spectra in the terahertz region[22,23]. For CH_4, C_2H_6, and other nonpolar gases, the absorption coefficient α is too little to show any fingerprint spectra in the THz region, in spite of the increase of partial pressure. According to the rule of the phase diagram, the sum of the coordinates of any point should equal 1. The aim of Figure 6 is to correctly describe the relationship between the THz response and the hydrocarbon components. Actually, pressure is another significant parameter in actual natural gas industry and has influence on THz response. In this research, pressure was not a constant throughout the experiment and was quantitatively determined along with the components' concentration by BPANN, which will be discussed later. Consequently, the amplitude ratio and Δt reflected a greatly different distribution in the whole phase projection area, indicating that the phase figures can be selected as the standard to identify or predict the classification and stability of gas mixtures. The results provided an appropriate suggestion for gas detection in the natural gas industry.

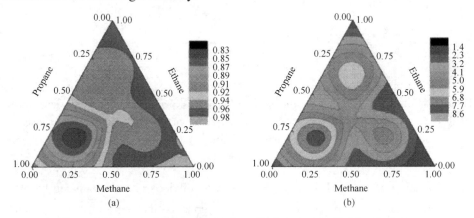

Figure 6 Phase projection picture of amplitude ratio (a) and delay deviation (b) between samples and reference in the $CH_4 + C_2H_6 + C_3H_8$ system

In addition, quantitative determination of component concentration and total pressure is always a necessary part to evaluate the quality and safety of gases in the natural gas industry. Also, synchronous characterization plays a very significant role to improve the detection efficiency because multiple properties are often difficult to be detected simultaneously. To realize the synchronous determination of the total pressure as well as the concentration of components, including CH_4, C_2H_6, and C_3H_8, BPANN was employed with the input of THz frequency domain spectra (THz-FDS), which was calculated from the THz-TDS of the three-component system in Figure 4 after the application of fast Fourier transform. Figure 7 shows the THz-FDS of selected samples in the three-component system, and the concentration of the components is also displayed in Figure 7. It is noted that the difference can be obviously observed in frequency dependent THz spectra among the mixtures with different pressure and components concentrations. The THz response reflect in Figure 7 is in good agreement with that in Figure 2~4.

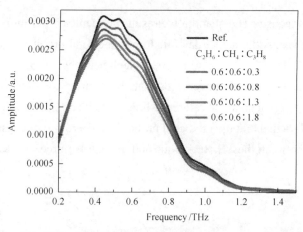

Figure 7　THz-FDS of the four selected samples in three-component systems

In addition, BPANN, a mathematical nonlinear dynamics system simulating structure and function of biological neural networks in the human brain, was also used. No priori models were required for ANN owing to capturing the inherent information from the considered variables and learning from the existing data, even when noise was present. Neurons, the elemental information processing units of the ANN structure, were linked up through synaptic weights to organize into several layers. Based on searching an error surface using gradient descent for a point with minimum error, the BP learning algorithm stored a lot of input-output mapping relationships without prior revealing of the mathematical equation and was composed of an input, hidden, and output layer. In the present study the THz frequency domain spectra data were introduced into the input layer and calculated in the hidden layer; then the results would be finally obtained in the output layer. The three-layer network is enough to simulate the complicated functions. In order to identify all of the subsequently given spectra correctly in the prediction set, the number of training sets should exceed half of the sample numbers. BPANN can store a large number of input-output mapping relationships without prior giving of the mathematical equation which describes the mapping relationships. Its learning rule is to use the method of steepest descent, to constantly adjust the network weights and threshold by backpropagation. More interesting is that multiple variables of the target samples can be determined simultaneously so that it is very promising for detecting natural gas in the petroleum industry[24]. In this research, BPANN was used to build a quantitative model between the THz technique and the gas properties with the input of THz-FDS over the range from 0.2THz to 1.5THz and without any spectral pretreatment. In order to efficiently determine the subsequently given spectra in the prediction set, the number in the training set should exceed that in the prediction set. Within all of the 60 samples, 15 groups were randomly selected as the prediction set and the remaining 45 samples were the training set. According to the significant response of both pressure and hydrocarbon components in the THz range, which are reflected in Figure 2~4, the pressure and components' concentrations are characterized simultaneously. The quantitative results are depicted in Figure 8, which represent the predicted values versus the actual data, including the total pressure and the concentration of CH_4, C_2H_6, and C_3H_8. The correlation coefficient between the actual and the predicted values is calculated as 0.9859. Moreover, most data both in the training set and the prediction set were found to be very close to the reference lines (two black lines), which represented zero residuals between the actual and the predicted properties. Therefore, the results show that there exists a special rule between the THz technique and the gas resource.

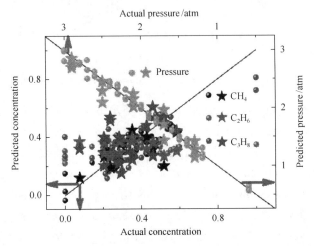

Figure 8 Quantitative model of CH_4, C_2H_6, and C_3H_8 concentrations as well as the total pressure built by BP-ANN. Circle and star points reflect the training set and prediction set, respectively

The research aimed to discuss the THz responses of principal ingredients mixtures in natural gas and realize the determination of the components' concentrations and the total pressure by the combination of THz technique and mathematical method. It was of interest that great differences were observed between any two different two or three-component systems. With the input of C_3H_8, the THz response gradually changed and the tendency can be obviously found in phase projection figures. Moreover, the concentration and pressure were quantitatively predicted with a very high correlation coefficient. The phase diagrams and the quantitative models provided a proof to identify the natural gas qualitatively and quantitatively. Actually, the more classification and larger numbers of gases that were employed to analyze with THz technique, more abundant qualitative and quantitative models can be obtained. Consequently, THz technique possesses a broadly promising application in the natural gas industry.

4 Conclusions

In summary, the practicability was demonstrated about the THz technique being applied to qualitatively and quantitatively detect the principal hydrocarbon components of natural gas, including CH_4, C_2H_6, and C_3H_8. Both two- and three- component systems were analyzed, and the amplitude ratio as well as delay deviation were extracted and related to the components' concentrations. The phase projection pictures obviously indicated the THz response of gas mixtures with different content ratio. Moreover, BPANN was employed to quantitatively determine the concentrations and the total pressure of the three-component system with the input of THz-FDS. Results showed that simple and efficient models were built between THz spectra and gas mixtures and evidently reflected in phase and BPANN pictures. Therefore, THz technique is a new selection to realize the rapid determination of principal hydrocarbon components in the natural gas industry.

Acknowledgements

We acknowledge the National Key Basic Research Program of China (Grant No. 2014CB 744302) and the Specially Founded Program on National Key Scientific Instruments and Equipment Development (Grant No. 2012YQ140005) for the financial support of this work.

References

[1] Ogunlowo O O, Bristow A L, Sohail M. Developing compressed natural gas as an automotive fuel in Nigeria: Lessons from international markets. Energy Policy, 2015, 76: 7-17.

[2] Yuan Z M, Deng Z B, Jiang M Z, et al. A modeling and analytical solution for transient flow in natural gas pipelines with extended partial blockage. J. Nat. Gas Sci. Eng., 2015, 22: 141-149.

[3] Sidding K, Grethe H. No more gas from Egypt? Modeling offshore discoveries and import uncertainty of natural gas in Israel. Appl. Energy, 2014, 136: 312-324.

[4] Fan S S, Zhang Y Z, Tian G L, et al. Natural gas hydrate dissociation by presence of ethylene glycol. Energy & Fuels 2006, 20: 324-326.

[5] Xu C G, Chen Z Y, Cai J, et al. Study on pilot-scale CO_2 separation from flue gas by the hydrate method. Energy & Fuels, 2014, 28: 1242-1248.

[6] Horiuchi N, Zhang X C. Searching for terahertz waves. Nat. Photonics, 2010, 4: 662.

[7] Zhan H L, Wu S X, Bao R M. et al. Qualitative identification of crude oils from different oil fields using terahertz time-domain spectroscopy. Fuel, 2015, 143: 189-193.

[8] Zhao H, Zhao K, Bao R. Fuel property determination of biodiesel-diesel blends by terahertz spectrum. J. Infrared, Millimeter, Terahertz Waves, 2012, 33: 522-528.

[9] Mittleman D M. Frontiers in terahertz sources and plasmonics. Nat. Photonics, 2013, 7: 666-669.

[10] Chen J, Zhao K, Zhao L J, et al. Probing disaggregation of crude oil in a magnetic field with terahertz time-domain spectroscopy. Energy & Fuels, 2014, 28: 483-487.

[11] Lu X F, Zhang X C. Generation of elliptically polarized terahertz waves from laser-induced plasma with double helix electrodes. Phys. Rev. Lett., 2012, 108: 123903.

[12] Jin W J, Zhao K, Yang C, et al. Experimental measurements of water content in crude oil emulsions by terahertz time-domain spectroscopy. Appl. Geophys., 2013, 10: 506-509.

[13] Yang Y H, Shutler A, Grischkowsky D. Measurement of the transmission of the atmosphere from 0.2 to 2THz. Opt. Express, 2011, 19: 8830-8838.

[14] Exter M V, Fattinger C, Grischkowsky D. Terahertz time-domain spectroscopy of water vapor. Opt. Lett., 1989, 14: 1128-1130.

[15] Mittleman D M, Jacobsen R H, Neelamani R, et al. Gas sensing using terahertz time-domain spectroscopy. Appl. Phys. B: Lasers Opt., 1998, 67: 379-390.
[16] Harde H, Zhao J. THz time-domain spectroscopy on ammonia. J. Phys. Chem. A, 2001, 105: 6038-6047.
[17] Foltynowicz R J, Allman R E, Zuckerman E. Terahertz absorption measurement for gas-phase 2,4-dinitrotoluene from 0.05THz to 2.7THz. Chem. Phys. Lett., 2006, 431: 34-38.
[18] Naftaly M, Foulds A P, Miles R E, et al. Terahertz transmission spectroscopy of nonpolar materials and relationship with composition and properties. Int. J. Infrared Millmeter Waves, 2005, 26: 55-64.
[19] Lattanzi V, Walters A, Pearson J C, et al. THz spectrum of monodeuterated methane. J. Quant. Spectrosc. Radiat. Transfer, 2008, 109: 580-586.
[20] Löffler T, Hahn T, Thomson M, et al. Large-area electro-optic ZnTe terahertz emitters. Opt. Express, 2005, 13: 5353-5362.
[21] Leng W X, Ge L N, Xu S S, et al. Pressure-dependent terahertz optical characterization of heptafluoropropane. Chin. Phys. B, 2014, 23: 107804.
[22] Lide D M. Microwave spectrum, structure, and dipole moment of propane. J. Chem. Phys., 1960, 33: 1514-1518.
[23] Drouin B J, Pearson J, Walters A, et al. THz measurements of propane. J. Mol. Spectrosc., 2006, 240: 227-237.
[24] Jiang S Y, Ren Z Y, Xue K M, et al. Application of BPANN for prediction of backward ball spinning of thin-walled tubular part with longitudinal inner ribs. J. Mater. Process. Technol., 2008, 196: 190-196.

Water adsorption dynamics in active carbon probed by terahertz spectroscopy

Honglei Zhan[1,3] Shixiang Wu[2] Rima Bao[1,3] Kun Zhao[1,3,4]
Lizhi Xiao[1] Lina Ge[1] Hongjie Shi[1]

(1.State Key Laboratory of Petroleum Resources and Prospecting, China University of Petroleum, Beijing 102249, China; 2.Petroleum Exploration and Production Research Institute, China Petroleum and Chemical Corporation, Beijing 100083, China; 3.Beijing Key Laboratory of Optical Detection Technology for Oil and Gas, China University of Petroleum, Beijing 102249, China; 4.Key Laboratory of Oil and Gas Terahertz Spectroscopy and Photoelectric Detection, China Petroleum and Chemical Industry Federation (CPCIF), Beijing 100723, China)

Abstract: It is vital to characterize the adsorption dynamics in oil-gas reservoirs and pollution control industry. Terahertz (THz) spectroscopy was used to study the adsorption of the water molecules in active carbon. The absorbance at selected frequencies and the first principal component scores over the whole THz range were related to the corresponding time lengths. The collective tendency expressly tracked the dynamics of water adsorbed in active carbon pores. Therefore, THz technique can be used as a promising tool to monitor the adsorption issues in petroleum and environment fields.

The adsorption phenomenon plays a significant role in many fields such as oil-gas reservoirs and wastewater treatment[1,2]. Shale gas, which is an unconventional natural gas hidden in the strata or mudstone layers in a free or an adsorbed state, is becoming an important force in the world market. The research regarding oil-gas adsorption in tight reservoirs has important practical significance[3,4]. Some studies are related to the adsorption as well as absorption of organic molecules or ions from water or other solutions along with the interaction between the adsorbed substances and porous materials[5-12]. The discussion regarding the dynamic adsorption process will be a key issue. Active carbon is used to adsorb water molecules in this study to simulate the adsorption dynamics in current oil-gas reservoirs and the environmental pollution industry. Scanning electron microscopy (SEM) and atomic force microscopy (AFM) are appropriate ways to describe the surface and structure of the holes; however, the dynamic process cannot be clearly observed by SEM and AFM. Most importantly, active carbon with liquid water was not appropriate during SEM measurements because a vacuum environment is necessary when measuring, and liquid water is not allowed because it may volatilize and harm the SEM setup. Moreover, the vacuum environment will greatly affect the adsorption dynamics because of the very large water concentration gradient between sample and vacuum environment. In addition, the vacuum, an extreme condition, is not normally found in the actual petroleum industry, thus the experiment is not appropriate to simulate the true adsorption dynamics. Finally, the water molecules cannot be clearly observed because of its nanometerscope size, which cannot be clearly distinguished using SEM. In addition, although AFM has a very large resolution of atom, its scanning velocity is very small and the time length is very long. AFM is also not a suitable online method to monitor the adsorption dynamics because the adsorption length is less than 20min[13].

Terahertz time-domain spectroscopy (THz-TDS), a technique to bridge the gap between the microwave and the infrared spectroscopy, has been rapidly developed over the last few decades[14-16]. This method can provide ample information on the intermolecular and intramolecular vibration modes and also simultaneously give the amplitude and phase information. The hydrogen bond collective network formed by water molecules changes on a picosecond (ps) timescale, thus causing the THz spectrum to be sensitive to fluctuations in the dipole moment of the water[17,18]. THz technique is an appropriate method for process monitoring because of its online properties and simple measurement conditions when detecting, which are necessary for the rapid identification of adsorption dynamics in the actual industry. Moreover, THz provides an indirect method for the characterization of adsorption dynamics depending on the absorption effect in THz frequency range. In this study, the THz measurements of water drops adhered on active carbon at various time frames were discussed. The research focuses on the process observation that the water molecules gradually moved into active carbon. First, the absorption spectra of the sample were obtained over the range of 0.1~1.45THz. Second, principal component analysis (PCA) was adopted to build a relationship between the THz adsorption and the timeframes. Finally, the THz absorbance at 0.5THz, 0.8THz, 1.0THz, 1.1THz, 1.2THz, 1.3THz, and 1.4THz were extracted along with the first principal component scores over the entire range and were associated with the corresponding timeframes. The results showed that the THz technique identified the different stages, especially the adsorption process of water molecules into the pores of active carbon.

Figure 1 shows the THz field amplitude as a function of time after the transmission of the THz pulse through the sample at different timeframes. The hydrogen bond network, which is a special intermolecular or intramolecular interaction and a type of strong molecular link, is formed with the mode of O–H···O in the water molecules. The ceaselessly forming and breaking of hydrogen bonds on the ps timescale, which is connected to the reorientation dynamics of the water molecules, are detected due to the sensitivity of THz-TDS. In this research, an air environment was selected, which was the most common condition. A relatively normal adsorption condition is better to simulate the adsorption phenomenon in actual petroleum and environment industries. After employing the fast Fourier transform to THz-TDS in Figure 1, the THz frequency-domain spectra (THz-FDS) was obtained. The absorbance (A) spectra of the sample was then calculated using $-\lg(Amp_{sam}/Amp_{ref})$, where Amp_{sam} and

Amp$_{ref}$ are the THz-FDS amplitudes of the sample and reference (air), respectively. The absorbance spectra of the samples only reflect the absorption effect of active carbon with or without water in the THz range. Figure 2 illustrates the frequency dependent absorbance as the water adhered to active carbon at various timeframes over the range of 0.1~1.45 THz and there is no characteristic peak because liquid was used in this experiment, which is consistent with the statement of liquid water absorption in the THz range in a previous report[19]. The results indicate the changes in the THz optical constants with the increasing adhering time.

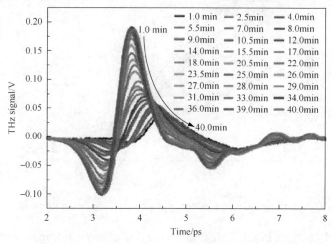

Figure 1   THz-TDS of the sample with different time frames from 1.0min to 40.0min

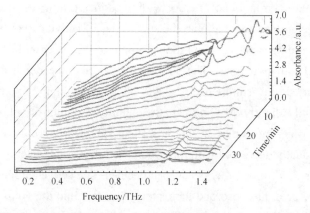

Figure 2   The frequency dependence of absorbance. The x-axis, y-axis and z-axis indicate the frequency range (0.1~1.45THz), the adhering time of water drop on the active carbons, and the absorbance, respectively

The dependence of the adsorption process on the adhering time was evaluated in detail at randomly selected frequencies. The corresponding results are displayed in Figure 3, which show the absorbance as a function of the adhered time after the transmission of the THz pulses. The absorbance basically remained unchanged in the range of 0~10.5min; then, decreased in the 10.5~29.0min range and remained invariant in the 29.0~40min range. The non-linear dynamics on the basis of collective tendency was brought into correspondence with each other at 0.5THz, 0.8THz, 1.0THz, 1.1THz, 1.2THz, 1.3THz and 1.4THz, indicating a special rule, in which a certain response existed for the water molecules that adhered onto and into the active carbon in the THz range.

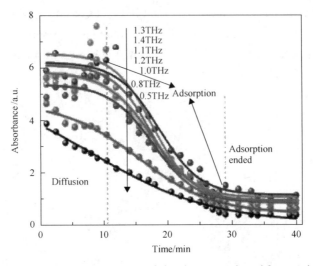

Figure 3 The time dependence of absorbance at selected frequencies

The observed changes in the THz absorbance were very significant during the adhering process. As mentioned above, the whole process can be divided into three stages: 1~10.5min, 10.5~29.0min and 29.0~40min, in accordance with the adhered time, as shown in Figure 3. In the first stage, the water drop adhered onto the superficies of active carbon and diffused to the surrounding area according to the concentration gradient of the water molecules. This diffusion process was divided into two parts: The surface motion and the depth diffusion. The two processes were carried out simultaneously and their competing actions made the THz absorbance change only slightly. At this stage, the quantity and intensity of hydrogen bonds do not change because water molecules are not adsorbed. However, during the second stage, the THz absorbance gradually decreased with the adherence time of the water drop. This stage was homologous with the adsorption process water molecules. The active carbon, which has a very large specific surface area, has adsorbed the water molecules into the voids. In the adsorption stage, with the increasing molecules adsorbed into active carbon, the water molecules were scattered in different holes, especially in inner ducts; thus, the intramolecular vibration changed and the THz response was weaker and weaker. Consequently, the THz absorbance spectra could trace the motion as the water molecules were gradually adsorbed into the voids of active carbon. In the third stage, when the adsorption ended, the intramolecular vibrations remained unchanged; thus, the absorbance values of the samples at different timeframes remained unchanged, indicating that the adsorption process ended and the water molecules were adsorbed in a stable state in the pores of the carbon. To highlight the difference between the non-adsorbed and adsorbed samples, a contrast of the amplitude and phase of THz-FDS, which were calculated by fast Fourier transform of THz-TDS, and optical parameters including refractive index (n) and absorbance (A) of the active carbon with water drop adsorbed at 40min and without water is illustrated in Figure 4. Although several peaks are found in Figure 4(a), they result from the vapor in air, which is selected as a measurement condition. The absorbance spectra in Figure 4(d) do not have any characteristic peaks because of the calculation of $-\lg(Amp_{sam}/Amp_{ref})$ and is consistent with the spectra shown in Figure 2. The significant differences that exist between the samples within the range of 0.1~1.45THz indicate that the water molecules were adsorbed into active carbon rather than volatilized into the air.

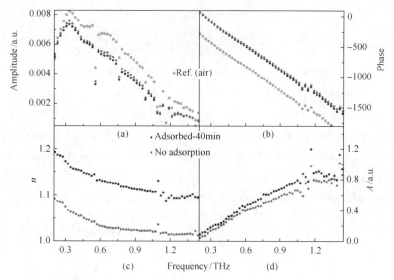

Figure 4 THz-FDS amplitude (a), phase (b) and optical parameters including refractive index n (c) and absorbance A (d) for active carbons with water at 40min and without water

In this study, a PCR method was used, which was able to narrow multiple variables to a few principal components (PCs) with dimension reduction technology, to analyze the motion process with the input of the THz absorbance spectra of the samples in the 0.1~1.45THz range; however, none of the spectral pretreatments were used[20]. As shown in Figure 5, the x-axis and y-axis indicate the first and the second PC scores, of which the contribution rates were 97.5% and 2.0%, respectively. Therefore, the first two PCs, particularly PC1, represent the majority of the sample information. The sample occupies different positions in the coordinate system at different adhering times. In regards to the adhered time, the sample had similar PC1 scores and different PC2 scores at 1.0~10.5min, different PC1 and PC2 scores at 10.5~29.0min, and the same PC1 and PC2 scores at 29.0~40.0min. The adsorption process with different adhered timeframes reflects the different PC scores; consequently, it can be classified as having three stages.

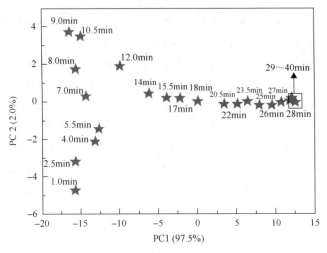

Figure 5　PC1 versus PC2 for absorbance data over the range of 0.1~1.45THz

Because PC1 presented most of the information on the original variables of the system due to its high contribution rate (97.5%), the PC1 scores were extracted and associated with the corresponding time, as shown in Figure 6. An analogous trend was obtained and its time intervals of transition points were similar to that shown in Figure 3. Each time interval was evident. Therefore, based on the regular curves from Figure 3 and 6, a conclusion was drawn that the THz technique can be used as an effective and promising tool to track the adsorption process of the fluid adsorbed into the porous structure.

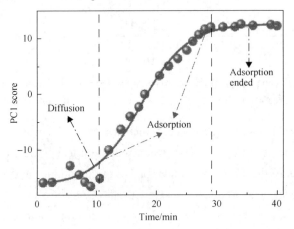

Figure 6　The time dependence of PC1 scores in the PCA system

To test and verify the repeatability and accuracy of the abovementioned conclusion, another similar experiment was performed where three drops were used and the PCA was also used to calculate the PC scores. Only the PC1 scores were extracted due to its high contribution rate (96.33%) and associated with the corresponding timeframes, as shown in Figure 7. Although the time intervals of the transition points appeared different, the three stages were obviously distinguished and the second stage was the adsorption process over the time range of 11~42min. The longer adsorption time resulted from the increase in the quantity of water molecules. These results showed that the adsorption process was evident and accurately expressed by the THz technique.

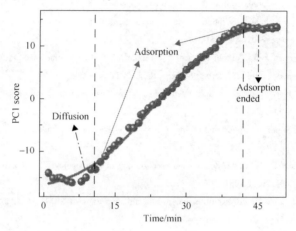

Figure 7　Dependence of PC1 scores on the corresponding time of three-drop-water experiment

In summary, the adsorption process of water into active carbon was monitored using THz-TDS. The absorbance at any frequency reflected the water adsorption dynamics into active carbon and the adsorption process was then validated by PCA calculation with the absorbance over the entire THz range. Therefore, THz-TDS represents a promising technique to monitor the adsorption dynamics; thus, THz technique might be greatly popularized in petroleum and environment industry in the future.

Acknowledgements

This work is supported by the National Key Basic Research Program of China (Grant No. 2014CB744302), the Specially Funded Program on National Key Scientific Instruments and Equipment Development (Grant No. 2012YQ140005), NSFC (Grant No. 61405259), and the Beijing National Science Foundation (Grant No. 4122064).

References

[1] Sheng J J, Chen K. Evaluation of the EOR potential of gas and water injection in shale oil reservoirs. Journal of Unconventional Oil and Gas Resources, 2014, 5: 1-9.

[2] Zietzschmann F, Altmann J, Ruhl A S, et al. Estimating organic micro-pollutant removal potential of activated carbons using UV absorption and carbon characteristics. Water Research, 2014, 56(2): 48-55.

[3] Gracceva F, Zeniewski P. Exploring the uncertainty around potential shale gas development-A global energy system analysis based on TIAM (TIMES Integrated Assessment Model). Energy, 2013, 57(1): 443-457.

[4] Vengosh A, Warner N, Jackson R, et al. The effects of shale gas exploration and hydraulic fracturing on the quality of water resources in the united states. Procedia Earth & Planetary Science, 2013, 7: 863-866.

[5] Moreno-Castilla C. Adsorption of organic molecules from aqueous solutions on carbon materials. Carbon, 2004, 42(1): 83-94.

[6] Namasivayam C, Kavitha D. Removal of congored from water by adsorption onto activated carbon prepared from coir pith, an agricultural solid waste. Dyes & Pigments, 2002, 54(1): 47-58.

[7] Bautista-Toledo M I, Rivera-Utrilla J, Ocampo-Pérez R, et al. Cooperative adsorption of bisphenol-A and chromium (III) ions from water on activated carbons prepared from olive-mill waste. Carbon, 2014, 73(14): 338-350.

[8] Tsai W T, Chang C Y, Lin M C, et al. Adsorption of acid dye onto activated carbons prepared from agricultural waste bagasse by $ZnCl_2$ activation. Chemosphere, 2001, 45(1): 51-58.

[9] Senthilkumaar S, Varadarajan P R, Porkodi K, et al. Adsorption of methylene blue onto jute fiber carbon: Kinetics and equilibrium studies. Journal of Colloid & Interface Science, 2005, 284(1): 78-82.

[10] Groszek A J. Study of the active carbon-water interaction by flow adsorption microcalorimetry. Langmuir, 1999, 15(18): 5956-5960.

[11] Ozcelik Y, Ozguven A. Water absorption and drying features of different natural building stones. Construction & Building Materials, 2014, 63(2): 257-270.

[12] Zhang J Q, Grischkowsky D. Terahertz time-domain spectroscopy of submonolayer water adsorption in hydrophilic silica aerogel. Optics Letters, 2004, 29(9): 1031-1033.

[13] Sujka M, Jamroz J, et al. α-Amylolysis of native potato and corn starches-SEM, AFM, nitrogen and iodine sorption investigations. LWT-Food Science and Technology, 2009, 42(7): 1219-1224.

[14] Horiuchi N, Zhang X C. Searching for terahertz waves. Nature Photonics, 2010, 4(9): 662.

[15] Bao R M, Wu S X, Zhao K, et al. Applying terahertz time-domain spectroscopy to probe the evolution of kerogen in close pyrolysis systems. Science China:Physics, Mechanics & Astronomy, 2013, 56(8): 1603-1605.

[16] Lundholm I, Wahlgren W Y, Piccirilli F, et al. Terahertz absorption of illuminated photosynthetic reaction center solution: A signature of photoactivation?. RSC Advances, 2014, 4(49): 25502-25509.

[17] Castro-Camus E, Palomar M, Covarrubias A A. Leaf water dynamics of Arabidopsis thaliana monitored in-vivo using terahertz time-domain spectroscopy. Scientific Reports, 2013, 3(10): 2910.

[18] Choi D H, Son H, Jung S, et al. Dielectric relaxation change of water upon phase transition of a lipid bilayer probed by terahertz time domain spectroscopy. Journal of Chemical Physics, 2012, 137(17): 1441.

[19] Xu J, Plaxco K W, Allen S J. Absorption spectra of liquid water and aqueous buffers between 0.3 and 3.72 THz. Journal of Chemical Physics, 2006, 124(3): 036101.

[20] Burnett A D, Fan W, Upadhya P C, et al. Broadband terahertz time-domain spectroscopy of drugs-of-abuse and the use of principal component analysis. Analyst, 2009, 134(8): 1658-1668.

Terahertz double-exponential model for adsorption of volatile organic compounds in active carbon

Jing Zhu[1,2]　Honglei Zhan[1,2]　Xinyang Miao[1,2]　Kun Zhao[1,2]　Qiong Zhou[1]

(1. State Key Laboratory of Petroleum Resources and Prospecting, China University of Petroleum, Beijing 102249, China; 2. Beijing Key Laboratory of Optical Detection Technology for Oil and Gas, China University of Petroleum, Beijing 102249, China)

Abstract: In terms of the evaluation of the diffusion-controlled adsorption and diffused rate, a mathematical model was built on the basis of the double-exponential kinetics model and terahertz(THz) amplitude in this letter. The double-exponential-THz model described the two-step mechanism controlled by diffusion. A rapid step involved external and internal diffusion, followed by a slow step controlled by intraparticle diffusion. The concentration gradient of the molecules promoted the organics molecules rapidly diffusing to the external surface of adsorbent. The solute molecules then transferred across the liquid film. Intraparticle diffusion began and was determined by the molecular sizes as well as affinities between organics and activated carbon.

Keywords: terahertz; double exponential model; adsorption; diffusion

1　Introduction

Volatile organic compounds (VOCs) are carbon-based compounds that exhibit a high vapor pressure at room temperature. Many types of VOCs are highly toxic, and can be damaging to the environment[1,2]. A number of studies have indicated the effectiveness of VOC removal by activated carbon adsorption[3-6]. Activated carbons are commonly used as the adsorbents for organic compounds, as they possess high adsorption capacity and fast adsorption kinetics[7,8]. Research on the time dependency of adsorption processes is crucial to predict kinetic parameters and to design operating conditions rationally[9].

A variety of theoretical models have been proposed to describe adsorption data, which can be generally classified as adsorption reaction models and adsorption diffusion models. Both models are applied to describe the kinetic process of adsorption. Adsorption diffusion models are typically constructed on the basis of three consecutive steps[10]: (i) external diffusion or film diffusion; (ii) internal diffusion or intra-particle diffusion; (iii) adsorption and desorption. Adsorption reaction models are based on the whole process of adsorption without considering these steps mentioned above. Generally, the overall rate of the sorption process may be controlled by any of these steps or in some cases by combination of two steps[9]. Therefore, diffusion rate constant, which could be used to estimate adsorption efficiency of VOCs, was

---

Originally published in *Journal of Physics D: Applied Physics*, 2017, 50(23): 235103.

determined to evaluate the adsorption behavior under different operation conditions. The double exponential is an analytical solution that highlight a two-step mechanism[11].

Terahertz time-domain spectroscopy (THz-TDS) could provide abundant information on intermolecular and low-frequency intramolecular modes of chemicals and give amplitude and phase involving the sample simultaneously[12-14]. Thus, THz techniques were effective means of evaluating adsorption. Different absorption properties of the adsorbed water were shown in hydrophilic silica aerogels, and the overall tendency expressly tracked the dynamics of water adsorption within active carbon pores[15,16]. THz emission spectroscopy can image molecular adsorption and desorption dynamics on graphene[17]. In addition, THz-TDS was applied to characterize the adsorption dynamics. Based on the pseudo-second-order kinetic model and the relationship between THz signal peak amplitudes (E_P) and time for adsorption, a mathematical model was built in terms of the adsorption rate and the THz parameter. Thus, the adsorption rates of isooctane, ethanol, and butyl acetate were assessed by the THz measurement, respectively[18]. In this research, the impacts of diffusion on adsorption of VOCs were investigated. We built a double-exponential kinetics model with respect to THz parameter, based on which the diffusion rate can be characterized. Therefore, THz-TDS was applied to characterize the adsorption quantitatively with a diffusion controlled model.

2 The mathematical model

The double-exponential function, proposed by Wilczak and Keinath, was used to describe lead and copper adsorption onto activated carbon[19]. It turns out that the two-step mechanism can be described precisely with the double-exponential model by Chiron et al.[11]. In addition, it is proposed that the double-exponential model describes a two-step mechanism that may be interpreted as a diffusion controlled process or a two-site adsorption mechanism, or both of them simultaneously.

The double-exponential model has the form

$$q(t) = q_e - \frac{D_1}{m_a}e^{(-K_1 t)} - \frac{D_2}{m_a}e^{(-K_2 t)} \tag{1}$$

where D_1 and D_2 are adsorption rate parameters of rapid and slow step, respectively. K_1 and K_2 are diffusion parameters which control the adsorption process of rapid and slow step, respectively, and m_a is the mass of adsorbent. If case that $K_1 \gg K_2$, rapid process can be neglected for overall kinetics.

Equation (1) can then be simplified as

$$q(t) = q_e - \frac{D_2}{m_a}e^{(-K_2 t)} \tag{2}$$

To quantify the initial adsorption rate v, the double-exponential equation is derived as follows:

$$v = \frac{dq(t)}{dt} = K_1 \frac{D_1}{m_a} e^{(-K_1 t)} + K_2 \frac{D_2}{m_a} e^{(-K_2 t)} \tag{3}$$

For initial conditions:

$$t = 0, \quad v_0 = K_1 \frac{D_1}{m_a} + K_2 \frac{D_2}{m_a} \tag{4}$$

3 Experimental

The real-time measurement was carried out to determine the adsorption behaviors of VOCs into active carbon. The active carbon, which mainly composes of carbon, hydrogen and oxygen element, is a kind of porous material whose void radius roughly varies from ~100nm to ~50μm. In our case the active carbon fiber cloth (ACFC), to which the active carbon pellets adhere tightly, was used to perform the adsorption experiment. The adsorption behaviors of ACFC fixed active carbon to isooctane, ethanol, and butyl acetate were studied by a typical THz-TDS setup with transmission geometry, respectively[20-22]. The polyethylene (PE) slices are transparent in THz range and are used to seal the ACFC with a single drop of chemical reagent on the middle of the ACFC[23].

The experimental setup was comprised of a conventional THz-TDS system. The THz pulse, which was generated by a p-type GaAs wafer and detected by a ZnTe sensor, carried sample's information concerning intermolecular and intramolecular vibration modes. The THz pulses of the reference and the sample were measured by scanning nitrogen and the ACFC with VOC droplet respectively at room temperature. The measurement was terminated on condition that E_P signal remains constant in the period over 30 min. The samples' measurement and THz-TDS data was derived from Reference[18].

4 Results and discussion

Figure 1 shows how the THz waveform evolves over in time under continuous excitation by femtosecond optical pulses from a Ti-sapphire laser to characterize the adsorption of organics including isooctane, ethanol and butyl acetate. The THz-TDS waveforms of butyl acetate at 1.6min, 77.5min and 104.3min were shown in Figures1(a) ~ (c), respectively. The green curves in Figures 1 (a) ~ (c) represent reference spectra obtained by probing the ACFC without any organics. The difference of E_P between reference and sample (ΔE_P) decreased in the order of $\Delta E_{P1.6} > \Delta E_{P77.5} > \Delta E_{P104.3}$ ($\Delta E_{P1.6}$, $\Delta E_{P77.5}$ and $\Delta E_{P104.3}$ refer to ΔE_P in 1.6min, 77.5min and 104.3min, respectively). As time increases, the signal E_P continuously augments. The frequency dependent THz power spectra of isooctane, ethanol and butyl acetate were calculated using fast Fourier transform (FFT) as shown in Figure 2. As adsorption proceeds, with detecting time increase, the augment of THz power was observed at many entire frequency range frequencies. According to our previous research, the change of E_P with time was caused by volatilization, diffusion and adsorption dynamics of organics on the surface activated carbon. Indeed, ethanol and butyl acetate were composed comprised of the hydrogen bond networks. The hydrogen bond collective network is formed by molecular changes on a picosecond (ps) timescale, leading to thus causing the sensitivity of THz spectrum to be sensitive to fluctuations of in the dipole moment of ethanol and butyl acetate, which is in agreement with the THz response in Figure 1[18].

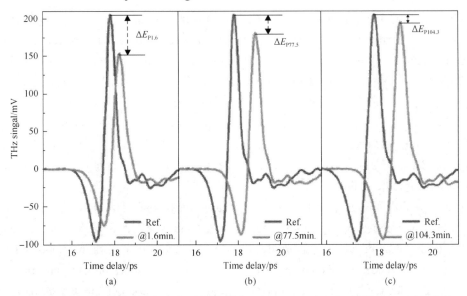

Figure 1 The THz-TDS of selected ACFC with butyl acetate at 1.6min, 77.5min and 104.3min

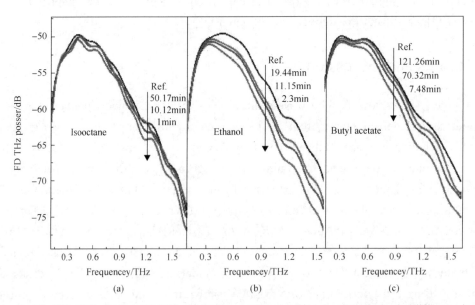

Figure 2 The THz-FDS of selected ACFC with isooctane, ethanol and butyl acetate obtained at different minutes

Previously, based on the pseudo-second order model and the assumption of the proportional relationship between the adsorbed amount and E_P, the adsorption equation has been proposed to describe adsorption kinetics. Itcan be generally classified as adsorption reaction model, and the adsorption rate of organics could be assessed by the THz measurement. It is noticeable that the double-exponential model, compared with the pseudo-second order model, is based on the whole process of adsorption in which diffusion process is considered. It is a model presented on the determination of adsorption stages[24]. More importantly, the rapid and slow adsorption can be described by the double-exponential model that has been employed in many fields such as optimization of wastewater treatment[25, 26].

Based on the double-exponential model and the assumption of proportional relationship between the adsorbed amount and E_P[18], the adsorption equation has been proposed to describe adsorption kinetics. The terahertz double-exponential model is described as

$$E_{P2}(t) = E_{PE1} - A_1 e^{(-K_{A_1} t)} - A_2 e^{(-K_{A_2} t)} \tag{5}$$

where $E_{P2}(t)$ and E_{PE1} are THz signal peak amplitudes at the t and at the equilibrium state causing by adsorption process. A_1 and A_2 are adsorption rate parameters of rapid and slow step, respectively. K_{A_1} and K_{A_2} are diffusion parameters which control the adsorption process of rapid and slow step.

On account of VOCs' volatilization fitted by the exponential model, the whole process is presented as follows:

$$E_{P1}(t) = E_{PE0} - A_0 e^{(-K_{A_0} t)} \tag{6}$$

$$E_P(t) = E_{P1}(t) + E_{P2}(t) \tag{7}$$

$$\begin{aligned} 0 \leqslant t < T_1, \quad & E_P(t) = E_{P0} - A_0 e^{(-K_{A_0} t)} \\ t \geqslant T_1, \quad & E_P(t) = E_{PE} - A_1 e^{(-K_{A_1} t)} - A_2 e^{(-K_{A_2} t)} \\ & E_P = E_{P1} + E_{P0} - A_0 e^{(-K_{A_0} T_1)} \end{aligned} \tag{8}$$

where A_0 and K_{A_0} are volatilization parameters. $E_{P1}(t)$ and $E_P(t)$ are THz signal peak amplitudes at the t causing by volatilization and whole process, respectively. E_{PE} are the values of E_P of the samples at the equilibrium state. E_{PE0} is used on diffusion equilibrium state.

Volatilization is described with exponential model and the terahertz double-exponential model is employed to describe diffusion and adsorption during whole process. The time dependent $E_{P1}(t)$, $E_{P2}(t)$ and $E_P(t)$ of ACFC with isooctane, ethanol and butyl acetate are shown in Figure 3 (a) ~ (c). T_1 represents the time when the adsorption starts. The exponential model indicated by fine line in Figure 3 (a) ~ (c) illustrates volatilization process of isooctane, ethanol and butyl acetate, respectively. The double-exponential model, suggested by blue curves in Figure 3 (a) ~ (c), is obtained when t exceeds T_1. Volatilization and diffusion of organics begins simultaneously in the sealed space. When t exceeds T_1, the molecules diffuse into holes and then were gradually adsorbed in walls of these holes.

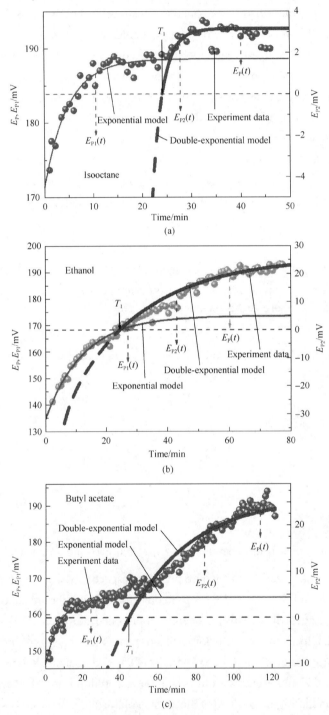

Figure 3　The thick and fine curves represent simulation of the dynamic processes for the exponential model and the double-exponential model of isooctane, ethanol and butyl acetate onto active carbon, respectively

Based on double-exponential theory, two steps indicated by $A_1 e^{(-K_{A_1} t)}$ and $A_2 e^{(-K_{A_2} t)}$, are both related to diffusion and adsorption. Diffusion dominates over kinetics since parameters that involve diffusion are exponential terms. K_{A_1} suggests rapid external and internal diffusion whereas K_{A_2} stands for a slow step controlled by intraparticle diffusion[11,19,27,28]. On the other hand, the double-exponential model also takes account of adsorption process. As porous absorbent with inhomogenous surface, activated carbon offers two different types of adsorption sites including oxygen-containing functional groups and carbon adsorption site[11]. Two adsorption sites contribute to values of A_1 and A_2 but fail to have impacts on exponential terms that are determined by diffusion process. Energy dispersive spectroscopy (EDS) was used to confirm the morphology and chemical composition of the active carbon. According to the EDS spectra, the mass fraction and atom fraction of the chemical element C are 68.00% and 91.27%, while those of O are 6.61% and 6.66%. The mass fraction of the chemical element O are 32/45, 32/44, 16/17 and 1 in the carboxyl, lactone group, phenolic hydroxyl and carbonyl which are the common oxygen-containing functional group on active carbon. Then, the mass fractions of four oxygen-containing functional groups are 9.3%, 9.09%, 7.02% and 6.61%, and all less than 10%. Therefore, the surface of active carbon is mainly carbon adsorption site. It indicates that two-site mechanisms can be ignored, and the organics adsorption is diffusion controlled mechanism. What's more, according to our previous research, the pseudo-second order model fits adsorption dynamics of organics such as ethanol and shows good linearity[18]. For the ethanol, there is a deviation from linearity as a downward curvature at initial minutes of whole process including adsorption and diffusion in usual t/E_P versus time plot. It agrees with the report that there is a deviation from linearity as a downward curvature at initial minutes of adsorption in usual t/q (time/the adsorbed amount) versus time plot, when diffusion contributes to the rate-controlling step of adsorption[9]. Therefore, the organics adsorption is diffusion controlled mechanism. An overall one-site mechanism was considered for adsorption of the organics on activated carbon and the double-exponential model can therefore be considered as diffusion controlled model.

During the rapid step, the organics adhere onto the superficies of active carbon and diffuse into the surrounding area. Due to the concentration gradient of the molecules, organics molecules diffuse rapidly through the solution to the external surface of adsorbent or the film diffusion of solute molecules. In this part, the solute molecules transfer across the liquid film according to the mass balance law[27]. The calculated parameters for the exponential model are shown in Table 1. The calculated values of E_{PE} are 191.8mV, 194.39mV and 190.86mV for isooctane, ethanol and butyl acetate, respectively. T_1 was 44.8min, 24.9min and 24.2min and decreased in the order of butyl acetate > ethanol > isooctane, which indicates the order of diffusion mean velocity and is consistent with the result of Reference[18]. The diffusion parameters K_{A_1} for isooctane, ethanol and butyl acetate are 0.72min^{-1}, 0.42min^{-1} and 0.33min^{-1}, respectively. The parameters including T_1 and K_{A_1} both reflected the rate relationship of rapid step for three organics, respectively.

Table 1 Parameters for terahertz double-exponential model

	Isooctane	Ethanol	Butyl acetate
T_1/min	24.2	24.9	44.8
E_{PE}/mV	191.8	194.39	190.86
A_1	153631.4	68.2	100.4
A_2	76815.9	69.19	101.9
K_{A_1}/min^{-1}	0.72	0.42	0.33
K_{A_2}/min^{-1}	0.45	0.0402	0.0303

When a slow step prevails, the organic molecules penetrate into the inner grafted layer and the external film resistance can be neglected[26]. In Table 1, the values of the adsorption parameters A_1 and A_2 are both increased in the order of isooctane, ethanol and butyl acetate. The order is in agreement with the adsorption rate relationship of three organics. In addition, the values of the diffusion parameter K_{A_2} are 0.45min^{-1}, 0.0402min^{-1} and 0.0303min^{-1}, respectively. In this stage, carbon adsorption sites on the surface of activated carbon has different affinities for different organics, and due to hydrophobic interaction, activated carbon has a greater affinity for hydrocarbons compared with ethanol and butyl acetate[20]. Therefore, the diffusion rate of isooctane is largest. The diffusion rate of ethanol and butyl acetate is limited because of adsorption. In addition, the molecules with bigger size require more time to diffuse into internal holes, therefore, the diffusion rate of ethanol is higher than that of butyl acetate whose molecules are larger. Adsorption rate is limited by diffusion, thus the intraparticle diffusion controls the adsorption rate.

5 Conclusions

In summary, according to our previous research, the change of E_P with time can characterize volatilization, diffusion and adsorption dynamics of organics on activated carbon. Based on the previous adsorption equation including THz parameter, a new adsorption equation has been proposed to describe adsorption and diffusion processes. Oxygen-containing functional groups on active carbon will not exceed 10% of the total adsorption so that the double-exponential model can be considered as diffusion controlled model. Therefore, the adsorption process can be divided into two steps, namely a rapid step involving external and internal diffusion, followed by a slow step controlled by intraparticle diffusion. The parameter T_1, K_{A_1} reflected the rate relationship of rapid step for three organics, respectively. According to analysis of the parameters K_{A_2}, A_1 and A_2, the process of diffusion and adsorption was investigated. In addition, the diffusion and adsorption rate relationship of three organics was quantitative characterization by the THz measurement. Based on the double-exponential kinetics model and terahertz amplitude infer the diffusion-controlled adsorption mechanism of the VOCs.

Acknowledgements

This work was supported by the National Nature Science Foundation of China (Grant No. 11574401), National Basic Research Program of China (Grant No. 2014CB744302), the Specially Founded Program on National Key Scientific Instruments and Equipment Development, China (Grant No. 2012YQ140005) and the China Petroleum and Chemical Industry Association Science and Technology Guidance Program (Grant No. 2016-01-07).

References

[1] Dwivedi P, Gaur V, Sharma A, et al. Comparative study of removal of volatile organic compounds by cryogenic condensation and adsorption by activated carbon fiber. Sep. Purif. Technol., 2004, 39:23-27.

[2] Irigaray P, Newby J A, Clapp R, et al. Lifestyle-related factors and environmental agents causing cancer: An overview. Biomed. Pharmacother, 2007, 61: 640-658.

[3] Li M S, Wu S C, Peng Y H, et al. Adsorption of volatile organic vapors by activated carbon derived from rice husk under various humidity conditions and its statistical evaluation by linear solvation energy relationships. Sep. Purif. Technol., 2016, 170: 102-108.

[4] Izquierdo M T, de Yuso A M, Valenciano R, et al.Influence of activated carbon characteristics on toluene and hexane adsorption: Application of surface response methodology. Appl. Surf. Sci., 2013, 264: 335-343.

[5] Kim D J, Shim W G, Moon H. Adsorption equilibrium of solvent vapors on activated carbons. Korean J. Chem. Eng., 2001, 18: 518-524.

[6] Kim K J, Kang C S, You Y J, et al. Adsorption-desorption characteristics of VOCs over impregnated activated carbons. Catal. Today, 2006, 111: 223-228.

[7] Franz M, Arafat H A, Pinto N G. Effect of chemical surface heterogeneity on the adsorption mechanism of dissolved aromatics on activated carbon. Carbon, 2000, 38: 1807-1819.

[8] Roostaei N, Tezel F H. Removal of phenol from aqueous solutions by adsorption. J. Environ. Manage., 2004, 70: 157-164.

[9] Haerifar M, Azizian S. Mixed surface reaction and diffusion-controlled kinetic model for adsorption at the solid/solution interface. J. Phys. Chem. C., 2013, 117: 8310-8317.

[10] Lazaridis N K, Asouhidou D D. Kinetics of sorptive removal of chromium(Ⅵ) from aqueous solutions by calcined Mg-Al-CO_3 hydrotalcite. Water Res., 2003, 37: 2875-2882.

[11] Chiron N, Guilet R, Deydier E. Adsorption of Cu(Ⅱ) and Pb(Ⅱ) onto a grafted silica: Isotherms and kinetic models. Water Res., 2003, 37: 3079-3086.

[12] Zhan H L, Zhao K, Zhao H, et al. The spectral analysis of fuel oils using terahertz radiation and chemometric methods. J. Phys. D Appl. Phys., 2016, 49: 395101.

[13] Zhan H L, Wu S X, Zhao K, et al. $CaCO_3$, its reaction and carbonate rocks: Terahertz spectroscopy investigation. J. Geophys. Eng., 2016, 13: 768-774.

[14] Jiang C, Zhao K, Zhao L J, et al. Probing disaggregation of crude oil in a magnetic field with terahertz time-domain spectroscopy. Energy & Fuels, 2014, 28: 483-487.

[15] Zhang J Q, Grischkowsky D. Terahertz time-domain spectroscopy of submonolayer water adsorption in hydrophilic silica aerogel. Opt. Lett., 2004, 29: 1031-1033.

[16] Zhan H L, Wu S X, Bao R M, et al. Water adsorption dynamics in active carbon probed by terahertz spectroscopy. RSC Adv., 2015, 5: 14389-14392.

[17] Yamaguchi A, Arafune H, Hotta K, et al. Adsorption and desorption dynamics of sodium dodecyl sulfate at the octadecylsilane layer on the pore surface of a mesoporous silica film observed *in-situ* by optical waveguide spectroscopy. Anal. Sci., 2011, 27: 597-603.

[18] Zhu J, Zhan H L, Miao X Y, et al. Adsorption dynamics and rate assessment of volatile organic compounds in active carbon. Phys. Chem. Chem. Phys., 2016, 18: 27175-27178.

[19] Wilczak A, Keinath T M. Kinetics of sorption and desorption of copper(II) and lead(II) on activated carbon. Water Environ. Res., 1993, 65: 238-244.

[20] Zhan H L, Wu S X, Bao R M, et al. Qualitative identification of crude oils from different oil fields using terahertz time-domain spectroscopy. Fuel, 2015, 143: 189-193.

[21] Leng W, Zhan H, Ge L, et al. Rapidly determinating the principal components of natural gas distilled from shale with terahertz spectroscopy. Fuel, 2015, 159: 84-88.

[22] Feng X, Wu S X, Zhao K, et la. Pattern transitions of oil-water two-phase flow with low water content in rectangular horizontal pipes probed by terahertz spectrum. Opt. Express, 2015, 23: A1693-A1699.

[23] Jin Y S, Kim G J, Jeon S G. Terahertz dielectric properties of polymers. J. Korean Phys. Soc., 2006, 49: 513-517.

[24] Fletcher A J, Yaprak U, Thomas K M. Role of surface functional groups in the adsorption kinetics of water vapor on microporous activated carbons. J. Phys. Chem. C, 2007, 111: 8349-8359.

[25] İsmail T. Ammonium removal from aqueous solutions by clinoptilolite: Determination of isotherm and thermodynamic parameters and comparison of kinetics by the double exponential model and conventional kinetic models. Int. J. Env. Res. Pub. He., 2012, 9: 970-984.

[26] Mason I G, Mclachlan R I, Gérard D T. A double exponential model for biochemical oxygen demand. Bioresour Technol., 2006, 97: 273-282.

[27] Zhou X, Wei J, Liu K, et al. Adsorption of bisphenol A based on synergy between hydrogen bonding and hydrophobic interaction. Langmuir, 2014, 30: 13861-13868.

[28] Cheung W H, Szeto Y S, McKay G. Intraparticle diffusion processes during acid dye adsorption onto chitosan. Bioresour Technol., 2007, 98: 2897-2904.

Adsorption dynamics and rate assessment of volatile organic compounds in active carbon

Jing Zhu   Honglei Zhan   Xingang Miao   Yan Song   Kun Zhao

(Beijing Key Laboratory of Optical Detection Technology for Oil and Gas, China University of Petroleum, Beijing 102249, China)

Abstract: In this paper, an investigation was presented about the terahertz time-domain spectroscopy (THz-TDS) as a novel tool for the characterization of the dynamic adsorption rate of volatile organic compounds, including isooctane, ethanol, and butyl acetate, in the pores of active carbon. The THz-TDS peak intensity (E_P) was extracted and corresponded with the measurement time frames. By analyzing E_P with time, the entire process could be divided into three physical parts including volatilization, adsorption and stabilization so that the adsorption dynamics could be clearly identified. In addition, based on the pseudo-second-order kinetic model and the relation ship between E_P and time in adsorption process, a mathematical model was built in terms of the adsorbed rate parameter and the THz parameter. Consequently, the adsorption rate of isooctane, ethanol, and butyl acetate could be assessed by the THz measurement, indicating that THz spectroscopy could be used as a promising selection tool to monitor the adsorption dynamics and evaluate adsorption efficiency in the recovery of pollutants.

1 Introduction

Volatile organic compounds (VOCs) are important precursors and participants of complex regional atmospheric pollution, while VOCs can cause mutagenic and carcinogenic effects on human beings[1,2]. It has been shown that prolonged exposure to VOCs could induce eye and throat irritation, cause damage to the liver and central nervous system and increase relative rates of leukemia and lymphoma[3]. To date more and more attention had been paid to control VOCs. Methods used for removal and recovery of VOCs include condensation, absorption, adsorption, thermal, catalytic, and photocatalytic oxidation[4-6], among which adsorption using activated carbon with high specific surface area and large pore volume was one of the most widely used and mature methods to recycle VOCs[7-9].

The common methods for adsorption characterization included calorimetric measurement, temperature-programmed desorption(TPD), quartz crystal microbalance (QCM), gravimetric and optical methods[10-17]. TPD experiments had been carried out to quantify the adsorbed-desorbed amount of each adsorbate over different adsorbent materials, and to conclude about the affinity between these materials and the studied adsorbates[15]. QCM required only a small amount of absorbent (μg level), offered short response time and allowed detection of mass change (ng level)[14]. According to the sensitivity to the intra- and inter-molecular vibration, terahertz (THz) wave was also an effective way to evaluate the adsorption. Different absorption properties of the adsorbed water was shown in hydrophilic silica aerogel, and the collective tendency expressly tracked the dynamics of water adsorbed into active carbon pores[18,19].

Several adsorption kinetic models have been established to understand the adsorption kinetics and rate-limiting step[20]. These included pseudo-first and -second-order rate model, Weber and Morris sorption kinetic model, Adam-Bohart-Thomas relation, first-order reversible reaction model, external mass transfer model, first-order equation of Bhattacharya and Venkobachar, Elovich's model and Ritchie's equation. The pseudo-first and -second-order kinetic models are the most well liked model to quantify the extent of uptake in sorption kinetics[21-25]. Generally, adsorption rate constant, which could be used to estimate removal efficiency of VOCs, was determined to evaluate the adsorption behavior under different operational conditions.

Indeed, dynamic recycling process of VOCs was complex including diffusion, adsorption and volatilization[26]. In order to simulate and investigate the adsorption law of VOCs in pollutants recovery, the isooctane, ethanol, and butyl acetate drops were employed to adhere on active carbon in this research. THz-TDS was applied to characterize the adsorption dynamics and the relationship was established between the adsorbed rate parameter k_0 and the THz response.

2 Results and discussion

The selected THz-TDS of ACFC with isooctane, ethanol, and butyl acetate at different time were displayed in Figure 1. As time increases, the THz signal peak amplitudes (E_P) continuously increase. In addition the lines in Figure 1 show the E_P of the first detection at $t = 0$ equaling 0.14131V, 0.14789V and 0.17368V for isooctane, ethanol, and butyl acetate, respectively, indicating that the isooctane has weaker absorption than ethanol and butyl acetate in THz range. Indeed, ethanol and butyl acetate were composed of the hydrogen bond network. The hydrogen bond collective network formed by molecules changes on a picosecond (ps) timescale, thus causing the THz spectrum to be sensitive to fluctuations in the dipole moment of the ethanol and butyl acetate, which is in agreement with the THz response in Figure 1[19].

In order to characterize the adsorption process of three samples on active carbon, the E_P was extracted to correspond with time as shown in Figure 2. t_{Ads} and t_{End} represented the time when the adsorption started and ended. The whole process could be distinctly divided into three parts: $0 < t < t_{Ads}$, $t_{Ads} < t < t_{End}$ and $t_{End} < t$. For isooctane, ethanol and butyl acetate, the values of the parameter t_{Ads} are equal to 22.01min, 22.2min and 41.16min; the values of the parameter t_{End} are equal to 32.49min, 58.34min and 105.11min; the time lengths Δt from t_{End} to t_{Ads} are 10.48min, 36.14min and 63.94min, respectively.

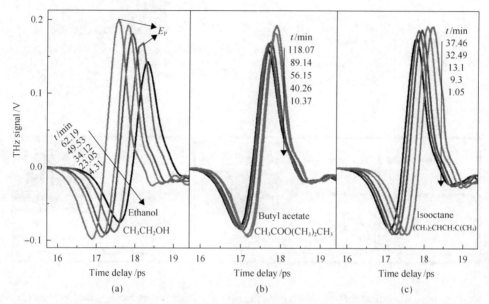

Figure 1 THz-TDS of selected ACFC with isooctane, ethanol, and butyl acetate at different time

According to our previous research, the three parts in Figure 2 were revealed as volatilization, diffusion and adsorption dynamics of organics on activated carbon. To further display the dynamics mechanism in the three processes, the motion sketch of organics on ACFC was shown in Figures 3 (a) ~ (c). The curves in Figure 3 (a) represented E_P as a function of time contributed by volatilization, diffusion and adsorption, respectively, and the black curve showed the superimposition of three models.

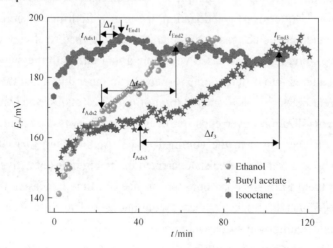

Figure 2 The time dependent E_P of ACFC with isooctane, ethanol and butyl acetate

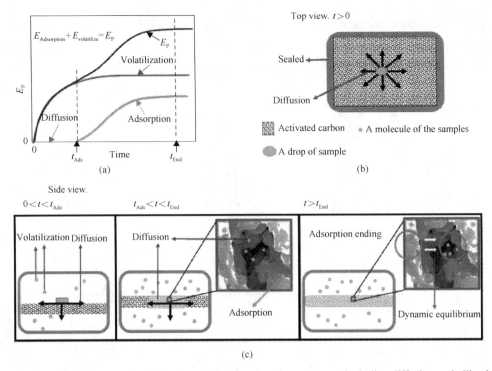

Figure 3 (a) The theoretical analysis sketch of E_P in respective processes including diffusion, volatilization and adsorption; (b) sketch map of diffusion in horizontal and vertically direction; (c) sketch map of diffusion in volatilization and diffusion, and the detailed adsorption state in the pores of active carbon

At the outset, the organics adhered onto the superficies of active carbon and diffused to the surrounding area due to the concentration gradient of the molecules. This diffusion process was divided into two parts: The surface motion [Figure 3 (b)] and the depth diffusion [the first portion of Figure 3 (c)]. In addition, diffusion has always existed before t_{End} and the THz signal (E_P) change only slightly. The quantities of organic molecules do not change because they are not adsorbed. In terms of volatilization, it began with diffusion simultaneously in the sealed space and made molecules gasification. Gas molecules had relatively weak response to THz radiation, therefore, the E_P increased with time going on. With the augment of time, the molecules were scattered into different holes especially in inner ducts, and then were gradually adsorbed in different holes[27-29]. The second inset of Figure 3 (c) described an early adsorption stage where the volatilization never ends but eventually achieves a dynamic equilibrium. In this stage, gas molecules filled to the confined space. At the same time, small part of the sample molecules which were in a free state, arrived in the big holes of active carbon and then a few amounts of them were adsorbed onto the surface. As time increased, the number of the molecules adsorbed on the surface increased continuously. The third inset of Figure 3 (c) showed the dynamic equilibrium stage, where most molecules were adsorbed and stabilized. A few of molecules were in dynamic equilibrium that adsorption and desorption took place with identical rate.

Based on pseudo-second order model and assumption of the proportional relation between the adsorbed amount and E_P, the adsorption equation can be presented as

$$\frac{t'}{E_P(t')} = \frac{1}{kE_{P0}^2} + \frac{t'}{E_{P0}}$$

where $t' = t - t_{Ads}$, k was the adsorption rate constant, and E_{P0} and $E_P(t')$ were the values of E_P of samples adsorbed onto active carbon at equilibrium state and t', respectively[20]. The data obtained in experiment were used to discuss the characteristics of the adsorption process. The linear positive correlation model between t'/E_P and the time t' of three samples was shown in Figure 4. The slope of three regressions equaled 0.00518, 0.00524 and 0.00544 for isooctane, ethanol and butyl acetate. The results indicated that E_{P0} calculated using the slope was roughly the same and was in accordance with that in Figure 2. In addition, the intercept d, which reflected the information of adsorption rate, was 0.00036min/mV, 0.00448min/mV and 0.00879min/mV for isooctane, ethanol and butyl acetate, respectively. Based on $k=1/(dE_{P0}^2)$, the k was 0.00613, 0.00337 and 0.73997 and increased in the order of butyl acetate < ethanol < isooctane, which indicated the order of adsorption rates.

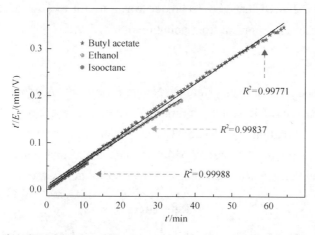

Figure 4　t'/E_P as a function of the time difference t' for adsorption of isooctane, ethanol and butyl acetate

Energy dispersive spectroscopy (EDS) was used to confirm the morphology and chemical composition of the active carbon. According to the EDS spectrum, the mass fraction and atom fraction of the chemical element C are 68.00% and 91.27%, while those of O are 6.61% and 6.66%. Activated carbon is constituted by blocks of small size imperfect graphene layers. The surface of the carbon is consequently rather hydrophobic since the interaction with water molecules is weak. Besides, hydrogen bond was formed between the element of O from hydroxyl group at the surface of active carbon and adsorbates including ethanol and butyl acetate. On the one hand, the adsorption rate improved due to the influence of hydrophilic microdomain and the participation of hydrogen bonding adsorption. However, hydrophobic interaction was the predominant driving force compared with hydrogen bonding in adsorption, and activated carbon had a greater affinity for hydrocarbons[30]. On the other hand, the molecules with bigger size demanded more time to scatter into inner holes and be adsorbed[31], therefore, adsorption rate of ethanol was higher than that of butyl acetate whose molecules were larger and it was isooctane that had the largest adsorption rate.

3 Conclusions

In summary, the research focused on the THz measurements of isooctane, ethanol, and butyl acetate drops adhered on active carbon at various time frames. By analyzing the THz signal peak amplitude E_P, the process in the whole range could be distinctly divided into three steps. Diffusion and volatilization played an important role in the first stage. Then the molecules were scattered in different holes especially in inner ducts, and gradually adsorbed in different holes in the second portion. With time increasing, dynamic equilibrium stage was shown in the third portion. What's more, a correlation between the adsorbed rate parameter k and the E_P of samples was obtained from the adsorption data and pseudo-second-order equation. This work showed that THz technique was a useful tool for adsorption characterization in protecting the environment from pollution of the VOCs.

Acknowledgements

This work was supported by the National Nature Science Foundation of China (Grant No. 11574401), National Basic Research Program of China (Grant No. 2014CB744302), the Specially Founded Program on National Key Scientific Instruments and Equipment Development, China (Grant No. 2012YQ140005) and the China Petroleum and Chemical Industry Association Science and Technology Guidance Program (Grant No. 2016-01-07).

References

[1] Gianluigi D G, Genoveffa F, Annalisa M, et al. Indoor and outdoor monitoring of volatile organic compounds in school buildings: Indicators based on, health risk assessment to single out critical issues. International Journal of Environmental Research and Public Health, 2013, 10(12): 6273-6291.

[2] Dwivedi P, Gaur V, Sharma A, et al. Comparative study of removal of volatile organic compounds by cryogenic condensation and adsorption by activated carbon fiber. Separation Purification Technology, 2004, 39(1-2): 23-37.

[3] Irigaray P, Newby J A, Clapp R, etal. Life style-related factors and environmental agents causing cancer: An overview. Biomedicine Pharmacotherapy, 2007, 61(10): 640-658.

[4] Chuang C L, Chiang P C, Chang E E. Modeling VOCs adsorption onto activated carbon. Chemosphere, 2003, 53(1): 17-27.

[5] Das D, Gaur V, Verma N. Removal of volatile organic compound by activated carbon fiber. Carbon, 2004, 42(14): 2949-2962.

[6] Kawasaki N, Kinoshita H, Oue T, et al. Study on adsorption kinetic of aromatic hydrocarbons onto activated carbon in gaseous flow method. Journal of Colloid Interface Science, 2004, 275(1): 40-43.

[7] Izquierdo M T, de Yuso, A M, Valenciano R, et al. Influence of activated carbon characteristics on toluene and hexane adsorption: Application of surface response methodology. Applied Surface Science, 2013, 264(1): 335-343.

[8] Kim D J, Shim W G, Moon H. Adsorption equilibrium of solvent vapors on activated carbons. Korean Journal of Chemical Engineering, 2001, 18(4): 518-524.

[9] Kim K J, Kang C S, You Y J, et al. Adsorption–desorption characteristics of VOCs over impregnated activated carbons. Catalysis Today, 2006, 111(3-4): 223-228.

[10] Dutour S, Nokerman J, Limborgnoetinger S, et al. Simultaneous determination of mass and calorimetric adsorption data of volatile organic compounds on microporous media in the low relative pressure range. Measurement Science Technology, 2003, 15(1): 185-194.

[11] Ulmer H, Mitrovics J, Noetzel G, et al. Odours and flavours identified with hybrid modular sensor systems. Sensors and Actuators B: Chemical, 1997, 43(1-3): 24-33.

[12] Serrano D P, Calleja G, Botas J A, et al. Adsorption and hydrophobic properties of mesostructured MCM-41 and SBA-15 materials for volatile organic compound removal. Industrial Engineering Chemistry Research, 2004, 43(22): 7010-7018.

[13] Zhao X S, Lu G Q, Hu X. Organophilicity of MCM-41 adsorbents studied by adsorption and temperature-programmed desorption. Colloids and Surfaces A: Physicochemical and Engineering Aspects, 2001, 179(2): 261-269.

[14] Huang C Y, Song M, Gu Z Y, et al. Probing the adsorption characteristic of metal–organic framework MIL-101 for volatile organic compounds by quartz crystal microbalance. Environmental Science and Technology, 2011, 45(10): 4490-4496.

[15] Shim W G, Lee J W, Moon H. Adsorption equilibrium and column dynamics of VOCs on MCM-48 depending on pelletizing pressure. Microporous and Mesoporous Materials, 2006, 88(1): 112-125.

[16] Wang C M, Kueisen C A, Chung T W, et al. Adsorption equilibria of aromatic compounds on activated carbon, silica gel, and 13X zeolite. Journal of Chemical and Engineering Data, 2004, 50(3): 527-531.

[17] Alvarezherrero A, Heredero R L, Bernabeu E, et al. Adsorption of water on porous vycor glass studied by ellipsometry. Applied Optics, 2001, 40(4): 527-532.

[18] Zhang J, Grischkowsky D. Terahertz time-domain spectroscopy of submonolayer water adsorption in hydrophilic silica aerogel. Optics Letters, 2004, 29(9): 1031-1033.

[19] Zhan H, Wu S, Bao R, et al. Water adsorption dynamics in active carbon probed by terahertz spectroscopy. Rsc Advances, 2015, 5(19): 14389-14392.

[20] Febrianto J, Kosasih A N, Sunarso J, et al. Equilibrium and kinetic studies in adsorption of heavy metals using biosorbent: A summary of recent studies. Journal of Hazardous Materials, 2009, 162(2-3): 616-645.

[21] Pavasant P, Apiratikul R, Sungkhum V, et al. Biosorption of Cu^{2+}, Cd^{2+}, Pb^{2+}, and Zn^{2+} using dried marine green macroalga Caulerpa lentillifera. Bioresource Technology, 2006, 97(18): 2321-2329.

[22] Djeribi R, Hamdaoui O. Sorption of copper(II) from aqueous solutions by cedar sawdust and crushed brick. Desalination, 2008, 225(1-3): 95-112.

[23] Baral S S, Das S N, Rath P. Hexavalent chromium removal from aqueous solution by adsorption on treated sawdust. Biochemical Engineering Journal, 2006, 31(3): 216-222.

[24] Apiratikul R, Pavasant P. Batch and column studies of biosorption of heavy metals by Caulerpa lentillifera. Bioresource Technology, 2008, 99(8): 2766-2777.

[25] Sağ, Yeşim, Aktay, et al. Kinetic studies on sorption of Cr(VI) and Cu(II) ions by chitin, chitosan and Rhizopus arrhizus. Biochemical Engineering Journal, 2002, 12(2): 143-153.

[26] Russo V, Tesser R, Trifuoggi M, et al. A dynamic intraparticle model for fluid–solid adsorption kinetics. Computers & Chemical Engineering, 2015, 74(4): 66-74.

[27] Feng X, Wu S X, Zhao K, et al. Pattern transitions of oil-water two-phase flow with low water content in rectangular horizontal pipes probed by terahertz spectrum. Optics Express, 2015, 23(24): A1693.

[28] Jin Y S, Kim G, Jeon S G. Terahertz dielectric properties of polymers. Journal of the Korean Physical Society, 2006, 49(2): 513-517.

[29] Ho Y S. Review of second-order models for adsorption systems. Cheminform, 2006, 136(3): 681-689.
[30] Bittner E W, Smith M R, Bockrath B C. Characterization of the surfaces of single-walled carbon nanotubes using alcohols and hydrocarbons: A pulse adsorption technique. Carbon, 2003, 41(6): 1231-1239.
[31] Lu Y, Jiang M, Wang C, et al. Impact of molecular size on two antibiotics adsorption by porous resins. Journal of the Taiwan Institute of Chemical Engineers, 2014, 45(3): 955-961.

第四篇 油气产品及污染物的光学技术表征评价

A spectral-mathematical strategy for the identification of edible and swill-cooked dirty oils using terahertz spectroscopy

Honglei Zhan[1,2]   Jianfeng Xi[2]   Kun Zhao[1,2]   Rima Bao[2]   Lizhi Xiao[1]

(1.State Key Laboratory of Petroleum Resources and Prospecting, China University of Petroleum, Beijing 102249, China; 2.Beijing Key Laboratory of Optical Detection Technology for Oil and Gas, China University of Petroleum, Beijing 102249, China)

Abstract: Effective technique and procedure based on the terahertz (THz) spectroscopy and mathematical methods, named THz Mathematics (T-Math), have been developed for the precise and rapid identification of edible oil from a series of typical swill-cooked dirty oils in daily life. Differences can be observed in the absorbance spectra of edible and dirty oils, while the waveforms were very similar. Principal component analysis (PCA) and support vector machine (SVM) were employed for the identification of edible or swill-cooked oil classification. Compared with dirty oils, normal edible oil had the largest first principal component (PC1) score with a contribution rate of 97.4%. The PC1 score deviations of edible oil and dirty oils varied from ~0.5 to 2.3. Meanwhile, by using SVM leave-one-out cross-validation, the swill-cooked oils can be directly identified with the precision of 100%. This study proved that the developed method was very suitable for the rapid determination of swill-cooked oils in food safety field.

Keywords: terahertz spectroscopy; swill-cooked oil; edible oil; principal component analysis; support vector machine

1 Introduction

Recently, research about swill-cooked dirty oil drew a wide attention because of the increasing events of food safety in many countries. Swill-cooked dirty oil referred to all kinds of inferior oils, such as recuperated cooking oil and repeated frying oils. Dirty oil often contained bacteria, heavy metal and harmful chemicals caused by the process of picking, decoloration, and deodorization so that it seriously jeopardized public health with an indirect way[1]. Thus, the governments in many countries widely called for the reliable methods to identify waste oil. Precise and rapid techniques can undoubtedly control waste oil diffusion, improve the food safety and really protect the interest of the consumers[2].

Actually, it is difficult to distinguish the edible oil and swill-cooked dirty oil not only because of the similar exterior, but also due to the analogous principal components. Glycerin trimyristate was the basis of edible and waste oils. Physical-chemical indicators such as acid value, solid fat, heavy metals etc., can detect dirty oils to some extent, but there were still relevant limits[3]. Spectral technologies have drawn wide attention due to the on-line detection and have been gradually applied in oil identification. Near infrared spectroscopy (NIR) was used to qualitatively identify the edible oil and dirty oils[4]. Recently THz spectroscopy has been developed rapidly and applied in many fields[5-8]. Some studies can be found about THz-organics analysis in oil-gas detection[9-16]. Some of the molecular vibration modes can be reflected in THz frequency range. However, there still existed some difficulties of organics determination due to the absence of absorption peaks or similar spectral waveform of different samples. Similar to NIR, THz detection of oils also need data-analysis methods to realize the clear identification[17]. In this study, THz time-domain spectroscopy (THz-TDS) was employed to scan eight kinds of swill-cooked oils and normal edible oil. The absorbance spectra reflected both the similarity and difference at selected frequencies. In order to classify the dirty oils more clearly, Principal component analysis (PCA) and support vetor machine (SVM) were utilized to build two kinds of models. Based on the PCA and SVM models, swill-cooked dirty and edible oils were clearly identified respectively. Such procedure can supply a new way for the food safety regulators to rapidly identify dirty oils on table and really protect consumers' health.

2 Experimental methods

2.1 Measurement setup and sample preparation

The measurement setup was based on a conventional THz-TDS system with transmission geometry. The THz pulse was generated by a p-type GaAs wafer with <100> orientation pumped by a Ti-sapphire laser with a center wavelength of 800 nm, a pulse width of 100 fs and a repetition rate of 80 MHz. A 2.8mm thickness <110>ZnTe was employed as the sensor and a standard lock-in technology was used in this setup. As shown in Figure 1, the laser beam was split into two beams by the S1 (splitter) after being reflected by M1 and M2 (mirror). The pump beam was used to generate THz radiation, which was focused onto a sample through optical lens (L1) and M7. The probe beam initially reached and transmitted through the automatic delay stage. After reflected by M4 and M5, the probe beam reached the detector. Meanwhile, the THz beam carrying sample's information transmitted L2 and was reflected by M6 and then met the probe laser beam in THz detector. In the detection system, the probe and THz pulse were collinear and passed through hyper spherical lens and the hybrid beams were focalized onto ZnTe crystal, whose index ellipsoid could be changed by monitoring THz electric field. Therefore, the polarization state of probe beam with linear polarization was altered due to the electro-optic crystal ZnTe. The signal from detector was amplified by a lock-in amplifier in the controller, which was also used to control the move of delay stage. A computer was employed to give the instructions and set the parameters. To dwindle the moisture absorption in the air and enhance the signal noise ratio (SNR), the setup was covered with dry nitrogen. The measurement temperature and relative humidity can be tested by a sensor in THz setup and displayed in the computer, and were 294.0K±0.3K and 3%±0.5%, respectively.

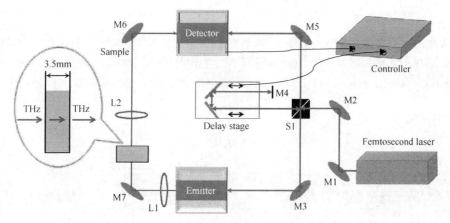

Figure 1   Measurement system of swill-cooked and edible oils based on THz-TDS

In this study, a common edible oil from a supermarket was measured, which was certified by Administration of Quality Supervision, Inspection and Quarantine (AQSIQ). Eight kinds of other oil were collected from the frying boilers of some vendors. These oils can be classified as swill-cooked oil due to the repeated frying of different foods, such as cake, gluten, egg, tofu, etc. The dirty oils initially cooled down in air and were then filtrated by a filter paper. All the nine kinds of oils were put into polytene (PE) samplers, which had little absorption in THz frequency range so that it was an ideal sampler for THz measurement. In Figure 1, the left illustration showed the enlarged view of sample. The width and height of PE sampler equaled 3.5mm and 30mm. The THz pulse passed through PE and oils and then was detected by the detector.

Herein, THz-TDS of the reference and samples were measured by scanning the empty PE sampler and the PE cells holding different kinds of swill-cooked oils or edible oil. After being processed by the fast Fourier transform (FFT), THz frequency-domain spectra (THz-FDS) of samples can be obtained in the efficient frequency range. Based on THz-FDS, the frequency dependent absorbance which is one of the THz parameters can be calculated by $-\lg(E_S/E_R)$, where E_S and E_R were the THz frequency-domain amplitudes of sample and reference, respectively.

2.2 Mathematical methods

PCA is a mathematical method which reduces the number of dimensions within the data while retains as much of the overall variations as possible based on uncorrelated projections. The calculation of PCA results in several variables called principal components (PCs). PCs are a set of new maximized variables which are uncorrelated and expressed as linear combination of original variables. PCs can be calculated via eigenvalue decomposition of the following scatter matrix S:

$$S = \sum_i (d_i - \eta)[(d_i - \eta)]^T \quad (1)$$

where d_i is an ith input pattern and η is the average value of d_i. If we let D be a diagonal matrix of eigenvalues in a descending order and E be an orthogonal matrix whose columns are the corresponding eigenvectors, the principal components P_i can be obtained as

$$S = EDE^T \quad (2)$$

$$P_i = E^T d_i \quad (3)$$

Dimension reduction is achieved by discarding the unimportant elements of d_i. The number of retained principal components would be determined according to the classification results[18].

SVM is one of the best classifier methods and has been applied in a wide variety of fields. The main applications are in regression analysis, classification, forecasting, and pattern recognition. It is based on statistical learning theory and the structural risk minimization principle. Structural risk minimization can reduce the upper-bound generalization error instead of traditional local training error. As shown in Figure 2, an input space of data can be transformed into nonlinear and high dimensional space by the use of kernel. The algorithm will generate a sparse prediction function by choosing a selected number of training points called support vectors. SVM has no hypothesis in functional transformation, making it essential to have linearly divisible data. The SVM calculation is based on the theory of Vapnik. The formula can be expressed as $f(x)=\omega\varphi(x)+b$, where ω and b represent a normal vector and the bias term, and $\varphi(x)$ is the high-dimensional space feature which can be described as a nonlinear mapping function[19].

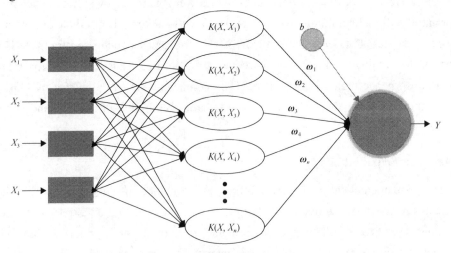

Figure 2　Principle of SVM method

3 Results and discussion

Initially, the reference pulse was obtained by scanning the empty PE sampler and then the PE cells holding oils were scanned and described as samples. As shown in Figure 3 the effective range of the frequency dependent absorbance spectra was selected from 0.2THz to 1.3THz. The temporal waveforms were obtained from THz-TDS and plotted in the inset of Figure 3. The THz-TDS of edible oil was different from that of fried tofu oil which can be classified as dirty oil. The differences referred to not only the time delay but the peak intensity, indicating that there were various effects for edible and dirty oils in THz range. Common characteristic absorption band can be found at ~1.00THz. Another two absorption features of swill-cooked oil were located at 0.70THz and 1.23THz, while the frequency of absorption feature of edible oil referred to 0.82THz in addition to 1.00THz. The absorption characteristics in Figure 3 indicated that there existed common chemical components, and the differences were also observed between edible and dirty oil, which was agreement with the actual conditions[20].

Swill-cooked oil was involved with kinds of inferior oils in daily life, including the waste cooking oil, repeated fried oil, the oil extracted from sewer trash, leftovers and inferior animal, etc. The glycerin trilaurate was the primary ingredients of both edible and waste oil. Compared to the edible oil, swill-cooked oils contain a number of fat oxidation products, trans-fatty-acids, cholesterol, as well as some toxic substance causing cancer.

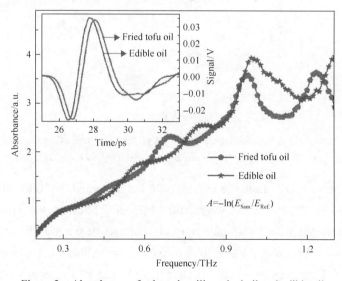

Figure 3 Absorbance of selected swill-cooked oil and edible oil

In order to discuss the comprehensive absorption response in the whole frequency range, PCA was employed to classify the edible and swill-cooked oils with the absorbance spectra over the whole range as the input. PCs, such as PC1, PC2 and PC3 were related to the original variables. PC1 reflected the most information due to the largest contribution rate which was used to judge the significance of PCs to the oils. Similarly, PC2 represented the second most information. In Figure 4, a two-dimensional space was plotted because of the large value of total contribution rate of PC1 and PC2. The contribution rate described the importance of PCs and was employed to evaluate the percentage of PCs representing samples' information. Concretely, the first two PCs of the absorbance spectra were found to describe 97.4% and 2.3% of the variance, with the total contribution rate equaling 99.7% in all deviations. The eight waste oils were simplified into eight numbers. The details were as following: Leftover vegetable oil (1), roll oil (2), fish-fried oil (3), Chuanlu sesame oil (4), gluten-baked oil (5), cake-egg oil (6), fried tofu oil (7) and hot-spicy pot oil (8). In the two-dimensional system, the edible oil was located at the right, while the swill-cooked oils were at the left and the middle locations. Consequently, PC1 versus PC2 system can directly distinguish the classification of oils.

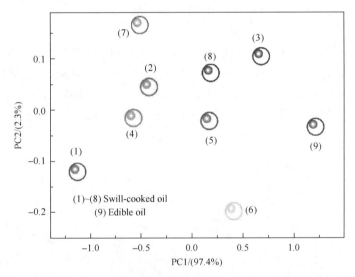

Figure 4　PC1 versus PC2 in the two-dimensional PCA system. PC1 and PC2 indicate 97.4% and 2.3% information of oils, respectively. (1)~(8) represent eight kinds of swill-cooked oils and (9) is an edible oil

Due to the high contribution rate (97.4%), PC1 represented the most information of the original variables in PCA system. Meanwhile, PC1 could reflect the primary absorption response in the selected frequency range not only because of the high contribution rate, but because that the input data was selected over the whole range. Such calculation can reduce the random error and improve the precision. Herein, only the PC1 scores were extracted and related to these oils. As shown in the PC1 score histogram of eight kinds of swill-cooked oils and edible oil in Figure 5, the edible oil had the largest PC1 score of ~1.2, while the scores of all the waste oils varied from ~−1.1 to ~0.7. Four kinds of dirty oils had positive scores and another four kinds had negative scores. Consequently, the PC1 score deviations of edible oil and swill-cooked oils equaled from ~0.5 to ~2.3, indicating a large difference between them in PCA model. The combination of THz absorbance spectra and PCA calculation would be an effective selection to rapidly distinguish the swill-cooked oils in our daily life.

In order to classify the different oils precisely and identify the swill-cooked oil rapidly, the classification analysis is necessary. Based on the absorbance of different oils, the assessment of oil quality was performed. SVM was used for both classification and regression analysis to build a qualitative model between the absorbance and oil types. As there was little difference about absorbance between edible and waste oils at some frequencies, the whole range of 0.2~1.3THz was selected for modeling. The classification plots of the leave-one-out cross-validation were shown in Figure 6. One of the swill-cooked oils was removed from the eight dirty oils and employed as prediction set. The left 7 waste oils and edible oil were set as two clusters (Nos. 1 and 2) and used for calibration. Such model building was processed 8 times due to 8 kinds of swill-cooked oil. It was obvious that all the prediction oils were judged into Nos. 1 cluster. Consequently, the direct identification of edible and swill-cooked oils can be realized by utilizing SVM algorithm based on the absorbance spectra over the whole range. In terms of new oil which cannot be recognized through the extrinsic features, the THz absorption spectra would be obtained by scanning the samples. Then SVM model will be built based on absorbance spectra. According to the database pre-built, the classification of the unknown oil can be directly and rapidly predicted due to the intelligent process composed of computing programs.

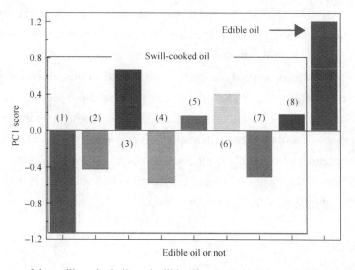

Figure 5　PC1 scores of the swill-cooked oils and edible oil extracted in Figure 4. PC1 scores vary from −1.13 to 1.21. Compared with swill-cooked oils, edible oil has largest PC1 score

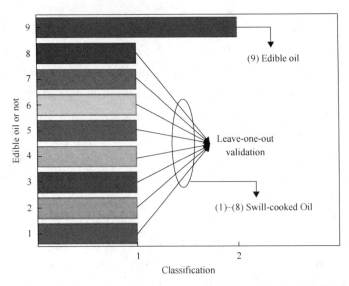

Figure 6 Leave-one-out validation of classification results in SVM calculation. In SVM calculation, one of the swill-cooked oils was employed as prediction set and the left 7 waste Oils as well as edible oil were used for calibration. Eight kinds of swill-cooked oil were calculated eight times and all were predicted as same classification

The present THz-TDS-based classification method was combined with PCA and SVM to quantify edible oil quality in a relatively scientific way. When applied to 9 kinds of oils including normal edible and swill-cooked oils collected from market, the THz-PCA model with a contribution rate of 97.4% showed that there was distinct difference of PC1 scores between edible and dirty oils. Meanwhile, when using leave-one-out cross-validation in calculation process, the THz-SVM model with an accuracy of 100% of classification prediction indicated that the eight samples belonged to cluster No. 1, which was in agreement with actual conditions. Overall, the absorbance spectra reflected the diversity of absorption, proving the different response caused by diverse components in oils. However, it was hard as a standard to directly identify the edible oil from swill-cooked oils due to the similarity at a number of frequencies. The utilization of mathematical methods, including PCA and SVM, supply a comprehensive tool to classify the edible oil and waste oils more clearly. The results reflected in this research were intuitive for relevant operation staff to identify unknown oils. This study will enrich the theory of edible oil identification and supply a new selection to directly check the oil quality in the food safety field.[21]

4 Conclusions

Both edible and swill-cooked oil were investigated by using the combination of THz-TDS and mathematical methods. Generally, swill-cooked oil and edible oil were hard to distinguish due to the similar exterior and physical properties. In THz system, two kinds of oils showed different absorption features at several frequencies while the main pulses of waveforms were basically same. The use of PCA and SVM can greatly amplify the difference of edible and dirty oils. Edible oil can be rapidly and directly identified from a series of swill-cooked oils according to the PC1 scores and SVM classification. The precise determination of swill-cooked oils is closely related to public health and interests. THz mathematics may be a promising selection for the food regulators and the public to protect the food safety.

Acknowledgements

This work was supported by the National Basic Research Program of China (Grant No. 2014CB744302), the Specially Funded Program on National Key Scientific Instruments and Equipment Development (Grant No. 2012YQ140005), and the National Nature Science Foundation of China (Grant Nos. 11574401 and 61405259).

References

[1] Ramani K, Karthikeyan S, Boopathy R, et al. Surface functionalized mesoporous activated carbon for the immobilization of acidic lipase and their application to hydrolysis of waste cooked oil: Isotherm and kinetic studies. Process Biochemistry, 2012, 47, 435-445.

[2] He J, Xu W, Shang Y, et al. Development and optimization of an efficient method to detect the authenticity of edible oils. Food Control, 2013, 31, 71-79.

[3] Maggio R M, Cerretani L, Chiavaro E, et al. A novel chemometric strategy for the estimation of extra virgin olive oil adulteration with edible oils. Food Control, 2010, 21, 890-895.

[4] Zhou Y, Liu T, Li J. Rapid identification between edible oil and swillcooked dirty oil by using a semi-supervised support vector machine based on graph and near-infrared spectroscopy. Chemometrics and Intelligent Laboratory Systems, 2015, 143, 1-6.

[5] Mittleman D M. Frontiers in terahertz sources and plasmonics. Nature Photonics, 2013, 7, 666-669.

[6] Ok G, Kim H J, Chun H S, et al. Foreign-body detection in dry food using continuous sub-terahertz wave imaging. Food control, 2013, 42, 284-289.

[7] Siegel P H. Terahertz technology in biology and medicine. IEEE Transactions on Microwave Theory and Techniques, 2004, 52, 2438-2447.

[8] Zhao H, Zhao K, Bao R. Fuel property determination of biodiesel-diesel blends by terahertz spectrum. Journal of Infrared, Millimeter, and Terahertz Waves, 2012, 33, 522-528.

[9] Bao R, Wu S, Zhao K, et al. Applying terahertz time-domain spectroscopy to probe the evolution of kerogen in close pyrolysis systems. Science China: Physics, Mechanics & Astronomy, 2013, 56, 1603-1605.

[10] Ge L, Zhan H, Leng W, Zhao, et al. Optical characterization of the principal hydrocarbon components in natural gas using terahertz spectroscopy. Energy & Fuels, 2015, 29, 1622-1627.

[11] Leng W, Zhan H, Ge L, et al. Rapidly determinating the principal components of natural gas distilled from shale with terahertz spectroscopy. Fuel, 2015, 159, 84-88.

[12] Zhan H, Li Q, Zhao K, et al. Evaluating $PM_{2.5}$ at a construction site using terahertz radiation. IEEE Transaction on Terahertz Science and Technology, 2015, 5, 1-6.

[13] Zhan H, Qin F, Jin W, Ge, et al. Quantitative determination of n-heptane and n-octane using terahertz time-domain spectroscopy with chemometrics methods. Frontiers of Optoelectronics, 2015, 8, 57-61.

[14] Zhan H, Wu S, Bao R, et al. Qualitative identification of crude oils from different oil fields using terahertz time-domain spectroscopy. Fuel, 2015, 143, 189-193.

[15] Zhan H, Wu S, Bao R, et al. Water adsorption dynamics into active carbon probed by terahertz spectroscopy. RSC Advances, 2015, 5, 14389-14392.

[16] Zhan H, Zhao K, Xiao L. Spectral characterization of the key parameters and elements in coal using terahertz spectroscopy. Energy, 2015, 93, 1140-1145.

[17] Qin F, Li Q, Zhan H, et al. Probing the sulfur content in gasoline quantitatively with terahertz time-domain spectroscopy. Science China: Physics, Mechanics & Astronomy, 2014, 57, 1404-1406.

[18] Burnett A D, Fan W, Upadhya P C, et al. Broadband terahertz time-domain spectroscopy of drugs-of-abuse and the use of principal component analysis. Analyst, 2009, 134, 1658-1668.

[19] Wang J, Hu J. A robust combination approach for short-term wind speed forecasting and analysis-combination of the ARIMA (Autoregressive Integrated Moving Average), ELM (Extreme Learning Machine), SVM (Support Vector Machine) and LSSVM (Least Square SVM) forecasts using a GPR (Gaussian Process Regression) model. Energy, 2015, 93, 41-56.

[20] Xu L, Xu X, Xiong H, et al. Rapid detection of vegetable cooking oils adulterated with inedible used oil using fluorescence quenching method with aqueous CTAB-coated quantum dots. Sensors and Actuators B: Chemical, 2014, 203, 697-704.

[21] Jedrkiewicz R, Glowacz A, Gromadzka J, et al. Determination of 3-MCPD and 2-MCPD esters in edible oils, fish oils and lipid fractions of margarines available on polish market. Food Control, 2016, 59, 487-492.

Qualitative identification of crude oils from different oil fields using terahertz time-domain spectroscopy

Honglei Zhan[1,3]　Shixiang Wu[2]　Rima Bao[3,4]　Lina Ge[3,4]　Kun Zhao[1,3,4]

(1.State Key Laboratory of Petroleum Resources and Prospecting, China University of Petroleum, Beijing 102249, China; 2.Petroleum Exploration and Production Research Institute, China Petroleum and Chemical Corporation, Beijing 100083, China; 3.Beijing Key Laboratory of Optical Detection Technology for Oil and Gas, China University of Petroleum, Beijing 102249, China; 4. Key Laboratory of Oil and Gas Terahertz Spectroscopy and Photoelectric Detection, China Petroleum and Chemical Industry Federation (CPCIF), Beijing 100723, China)

Abstract: The purpose of this article is to demonstrate the practicability to classify the sources and types of crude oils from different oil fields in several countries and regions with terahertz time-domain spectroscopy (THz-TDS). THz parameters spectra, such as refractive index and absorption coefficient, were calculated. Multivariate statistical methods, including cluster analysis (CA) and principal component analysis (PCA), were used to build models between THz parameters and crude oils from different countries and regions. The distances of CA between oils and first principal component (PC1) scores of oils in PCA method reflected the oil-dependent differences, indicating that there existed consistency between CA and PCA. Consequently, the combination of THz technology as well as multivariate statistical methods could be an effective method for rapid identification of crude oils with different properties and geographical locations.

Keywords: crude oil; terahertz; cluster analysis; principal component analysis

1 Introduction

Crude oil, an unprocessed oil exploited from oil well, is a brownish black or dark green viscous liquid or semi-solid, which is a combustible substance composed of various hydrocarbons and other elements such as oxygen, sulfur, nitrogen and so on. After exploited from oil wells and pretreated preliminarily, crude oil will be transported to refineries using pipelines, oilers or some other oil tools. The classification of crude oils in pipelines is of great importance to determine the sources, types and properties of various crude oils for the sake of subsequent processing. One of the effective tools wis true boiling point (TBP) method which is a batch distillation process widely used for the characterization of crude oils[1]. Near infrared and mid-infrared spectroscopy methods are also validated to be a fast and effective tool to determine crude oil and oil products[2-5].

Terahertz time-domain spectroscopy (THz-TDS), which ranges from 0.1THz to 10THz, bridges the gap between microwave and infrared spectroscopy. THz-TDS is based on the THz electric field with time resolution, which is generated using a femtosecond laser pulse. After applying the fast Fourier transform (FFT), the spectral information can be obtained. Spectral features are often employed to qualitatively analyze the material structures and physical properties of tested samples. THz technique was considered for crude oil characterization because of its advantages. It was highly sensitive in both time and frequency domains and was often acted as a nondestructively on-line method due to its low photon energy. In addition, THz-TDS could provide abundant information of intermolecular and low-frequency intramolecular modes of chemicals and give the amplitude and phase information of sample simultaneously[6-9]. As a contactless method, THz technique was little sensitive to the thermal background radiation and did not need sample pretreatment, such as dilution, which was always necessary to some chemical methods because the properties of crude oils were too complicated, for example, the deep color would result in a difficult observation of experimental phenomenon. Meanwhile, THz-TDS method was proved to be an effective tool to qualitatively and quantitatively detect the properties in crude oil[10,11], the particle aggregation state in crude oil[12], and the inflammable oil products such as gasoline, biodiesel-diesel, lubricating oil, and so on[13-19]. In this research, different crude oils from several oil fields in several countries and regions were determined by THz-TDS technique. Refractive index and absorption coefficient spectra of crude oils were calculated, which obviously reflected the differences of corresponding oil-dependent properties. Multivariate statistical methods, including cluster analysis (CA) and principal component analysis PCA, were then adopted to classify crude oils with the input of refractive index and absorption coefficient spectra. The consistency was observed between the two methods from the CA dendrograms and PCA histograms, indicating that THz-TDS technique was simple and sensitive to realize the classification of crude oils from different oil fields.

2 Materials and methods

2.1 Experimental method

Seven kinds of crude oils were supplied from different oil fields in several countries and regions. Detailed information about samples, including viscosity v, H/C, American Petroleum Institute (API) Gravity γ, content of paraffin η_p and content of asphaltene η_a, was listed in Table 1 and identifier numbers will be used on behalf of oil's appellation in the following discussion. The seven kinds of crude oils had the representativeness to some extent because of the different geographical locations, not only China, but also overseas, such as Middle East, Sultan, and Brazil. On the other hand, clean oil ($\gamma>30$), heavy oil ($\gamma<20$), and intermediate oil ($20<\gamma<30$) were involved according to the density. Also, low- ($\eta_p<2.5\%$), high- ($\eta_p>10\%$), and intermediate-paraffin ($2.5\%<\eta_p<10\%$) oils can be observed according to the paraffin content. In addition, the representativeness can also be proved according to other properties such as v, H/C, η_a and so on. All the properties reflected the big variations so that the seven crude oils covered abundant information. Therefore, these crude oils were typical in the actual petroleum industry of China.

Table 1 The physical properties of seven crude oils from different oil fields

	Crude oil 1 (Middle East)	Crude oil2 [Bohai Gulf (China)]	Clean oil3 [Liaohe (China)]	Heavy oil4 [Liaohe (China)]	Crude oil5 (Sultan)	Crude oil6 (Brazil)	Crude oil7 [Banqiao (China)]
v (cP/50℃)	3.65	47.2	8.02	63.2	16.2	26.1	4.58
H/C	1.80	1.89	1.87	1.62	1.85	1.60	1.90
γ	32.84	22.78	31.83	17.72	33.77	23.39	48.21
η_v/%	3.82	11.76	13.52	7.86	24.53	1.24	18.49
η_a/%	1.42	0.12	0.17	0.26	0.63	2.55	1.14

As shown in Figure 1, the experimental setup we used was comprised of a conventional THz-TDS system with transmission geometry and a diode-pump mode-locked Ti-sapphire laser (MaiTai, Spectra Physics) from Zomega Terahertz Corporation. An 800nm femtosecond laser beam was split into two beams. The pump beam was used to generate THz radiation through an emitter composed of photo conductive antenna. THz pulses were focused onto a sample by optical lens and the THz beam carrying sample's information met the probe laser beam at the ZnTe crystal in THz detector[20]. The signal was amplified by a lock-in amplifier and sent to computer for next processing. To avoid moisture absorption in the air and enhance the signal noise ratio (SNR), the setup was covered with dry nitrogen. As shown in the left illustration of Figure 1, 10mm thick polystyrene (PS) was selected as the sampler, whose width and height equaled 10mm, and 45mm, respectively. PS had little absorption in THz range and the PS cell would be an ideal sampler for THz measurement.

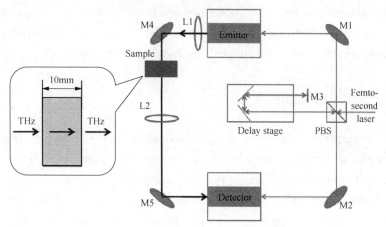

Figure 1 Diagram of transmission THz-TDS setup

The THz-TDS of the reference and samples were measured by scanning the empty PS cell and the PS cells holding the crude oils. After the calculation of FFT, THz frequency-domain spectra (THz-FDS) of samples were obtained. In addition, samples' THz parameters, such as the frequency dependence of refractive index (n) and absorption coefficient (α) can be calculated by $1+[\varphi c/(\omega d)]$ and $(2/d)\ln\{4n/[\rho(n+1)^2]\}$, respectively, where ω is the frequency, ρ and φ are the FFT amplitude ratio and phase difference of the sample and reference signal, c equals the velocity of light in vacuum, and d is the thickness of sample[21, 22].

2.2 Modeling methods

As a multivariate statistical method to study the classification problem, CA is used to divide objects into several classes, which are not given in advance and whose structures are also not assumed but only determined by the feature of the data. Objects in same class trend to resemble each other, but objects in different classes hardly resemble. CA is an exploratory analysis owing to the process of classification without prior to be given. It can simplify the complex data of the samples by making a logistic calculation to build a simple model and extract the significant information which can be not reflected directly in the original data or spectra[4]. In addition, PCA, a multivariate statistical method, reduces the number of dimensions within the data and identifies potential structure of large spectral data as well as groups. PCA method has great advantages and has been applied widely to the fields involving a lot of variables. The results calculated by PCA are known as principal components (PCs), which can reflect as much of the overall variation as possible and are usually expressed as linear combination of original variables. PCs are uncorrelated from each other so that the information contained in PCs is not overlapping. By using PC scores, the information of samples can be qualitatively and quantitatively analyzed. For example, by plotting the scores of the early PCs against each other, samples which are closely related would cluster together in a two- or three-dimensional space, or obtain the most similar scores by comparing all the PC1 scores, thus qualitative identification can be achieved[23,24]. Also, the relationship among the crude oils can be displayed more intuitively by using CA and PCA methods.

3　Results and discussions

The reference pulse was obtained by scanning the empty PS cell firstly, and the waveforms of all the samples' spectra were variant from each other as shown in Figure 2(a), not only the amplitude, but the peak time, indicating that THz technique could give the amplitude and phase information of samples simultaneously. After the application of FFT, the THz-FDS were calculated in Figure 2(b), which depended on the place where the oil was made. Due to the strong absorption of oil, the effective frequency range was reduced to 0.4~2.0THz.

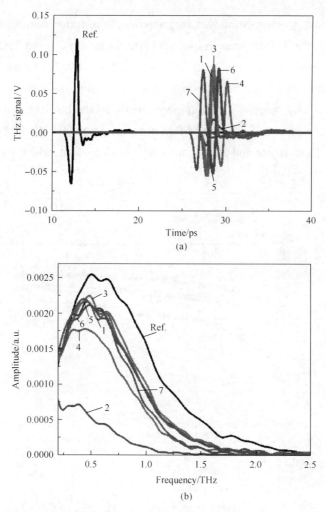

Figure 2 (a) THz-TDS and (b) THz-FDS of samples and reference

Seven kinds of oils exhibited distinctive refractive index n and absorption coefficient α spectra in Figure 3. n kept the maximum for Liaohe heavy oil compared to others over the hole range, and subsequently ranked the oils of Brazil, Sultan, Liaohe clean, Bohai Gulf, Middle East and Banqiao in accordance with the order of decreasing; however, overlapping phenomenon was also found between two specific oils at several frequencies. It was observed that at frequency of 1THz, two samples crossed and the refractive index changed, and the crossing phenomenon was resulted from the frequency dependence of dispersion. Electromagnetic parameters changed with the frequency variation due to the hysteresis of dielectric polarization in some special systems. Herein, the crude oils from Middle East and Banqiao (China) reflected the similar refractive spectra but different dispersion effect, which may be caused by the different ingredient species and the different contents of the same components between the two crude oils. Differences could also be observed from the absorption coefficient spectra. Moreover, characteristic absorption bands, such as peaks of Banqiao oil at 1.87THz, Liaohe heavy oil at 1.92THz, Bohai Gulf oil at 1.45THz, and Brazil oil at 1.73THz, were acted as the base of determination of oil sources. Therefore, THz technique could serve as an easy tool to detect crude oils from different regions due to their special responses in THz range. To build a more precise model between oils and THz spectra and realize the determination of oils from different regions more directly and efficaciously, CA and PCA were used to analyze and display the classification of oil, which would be detailedly discussed in the next section.

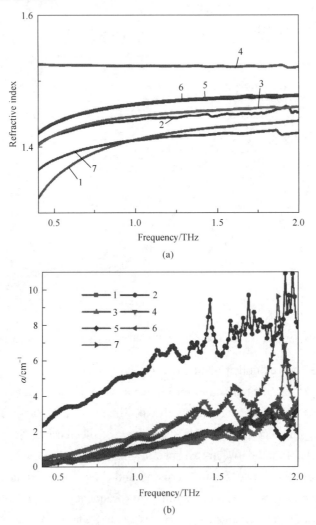

Figure 3 (a) Refractive index n and (b) absorption coefficient α spectra of seven kinds of oils

System cluster analysis, a kind of CA method, was used to classify crude oils from different regions. Every sample was set as a class at the beginning of calculation and then two classes with minimum Euclidean distance, a variable for measuring the similitude between samples, would be classified as a new one. All samples were calculated based on the distances between them and finally clustered as one class. The relationship of samples calculated by CA would be plotted in a dendrogram. The cophenetic correlation coefficient (CCC), which was defined as a linear correlation coefficient between the cophenetic distances obtained from the tree and the original distances used to construct the tree, represented the reliability of the dendrogram in CA model. The importance of PCs was described by corresponding contribution rate. The larger contribution rate was, the more information PC reflected. According to the great differences between dispersion and absorption effect shown in Figure 3, the next calculation with CA and PCA methods was obtained with the data of refractive index, and then absorption coefficient over the range of 0.4~2.0THz, respectively. As shown in Figure 4, the refractive index spectra was firstly calculated with CA and PCA methods obtaining dendrogram of CA and PC1 score histogram, respectively. All the samples were clustered into one class with six steps, where the final CCC value equaled 0.7894. The first PC (PC1) of the data set was found to describe 98.6% of the variances within the data, with the first three PCs describing 99.9% of the variances. The more similar PC1 scores of any two samples were, the more similitude they provided. On the contrary, the larger PC1 scores deviation of samples was, the more differences they had. Results showed that Liaohe heavy oil was the most special oil shown in both CA dendrogram and PCA histogram. The distance and deviation between Liaohe heavy oil and others were the largest, on the contrary, the distance between Sultan and Brazil was the smallest in Figure 4, which was consistent with the deviation of corresponding PC1 scores. Oils from other regions in this research were clustered, reflecting similarities among them.

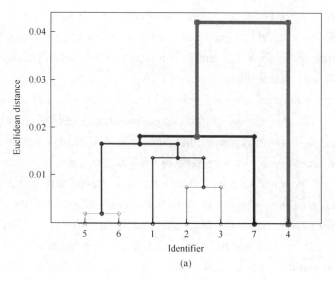

(a)

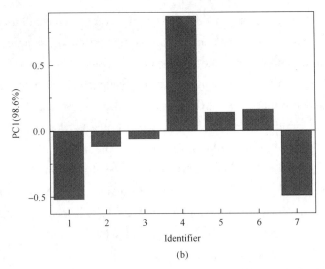

Figure 4 (a) The Euclidean distance dendrogram and (b) PC1 score histogram calculated with refractive index data

    The results calculated by CA and PCA methods with the input of absorption coefficient spectra data were depicted in Figure 5, including CA dendrogram and PC1 score histogram, respectively. Herein CCC value equaled 0.9561 and PC1 was found to describe 95.1% of the variances within the data. Bohai Gulf oil reflected a great particularity both in CA and PCA calculation progress. The distance between Bohai Gulf oil and others reached a maximum, and its PC1 score was in the lead compared to other oils. From Figure 5, we inferred that Middle East oil and Liaohe clean oil had the some similarities because of the small distance and analogous PC1 scores. Similar phenomenon was found between the crude oils from Sultan and Brazil. These were in accord with preceding results calculated with refractive index data.

    Different responses in THz range were observed concerning seven representative kinds of crude oils from different regions. CA and PCA methods were used to classify the similarity of oils and results showed that the distances and PC1 scores matched well with certain properties of different oils. In spite of existing differences in refractive index and absorption coefficient spectra of crude oils, the use of multivariate statistical methods greatly improved the identification effect and provided a more direct and effective display way to identify different oils. Actually, the more crude oils were used to analyze, the more abundant models could be obtained. Based on the results above, more and more crude oils could be measured by THz-TDS and clustered with CA and PCA to build an abundant data base, which would be acted as the identification standard of crude oils from different regions. Consequently, THz-TDS is a method for rapid identification of crude oils and can be used as a supplementary mean of traditional methods in crude oils' pipeline transportation fields.

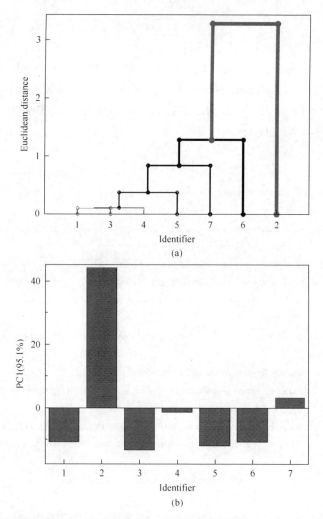

Figure 5 (a) The Euclidean distance dendrogram and (b) PC1 score histogram calculated with absorption coefficient data

4 Conclusions

In summary, THz technique was proved to be an effective tool to realize the identification of crude oils from different regions. There existed obvious differences among all kinds of oils both in refractive index and absorption coefficient spectra in THz range. Two multivariate statistical methods, CA and PCA, were used to build qualitative models between oils and their THz spectra. The distances in CA dendrogram and PC1 scores in PCA histogram directly displayed the similarity or differences between any two oil from different countries and regions. Therefore, THz technique plus multivariate statistical methods is a potential selection for the further identification of crude oils.

Acknowledgements

This work was supported by the National Key Basic Research Program of China (Grant No. 2014CB744302), the Specially Founded Program on National Key Scientific Instruments and Equipment Development (Grant No. 2012YQ140005), and the Beijing National Science Foundation (Grant No. 4122064).

References

[1] Behrenbruch P, Dedigama T. Classification and characterisation of crude oils based on distillation properties. J. Pet. Sci. Eng., 2007, 57: 166-180.

[2] Aske N, Kallevik H, Sjöblom J. Water-in-crude oil emulsion stability studied by critical electric field measurements. Correlation to physico-chemical parameters and near-infrared spectroscopy. J. Pet. Sci. Eng., 2002, 36: 1-17.

[3] Chu X L Xu, Y P Tian, S B, et al. Rapid identification and assay of crude oils based on moving-window correlation coefficient and near infrared spectral library. Chemom. Intell. Lab. Syst., 2011, 104: 44-49.

[4] Li L, Lomov S V, Yan X, et al. Cluster analysis of acoustic emission signals for 2D and 3D woven glass/epoxy composites. Compos. Struct., 2014, 116: 286-299.

[5] Bassbasi M, Hafid A, Platikanov S, et al. Study of motor oil adulteration by infrared spectroscopy and chemometrics methods. Fuel, 2013, 104: 798-804.

[6] Horiuchi N, Zhang X C. Searching for terahertz waves. Nat. Photonics, 2010, 4: 662.

[7] Bao R M, Wu S X, Zhao K, et al. Applying terahertz time-domain spectroscopy to probe the evolution of kerogen in close pyrolysis systems. Sci. China: Phys. Mech. Astron., 2013, 56: 1603-1605.

[8] Siegel P H. Terahertz technology in biology and medicine. IEEE T. Microw. Theory, 2004, 52: 2438-2447.

[9] Takenori T, Takahiro O, Ikumi K, et al Estimation of water content in coal using terahertz spectroscopy. Fuel, 2013, 105: 769-770.

[10] Jin W J, Zhao K, Yang C, et al. Experimental measurements of water content in crude oil emulsions by terahertz time-domain spectroscopy. Appl. Geophys., 2013, 10: 506-509.

[11] Bao R M, Tian L, Zhao K, et al. Spectroscopy studies on the Tuha crude oil in the terahertz range. SPIE Proceedings, 2011, 8195: 81951J.

[12] Jiang C, Zhao K, Zhao L J, et al. Probing disaggregation of crude oil in magnetic field with terahertz time-domain spectroscopy. Energy Fuels, 2014, 28: 483-487.

[13] Ikeda T, Matsushita A, Tatsuno M, et al. Investigation of inflammable liquids by terahertz spectroscopy. Appl. Phys. Lett., 2005, 87: 034105.

[14] Jin Y S, Kim G J, Shon C H, et al. Analysis of petroleum products and their mixtures by using terahertz time domain spectroscopy. J. Korean Chem. Soc, 2008, 53: 879-1885.

[15] Qin F L, Li Q, Zhan H L, et al. Probing the sulfur content in gasoline quantitatively with terahertz time-domain spectroscopy. Sci. China: Phys. Mech. Astron., 2014, 57: 1404-1406.

[16] Tian L, Zhou Q L, Jin B, et al. Optical property and spectroscopy studies on the selected lubricating oil in the terahertz range. Sci. China: Phys. Mech. Astron., 2009, 39: 1938-1943.

[17] Zhao H, Zhao K, Bao R M. Fuel property determination of biodiesel-diesel blends by terahertz spectrum. J. Infrared Milli. Terahertz Waves, 2012, 33: 522-528.

[18] Al-Douseri F M, Chen Y Q, Zhang X C. THz wave sensing for petroleum industrial application. Int. J. Infrared Milli., 2006, 27: 481-503.

[19] Leng W X, Ge L N, Xu S S, et al. Pressure-dependent terahertz optical characterization of heptafluoropropane. Chin. Phys. B, 2014, 23:107804.
[20] Löffler T, Hahn T, Thomson M, et al. Large-area electro-optic ZnTe terahertz emitters. Opt. Express, 2005, 13:5353-5362.
[21] Mickan S P, Zhang X C. T-ray sensing and imaging. Int. J. High Speed Electron. Syst, 2003, 13:601-676.
[22] Dorney T D, Baraniuk B G, Mittleman D M. Material parameter estimation with terahertz time-domain spectroscopy. J. Opt. Soc. Am. A, 2001, 18:1562-1571.
[23] Hwang J, Choi N, Park A, et al. Fast and sensitive recognition of various explosive compounds using Raman spectroscopy and principal component analysis. J. Mol. Struct., 2013, 1039:130-136.
[24] Burnett A D, Fan W, Upadhya P C, et al. Broadband terahertz time-domain spectroscopy of drugs-of-abuse and the use of principal component analysis. Analyst, 2009, 134:1658-1668.

The spectral analysis of fuel oils using terahertz radiation and chemometric methods

Honglei Zhan[1,2] Kun Zhao[1,2] Hui Zhao[2] Qian Li[2]
Shouming Zhu[2] Lizhi Xiao[1]

(1.State Key Laboratory of Petroleum Resources and Prospecting, China University of Petroleum, Beijing 102249, China; 2.Beijing Key Laboratory of Optical Detection Technology for Oil and Gas, China University of Petroleum, Beijing 102249, China)

Abstract: The combustion characteristics of fuel oils are closely related to both engine efficiency and pollutant emissions, and the analysis of oils and their additives is thus important. These oils and additives have been found to generate distinct responses to terahertz (THz) radiation as the result of various molecular vibrational modes. In the present work, THz spectroscopy was employed to identify a number of oils, including lubricants, gasoline and diesel, with different additives. The identities of dozens of these oils could be readily established using statistical models based on principal component analysis. The THz spectra of gasoline, diesel, sulfur and methyl methacrylate (MMA) were acquired and linear fittings were obtained. By using chemometric methods, including back propagation, artificial neural network and support vector machine techniques, typical concentrations of sulfur in gasoline (ppm grade) could be detected, together with MMA in diesel below 0.5%. The absorption characteristics of the oil additives were also assessed using 2D correlation spectroscopy, and several hidden absorption peaks were discovered. The technique discussed herein should provide a useful new means of analyzing fuel oils with various additives and impurities in a non-destructive manner and therefore will be of benefit to the field of chemical detection and identification.

Keywords: THz spectroscopy; fuel oils; additive; chemometric

1 Introduction

At present, environmental pollution is a global issue, largely due to increasing concerns regarding climate change, and the negative impacts of pollution pose a threat to long-term economic sustainability[1]. Much of this environmental degradation stems from the combustion of fuel oils such as gasoline and diesel, which is currently a necessity for both industry and daily life[2]. As important power sources, gasoline and diesel engines are widely used in motorcycles, cars, and off-road vehicles, and the emissions such as particulate matter or gaseous emissions from the combustion in engines are closely related to their energy requirements[3]. In terms of common fuel oils, the lubricants' chemical compositions are heavy hydrocarbons with high boiling points and non-hydrocarbon mixtures; the main components of gasoline include aliphatic hydrocarbon (C_5~C_{12}) as well as naphthenic and aromatic hydrocarbons[4], while the principal ingredients of diesel are C_{10}~C_{22} complex hydrocarbons[5]. The different intramolecular forces, chemical structures and vibrational modes in gasoline are not only a direct contributor to SO_x emissions, but can also affect the low-temperature activity of automotive catalytic converters. For this reason, the Euro III standard allows a maximum of 150ppm sulfur in gasoline[6], and regulations in China impose the same limit[7,8]. Diesel is more commonly utilized in trains, tractor, and ships under comparatively extreme conditions, such as low-temperature environments, because it doesn't tend to form solid precipitates, thus allowing the continued operation of diesel engines under these conditions[9-11].

The performance characteristics of various fuels, such as density, viscosity and heating value, are sometimes improved through the use of chemical additives having relatively low flash and auto-ignition temperatures. The use of oil additives is based on considering the tradeoff between improved oil quality and increased particulate emissions[12-15]. Taking diesel as an example, a pour point depressant (PPD) can be used to lower the freezing point. One such PPD is methyl methacrylate (MMA), which is an ester. In order to obtain the optimal benefit from MMA in oil, the real-time monitoring of the MMA concentration during processing or use of the fuel would be helpful.

Fuel oils are primarily composed of hydrocarbons, particularly *n*-alkanes. These alkanes, the sulfur in gasoline, and MMA all have very different molecular structures, polarities and vibrational modes, and these differences can be used to monitor the levels of these additives or contaminants in gasoline and diesel oils. The newly developed spectroscopic method known as THz spectroscopy is very sensitive to inter- and intra-molecular interactions, including van der Waals forces and especially hydrogen bonds[16]. THz spectroscopy operates over the vibrational range from 0.1THz to 10THz, which bridges the gap between infrared and microwave radiation. Due to the low photon energy employed and the sensitivity to molecular vibration modes, THz spectroscopy offers some unique advantages and has been used in biology and medicine, and in the analysis of various materials, such as coal, petroleum and pollutants[17-20].

As noted, sulfur in gasoline and MMA in diesel are closely related to pollutant emissions and engine efficiency, both of which are very significant with regard to controlling pollution and protecting the environment. The attenuation of THz radiation by oils means that radiation in this spectral band is highly sensitive during non-contact or on-line analysis of oil additives. In addition, the strong absorption of additives such as MMA allows us to determine additive contents in oils based on their THz spectra. As the principles behind this new technique are quite different from those of traditional chemical test methods, this new method represents a significant theoretical and technological innovation[21]. An additional advantage of THz spectroscopy is that it requires neither contact with the oils nor any sample pretreatment. The application of this technique to the analysis of additives or impurities in oils would therefore allow real-time monitoring of oils, and continuous data acquisition could even permit the assessment of additive concentrations based on statistical models immediately prior to combustion if the engine has code to change injection/ignition parameters based on sulfur or MMA content. In addition, this method is sensitive enough to detect low levels of additives or contaminants when used in conjunction with spectroscopic analysis techniques. Therefore, the THz technique can act as a supplementary tool for traditional approaches to oil analysis such as chemical methods, due to its sensitivity to molecular vibrational modes and little sample pretreatment, and has been examined for applications in petroleum analysis, security screening, the assessment of cultural relics and environmental monitoring[22-24].

Terahertz time-domain spectroscopy (THz-TDS) system has previously been used to analyze oil and gas reservoirs and similar resources in either the solid, liquid or gas states[25-28]. Various oils and rocks have also been assessed by employing different sample holders. In additions, previous investigations have focused on the spectral features of commercial fuel oils and on predicting the cold flow properties of diesel by THz-TDS[29,30], while this study concentrated on the direct identification of oil types, quantification of additives in oils and the properties of fuels onboard a vehicle. In the present research, we investigated the THz spectroscopy of various standard fuel oils in conjunction with different additives. Mathematical methods, including principle component analysis (PCA), back propagation artificial neural network (BPANN), support vector machine (SVM) and two-dimensional correlation spectroscopy (2DCOS) were employed to interpret the spectra of the fuel oils and additives. The utilization of this technique is for the precise identification of lubricants, gasoline and diesel, and for the quantitative determination of the additives' levels in oils. This work demonstrates that THz spectroscopy combined with mathematical methods is an appropriate tool for the characterization of fuel oil and therefore has applications in the field of petrochemical engineering.

2 Experimental methods

The experimental setup was comprised of a conventional THz-TDS system functioning in transmission mode and employing an fs Ti-sapphire laser (Spectra Physics Company, Santa Clara, California, USA). As shown in Figure 1, laser pulses were generated from the diode-pump mode-locked laser with a repetition rate of 80 MHz, a 100 fs duration and a central wavelength of 800 nm. Prior to connecting the laser to the THz system, the output was attenuated so that the average power was less than 100 mW. The input laser beam was split into two beams: a pump beam with more power and a probe beam with less. The pump beam generated THz radiation while passing through the 2.8mm thickness p-type GaAs with a <100> orientation. After transmission through the sample, the THz pulses carrying information regarding the specimen were focused onto a 2.8mm <110> ZnTe substrate together with the probe laser beam. The

THz signals were detected and amplified using a lock-in amplifier, and the data were recorded by computer. During each measurement, a delay stage was employed to change the time delay between the THz pulse and probe laser so that THz-TDS could be obtained[31].

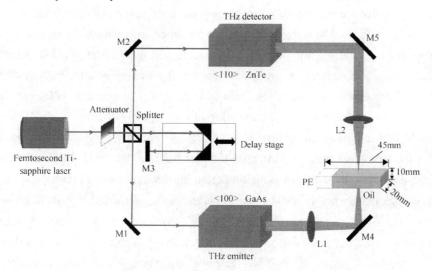

Figure 1   THz-TDS system

In the present research, three types of fuel oils (gasoline, diesel and lubricants) were analyzed based on their THz-TDS, comprising 6 lubricants, 20 gasolines and 15 diesel specimens. These specimens represented multiple grades; the lubricating oils used in this work included SMA engine oil SG 5W-30, GEELY engine oil SG 10W-30, SMA engine oil SG 10W-30, Shell HELIX 10W-40, SMA engine oil SG 75W-90 and Mobilude GX 80W-90. In terms of gasoline, 20 types were assessed, based on the GWB(E)06110 standard for determining the sulfur content in oils[8]. These were obtained from the CRM/RM Information Department of China and had sulfur concentrations ranging from 0.2ppm to 50ppm. The present study focused on the properties of gasoline with varying sulfur contents that had been investigated by different mathematical methods. The diesel samples used in this study were produced by blending MMA with 0# diesel to generated 15 different samples, with MMA concentrations ranging from 0.2% to 3%. Prior to analysis, each oil sample was transferred into a polyethylene (PE) cell 10mm in thickness and 45mm in height, which had very little absorption in the THz range. To minimize the absorption of atmospheric moisture and thereby to improve the signal-to-noise ratio, the measurement apparatus was blanketed with dry nitrogen gas (99.9%). All measurements were performed at room temperature (294.1 K ± 0.3 K).

The THz-TDS of fuel oils were obtained by scanning each 10mm thickness PE cell once filled with lubricant, gasoline or diesel samples, employing an empty PE cell as a reference. To minimize the systematic error in the measurement process, a reference spectrum was collected prior to each sample filling of the cell. Following the fast Fourier transform (FFT) of the reference and sample spectra, the THz frequency-domain spectra (FDS) were obtained. Based on these THz-FDS, the frequency dependent absorbance (A) spectra of each sample were generated based on the calculation $A=\lg[E^2_{ref}(\omega)/E^2_{sam}(\omega)]$, where ω is the frequency and $E_{ref}(\omega)$ and $E_{sam}(\omega)$ are the amplitudes of the sample and reference, respectively[32]. The resulting THz absorbance spectra were then employed in the subsequent statistical analyses.

PCA is typically used to reduce the number of dimensions within the sample data while retaining as much of the overall variation as possible and identifying potential structures in large spectral data sets, as well as groups within the absorbance spectra. PCA results in the generation of a number of variables, termed principal components (PC), representing maximized variables based on linear combination of the original variables[33].

BPANN is a mathematical nonlinear dynamics system that simulates the structure and function of the biological neural networks in the human brain. No priori model is required for ANN because this system is able to extract inherent information from variables and learn from the existing data, even when noise is present. Neurons, the elemental information processing units within the ANN structure, are linked through synaptic weighting and organized into several layers. Based on searching an error surface using a gradient descent to determine points with minimum error, the BPANN learning algorithm stores numerous input-output mapping relationships without prior knowledge of any mathematical equations and generates input, hidden and output layers[34]. The THz absorbance spectra of gasoline and diesel samples were introduced into the input layer and calculated in the hidden layer, and the results were generated within the output layer.

SVM is a class of learning models that can be trained by utilizing convex optimization techniques and a new soft computing learning algorithm that has been previously applied to the fields of hydrology and environmental studies. This process, which is originally developed by Vapnik, can be employed for regression analysis, classification, forecasting and pattern recognition, and exhibits improved performance relative to former methods[35]. The basis of SVM is similar to that of ANN in that it uses a set of input data, called the training data, and computes suitable weights to learn the governing function. In the process of calculation, only a certain number of input sample vectors, called the support vectors, are selected, and these are always less than the total number of samples. Through the input of a kernel, an input data space can be transformed into a nonlinear, high-dimensional space. The algorithm eventually generates a sparse prediction function by choosing support vectors[36]. SVM is particularly suitable for both qualitative and quantitative analysis of fuel oils based on THz absorbance over the entire frequency range.

2DCOS, which has been used for vibrational spectroscopy analysis in both the infrared and near infrared regions, is generally employed with dynamic spectra undergoing perturbations. These perturbations can include various physical or chemical parameters, such as temperature, electric field and concentration. A 2D correlation spectrum is calculated based on the equation $X(v_1, v_2)=[y(v_1, t), y(v_2, t')]=\Phi(v_1, v_2)+i\Psi(v_1, v_2)$, where v is the frequency, $y(v, t)$ represents the spectral responsivity of the output, and $\Phi(v_1, v_2)$ and $\Psi(v_1, v_2)$ are the synchronous and asynchronous correlation intensities. $\Phi(v_1, v_2)$ indicates the similarity while $\Psi(v_1, v_2)$ reflects the dissimilarity. In the 2D plot of a synchronous correlation function, if the sign of a cross peak is positive, intensities at the corresponding frequency are increasing or decreasing together; if negative, one is increasing while the other is decreasing. In an asynchronous correlation spectrum, the sequential changes of two peaks are depicted as a pair of crossed peaks with positive and negative signs. This method can amplify slight differences in the spectral changes of overlapped peaks[37]. Based on the principles above, THz absorption spectra can be related to the levels of additives in oils.

3 Results and discussion

3.1 Identification of oils with different types

The direct identification of different types of fuel oils was initially investigated based on the THz dielectric effect of the oils. In order to show the absorption of various lubricants, gasoline and diesel mixtures in THz range, the THz absorbance spectra were depicted separately in Figures 2 (a)~(c). The absorbance A values augmented with the increasing of frequency for all lubricants, gasoline and diesel samples. As shown in the inset to Figure 3, lubricants, diesel and gasoline have different degrees of THz absorbance, such that diesel demonstrates the greatest absorbance while gasoline shows the lowest at relatively high frequencies. Meanwhile, there are various cross features among the oil spectra in the range of 0.6~0.9THz, possibly resulting from the different absorption intensities of hydrocarbons and variations in the absorption tendency with increasing frequency. Therefore, THz absorption is determined by the chemical structures of the oils, indicating that the THz technique is a promising tool for the rapid identification of fuel oils.

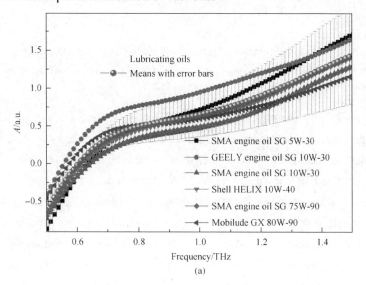

(a)

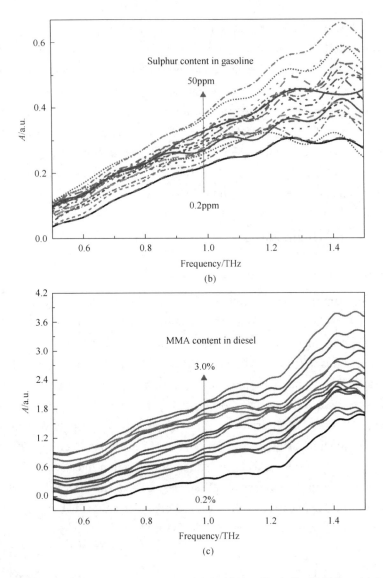

Figure 2　Frequency dependent absorbance (*A*) spectra of three kinds of fuel oils, including: (a) 6 lubricants; (b) 20 gasoline with the sulfur content from 0.2ppm to 50ppm; (c) 15 diesels with the MMA content from 0.2% to 3%

　　To more effectively identify a series of oils having different characteristics, PCA was employed to generate clusters corresponding to the 6 lubricant, 20 gasoline and 15 diesel samples. By plotting the scores of the first two PCs (those having the greatest contributions), a 2D space was obtained, as shown in Figure 3. The three types of oils were well separated by the PC analysis, and the gasoline, diesel and lubricant clusters are located at the top left, top right and bottom left, respectively. In the gasoline cluster, the greatest variations (maximum to minimum) in the PC1 and PC2 values were below 1.3 and 0.7, indicating only minimal variation between gasoline samples. In the case of the lubricants, the six individual samples exhibit some differences in their PC1 and PC2 values but the deviations are still relatively small, while the diesel samples, although all in the same general area, show deviations between the various samples that are much larger than those in the gasoline and lubricant clusters. This phenomenon is caused by the strong absorption of MMA in the THz range which can be revealed in Figures 2 (c) and 5 (d) due to the ester functional group. The ester group has a dissymmetrical structure and is highly polar, leading to significant absorption in the THz range[38]. These data demonstrate that THz spectroscopy can distinguish between different oils, and that the combination of THz and PCA can precisely identify fuel oils with different properties in a data set with hundreds of samples.

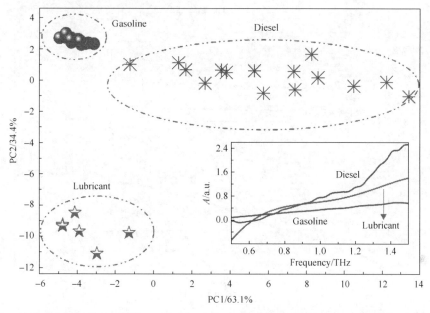

Figure 3   2D scatter plot generated by the PCA analysis of THz absorbance spectra. The x and y axes represent PC1 and PC2, whose contribution rates are 63.1% and 34.4%, respectively. The scatters are composed of 6 lubricants, 20 gasoline and 15 diesel, whose PC1- PC2 clusters are located at bottom left, top left and top right, respectively. The inset plots the frequency dependence of the absorbance of three selected oils

3.2 Spectral determination of fuel oils' additives

As noted, the sulfur in gasoline can generate atmospheric pollutants after a series of reactions and transportation in air. The frequency dependence of the absorbance of gasoline over the range of 0.2~1.5THz at selected sulfur concentrations is plotted in Figure 4. Similar spectral values and profiles were found, indicating minimal experimental error and reproducible operation of the THz instrumentation. The absorbance values were shown to increase at higher frequencies, although no sharp absorption peaks were evident. The relationship between the THz response and the sulfur concentration in the gasoline was assessed by plotting the absorbance against sulfur content for all 20 gasoline samples at randomly selected wavelength values of 0.5THz, 0.8THz, 1.1THz and 1.4THz, as shown in the inset of Figure 4. In each case, a linear relationship was built with the residual sum of squares (RSS) of 0.01034, 0.01635, 0.04557 and 0.13912, and the slope of the best fit line increased with the frequency. The linear relationships at selected frequencies prove the monotonous increasing trend of the sulfur content and the THz absorption, but are not enough to precisely determine the sulfur content in gasoline due to partially indistinguishable absorbance of the >40ppm and <10ppm-sulfur gasoline. In order to quantitatively determine the sulfur content, the chemometric methods will be employed with the absorbance over the entire range as the input, which will be discussed in the next section.

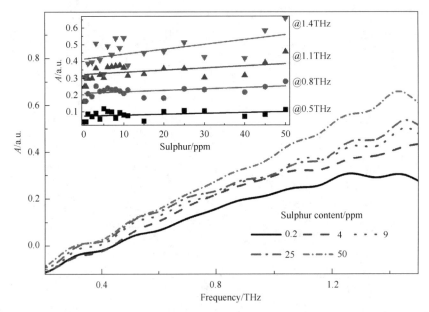

Figure 4 The frequency dependent absorbance spectra of gasoline with sulfur concentrations ranging from 0.2ppm to 50ppm. The inset summarizes the absorbance at various sulfur levels and at selected frequencies of 0.5THz, 0.8THz, 1.0THz and 1.4THz, together with linear fits

MMA is a common diesel PPD, as it can significantly improve the performance of diesel at low temperatures. In the present work, MMA was blended with diesel samples over the range of 0.2%~3% at intervals of 0.2%. The THz field signals of various samples as functions of time are shown in Figure 5 (a), obtained after transmission of the THz pulses through a blank sample holder and through cells holding the diesel samples. Compared with the blank reference, the sample THz-TDS signals exhibit obvious time delay and attenuation. The THz peak in the TDS contains significant information regarding the refraction and absorption, and the time delay, τ, and amplitude, E_p, of the THz peaks were extracted from the THz-TDS. E_p and τ values of 0.0892V and 14.44ps, 0.0837V and 14.48ps, and 0.0754V and 14.53ps, respectively, were obtained from samples having MMA concentrations of 0.2%, 1.2% and 3.0%. All 15 diesel fuels were employed to build linear models and, as shown in

Figure 5 (b), the τ and E_p values of all 15 samples were determined and plotted against the respective MMA concentrations. Linear relationships were established between MMA concentrations and both τ and E_p, exhibiting increasing and decreasing trends. Accordingly, the first derivatives of these lines were positive and negative, respectively. All of the experimental data points were quite close to the best fit lines with the R-square (R^2) of 0.737 as well as 0.975 and in each case the error bar intersects with the line, demonstrating the good precision of the linear models.

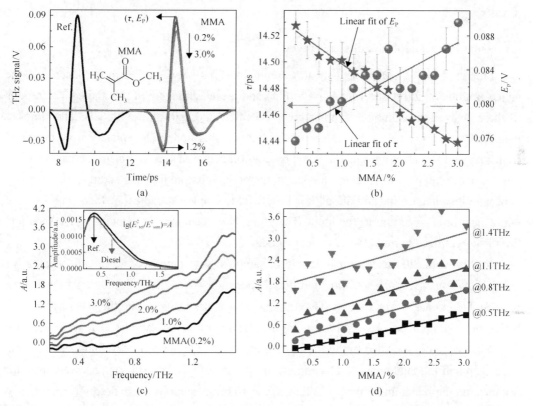

Figure 5 (a) THz spectra of a blank PE cell as the reference signal and of selected diesel samples with different MMA concentrations. E_p and τ represent the intensity and time delay of the THz peaks. (b) τ and E_p as functions of the MMA concentration with linear fits. (c) Frequency dependent absorbance of diesel samples with different MMA concentrations, based on absorbance spectra. The inset image shows the THz-FDS of a reference and a selected diesel sample. (d) Absorbance as a function of MMA concentration for diesel samples at 0.5THz, 0.8THz, 1.1THz and 1.4THz

The THz-FDS of the reference cell and of various diesel samples are presented in the inset of Figure 5 (c), and the frequency dependent absorbance spectra are presented in Figure 5 (c). The effective frequency range was from 0.2THz to 1.5THz, over which range the absorbance together with the slope gradually increased as the frequency increased. The absorbance values of diesel samples with different MMA concentrations were also extracted at selected frequencies. As shown in Figure 5 (d), the scatter plots of absorbance at 0.5THz, 0.8THz,

1.1THz and 1.4THz demonstrate a linear increase with increasing MMA concentration. These linear relationships were evident at each frequency and all data points were found to lie close to the best fit lines with the R^2 of 0.973, 0.922, 0.821 and 0.677, respectively. In order to build a more effective model of oils with high reproducible operation, more gasoline-sulfur or diesel-MMA samples can be employed and measured by THz system. These results confirm that THz spectroscopy is an effective means of analyzing additives in diesel samples and could potentially be applied to the online quality inspection of the physical properties of oils during processing.

3.3 Quantification of fuel oils' additives

Figures 4 and 5 (d) present the linear relationships between THz absorbance and the concentrations of sulfur and MMA in gasoline and diesel, and suggest that the concentrations of these analytes in unknown samples could be calculated using these models. The simplest means of doing so is to apply Beer's law, although this approach typically uses only one frequency and is thus more easily affected by environmental factors and instrumental noise. To build a more robust and precise model to quantitatively predict the additive or impurity concentrations in gasoline and diesel, it is beneficial to employ statistical methods that uses the entire absorption spectrum as the input. Therefore, two chemometric techniques, BPANN and SVM, were utilized to build the statistical models.

In this work, 20 gasoline and 15 diesel samples were employed to obtain quantitative analysis systems. Both the gasoline and diesel specimens were divided into two subsets, one for training and one for prediction. The training set was used to generate the statistical model, while the prediction set was employed for validation, such that samples that were used for prediction were not included in the training. In each case, the number of training samples was more than half the total quantity of samples. In the case of gasoline, 14 samples were used for training and the remaining 6 for prediction, while 10 diesel samples were used for training and 5 for prediction. Figure 6 shows plots of the predicted values against the actual values for both the training and prediction sets, using: (ⅰ) the BPANN model for gasoline, (ⅱ) the SVM model for gasoline, (ⅲ) the BPANN model for diesel and (ⅳ) the SVM model for diesel. In each case, the data points are situated close to the dashed line labeled as "Ref.", which represents zero residuals between the actual and predicted values. To evaluate the reliability and precision of this method, the correlation coefficient, R, and the root-mean square error, RMSE, were calculated for the data in both the training and prediction sets. These values are related to the covariance, $cov(x, y)$, and variance matrix, $v(x)$, according to the formulae $R=cov(x, y)/[v(x)v(y)]^{1/2}$ and $RMSE=[\Sigma(x-y)^2/n]^{1/2}$, where n is the number of samples in a set, x and y represent actual and predicted concentrations, respectively. In the case of gasoline, R exceeded 0.98 and 0.92, while the RMSE were below 2.1 and 6.5 when using the BPANN and

SVM models. The *R* and RMSE of the diesel samples were all above 0.99 and below 0.2 for both models. Compared to the gasoline, the diesel set showed better correlation. This may be because MMA has significant molecular polarity, resulting in strong absorption in the THz range. In short, BPANN and SVM combined with THz spectroscopy are appropriate to the characterization of trace additives or contaminants in fuel oils.

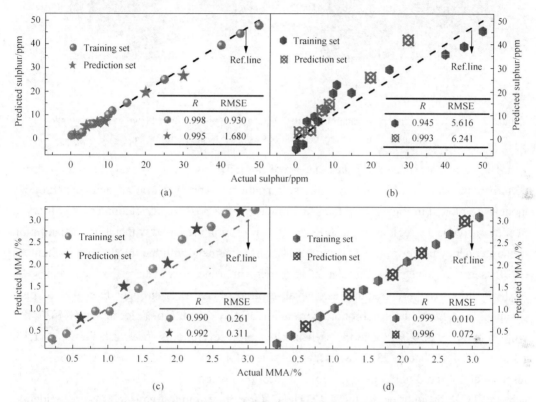

Figure 6 Quantification of sulfur in gasoline and MMA in diesel based on BPANN and SVM. Predicted versus actual sulfur concentrations in gasoline using the BPANN (a) and the SVM model (b). In each case, 14 samples were used as the training set and the remaining 6 as a prediction set. (c) and (d) show the predicted versus actual MMA concentrations in diesel using the BPANN and the SVM model. Here 10 samples were selected as the training set and the remaining five were used as the prediction set. The input data consisted of the absorbance spectra from 0.2THz to 1.5THz

3.4 Dynamics spectra analysis of fuel oils

As shown in Figures 2 (b) and (c), the absorbance values increased for both oils at higher additive concentrations. To further investigate the interactions of THz radiation with oils and their additives, 2DCOS was employed to evaluate the *A* spectra.

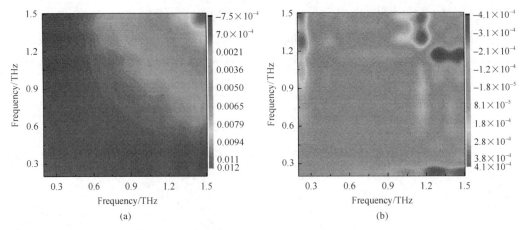

Figure 7 Synchronous (a) and asynchronous (b) 2D correlation plots obtained from gasoline samples with increasing sulfur concentrations as the external perturbation

The dynamic spectra of gasoline were initially analyzed by varying the external perturbation, which in this case was the sulfur content. The synchronous and asynchronous models are plotted in Figure 7. In the synchronous plot, a positive correlation can be observed over most frequency regions, indicating that the absorption increased with sulfur concentration in the frequency range from 0.5THz to 1.5THz, in agreement with the trends seen in Figure 2 (b). In addition, a negative correlation is evident in some areas where the frequencies are relatively low. At these frequencies, the absorption of sulfur is stronger than that of pure gasoline. In addition, the synchronous correlation displays an auto-peak located at (1.48, 1.48), with an intensity of 0.01135. In the asynchronous plot, several cross peaks are found, at (1.18, 1.31), (1.17, 1.47) and (0.23, 1.28), with intensities of 4.10×10^{-4}, 3.78×10^{-4} and 3.68×10^{-4}, respectively. Accordingly, there are four absorption features at 0.23THz, 1.17THz, 1.31THz and 1.47THz. In addition to the 1.47THz feature, three other frequencies were evidently hidden in the raw spectra. In a previous research, a split-ring-resonator (SRR) with the gasoline of 93# had the absorption features at 0.346THz and 1.173THz, respectively[39]. Herein, gasoline was blended with sulfur, and the mixtures exhibited not only similar absorption features to the results in Reference[39], but the different peaks as a result of the absorption of sulfur.

Similarly, after 2DCOS processing of the dynamic spectra, synchronous and asynchronous correlation intensities were obtained for diesel, as shown in Figure 8. In the synchronous image, a positive correlation can be observed over the entire frequency range, indicating that the absorption increased with the MMA concentration. This synchronous plot also displays a correlation square, with four vertexes located at (1.46, 1.46), (1.46, 1.18), (1.18, 1.18) and (1.18, 1.46). The high correlation value (0.38694) at 1.46THz reflects the significant extent of MMA-induced dynamic fluctuations in the absorbance spectra. The synchronous correlations seen here are in agreement with the results in Figure 2 (c), where absorption was found to vary with MMA concentration in the THz range. In the asynchronous image, three cross peaks are observed at (1.13, 1.49), (1.13, 0.63) and (1.13, 0.41), with intensities of 0.02100, 0.01515 and 0.01566, respectively. Herein, one of the peaks found are located at the end of the bandwidth such as 1.49THz, thus it will be not considered in next discussion due to the relatively low signal to noise level. The 1.13THz feature is related to that at 1.18THz in the synchronous plot. The feature at 1.13THz changes before those at 0.63THz and 0.41THz, indicating that the 1.13THz frequency is related to the absorption of the diesel itself. As the 1.13THz and 1.18THz features as well those at 1.46 and 1.48THz can be considered as equal to the 1.13THz and 1.46THz features with small deviations, the gasoline and diesel exhibit common features at these two wavelengths. In a previous report, the absorption spectra of the gasoline and diesel collected from a gas station in Turkey were determined in the range from 0.15THz to 1.1THz, and both of them had absorption features at ~1.05THz[40]. The results herein are in agreement with those in Reference[40]. The features at 0.63THz and 0.41THz are attributed to the MMA. Based on the results in Figures 7 and 8, the combination of THz spectroscopy and 2DCOS is evidently a practical means of analyzing fuel oils and their additives, and could be applied to monitor the properties of fuels onboard a vehicle.

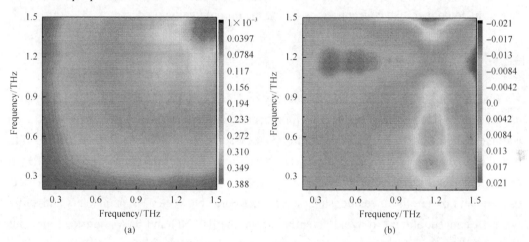

Figure 8 Synchronous (a) and asynchronous (b) 2D correlation plots obtained from diesel samples with increases in the MMA concentration as the external perturbation

3.5 Discussion

Generally, clear differences of gasoline and diesel in optical properties offer a simple yet effective way to discriminate fuel oils from each other by using THz spectra without any danger of combustion or decomposition[40]. Gasoline with three grades have different refractive index and absorption coefficient so that THz-TDS would be an alternative for evaluation of oil viscosity[41]. In order to enhance the efficiency, the identification of fuel oils with THz spectra was realized by using a split-ring-resonator as the sensor, which made a 0.027THz-redshift between 93# and 97# gasoline[39]. According to a previous report, the THz absorption increased with the 97# gasoline in 93#-97# gasoline mixtures, and the multiparameter-combined analysis allowed the determination of 97# gasoline content with a low absolute error (6%), resulting in predictions more accurate and precise than those obtained by absorption spectra and dualistic linear regression fitting[42,43]. As a common additive, ethanol was blended with gasoline. Similar to the spectra in this research, the gasoline samples had no obvious peaks in THz range, and the absorption augmented with the increasing of ethanol content in the absorption coefficient spectra. A multiple linear regression can predict ethanol content ranging from 0.5% to 50%[44]. In this research, the combination of THz absorbance spectra and chemometric methods can predict ppm grade sulfur and 0.2% MMA in oils. Therefore, the use of statistics can improve the precision and prediction limitation by using the absorption spectra over the entire range as the input. Particularly, 2DCOS, which is generally used in near-infrared analysis, has been applied to the crystallization of poly(3-hydroxybutyrate) at 90℃[45] and not been applied to the analysis of fuel oils yet. In the present research, a comparison between additive concentration results and the absorption features identified by 2DCOS shows that this procedure could allow sensitive and precise determination of additive concentrations during the oil production process.

4 Conclusions

Our analysis suggests that THz technology can be applied to the efficient characterization of fuel oils, as it reflects the molecular vibrational modes and intensities of both oils and their additives. In this work, we compared the THz absorption responses of three typical oils, including lubricant, gasoline and diesel samples. Based on the absorption spectra and PCA calculations, each series of oils could be correctly clustered by type. In the case of both gasoline and diesel, linear relationships were determined between THz parameters and sulfur or MMA concentrations. To avoid potential errors, the BPANN and SVM statistical methods were employed to quantitatively determine the additive contents in gasoline and diesel specimens. The resulting statistical models were highly precise due to $R > 0.94$ in both BPANN and SVM models. In addition, 2DCOS was shown to be an appropriate method for the extraction of hidden peaks in absorption spectra. The results presented herein provide a new perspective on the analysis of fuel oils, in particular the monitoring of additives or impurities. It is expected that this THz technique will prove extremely helpful in the assessment of additives in oils in the future. The development of miniaturized, integrated and higher-resolution THz instrumentation will enable this method to be applied in the field of petrochemical engineering.

Acknowledgements

This work was partially funded by the National Basic Research Program of China (Grant

No. 2014CB744302), the Specially Funded Program on National Key Scientific Instruments and Equipment Development (Grant No. 2012YQ140005), the China Petroleum and Chemical Industry Association Science and Technology Guidance Program (Grant No. 2016-01-07) and the National Nature Science Foundation of China (Grant No. 11574401).

References

[1] Zheng S M, Yi H T, Li H. The impacts of provincial energy and environmental policies on air pollution control in China. Renew. Sust. Energ Rev, 2015 49:386-394.

[2] Azimi S S, Namazi M H. Modeling of combustion of gas oil and natural gas in a furnace: Comparison of combustion characteristics. Energy, 2015, 93:458-465.

[3] Geng P, Zhang H, Yang S C. Experimental investigation on the combustion and particulate matter (PM) emissions from a port-fuel injection (PFI) gasoline engine fueled with methanol-ultralow sulfur gasoline blends. Fuel, 2015, 145:221-227.

[4] Annable M D, Wallace R B, Hayden N J, et al. Reduction of gasoline component leaching potential by soil venting. J. Contam. Hydrol. 1993, 12: 151-170.

[5] Tsien A, Diaz-Sanchez D, Ma J, et al. The organic component of diesel exhaust particles and phenanthrene, a major polyaromatic hydrocarbon constituent, enhances ige production by ige-secreting ebv-transformed human b cells in vitro. Toxicol. Appl. Pharm., 1997, 142(2): 256-263.

[6] Zhang K, Hu J, Gao S, et al. Sulfur content of gasoline and diesel fuels in northern China. Energy Policy, 2010, 38:2934-2940.

[7] Zhu L F, Qian Y, Wang X L, et al. Effects of direct injection timing and premixed ratio on combustion and emissions characteristics of RCCI (Reactivity Controlled Compression Ignition) with N-heptane/gasoline-like fuels. Energy, 2015, 93: 383-392.

[8] Qin F L, Li Q, Zhan H L, et al. Probing the sulfur content in gasoline quantitatively with terahertz time-domain spectroscopy. Sci. China: Phys. Mech. Astron., 2014, 57: 1404-1406.

[9] Li J, Yang W M, An H, et al. Modeling on blend gasoline/diesel fuel combustion in a direction injection diesel engine. Appl. Energy, 2015, 160: 777-783.

[10] Du J K, Sun W C, Guo L, et al. Experimental study on fuel economies and emissions of direct-injection premixed combustion engine fueled with gasoline blends. Energ. Convers. Manage., 2015, 100: 300-309.

[11] Benajes J, Novella M R, Thein K. Understanding the performance of the multiple injection gasoline partially premixed combustion concept implemented in a 2-stroke high speed direct injection compression ignition engine. Appl. Energy, 2016, 161: 465-475.

[12] Keskin A, Ocakoglu K, Resitoglu I A, et al. Using Pd (II) and Ni (II) complexes with N,N-dimethyl-N'-2-chlorobenzoylthiourea ligand as fuel additives in diesel engine. Fuel, 2015, 162:202-206.

[13] Broatch A, Tormos B, Olmeda P, et al. Impact of biodiesel fuel on cold starting of automotive direct injection diesel engines. Energy, 2014, 73:653-660.

[14] Ali O M, Mamat R, Masjuki H H, et al. Analysis of blended fuel properties and cycle-to-cycle variation in a diesel engine with a diethyl ether additive. Energ. Convers. Manage., 2016, 108: 511-519.

[15] Wang Y S, Liang X Y, Shu G Q, et al. Effect of lubricating oil additive package on the characterization of diesel particles. Appl. Energy, 2014, 136:682-691.

[16] Wu J. Tunable ultranarrow spectrum selective absorption in a grapheme monolayer at terahertz frequency. J. Phys. D: Appl. Phys., 2016, 49: 215108.

[17] Siegel P H. Terahertz technology in biology and medicine. IEEE Trans. Microw. Theory Techn., 2004, 52: 2438-2447.

[18] Zhan H L, Zhao K, Xiao L Z. Spectral characterization of the key parameters and elements in coal using terahertz spectroscopy. Energy, 2015, 93:1140-1145.

[19] Ge L N, Zhan H L, Leng W X, et al. Optical characterization of the principal hydrocarbon components in natural gas using terahertz spectroscopy. Energy & Fuels, 2015, 29: 1622-1627.

[20] Zhan H L, Li Q, Zhao K, et al. Evaluating PM_{25} at a construction site using terahertz radiation. IEEE Trans. Sci. Technol, 2015, 5: 1028-1034

[21] Camus E C, Palomar M, Covarrubias A A. Leaf water dynamics of Arabidopsis thaliana monitored in-*vivo* using terahertz time-domain spectroscopy. Sci. Rep., 2013, 3:2910.

[22] Lu X F, Zhang X C. Investigation of ultra-broadband terahertz time-domain spectroscopy with terahertz wave gas photonics. Front. Optoelectron., 2014, 7:121-155

[23] Hong B B, Huang L P, Xu X L, et al. Hollow core photonic crystal for terahertz gyrotron oscillator. J. Phys. D: Appl. Phys., 2015, 48: 045104.

[24] Mittleman D M. Frontiers in terahertz sources and plasmonics. Nat. Photonics, 2013, 7:666-669.

[25] Zhan H L, Wu S X, Bao R M, et al. Qualitative identification of crude oils from different oil fields using terahertz time-domain spectroscopy. Fuel, 2015, 143: 189-193.

[26] Zhao H, Zhao K, Bao R M. Fuel property determination of biodiesel-diesel blends by terahertz spectrum. J Infrared Milli. Terahz. Waves, 2012, 33: 522-528.

[27] Leng W X, Zhan H L, Ge L N, et al. Rapidly determinating the principal components of natural gas distilled from shale with terahertz spectroscopy. Fuel, 2015, 159: 84-88.

[28] Zhan H L, Wu S X, Bao R M, et al. Water adsorption dynamics into active carbon probed by terahertz spectroscopy. RSC Adv., 2015, 5:14389-14392.

[29] Zhao H, Zhao K, Tian L, et al. Spectrum features of commercial derv fuel oils in the terahertz region. Sci. China: Phys. Mech. Astron., 2012, 55: 195-198.

[30] Zhao H, Zhao K, Bao R M. Predicting cold flow properties of diesel by terahertz time-domain spectroscopy. ISRN Spectroscopy, 2012: 876718.

[31] Zhan H L, Sun S N, Zhao K, et al. Less than 6 GHz resolution THz spectroscopy of water vapor. Sci. China: Tech. Sci., 2015, 58: 2104-2109.

[32] Liu H B. Terahertz spectroscopy for chemical and biological sensing applications. Troy, NY: Dept. Phys., Rensselaer Polytechnic Inst. 2006.

[33] Burnett A D, Fan W, Upadhya P C, et al. Broadband terahertz time-domain spectroscopy of drugs-of-abuse and the use of principal component analysis. Analyst, 2009, 134: 1658-1668.

[34] Jiang S Y, Ren Z Y, Xue K M, et al. Application of BPANN for prediction of backward ball spinning of thin-walled tubular part with longitudinal inner ribs. J. Mater. Process. Tech. 2008, 196: 190-196.

[35] Abu-Mostafa Y. The Vapnik-Chervonenkis dimension: information versus complexity in learning. Neural. Comput. 1989, 1: 312-317.

[36] Wang J, Hu J. A robust combination approach for short-term wind speed forecasting and analysis e-Combination of the ARIMA (autoregressive integrated moving average), ELM (extreme learning machine), SVM (support vector machine) and LSSVM (least square sVM) forecasts using a GPR (gaussian process regression) model. Energy, 2015, 93: 41-56.

[37] Ozaki Y, Liu Y, Noda I. Two-dimensional near-infrared correlation spectroscopy study of premelting behavior of Nylon. Macromolecules, 1997, 30: 2391-2399.

[38] Cunningham P D, Hayden L M. Carrier dynamics resulting from above and below gap excitation of p3ht and p3ht/pcbm investigated by optical-pump terahertz-probe spectroscopy. J. Phys. Chem. C, 2008, 112(21): 722-728.

[39] Hu F R, Zhang L H, Xu X L, et al. Study on split-ring-resonator based terahertz sensor and its application to the identification of product oil. Opt. Quant. Electron., 2015, 47: 2867-2879.

[40] Arik E, Altan H, Esenturk O. Dielectric properties of diesel and gasoline by terahertz spectroscopy. J. Infrared Milli. Terahz. Waves, 2014, 35: 759-769.

[41] Adbul-Munaim A M, Reuter M, Koch M, et al. Distinguishing gasoline engine oils of different viscosities using terahertz time-domain spectroscopy. J. Infrared Milli. Terahz. Waves, 2015, 36:687-696.

[42] Li Y N, Zeng Z M, Li J, et al. Terahertz quantitatively distinguishing gasoline mixtures using multiparameter-combined analysis. Appl. Opt., 2013, 52:7382-7388.

[43] Li J. Terahertz spectroscopy plus analysis distinguishes gasoline mixtures. Laser Focus World, 2013, 49:9.

[44] Arik E, Altan H, Esenturk O. Dielectric properties of ethanol and gasoline mixtures by terahertz spectroscopy and an effective method for determination of ethanol content of gasoline. J. Phys. Chem. A, 2014, 118:3081-3089.

[45] Hoshina H, Ishii S, Morisawa Y, et al. Isothermal crystallization of poly(3-hydroxybutyrate) studied by terahertz two-dimensional correlation spectroscopy. Appl. Phys. Lett., 2012, 100: 011907.

Evaluating PM$_{2.5}$ at a construction site using terahertz radiation

Honglei Zhan[1,2,3] Qian Li[1,2,3] Kun Zhao[1,2,3] Leiwei Zhang[4] Zhenwei Zhang[4]
Cunlin Zhang[4] Lizhi Xiao[1,2,3]

(1.State Key Laboratory of Petroleum Resources and Prospecting, China University of Petroleum, Beijing 102249, China; 2.Beijing Key Laboratory of Optical Detection Technology for Oil and Gas, China University of Petroleum, Beijing 102249, China; 3.Key Laboratory of Oil and Gas Terahertz Spectroscopy and Photoelectric Detection, China Petroleum and Chemical Industry Federation (CPCIF), Beijing 100723, China; 4.Department of Physics, Capital Normal University, Beijing 100048, China)

Abstract: Economic and industrial development has led to increasing problems with particulate pollution in many countries. Particulate matter with the diameter of less than 2.5μm (PM$_{2.5}$) is of concern in many cities. In this study, a total of 70 samples of PM$_{2.5}$ were collected from dusty environments and analyzed using terahertz (THz) radiation. The transmission spectrum of PM$_{2.5}$ had two distinct absorption bands between 2.5THz and 7.5THz. Their center frequencies were 3.36THz and 6.91THz, respectively. Based on the THz absorbance spectra, the elemental compositions were studied by monitoring PM$_{2.5}$ masses in conjunction with two-dimensional correlation spectroscopy. Correlations between absorption bands and cross peaks in the synchronous and asynchronous plots indicated that the metallic oxides showed absorption features in the range from 2.5THz to 7.5THz. The vibration modes of anions and cations were located at relatively low and high frequencies, respectively. Statistical methods, including partial least squares and back propagation artificial neural network, were used to quantitatively characterize the PM$_{2.5}$ content with the input of absorbance over the whole frequency range. This research indicates that THz spectral analysis of PM$_{2.5}$ is a promising tool for investigating the composition and mass of pollutants.

Keywords: PM$_{2.5}$; statistical methods; terahertz (THz); two-dimensional correlation spectroscopy

1 Introduction

City air pollution has become a global issue because of rapid industrialization and urbanization. Consequently, techniques for monitoring this pollution are of importance. Particulate matter (PM), especially fine particles with diameters of ≤2.5 μm (PM$_{2.5}$), is a major contributor to air pollution. It was suggested that PM$_{2.5}$ could be used to monitor air quality, which is important to protect public health[1,2].

The adverse effects of PM exposure are marked by oxidative stress, inflammation, and tissue damage. Therefore, it is important to monitor air for the content of PM (size and mass)[3].

Organic and nitrogen oxide emissions of PM are clearly related to ozone formation. This link is of great importance for government to establish new ozone and PM standards[4]. $PM_{2.5}$ pollution in agricultural settings impacts the human health, and especially affects children[5]. PM can be composed of polyaromatic hydrocarbons, metals, and polar organic compounds, and PM compositions are important in increasing oxidative stress burdens[6]. In hospital emergency room visits, $PM_{2.5}$ is often linked to adverse health effects. Major contributors to $PM_{2.5}$ include organic carbon, elemental carbon, sulfate, nitrate, and ammonium[7]. Therefore, efforts need to be made to prevent and control PM pollution.

As a developing country that is undergoing rapid economic development, China faces serious environmental problems that are of concern to both the government and the public. Many efforts are being made by the government towards sustainable development to reduce $PM_{2.5}$ pollution and protect the environment. Many studies have focused on determining the sources and constituents of PM, and on methods for detection of PM[5,8,9]. In the large cities in China, such as Beijing, and Shanghai, haze occurs that is driven to a large extent by secondary aerosol formation. To control China's $PM_{2.5}$ levels, emissions of both primary particulates and secondary aerosol precursors need to be monitored[10]. Although the level of exposure to $PM_{2.5}$ and its elemental composition vary from region to region, the exposure level and composition are always significantly correlated with each other[11].

Dust is a common source of pollution in China, and this can arise from dust storms or dust pollution produced at construction sites. Some studies have investigated the constituents of dust pollution, and its effects on public health. Metallic oxides are usually the dominant component of $PM_{2.5}$ in dusty environments, but this can vary with the location, and in some places, water-soluble ions like Al, Fe, and Ca are the principal components. The characterizeation of $PM_{2.5}$ is usually performed by chemical methods with good precision[12-14].

Direct $PM_{2.5}$ characterization methods include optical techniques, such as spectral methods. THz spectroscopy covers the range from 0.1THz to 10THz and bridges the gap between microwave and infrared in the electromagnetic spectrum. This technique provides higher penetrability of nonpolar compounds than infrared and better spatial resolution than microwave. Many compounds, such as proteins and macromolecules, have spectral fingerprints in this range. Compared to infrared and microwave techniques, THz radiation is safe and reliable for online and real-time monitoring because of its lower photon energy and higher signal-to-noise ratio (SNR)[15-19].

In an earlier study, we reported on the application of THz radiation to analysis of $PM_{2.5}$ in the atmosphere[20]. In the present study, the analysis of $PM_{2.5}$ in air samples collected from a dusty environment was performed using THz radiation. Elemental analysis of the PM was conducted using X-ray fluorescence (XRF). Then, two-dimensional correlation spectroscopy (2DCOS) was used to analyze in detail the interactions between the THz radiation and the PM components. In addition, statistical methods, including partial least squares (PLS) and back propagation artificial neural network (BPANN), were used to quantify the PM$_{2.5}$. Here, we focus on the application and basic science of the THz spectrum in environmental science. PM$_{2.5}$ has distinct absorption features in THz range. Combining mathematical methods, the overlapped peaks of the constituents can be identified and the PM$_{2.5}$ content can be rapidly determined. The results are very useful for the characterization of pollutant sources and content. Such research cannot only improve the standard of PM$_{2.5}$ monitoring, but also promote the development of a THz technique.

2 Methods and procedures

An air sampler (Minivol Tactical Air Sampler) was used to collect the PM samples in a dusty environment. Air (flow rate N = 9 L/min) was drawn through a particle size separator, which had a 10 μm cut-off and a 2.5 μm cut-off, and the PM$_{2.5}$ was collected on a quartz filter membrane (a diameter of 47mm and a mass of ~ 0.1450 g). A total of 70 samples were collected from a construction site in Shijiazhuang, China, in June and July, 2014. All the blank filters were weighed before use, and then reweighed after sample collection to calculate the weight of PM$_{2.5}$. The PM$_{2.5}$ concentration ρ was obtained using the expression $\rho = mN^{-1}t^{-1}$, where m, N and t are the mass of PM$_{2.5}$, flow rate and sampling time, respectively.

A Fouriertransform (FT) infrared spectrometer was used to measure the absorption spectra of the PM$_{2.5}$ samples in THz range. Its optical path is referred to Figure 1. The optical source, a silicon carbide, can emit steady and strong emissions with continuous wavelength including THz range. After being reflected by CM1 and CM2 (concave mirror), the pulses are parallel to each other and then entered a Michelson interferometer where interference pulses can be obtained by interferometer. The pulses reach sample cavity after being reflected by tunable M1 (mirror) and CM3. Herein, the sample needs to be located at the focal point of the pulses. Finally, the pulses carrying the sample information are reflected by CM4 and accepted by the detector, and the interference spectrum is observed. A computer is employed to process the spectra by using FT. Then the FT infrared spectrum can be obtained. In this study, the blank filters are used as references. The absorbance spectra are calculated using $\ln(I_0/I)$, where I_0 is the signal intensity of the reference and I is the signal intensity of the sample[20].

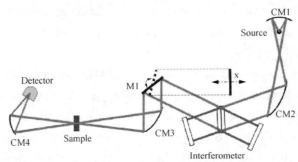

Figure 1　Diagram of the optical path in FT infrared spectrometer

2DCOS is a generally applicable to a broad range of spectroscopic technique based on a set of spectral data from a system under some perturbation. It can enhance spectral resolution by spreading peaks along the second dimension. Selective development of two-dimensional (2D) peaks provides better access to information which is not readily observed in conventional one-dimensional (1D) spectra. Sign of cross peaks is used to determine relative direction of intensity changes and sequential order of events. The 2D correlation spectrum can be calculated by

$$X(v_1,v_2)=[y(v_1,t),y(v_2,t')]=\Phi(v_1,v_2)+i\Psi(v_1,v_2)$$

where v is the frequency, $y(v,t)$ represents the spectral responsivity of output, $\Phi(v_1,v_2)$ and $\Psi(v_1,v_2)$ are the synchronous and asynchronous correlation intensities. $\Phi(v_1,v_2)$ (synchronous) indicates the similarity of signal dependence on time t, and $\Psi(v_1,v_2)$ reflects the dissimilarity. In the synchronous plot, auto-peaks at diagonal positions indicate the extent of perturbation-induced dynamic fluctuations of spectral signals. Cross peaks represent simultaneous change of spectral signals at two different frequencies. Herein, if the sign of a cross peak is positive, intensities at corresponding frequency are increasing or decreasing together; if negative, one is increasing while another is decreasing. In asynchronous correlation spectrum, cross peaks develop only if the intensity varies out of phase with each other for some Fourier frequency components of signal fluctuations. Here, the sign is positive if the intensity changes at v_x occurs before v_y and negative if v_x after v_y. The above sign rules are correct only when $\Phi(v_1,v_2) > 0$ and are reversed if $\Phi(v_1,v_2) < 0$[21].

To quantitatively determine the total mass of PM collected on the quartz filter membrane, PLS and BPANN were employed in this research. PLS is one of the most widely mathematical methods, which relates two data matrices by a linear multivariate model. It can extract the spectrum features from the absorbance spectra and then correlate these features to the physical properties by using all the absorbance values within a given frequency range. The extracted features remain the most useful information, and eliminate multi-collinearity, uncorrelated information and noise in the absorbance spectra[22]. BPANN is another non-linear chemometrics method imitating and simplifying the basic characteristic of human brain. It can store many input-output mapping relationships without prior revealing the mathematical equation. Also, it can process the considered variables and learn from the present data, even when the noise exists. Its learning rule is to use the method of steepest descent and to constantly adjust the network weights and threshold by back propagation. Generally, BPANN algorithm includes at least three layers (e.g., input, hidden and output layers). Input layer receives and distributes plenty of input information. Then hidden layer captures the mapping relationship between the inputs and outputs. Finally, output layer produces the calculated results from hidden layers. The hidden layer is the key part to create a more correct model and often involves a great many layers[23]. In this research, the THz absorbance spectra data over the whole frequency range were employed as the input variables.

3 Results and discussion

A Savitzky-Golay filter was applied to pre-process the absorption data to reduce instrument noise and smooth curve in the absorption spectrum, which did not distort the spectral waveforms and absorption features[24]. An FT infrared spectrometer has a larger SNR in relatively high frequency range above 3THz. In our case it is noted that the SNRs is not rather high in 1~2.5THz range and 7.5~10THz range. The frequency between 2.5~7.5THz was selected for the analysis of samples in this study. Figure 2 shows the smoothed absorbance spectra as a function of the frequency from 2.5THz to 7.5THz for the individual $PM_{2.5}$ samples. The $PM_{2.5}$ mass ranged from 0.2mg to 10.8mg. An absorption effect was observed, and the larger samples showed stronger absorption effects. Therefore, the optical response was linked to both the type and mass of the particles. Two broad absorption bands were observed at approximately 3.36THz and 6.91THz (Figure 2). Further study of the interaction between the pollutant components and the THz radiation is important for characterization of $PM_{2.5}$ in dusty environments.

The mass of $PM_{2.5}$ can be initially measured by the difference value of the quartz filter with and without $PM_{2.5}$. Also, the absorbance spectra were directly obtained by the FT infrared spectrometer. Based on the spectra, the absorbance intensities of all samples at two central frequencies can be extracted and acted as Y-axis. The mass of corresponding samples remained at the X-axis. Figure 3(a) and (b) show the linear regressions for the absorbance versus $PM_{2.5}$ mass at 3.36THz and 6.91THz, respectively. The correlation coefficients (R) are listed in Figure 3. An R value close to 1 indicates that the quantitation is accurate. Compared to that at 6.91THz, the regression at 3.36THz had a higher R value for the relationship. These results indicate that the $PM_{2.5}$ is sensitive to the THz frequency.

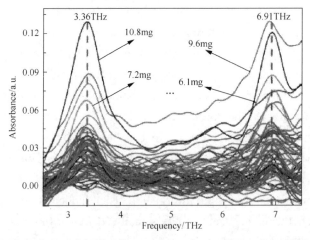

Figure 2 Frequency dependence of the absorbance spectra for the 70 samples of $PM_{2.5}$ collected in dusty environments. The mass of $PM_{2.5}$ ranged from 0.2mg to 10.8mg

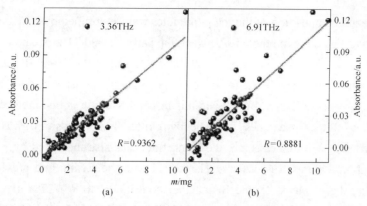

Figure 3 The $PM_{2.5}$ mass versus THz absorbance at (a) 3.36THz and (b) 6.91THz. The lines represent the linear fit between the absorbance and the mass. The correlation coefficients (R) were 0.9362 at 3.36THz and 0.8881 at 6.91THz

To investigate the relationship between the $PM_{2.5}$ and the absorption effect in the THz range, the dependence of absorbance (A) on mass (A/m) at 3.36THz and 6.91THz was calculated and displayed in Figure 4. The A/m values were basically constant (0.01mg^{-1}) over the $PM_{2.5}$ mass range from 1.5mg to 10.8mg. Some of the samples ($PM_{2.5}$ mass < 1.5mg) deviated from the trend observed for the rest of the samples, which may have been affected by weighing errors with electronic balance and systematic errors with the spectrometer. The trend in the data showed that all the $PM_{2.5}$ samples had the same principal components, which is to be expected, as they were sourced from the same building site. Pollutants from construction usually come from construction materials such as concrete, which contains cement, sand, stones and water. These materials are rich in silicon (Si), oxygen (O), and calcium (Ca). In the building process, light and fine concrete particles are produced and spread in the air. Other $PM_{2.5}$ sources include industrial emissions and automobile exhaust, which can contain sulfur (S), O, and titanium (Ti). The main components of the $PM_{2.5}$ were of great importance to the THz absorption effect shown in Figure 2.

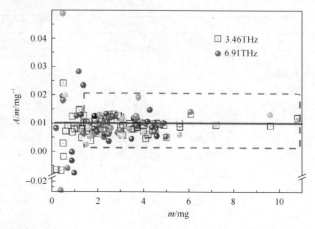

Figure 4 Dependence of the absorbance on the particulate matter mass at 3.36THz and 6.91THz. The dashed box represents the trend over the particulate matter mass range from 1.5mg to 10.8mg

The elemental composition of the $PM_{2.5}$ particles was determined by XRF, and the results were used to investigate the interactions between the particles and THz radiation. The samples studied by XRF had $PM_{2.5}$ masses of 0mg, 0.2mg, 2.4mg, 5.6mg and 10.8mg, and their elemental analysis results are listed in Table 1 and plotted against the total $PM_{2.5}$ mass in Figure 5. The elements measured by XRF in this research were in good agreement with those in similar studies of other researches[14,25]. The blank filter ($PM_{2.5}$ mass of 0) was composed of silicon dioxide (SiO_2). In this research, the blank filter was also measured by THz setup and none of absorption features was observed, in accord with the results in a previous report[26]. For this sample, the elements O and Si together contributed to 98.727% of the total mass (52.867%+45.860%), and the O/Si ratio (52.867/45.860=1.153) was close to the theoretical value of 1.143 (32/28). The slightly larger ratio indicated that this sample contained some other oxides, such as CaO or Al_2O_3. However, the amount of these oxides was small enough that it had little effect on the experiment. As the mass of $PM_{2.5}$ on the quartz filters increased, the contents of most elements changed. The content of Si gradually decreased, and the contents of Na and Zn remained the same, while those of all the other elements (Ca, Al, Mg, K, Fe, S, Ti, P, and Cl) increased. The total mass of the filter included that of the blank filter and $PM_{2.5}$, and changes in the O/Si ratio (1.143, 1.153, 1.157, 1.208, 1.270 and 1.398) indicated that the content of O increased as the $PM_{2.5}$ mass increased.

Therefore, some O must come from the $PM_{2.5}$ in addition to that from SiO_2 in the filter. The XRF results (Table 1) indicated that metallic oxides were the principal compounds in the $PM_{2.5}$. While the contents of most of the elements changed as the $PM_{2.5}$ mass increased, the ratios between the elements remained constant, which was in good agreement with the results in Figure 4. The following analysis is based on the metallic oxides being the major components, and the anions and cations being important in the THz absorption.

Table 1 The elemental compositions (XRF) of the filters containing particulate matter samples   (unit:%)

Elemental	$PM_{2.5}$mass				
	0mg	0.2mg	2.4mg	5.6mg	10.8mg
O	52.867	52.834	52.174	51.176	49.621
Si	45.860	45.661	43.174	40.287	35.505
Ca	0.276	0.365	1.459	2.949	5.659
Al	0.135	0.219	1.506	3.263	5.389
Mg	0.122	0.148	0.508	0.921	1.41
K	0.034	0.042	0.167	0.424	0.782
Fe	0.012	0.016	0.113	0.262	0.579
Na	0.621	0.602	0.548	0.501	0.464
S	0.024	0.086	0.271	0.085	0.291
Ti	0.003	0.005	0.026	0.039	0.104
P	0.032	0.012	0.024	0.047	0.077
Cl	0.01	0.01	0.022	0.035	0.075
Zn	0.002	0.002	0.003	0.003	0.016

As shown in Figure 2, two broad absorption bands were observed from 2.5THz to 4THz (center at 3.36THz) and from 6THz to 7.5THz (center at 6.91THz). Simple linear relationships were observed between the absorbance and the mass, and the XRF results and constant A/m showed that the main components of all the $PM_{2.5}$ samples were the same. The THz response depended on the molecular vibration modes and intensities[18], so the absorption features at 3.36THz and 6.91THz in Figure 2 were caused by the principal components, which can be analyzed from Table 1. Therefore, Table 1 was closely related to Figure 2. To characterize the optical responses of the particles collected from the dusty environment, 2DCOS was applied to the absorbance spectra of the 70 samples, containing the $PM_{2.5}$ with the mass from 0.2mg to 10.85mg. In 2DCOS models, the absorbance spectra from 2.5THz to 7.5THz were employed as the input and the mass of all samples were used as the perturbation. After the two-dimensional correlation calculation of dynamic spectrum, synchronous and asynchronous correlation intensities were calculated and the 2DCOS models can be obtained[27]. Positive correlation was observed over the entire frequency range (Figure 6), which indicated that the absorption increases with $PM_{2.5}$ mass over the entire frequency range (2.5~7.5THz). The synchronous 2D correlation displayed an obvious correlation square, the four vertexes of which were located at (3.36, 3.36), (3.36, 6.91), (6.91, 6.91) and (6.91, 3.36). The large correlation (4.56×10^{-4}) for first auto-peak at 3.36THz represents a high extent of mass-induced dynamic fluctuations of the absorbance spectra, and the even larger correlation (6.55×10^{-4}) for the auto-peak at 6.91THz shows an even higher extent of dynamic fluctuations. Thus, the two auto-peaks showed good agreement with Figure 2, in that the absorption response was very sensitive to the $PM_{2.5}$ mass at the selected frequencies. In addition, two symmetrical cross peaks at the other vertexes of the correlation square represented simultaneous changes of the spectral signals at 3.36THz and 6.91THz, and suggested a related origin for the intensity variations. Accordingly, the positive signs show the absorbance deviates in the same direction at these frequencies. Another cross peak was located at (3.36, 5.84), and showed that a weak

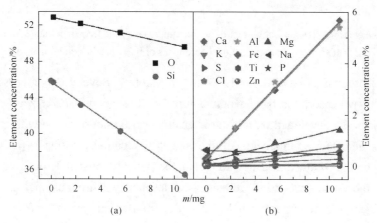

Figure 5　Dependence of the elemental analysis results (XRF) on the particulate matter mass

interaction existed between the corresponding groups whose characteristic vibration was located at these frequencies. Consequently, the group at 3.36THz had a synergistic reaction with those at 5.84THz and 6.91THz.

According to elemental analysis in Table 1, the top six elements of mass content belonged to O, Si, Ca, Al, Mg and K. In these elements, only O can be existed with the anion state; yet, others would be kept with cation form. S can form another complexion SO_4^{2-}, but the content is too few. Therefore, it can be concluded that anions contribute to the strong vibration at approximately 3.36THz because the anions O^{2-} can be related to many other elements, such as Ca, Fe, Al and Mg. Some metallic oxide, such as CaO, Al_2O_3 and MgO, would be the principal components of $PM_{2.5}$ collected from construction site. Accordingly, the cations contributed to the strong vibration at relatively high frequencies which included 5.84THz and 6.91THz as the central frequencies. As discussed formerly, the O content increased as the mass of the $PM_{2.5}$ sample increased, while the masses of S and Cl changed very little. The significance of O in $PM_{2.5}$ is reflected in the synchronous plot. The auto and cross peaks in the synchronous plot could be used to evaluate the elemental composition of pollutants in THz range.

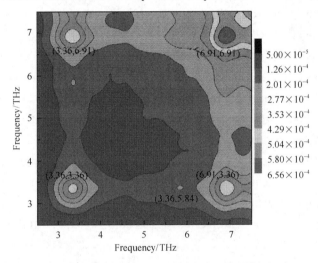

Figure 6 Synchronous 2D correlation plot of all the samples over the frequency range from 2.5THz to 7.5THz. The numbers represent the coordinates of the peaks in synchronous data

The asynchronous 2DCOS figure is plotted in Figure 7. Several strong cross peaks were observed in the two absorption bands, including at 2.5~4THz and 6~7.5THz as seen in Figure 2 and 6. Cross peaks developed only if the intensities varied out of phase with each other for some Fourier frequency components of the signal fluctuations. While there was positive correlation over the whole frequency range in the synchronous plot, positive or negative correlation in the asynchronous plot reflected the special asynchronous characteristics of the signal intensities at different frequencies, and the information revealed from positive and negative correlation was different. In terms of the absorption effect at approximately 3.36THz, a negative correlation was found between the bands at 2.69THz and 3.39THz. In the raw spectra, the absorbance at 2.69THz was difficult to distinguish. However, the cross peak in the asynchronous plot indicated that the absorption band centered at approximately 3.36THz was composed of overlapping peaks, which were located at 2.69THz and 3.39THz. The peak at 3.39THz was the main component because of the small distance between it and 3.36THz, and the peak at 2.69THz was a weak absorption feature. According to the data in Table 1, only O and S were listed among the top ten elements of content. O^{2-} and SO_4^{2-} were the principal anions of $PM_{2.5}$. Combined with the elemental composition information and the analysis about the anion absorption position from Figure 7, it can be ascertained that 3.39THz is the characteristic absorption frequency of O^{2-} and the absorption at 2.69THz is for SO_4^{2-}. Moreover, the intensity change of O^{2-} occurred after that of SO_4^{2-} because of the negative correlation. For the absorption band from 6THz to 7.5THz, more information that was hidden in the raw spectra could be observed in asynchronous plot. Similarly, the cross peaks for this band was consistent with three overlapping peaks centered at 7.43THz, 6.89THz and 6.36THz. The band at 6.89THz showed positive cross peaks with those at 6.36THz and 7.43THz in the asynchronous plot, which indicated that the intensity change at 6.89THz occurred before those at 6.36THz and 7.43THz. In addition, three cross peaks were also observed at (6.89, 5.01), (6.00, 7.41) and (5.29, 7.23) in Figure 7. The 7.41THz and 7.23THz bands could be classified as 7.43THz because of their small deviations from this frequency and the coherence reflected in the asynchronous plot. All the positive signs of correlation illustrated the sequencing of intensity changes at the two frequencies. The band at 6.89THz changed before that at 5.01THz, and the band at 7.43THz changed after the bands at 6.00THz and 5.29THz. These new peaks can be inferred to be caused by cations because the 3.36THz absorption band was caused by anions.

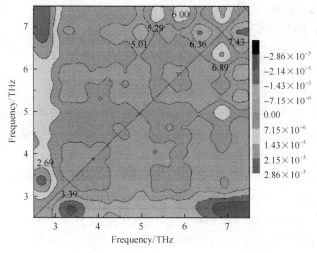

Figure 7　Asynchronous 2D correlation plot of all the samples over the frequency range from 2.5THz to 7.5THz. The numbers represent the horizontal ordinates of the peaks in asynchronous data

According to Table 1, Al, Ca and Mg ranked the first three elements, the Al content was larger than Ca due to the similar mass but smaller molecular weight of Al (Al-27, Ca-40), and Ca ranked second and Mg third. Accordingly, Al^{3+} corresponded to the 6.89THz peak because of the small distance from 6.91THz in the asynchronous plot and the largest correlation of 6.91THz in synchronous plot. Then, Ca^{2+} and Mg^{2+} corresponded to the 7.43THz and 6.36THz peaks. Thus, we could conclude that the absorption band from 6THz to 7.5THz in Figure 2 was overlapped by three absorption peaks which belonged to Al^{3+}, Ca^{2+} and Mg^{2+} at 6.89THz, 7.43THz and 6.36THz, respectively. In addition, several new peaks were observed which were not rather distinct in Figure 2. Their absorption frequencies were 5.01THz, 6.00THz and 5.29THz and could be attributed to K^+, Fe^{3+} and Fe^{2+} due to the relationship between the elemental mass in Table 1 and the correlation intensities in Figure 7. However, the amounts of these ions were very small and they did not feature largely in the raw absorbance spectra or the 2D spectroscopy. While the raw absorbance spectra only showed broad absorption bands from overlapping features, the asynchronous plot could be used to characterize the overlapping bands and obtain the individual absorption peaks.

Quantitation of $PM_{2.5}$ is of importance for environmental monitoring and measuring the mass is one of the common methods. Generally, measuring the mass to monitor $PM_{2.5}$ needs more manual operation, which is not conducive to the intelligentialization and standardization of $PM_{2.5}$ determination. In addition, it can not characterize the components and identify the main pollutant sources. However, in this study, the mass of $PM_{2.5}$ needs measuring to be a perturbation in 2DCOS calculation and an output index in PLS and BPANN calculation. Measuring the mass is used for the establishment of qualitative and quantitative models. After the models are built, THz spectra can be employed to characterize $PM_{2.5}$ without measuring the mass in the future. Herein, the mass of $PM_{2.5}$ collected from the dusty environment could be calculated by weighing the filter the sample was collected on. PLS was then used to build a model between the $PM_{2.5}$ mass and the absorption response with an input of the smoothed absorbance spectra (Figure 2) from 2.5THz to 7.5THz and an output of the $PM_{2.5}$ mass. A total of 70 samples were divided into two groups for calibration and validation. Here, validation set is used to evaluate the precision of the model built by calibration set. To construct a reliable model, more than half of the sample set should be used for calibration. Therefore, 47 samples were used for calibration and 23 were used for validation. Figure 8 shows the predicted $PM_{2.5}$ mass plotted against the actual values for both the calibration and validation sets. The reference (Ref.) line represents zero residuals between the predicted and actual data. All points for the calibration and validation sets were located close to the Ref. line, which indicated that the model is good quantitation. To evaluate the reliability and precision of the model, the correlation coefficient (R) and root-mean square error (RMSE) were calculated (Figure 8). R is related to the covariance $cov(x,y)$ and variance matrix $v(x)$ by $R=cov(x,y)/(v(x) \times v(y))^{1/2}$, and RMSE can be calculated by RMSE$=[\Sigma(x-y)^2/n]^{1/2}$, where n is the sum of samples in a set, and

x as well as y represent actual and predicted mass, respectively. R value close to 1 and a small RMSE indicate the model has high precision. Herein, the R of calibration set was 0.9749 and RMSE equaled 0.49mg, proving that the calibrated model was effective. Meanwhile, the R and RMSE of validation set equaled 0.9628 and 0.53mg. This indicated that the predicted data were very close to the actual values and the calibration model was quite correct. Therefore, PLS can be used to quantitatively determine $PM_{2.5}$ based on absorbance data in THz range.

BPANN was also used for quantitation of the $PM_{2.5}$ samples. Figure 9 shows the quantitative model constructed using the BPANN. Within the 70 samples, 47 were used in the training set and the remainder in the prediction set. The data points from both sets were close to the Ref. line. The R values of the training and prediction sets were 0.9987 and 0.9425, respectively, and the RMSE values were 0.10mg and 0.68mg, respectively. The high R and small RMSE of the two sets show that BPANN is also reliable for quantitative determination of $PM_{2.5}$. Therefore, in terms of the unknown $PM_{2.5}$, its content can be directly predicted by the THz absorption and the built models.

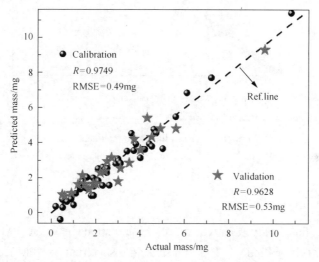

Figure 8 Predicted $PM_{2.5}$ mass versus actual mass from the PLS quantitative model. The black dotted line represents the reference (Ref.) line (zero residuals)

This study was performed to validate the application of THz radiation in the characterization of $PM_{2.5}$ from a dusty environment. Linear regressions, 2DCOS and statistical models were finally built to evaluate the constituents and content. Results show THz measurements can be employed and required for the application in characterization of the air pollution. Initially, the THz response of $PM_{2.5}$ depends on the molecular vibration modes and intensities of the constituents. Such principle is different from that in traditional chemical techniques. It will be important for the theoretical and technological innovation to utilize such a new optical method in $PM_{2.5}$ identification. Besides, THz spectra can be used to build a fingerprint database of $PM_{2.5}$ so that THz features are of great significance to the rapid, direct and correct identification of pollutant sources, which are always necessary and difficult to be detected. Finally, THz technique is a contactless method. This can greatly reduce the damage of some toxic and corrosive compounds in $PM_{2.5}$ to the instrument, which will save the cost and improve the efficiency. Many developing countries are experiencing severe particulate pollution, and effective methods for monitoring and controlling $PM_{2.5}$ are imperative. These results will be of importance for environmental monitoring and for predicting PM emissions from dust storms, and for controlling PM emissions in dusty environments.

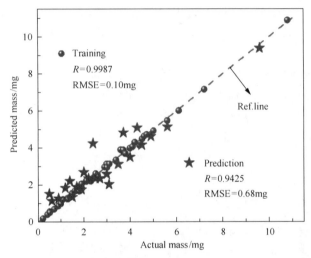

Figure 9   Predicted $PM_{2.5}$ mass versus actual mass with the BPANN model. The dotted line represents the reference (Ref.) line (zero residuals)

4 Conclusions

In summary, this study focused on the interaction between $PM_{2.5}$ from a dusty environment and THz radiation. The transmission spectrum of $PM_{2.5}$ had two distinct absorption bands between 2.5THz and 7.5THz. Their center frequencies were 3.36THz and 6.91THz, respectively. Linear correlations were observed between the deposited mass and the peak intensities. Besides, a number of constituents of $PM_{2.5}$ has been identified using XRF and the elemental variation versus deposited mass has been investigated. In order to explore the absorption features of constituents in THz range, 2DCOS was applied to analyze the THz absorbance spectra. Based on correlation between the absorption bands and the cross peaks in synchronous and asynchronous plots, the broad bands were found to contain several overlapping peaks. The vibration modes of anions and cations were located at relatively low and high frequencies, respectively. Quantitation of $PM_{2.5}$ mass based on a large statistics (70 spectra) was realized using statistical methods with the absorbance spectra as the input data. A good correlation was stated between predicted (from a single spectrum) and measured (by weighing the sample) data. According to this research, optical techniques, and especially spectral methods, should be considered for $PM_{2.5}$ monitoring. The results of this research could be used by environmental policymakers and experts to develop appropriate plans and techniques for PM control.

References

[1] Austen K.Step aside, fitness trackers. The next wave of personal sensors is giving people the ability to monitor the air they breathe. Nature, 2015, 517: 136-138.

[2] Akimoto H. Global air quality and pollution. Science, 2003, 302(5651): 1716-1719.

[3] Nel A. Air pollution-related illness: Effects of particles. Science, 2005, 308(5723): 804-806.

[4] Meng Z, Dabdub D, Seinfeld J H. Chemical coupling between atmospheric ozone and particulate matter. Science, 1997, 277(5322): 116-119.

[5] Loftus C, Yost M, Sampson P, et al. Regional $PM_{2.5}$ and asthma morbidity in an agricultural community: A panel study. Environmental Research, 2015, 136: 505-512.

[6] Wei Y J, Han I K, Min S, et al. $PM_{2.5}$ constituents and oxidative DNA damage in humans.. Environmental Science & Technology, 2009, 43(13): 4757-4762.

[7] Qiao L, Cai J, Wang H, et al. $PM_{2.5}$ constituents and hospital emergency-room visits in Shanghai, China. Environmental Science & Technology, 2014, 48(17): 10406-10414.

[8] Yuan Y, Liu S, Castro R, et al. $PM_{2.5}$ monitoring and mitigation in the cities of China. Environmental Science & Technology, 2012, 46(7): 3627-3628.

[9] Pui D Y H, Chen S C, Zuo Z. $PM_{2.5}$ in China: Measurements,sources,visibility and health effects,and mitigation. Particuology, 2014, 13(2): 1-26.

[10] Huang R J, Zhang Y, Bozzetti C, et al. High secondary aerosol contribution to particulate pollution during haze events in China. Nature, 2014, 514(7521): 218-222.

[11] Brokamp C, Rao M B, Fan Z, et al. Does the elemental composition of indoor and outdoor $PM_{2.5}$ accurately represent the elemental composition of personal $PM_{2.5}$?. Atmospheric Environment, 2015, 101: 226-234.

[12] Tao J, Zhang L, Engling G, et al. Chemical composition of $PM_{2.5}$ in an urban environment in Chengdu, China: Importance of springtime dust storms and biomass burning. Atmospheric Research, 2013, 122: 270-283.

[13] Zhang Q, Jian Z, Yang Z, et al. Impact of $PM_{2.5}$ derived from dust events on daily outpatient numbers for respiratory and cardiovascular diseases in Wuwei, China. Procedia Environmental Sciences, 2013, 18: 290-298.

[14] Han S, Youn J S, Jung Y W. Characterization of PM_{10} and $PM_{2.5}$ source profiles for resuspended road dust collected using mobile sampling methodology. Atmospheric Environment, 2011, 45(20): 3343-3351.

[15] Horiuchi N. Searching for terahertz waves. Nature Photonics, 2010, 4(9): 662.

[16] Zhan H, Wu S, Bao R, et al. Water adsorption dynamics in active carbon probed by terahertz spectroscopy. RSC Advances, 2015, 5(19): 14389-14392.

[17] Bidgoli H, Cherednichenko S, Nordmark J, et al. Terahertz spectroscopy for real-time monitoring of water vapor and CO levels in the producer gas from an industrial biomass gasifier. IEEE Transactions on Terahertz Science & Technology, 2017, 4(6): 722-733.

[18] Zhan H, Wu S, Bao R, et al. Qualitative identification of crude oils from different oil fields using terahertz time-domain spectroscopy. Fuel, 2015, 143: 189-193.

[19] Popovic Z, Grossman E N. THz metrology and instrumentation. IEEE Transactions on Terahertz Science & Technology, 2011, 1(1): 133-144.

[20] Qian L I, Zhao K, Zhang L W, et al. Probing PM$_{2.5}$ with terahertz wave. Science China: Physics, Mechanics & Astronomy, 2014, 57(12): 2354-2356.

[21] Zhang J M, Tsuji H, Noda I, et al. Structural changes and crystallization dynamics of poly(l-lactide) during the cold-crystallization process investigated by infrared and two-dimensional infrared correlation spectroscopy. Macromolecules, 2004, 37(17): 6433-6439.

[22] Cai C B, Yang H W, Wang B, et al. Using near-infrared process analysis to study gas-solid adsorption process as well as its data treatment based on artificial neural network and partial least squares. Vibrational Spectroscopy, 2011, 56(2): 202-209.

[23] Jiang S Y, Ren Z Y, Xue K M, et al.Application of BPANN for prediction of backward ball spinning of thin-walled tubular part with longitudinal inner ribs.Journal of Materials Processing Technology, 2008, 196: 190-196.

[24] Burnett A D, Fan W, Upadhya P C, et al. Broadband terahertz time-domain spectroscopy of drugs-of-abuse and the use of principal component analysis. Analyst, 2009, 134(8): 1658-1668.

[25] Shahsavani A, Naddafi K, Haghighifard N J, et al. The evaluation of PM$_{10}$, PM$_{2.5}$, and PM$_1$ concentrations during the Middle Eastern Dust (MED) events in Ahvaz, Iran, from April through September 2010. Journal of Arid Environments, 2012, 77: 72-83.

[26] Kojima S, Kitahara H, Nishizawa S, et al. Terahertz time-domain spectroscopy of low-energy excitations in glasses. Journal of Molecular Structure, 2005, 744(7): 243-246.

[27] Hoshina H, Ishii S, Morisawa Y, et al. Isothermal crystallization of poly(3-hydroxybutyrate) studied by terahertz two-dimensional correlation spectroscopy. Applied Physics Letters, 2012, 100(1): 011907.

Non-contacting characterization of PM$_{2.5}$ in dusty environment with THz-TDS

Honglei Zhan[1,2] Kun Zhao[2] Lizhi Xiao[1]

(1.State Key Laboratory of Petroleum Resources and Prospecting, China University of Petroleum, Beijing 102249, China; 2.Beijing Key Laboratory of Optical Detection Technology for Oil and Gas, China University of Petroleum, Beijing 102249, China)

Nowadays, many developing countries (such as China) are experiencing severe air pollution due to the rapid urbanization and industrialization. PM$_{2.5}$, which refers to the particulate matter with an aerodynamic diameter of 2.5μm or less, is the most important causation of air pollution in cities[1]. Compared with the coarse atmospheric particles, fine PM$_{2.5}$ has special characteristics including small size, large area, strong activity, and harmful substances (e.g., heavy metals microorganism, etc.). Meanwhile, PM$_{2.5}$ can reside long time and transmit large distance in air so that it has a greater impact on public health and the quality of air. The air containing PM$_{2.5}$ can be breathed into folks' respiratory tracks and cause kinds of respiratory illness such as asthma[2]. PM$_{2.5}$ in atmosphere is composed of many kinds of chemical materials originating from conditions, such as coal-fired power plant, vehicle exhaust gas, dust etc. Dust, an open pollution source driven by the wind, is an important part of the total suspended particulate matter in ambient air.

The precise monitoring of PM$_{2.5}$ is the basis to efficiently control PM$_{2.5}$ and protect air. Tapered element oscillating microbalance and β-ray absorption techniques have been employed in PM detection[3]. They have special advantages, but are still not enough to process PM$_{2.5}$ detection due to the high complexity of air pollution. Recently, the application of terahertz (THz) technique in oil-gas and pollutant detection drew a high attention due to a series of reports[4-10]. In this letter, THz time-domain spectroscopy (THz-TDS) combined with back propagation artificial neural network (BPANN) was utilized to detect PM$_{2.5}$ collected from dusty environment.

A special Minivol Tactical Air Sampler was used to collect PM$_{2.5}$ in dusty environment. The collection was finished at a construction site in Shijiazhuang, in June 2014. A quartz filter membrane with the diameter of 47mm was used and the air flow rate was set as 9 L/min. The blank filters were weighed and the filters with PM$_{2.5}$ were reweighed so that the PM$_{2.5}$ mass can be calculated. A typical TDS setup with transmission geometry was used. The light resource was generated by a Ti-sapphire laser with the wavelength of 800 nm, a pulse width of 100fs, and a repetition of 80MHz. The setup was employed to scan PM$_{2.5}$ and the THz-TDS can be obtained[10].

Originally published in *Science China: Physics, Mechanics & Astronomy*, 2016, 59(4): 644201.

In this research, a basic study was performed about the THz dielectric effect of 30 groups of PM$_{2.5}$ with different mass collected from dusty environment. The inset of Figure 1(a) shows the THz field signal as a function of time delay after the transmission of the THz pulse through the filters with PM$_{2.5}$. In order to get the relationship between the THz responses and the PM$_{2.5}$ mass in dusty conditions, we described the PM$_{2.5}$ content dependent peak intensity E_p, as shown in Figure 1(a), where the E_p linearly decreased with the increase of PM$_{2.5}$ mass and the linear function can be described as $y=0.16806-0.00277x$. The points were located at the two sides and close to the fitted line. Error bars of 5%-fluctuation indicated a high precision of the linear model. Therefore, the tendency and the linear model can be employed to rapidly and quantitatively evaluate the PM$_{2.5}$ in dusty environment.

Quantitation of PM$_{2.5}$ is an important part for the prediction of air pollution. Back propagation artigicial neutral network (BPANN) was used to build a quantitative model. BPANN is a mathematical nonlinear dynamics system simulating structure and function of biological neural networks in the human brain. No priori model was required for ANN owning to capturing the inherent information from the considered variables and learning from the existing data, even when noise was present. BP learning algorithm stored a lot of input-output mapping relationships without prior revealing the mathematical equation and composed of input, hidden and output layers. In the calculation, samples were divided into two subsets: The training and prediction set. In order to identify all the subsequently given spectra correctly in prediction set, the number of training set should exceed the half of sample numbers[8].

To build a robust and precise model to quantitatively predict PM$_{2.5}$ content, BPANN can be employed by using the absorption spectra in the whole range as the input. In this study, 10 samples were randomly selected and used for training and the left 20 were employed to predict. Figure 1(b) shows the predicted versus actual mass in both the training set and prediction set. The points in training and prediction were located close to the Ref. line, which represented zero residuals. To evaluate the precision of the BPANN model, we calculated the correlation

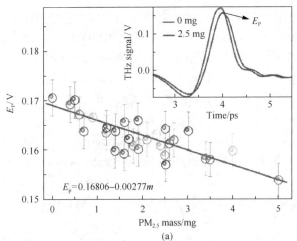

(a)

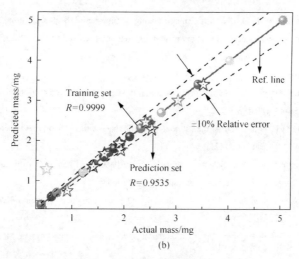

Figure 1 (a) PM$_{2.5}$ mass dependent E_P intensities. Error bars represent 5%-fluctuation. The inset graph shows the THz-TDS of selected PM$_{2.5}$; (b) predicted mass versus actual mass of PM$_{2.5}$ in dust environment of a construction site. The quantitatively model was built using BPANN. The Ref. line indicates zero errors and the dotted lines represent ±10% relative errors between actual and predicted values

coefficient (R). R referred to the covariance cov(x, y) and variance matrix $v(x)$ by R=cov(x, y)/$[v(x) \times v(y)]^{1/2}$, where x and y were actual and predicted mass. Here, R of training set and prediction set equaled 0.9999 and 0.9535, respectively. Also, two lines representing 10% relative errors showed that the predicted value were much close to the actual mass. Such model would be an efficient selection to monitor PM$_{2.5}$ in dusty environment.

In summary, the practicability was proved about the THz-TDS technique being applied to characterize PM$_{2.5}$ in dusty environment. A linear relationship was found between THz peak intensities and PM$_{2.5}$ mass. Moreover, BPANN was employed to build a quantitative model and characterize the PM$_{2.5}$ mass with high R and small relative errors. Therefore, THz-TDS is a promising tool to detect PM$_{2.5}$. The combination of THz and mathematical method will supply more selections in air pollution monitoring.

Acknowledgements

This work was supported by the National Key Basic Research Program of China (Grant No. 2014CB744302), the Specially Funded Program on National Key Scientific Instruments and Equipment Development (Grant No. 2012YQ140005) and the National Natural Science Foundation of China (Grant No. 11574401).

References

[1] Huang R J, Zhang Y, Bozzetti C, et al. High secondary aerosol contribution to particulate pollution during haze events in China. Nature, 2014, 514(7521): 218-222.

[2] Loftus C, Yost M, Sampson P, et al. Regional PM$_{2.5}$ and asthma morbidity in an agricultural community: A panel study. Environmental Research, 2015, 136: 505-512.

[3] Winkel A, Rubio J L, Veld J W H H, et al. Equivalence testing of filter-based, beta-attenuation, TEOM, and light-scattering devices for measurement of PM_{10} concentration in animal houses. Journal of Aerosol Science, 2015, 80: 11-26.

[4] Zhang Z W, Wang K J, Lei Y, et al. Non-destructive detection of pigments in oil painting by using terahertz tomography. Science China: Physics, Mechanics & Astronomy, 2015, 58(12): 124202.

[5] Zhan H L, Wu S X, Bao R M, et al. Qualitative identification of crude oils from different oil fields using terahertz time-domain spectroscopy. Fuel, 2015, 143: 189-193.

[6] Bao R M, Li Y Z, Zhan H L, et al. Probing the oil content in oil shale with terahertz spectroscopy. Science China: Physics, Mechanics & Astronomy, 2015, 58(11): 114211.

[7] Wang D C, Huang Q, Qiu C W, et al. Selective excitation of resonances in gammadion metamaterials for terahertz wave manipulation. Science China: Physics, Mechanics & Astronomy, 2015, 58(8): 84201.

[8] Zhan H L, Li Q, Zhao K, et al. Evaluating $PM_{2.5}$ at a construction site using terahertz radiation. IEEE Transactions on Terahertz Science & Technology, 2015, 5(6): 1028-1034.

[9] Li Q, Zhao K, Zhang L W, et al. Probing $PM_{2.5}$ with terahertz wave. Science China: Physics, Mechanics & Astronomy, 2014, 57(12): 2354-2356.

[10] Zhan H L, Sun S N, Zhao K, et al. Less than 6 GHz resolution THz spectroscopy of water vapor. Science China: Technological Sciences, 2015, 58(12): 2104-2109.

Terahertz assessment of the atmospheric pollution during the first-ever red alert period in Beijing

Honglei Zhan[1] Ning Li[1] Kun Zhao[1] Zhenwei Zhang[2] Cunlin Zhang[2]
Rima Bao[1]

(1.Beijing Key Laboratory of Optical Detection Technology for Oil and Gas, China University of Petroleum, Beijing 102249, China; 2.Department of Physics, Capital Normal University, Beijing 100048, China)

Haze or smog episodes, which are characterized by the presence of particulate matter at diameters less than 2.5μm ($PM_{2.5}$), have attracted increasing attention during the past few decades[1]. $PM_{2.5}$ has adverse effects on human respiratory health as well as on air visibility[2,3]. In the Beijing-Tianjin-Hebei (BTH) region of China, haze has become especially serious in recent years because of industrial expansion and traffic-related emissions[4,5] (http://news. xinhuanet.com/english/2016-01/07/c_134987525.htm;http://news.cenews.com.cn/html/2016-01/07/content_38614.htm). To protect public health, the first-ever red alert for smog in China was issued by the Beijing municipal environmental protection bureau (MEPB), implemented at 7 a.m. on 8 December 2015, and lifted at noon on 10 December 2015. During these two days, the government announced a series of counteractive measures in Beijing that included limitations on driving private cars, reductions in coal consumption, and controls on industrial vehicles and operations (http://en.people.cn/n/2015/1208/c90000-8987556.html; http://en. people. cn/n/2015/1210/c90000-8988321.html). These measures helped to cut concentrations of $PM_{2.5}$ by 10% on Tuesday. This success indicates that $PM_{2.5}$ monitoring of both concentrations and the sources of pollutants is of great significance for environmental departments and the public.

Normal PM monitoring approaches are very sensitive to $PM_{2.5}$ concentrations but identifying pollution sources, which is necessary for environmental protection, remains challenging. Optical (i.e., non-contact) methods have been applied to characterize $PM_{2.5}$ in different environments[6]. One of these, terahertz (THz) spectroscopy technology, has been useful in biomedicine, materials, safety, petroleum, and other areas[7-13]; it is a non-contact, real-time way to monitor $PM_{2.5}$. The THz signal represents substantial inter and intra-molecular vibration modes[14-16].

To improve air quality in Beijing, environmental departments will be making a series of emergency measures (EM) tests; to best assess the effect of counteractive measures, researchers will need proven techniques that can continuously monitor pollution levels and sources. It is known that THz absorption intensities and bands are closely related to $PM_{2.5}$

concentrations and components, and that the information hidden in absorption bands can be extracted by combining mathematical methods. In the present study, the concentration and pollution sources of $PM_{2.5}$ collected around Beijing between 8 December and 10 December 2015 were identified by the combination of THz absorption and 2D correlation spectroscopy (2DCOS). The results showed that EM can be highly effective in reducing pollution and improving air quality.

As shown in Figure 1(a), the spectra of a series of $PM_{2.5}$ in 3~10THz were obtained with a common Fourier transform infrared (FTIR) spectrometer, manufactured by Bruker; 64 scans at a resolution of $4cm^{-1}$ were co-added with strong apodization. Silicon carbide was used as the optical source because it can emit steady, strong emissions with continuous wavelengths in the THz range. Every measurement was processed in a vacuum environment. All of the $PM_{2.5}$ samples used in the present study were collected in the Changping district of Beijing. Two groups of $PM_{2.5}$ were investigated under different conditions, selected samples are shown in Figure 1(b). The first group of samples was obtained between 8 December and 10 December 2015, with a collection time length of 2h. The second group of samples was selectively collected in high-pollution conditions from 1 December to 5 December 2015, with a collection time length of 12.5h.

We used a standard air sampler (an Airmetrics Minivol) to collect $PM_{2.5}$ and designed two separators with respective cutoffs of 10μm and 2.5μm. Polytetrafluoroethylene (PTFE) filters with a diameter of 47mm were employed because of the small absorption of PTFE in the THz range. During the collection process, the air flow rate was set at 8 L/min. In our THz measurement, PTFE without and with $PM_{2.5}$ were used as reference and sample, respectively. All spectra were background-corrected using a blank filter spectrum, which was renewed after each scan. The sample holder and measurement area were carefully cleaned before every scan. The spectra were recorded as frequency-domain spectra (FDS) at each data point. All of the samples were measured in vacuum condition. The frequency dependence of absorbance (A) spectra was obtained by $\lg(E_{ref}/E_{sam})$, where E_{ref} and E_{sam} are the amplitudes of the reference and the sample in the FDS; these were obtained directly by FTIR measurement, as shown in Figure 2(a)[17].

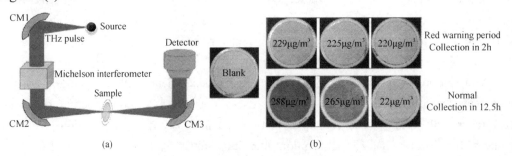

Figure 1 (a) Optical path of THz wave in FTIR; (b) images of a blank filter (reference) and a series filters with $PM_{2.5}$ collected under different conditions

2DCOS, which is generally applicable to a broad range of spectroscopic techniques, can analyze a set of spectral data and enhance spectral resolution from a system under some perturbation. 2DCOS can also enhance spectral resolution by spreading peaks along the second dimension. Crossed or overlapping peaks are used to determine the relative direction of intensity changes and the sequential order of events. The 2D correlation spectrum can be calculated by

$$X(v_1,v_2)=[y(v_1,t),y(v_2,t')]=\Phi(v_1,v_2)+i\Psi(v_1,v_2)$$

where v is the frequency; $y(v,t)$ represents the spectral responsivity of output; and $\Phi(v_1,v_2)$ and $\Psi(v_1,v_2)$ are the synchronous and asynchronous correlation intensities. $\Phi(v_1,v_2)$ (synchronous) indicates similarity of signal dependence on time t; $\Psi(v_1,v_2)$ reflects dissimilarity. Synchronous and asynchronous plots reflect the similarity and dissimilarity of signal dependence on t. In a synchronous correlation spectrum, auto-peaks at diagonal positions represent the extent of perturbation-induced dynamic fluctuations of signals, and crossed or overlapping peaks represent simultaneous changes of spectral signals at two wavelengths; the latter indicates a coupled or related origin of intensity variations. In an asynchronous correlation spectrum, crossed or overlapping peaks develop only if the variation of their intensities are out of phase, which indicates differences in the spectral changes of the crossed or overlapping peaks[18]. Generally, the sign is positive if the intensity changes at v_x occur before v_y, and is negative if the intensity changes at v_x occur after v_y. However, these rules are correct only when $\Phi(v_1,v_2)>0$; they are reversed if $\Psi(v_1,v_2)<0$.

During a red alert, the $PM_{2.5}$ content is the parameter of greatest concern, for both the government and the public, because it can directly indicate whether the EM made by MEPB effectively relieves the haze pollution level of smog. Our plots of selected FDS and A spectra from the first group samples, 3~10THz, are shown in Figures 2(a) and (b). We may see that the absorption of high $PM_{2.5}$ was larger and the central frequencies of the absorption band were located at ~6.8THz. Figure 2(c) shows the absorbance data at 6.0THz, 6.5THz, 6.8THz, 7.1THz, and 7.5THz, which were extracted and compared with the $PM_{2.5}$ content listed on a government website. As expected, the samples with high $PM_{2.5}$ are related to larger absorption at selected frequencies, and THz absorption of low $PM_{2.5}$ is similarly small under low pollution. Therefore, the information gained by EM during the red alert period was effective in relieving the haze pollution by gradually decreasing $PM_{2.5}$ concentrations.

We assessed the meliorating tendency of $PM_{2.5}$ concentration by measuring THz absorption during the red alert period, which was decided by a governmental agency. But another matter of public concern is whether the pollution sources indeed vary. In early December 2015, Beijing experienced haze pollution (a high-normal pollution condition) for

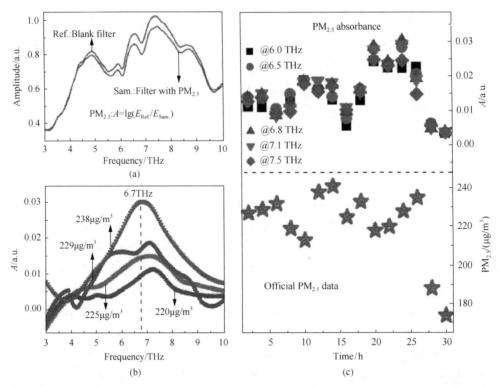

Figure 2 (a) Typical THz-FDS of reference and sample; (b) frequency-dependent absorbance A spectra of the selected $PM_{2.5}$; (c) time-dependent $PM_{2.5}$ absorbance (above) and corresponding $PM_{2.5}$ from government (below) during a selected period between 8 December and 10 December 2015

which no red alert was ordered. Thus we were able to apply 2DCOS to two kinds of $PM_{2.5}$, some collected during the red alert period and some collected during the red alert period and some collected under the normal high-pollution condition, using the $PM_{2.5}$ content as the external perturbation. The synchronous and asynchronous correlation intensities are shown in Figures 3 and 4. Positive correlations can be seen over the entire frequencies [Figure 3(a) and 4(a)]. Obvious auto-peaks occur at (6.7, 6.7) in the red alert condition and (6.1, 6.1) in the normal condition. The asynchronous 2DCOS plots are displayed in Figure 3(b) and 4(b). An asynchronous plot can ascertain whether the absorption band is due to non-peak absorption or to relative changes in the intensities of two or more overlapped peaks[19]. The intersecting peaks observed in these asynchronous plots indicate a series of vibrational modes at relative frequencies that undergo changes in intensity with $PM_{2.5}$.

In terms of $PM_{2.5}$ during the red alert [Figure 3(b)], a negative cross peak is located at (5.6, 6.8) (i.e., the absorption band whose central frequency equals 6.7THz is overlapped by peaks at 5.6THz and 6.8THz). The absorption band of $PM_{2.5}$ in normal high-pollution condition is overlapped by other single peaks as well. As shown in Figure 4(b), positive cross-peaks are located at (6.3, 3.4), (6.3, 5.2), (6.3, 7.0) and (6.3, 8.0). The 6.1THz center-frequency band is overlapped by five single peaks, at 3.4THz, 5.2THz, 6.3THz, 7.0THz and 8.0THz. The feature at 6.3THz changes before the peaks at 3.4THz, 5.2THz, 7.0THz, and 8.0THz.

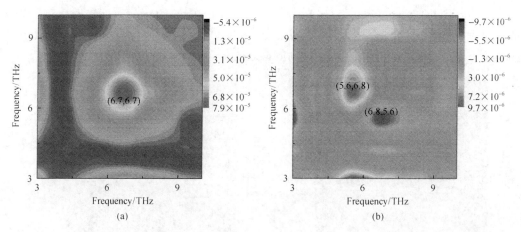

Figure 3 (a) Synchronous and (b) asynchronous 2D correlation plots of the selected $PM_{2.5}$ during the red alert period over the frequency range of 3~10THz

Based on the comparison of THz peaks of $PM_{2.5}$ collected in the periods with and without red alert, features can be attributed to two conditions. The absorption features around ~6.5THz are related to the PM's principal components in Beijing, which include sulfate emitted from vehicle exhaust and coal combustion. The peaks occurred at 5.6THz and 6.8THz during the red alert period, whereas the peaks at 5.2THz, 6.3THz, 7.0THz, and 8.0THz during the normal high pollution condition. According to the MEPB's requirement, private vehicle use was restricted in Beijing based on even- and odd-numbered license plates. In addition, 30% of the vehicles belonging to local governments, public institutions, social groups, and state-owned enterprises were allowed to be used. These uses were restricted because sulfates, such as Na_2SO_4, $Al_2(SO_4)_3$, $CaSO_4$, and $MgSO_4$, contributed greatly to air pollution in Beijing. As a result of these restrictions, the types and quantity of sulfates decreased. The peaks at 6.3THz and 8.0THz seen in normal high-pollution conditions were respectively assigned to the vibration modes of $MgSO_4$ and Na_2SO_4. Because the vibration modes of anions and cations had been previously located at relatively low and high frequencies, respectively, we assigned the peak at 3.4THz to the vibration of the O^{2-} of metallic oxides in dust[16]. During the red alert period, use of construction waste and regular garbage-transport vehicles, concrete tank trucks, and gravel transport vehicles were prohibited; outdoor construction was halted. The government also increased road sweeping and cleaning, and stopped or limited industrial enterprises. As a result, pollutants from construction sites were absent during this period.

In summary, it was known that THz absorption can directly monitor $PM_{2.5}$ in the atmosphere. We applied 2DCOS to absorbance spectra of $PM_{2.5}$ during China's first red-alert air pollution period and under normal high-pollution conditions (both in Beijing). By comparing features in synchronous and asynchronous plots of these spectra, we found that sulfate types were different under the two conditions, and that metallic oxides belonging to $PM_{2.5}$ sources in the normal high-pollution condition were absent during the red alert period. We concluded that it would be useful to develop a real-time monitoring system for air quality in Beijing, given that haze conditions vary with time and location. In the future, metamaterials should be considered to measure relatively low densities of $PM_{2.5}$ in real time[20].

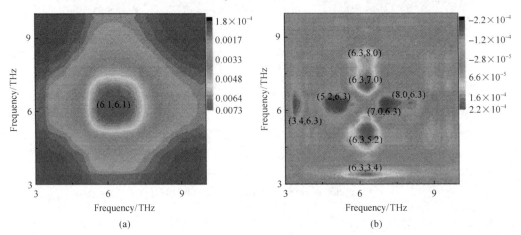

Figure 4 (a) Synchronous and (b) asynchronous 2D correlation plots of the selected $PM_{2.5}$ in a haze condition (normal high pollution) without red alert over the frequency range of 3~10THz

Acknowledgements

This work was funded by the National Basic Research Program of China (Grant No. 2014CB744302), the Specially Funded Program on National Key Scientific Instruments and Equipment Development (Grant No. 2012YQ140005), the China Petroleum and Chemical Industry Association Science and Technology Guidance Program (Grant No. 2016-01-07), and the National Nature Science Foundation of China (Grant No. 11574401).

References

[1] Huang R J, Zhang Y, Bozzetti C, et al. High secondary aerosol contribution to particulate pollution during haze events in China. Nature, 2014, 514(7521): 218-222.

[2] Nel A. Air pollution-related illness: Effects of particles. Science, 2005, 308(5723): 804-806.

[3] Yuan Y, Liu S, Castro R, et al. $PM_{2.5}$ monitoring and mitigation in the cities of China. Environmental Science & Technology, 2012, 46(7): 3627-3628.

[4] Zhang Y L, Cao F. Fine particulate matter ($PM_{2.5}$) in China at a city level.Scientific Reports, 2015, 5: 14884-14884.

[5] Zhang Q, Crooks R. (Asian Development Band, 2012).

[6] Winkel A, Llorens J, Rubio, I. V. J. W. H. Huis, et al. Equivalence testing of filter-based, beta-attenuation, TEOM, and light-scattering devices for measurement of PM_{10} concentration in animal houses. Journal of Aerosol Science, 2015, 80: 11-26.

[7] Wang D C, Huang Q, Qiu C W, et al. Selective excitation of resonances in gammadion metamaterials for terahertz wave manipulation. Science China: Physics, Mechanics & Astronomy, 2015, 58(8): 84201-084201.

[8] Zhan H L, Wu S X, Zhao K, et al. $CaCO_3$, its reaction and carbonate rocks: terahertz spectroscopy investigation. Journal of Geophysics & Engineering, 2016, 13(5): 768-774.

[9] Bao R M, Li Y Z, Zhan H L, et al. Probing the oil content in oil shale with terahertz spectroscopy. Science China: Physics, Mechanics & Astronomy, 2015, 58(11): 114211.

[10] Wang D D, Miao X Y, Zhan H L, et al. Non-contacting characterization of oil-gas interface with terahertz wave. Science China: Physics, Mechanics & Astronomy, 2016, 59(7): 1-2.

[11] Zhan H L, Wu S W, Bao B M, et al. Qualitative identification of crude oils from different oil fields using terahertz time-domain spectroscopy. Fuel, 2015, 143: 189-193.

[12] Bao R M, Miao X Y, Feng C J, et al. Characterizing the oil and water distribution in low permeability core by reconstruction of terahertz images. Science China: Physics, Mechanics & Astronomy, 2016, 59(6): 664201.

[13] Zhan H L, Zhao K, Xiao L Z. Spectral characterization of the key parameters and elements in coal using terahertz spectroscopy. Energy, 2015, 93: 1140-1145.

[14] Zhan H L, Zhao K, Xiao L Z. Non-contacting characterization of $PM_{2.5}$ in dusty environment with THz-TDS. Science China: Physics, Mechanics & Astronomy, 2016, 59(4): 644201.

[15] Li Q, Zhao K, Zhang L W, et al. Probing $PM_{2.5}$ with terahertz wave. Science China: Physics, Mechanics & Astronomy, 2014, 57(12): 2354-2356.

[16] Zhan H L, Li Q, Zhao K, et al. Evaluating $PM_{2.5}$ at a construction site using terahertz radiation. IEEE Transactions on Terahertz Science & Technology, 2015, 5(6): 1028-1034.

[17] He Y Y, Su Y J, Zhu M, et al. Effects of ion bombardment on microcrystalline silicon growth by inductively coupled plasma assistant magnetron sputtering. Science China: Physics, Mechanics & Astronomy, 2012, 55(11): 2070-2075.

[18] Hoshina H, Ishii S, Morisawa Y, et al. Isothermal crystallization of poly(3-hydroxybutyrate) studied by terahertz two-dimensional correlation spectroscopy. Applied Physics Letters, 2012, 100(1): 97-97.

[19] Zhan H L, Zhao K, Zhao H, et al. The spectral analysis of fuel oils using terahertz radiation and chemometric methods. Journal of Physics D: Applied Physics, 2016, 49(39): 395101.

[20] Chen Z C, Mohsen R, Gong Y D, et al. Realization of variable three-dimensional terahertz metamaterial tubes for passive resonance tunability. Advanced Materials, 2016, 24(23): 143-147.

Terahertz-dependent PM$_{2.5}$ monitoring and grading in the atmosphere

Xin yang Miao[1,2]   Honglei Zhan[1,2,3]   Kun Zhao[1,2]   Zhenwei Zhang[4]   Lei Xu[4]
Cunlin Zhang[4]   Lizhi Xiao[1]

(1.State Key Laboratory of Petroleum Resources and Prospecting, China University of Petroleum, Beijing 102249, China; 2.Beijing Key Laboratory of Optical Detection Technology for Oil and Gas, China University of Petroleum, Beijing 102249, China; 3.Department of Material Science and Engineering, China University of Petroleum, Beijing 102249, China; 4.Department of Physics, Capital Normal University, Beijing 100048, China)

Abstract: Rapid industrialization and economic development have led to serious pollution in the form of fine particulate matter (PM$_{2.5}$, particulate matter with a diameter of less than 2.5μm). In China, PM$_{2.5}$ has been one of the most debated topics in councils of government and issues of public concern. Terahertz (THz) radiation was employed to measure the PM$_{2.5}$ in the atmosphere from September 2014 to April 2015 in Beijing. Comparison of the PM$_{2.5}$ level from the website with THz absorbance revealed a significant phenomenon: THz radiation can be used to monitor PM$_{2.5}$ in the atmosphere. During Asia-Pacific Economic Cooperation (APEC) 2014, "APEC Blue" was also recorded in a THz system. The relationship between absorbance and PM$_{2.5}$ demonstrates that THz radiation is an effective selection for air pollution grading. Based on the absorbance spectra, the elemental compositions were studied by two-dimensional correlation spectroscopy (2DCOS) in conjunction with X-ray fluorescence. Several single absorption peaks were revealed and caused by sulphate from coal combustion, vehicle exhaust emissions and secondary reactions. Furthermore, mathematical algorithms, such as the BPANN and SVM, can process the THz absorbance data and greatly improve the precision of the estimation of PM$_{2.5}$ mass. Our results suggest that THz spectroscopy can not only reveal the component information for pollution source determination, but quantitatively monitor the PM$_{2.5}$ content for pollution level evaluation. Therefore, the use of THz radiation is a new method for future air pollution monitoring and grading systems.

Keywords: terahertz; PM$_{2.5}$; monitoring; grading

1 Introduction

The rapid economic development, industrial expansion and urbanization over the past few decades have caused increasing haze or smog episodes in many cities, especially in northern China, due to relatively high particulate matter (PM) emissions and unfavourable meteorological conditions for pollution dispersion[1]. In recent years, a high seasonal variability in PM$_{2.5}$ (PM with a diameter of less than 2.5μm) has been observed, with the highest emissions occurring during the winter and the lowest during the summer. The population-weighted mean of $PM_{2.5}$, an important parameter to evaluate the influence of pollution on folks, is 61 μg/m^3 in Chinese cities, approximately 3 times as high as the global population-weighted mean, often with a high daily concentration of 300 μg/m^3[2]. According to a report from Asian-development Back, only <1% of the 500 largest cities in China could meet the air quality guideline suggested by the World Health Organization, and several cities were ranked among the most polluted cities in the world. Such serious air pollution was accompanied by extremely poor visibility, poor air quality and a series of respiratory diseased, all of which has drawn considerable public attention. $PM_{2.5}$ mitigation has become a frequent issue in the National People's Congress (NPC) and Political Consultation Congress (PCC). The Chinese government implemented a series of major initiatives in response to the severe haze events. China established a mandatory monitoring range for $PM_{2.5}$ for the first time in 2012, and the new "National Ambient Air Quality Standards" were officially released by the Chinese Ministry of Environmental Protection (CMEP); subsequently, the Chinese State Council released the "Atmospheric Pollution Prevention and Control Action Plan (APPCAP)" in September 2013. This project aimed to reduce $PM_{2.5}$ emissions by up to 25% by 2017 relative to the 2012 levels and was backed by 1750 billion RMB in investments from the central government[3-7].

In November 2014, the issue of the rarity of blue skies in Beijing (APEC Blue) was widely discussed during APEC China 2014 due to the emissions reduction campaign directed by the Chinese government. According to data from the Beijing Municipal Environment Monitoring Centre, from November 1 to 12, the densities of $PM_{2.5}$, PM_{10}, SO_2, and NO_2 decreased by 55%, 44%, 57%, and 31%, respectively, from the same period in the previous year; in addition, the concentration of various pollutants was at the lowest level over the same period in the past 5 years[8,9]. To guarantee quality air for the military parade marking the 70th anniversary of the end of World War II (WWII V-Day) on September 3, many steel mills in Beijing and the neighbouring regions were shut down or reduced production starting in late August. According to the report from the Beijing Municipal Environmental Protection Bureau, the average density of $PM_{2.5}$ decreased by 73.2% after the efforts were implemented to ensure clean air for the International Association of Athletics Federations (IAAF) world championships and the WWII V-Day military parade[10]. For two weeks before the parade, the sky was blue and pollution levels were low; the day after, the sky was grey and pollution was back to normal.

Controlling air pollution requires optimized emission control and monitoring strategies[11-13]. Air pollution in China is extremely complex because of the comprehensive interactions between industrial structure, regional characteristics, climate influence and development levels. The factors governing $PM_{2.5}$ components and concentrations are extremely sophisticated and poorly constrained[14]. The pollution sources and levels change rapidly. Thus, it is necessary to develop new methods and build a comprehensive calibration system for the rapid identification and quantitative analysis of pollution sources and the systemic grading of the pollution levels. Such an investigation will provide innovative ideas and support for the control of air pollution. In this research, we combine a comprehensive set of theories and methods based on THz radiation to characterize $PM_{2.5}$ from the atmosphere.

2　Methods and procedures

To obtain the actual $PM_{2.5}$ levels in the atmosphere for THz measurement, we used a standard air sampler (Minivol Tactical Air Sampler from Airmetrics, America), shown in Figure 1 (a). In the inner machine, two separators with cut-offs of 10μm and 2.5μm were designed for $PM_{2.5}$ collection. Quartz filter membranes, which had a diameter of 47mm and a mass of 0.1420g, were employed so that $PM_{2.5}$ in the atmosphere could be collected on the filters. In order to eliminate water absorbed onto the filters, all filters were dried in a vacuum loft driver with 322.2K for 2h before collection. During the collection process, the air flow rate was set at 7L/min, and the time frame of every sample was 24h. The THz absorption effect is related to the samples' components and chemical structure, thus the $PM_{2.5}$ samples should be collected in a same region. A total of 128 $PM_{2.5}$ samples were obtained in a long-term collection effort from September 2014 to April 2015. The sampling took place on a bracket with 1.5 m from the ground at the Beijing Campus, China University of Petroleum. This site, located in Changping District, is about 36 km northwestern from Tian'anmen Square and ~28 km from the Olympic Bird Nest, Beijing, China. There was a specific pattern of days to collections over the time period in a way that seasonality could be important. However, some data were excluded because of the lack of collection in rainy days.

In terms of $PM_{2.5}$, their mass was proportional to the $PM_{2.5}$ content in the air due to the use of the same conditions in the collection process. Next, all $PM_{2.5}$ samples were scanned to obtain the absorption spectra in the THz range using a Fourier-transform infrared (FTIR) spectrometer. The optical path of the spectrometer is shown in Figure 1 (b). The optical source of the spectrometer is a silicon carbide, which can emit steady and strong emissions with continuous wavelength in THz range. After being reflected by concave mirrors 1 and 2 (CM1 and CM2, respectively), the pulses are parallel to each other and enter a Michelson interferometer, where interference pulses are obtained by an interferometer. The pulses reach the sample cavity after being reflected by tuneable mirror 1 (M1) and CM3. Herein, the sample must be located at the focal point of the pulses. Finally, the pulses carrying the sample information are reflected by CM4 and are accepted by the detector. Then, the interference spectrum is observed. A computer is used to process the spectra by using Fourier transform (FT). In this manner, the FTIR spectrum can be obtained. In this study, the blank filters are used as references shown in Figure 1 (a), and the filters were scanned as the operating principal of THz work, instead of ambient air. All of the samples were measured in vacuum condition at room temperature. The absorbance A spectra were obtained using $\ln(I_0/I)$, where I_0 and I are the signal intensity of the reference and the sample, respectively[15,16]. The frequency range between 1.5THz and 9THz is appropriately selected for an acceptable signal-to-noise ratio.

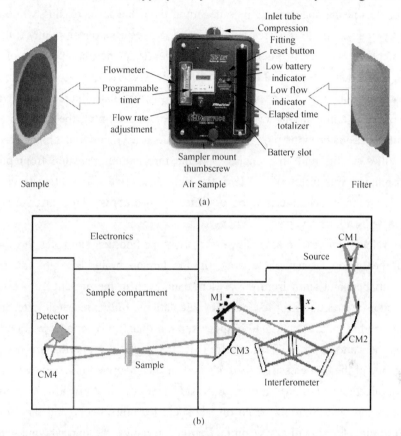

Figure 1　Sampling sketch (a) and measurement principle of the FTIR spectrometer (b)

Two-dimensional correlation spectroscopy (2DCOS) is generally applicable to a broad range of spectroscopic techniques based on a set of spectral data from a system under some perturbation. 2DCOS can enhance the spectral resolution by spreading peaks along the second dimension. Selective development of two-dimensional (2D) peaks provides better access to information that is not readily seen in conventional one-dimensional (1D) spectra. The sign of cross peaks is used to determine relative direction of the intensity changes and the sequential order of events. The 2D correlation spectrum can be calculated by

$$X(v_1,v_2)=[y(v_1,t),y(v_2,t')]=\Phi(v_1,v_2)+i\Psi(v_1,v_2) \tag{1}$$

where v is the frequency, $y(v, t)$ represents the spectral responsivity of the output, $\Phi(v_1,v_2)$ and $\Psi(v_1,v_2)$ are the synchronous and asynchronous correlation intensities, respectively. $\Phi(v_1,v_2)$ (synchronous) indicates the similarity of the signal dependence on time t, and $\Psi(v_1,v_2)$ reflects the dissimilarity. In the synchronous plot, auto-peaks at the diagonal positions indicate the extent of the perturbation-induced dynamic fluctuations of the spectral signals. The cross peaks represent simultaneous change of spectral signals at two different frequencies. Herein, if the sign of a cross peak is positive, then the intensities at the corresponding frequency are increasing or decreasing together; if it is negative, then one is increasing while another is decreasing. In an asynchronous correlation spectrum, cross peaks develop only if the intensity varies out of phase with each other for some Fourier frequency components of the signal fluctuations. Cross peaks developed only if the intensities varied out of phase with each other for some Fourier frequency components of the signal fluctuations. Whereas there was positive correlation over the entire frequency range in the synchronous plot, the positive or negative correlation in the asynchronous plot reflected the special asynchronous characteristics of the signal intensities at different frequencies, and the information revealed from positive and negative correlation was different[17]. Here, the sign is positive if the intensity changes at v_x occurs before v_y and is negative if v_x occurs after v_y. The above sign rules are correct only when $\Phi(v_1,v_2) > 0$ and are reversed if $\Phi(v_1,v_2) < 0$.

The BPANN is a normal mathematical statistical method that simulates the structure and function of the biological neural networks in the human brain. BPANN can process the considered variables and learn from the present data, even in the presence of noise. None of the priori models are required for the BPANN calculations. In this research, after the constant training between the input spectra and output values, a quantitative model was built and used to determine the samples whose THz spectra exist but physical properties are not given[18,19]. The SVM is one of the best classifier methods and has been applied in a wide variety of fields. The main applications of SVMs are in regression analysis, classification, forecasting, and pattern recognition. The SVM is based on statistical learning theory and the structural risk minimization principle. Structural risk minimization can reduce the upper-bound generalization error instead of the traditional local training error[20].

To evaluate the reliability and precision of the BPANN and SVM models, the absolute errors $(x-y)$ were initially calculated in terms of all 128 $PM_{2.5}$ samples and are listed in the inset of Figure 2, with less than 20 samples in the prediction set and less than 10 samples in the training set. The R and RMSE values (which are related to the covariance $cov(x, y)$ and variance matrix $v(x)$ by $R=cov(x, y)/[v(x)v(y)]^{1/2}$ and $RMSE=[\Sigma(x-y)^2/n]^{1/2}$, where n is the sum of samples in a set, and x as well as y represent actual and predicted mass, respectively) were calculated according to the present data in both the training and prediction sets.

In this research, considering the signal-noise ratio and the analysis of absorption features, the frequency between 1.5THz and 9.0THz was employed as the THz range. THz absorbance at 6.5THz was selected for mass measurement because of the common central frequency of absorption features, however, the absorbance data at all frequencies in THz range were used to build the models of 2DCOS, BPANN and SVM.

3 Results and discussion

3.1 THz parameter analysis of PM$_{2.5}$

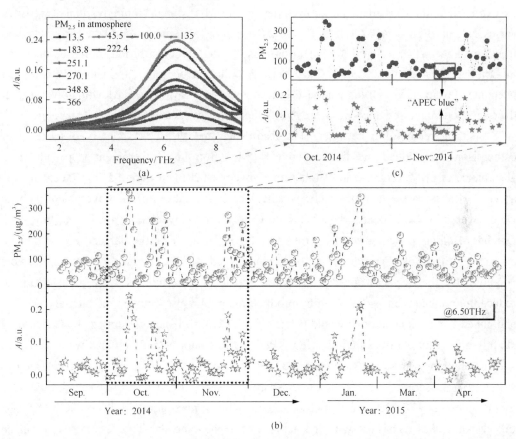

Figure 2 Spectral analysis of PM$_{2.5}$ content based on THz spectroscopy. (a) Frequency-dependent absorbance A spectra of PM$_{2.5}$ from 1.5THz to 9THz. The content of selected PM$_{2.5}$ ranges from 13.5μg/m^3 to 366μg/m^3. An absorption feature with the central frequency located at ~6.5THz is shown. (b) Time-dependent PM$_{2.5}$ from the government (circles) and THz absorbance of PM$_{2.5}$ (stars) from September 2014 to April 2015. A correlation can be seen between the two datasets. (c) Time-dependent PM$_{2.5}$ from the government (up) and corresponding absorbance in the selected period of October and November 2014. The pollution in the two months is highly representative. The Chinese National Day and APEC were held during this period. "APEC blue" attracted widespread attention from the public due to the long period of high air quality

The online monitoring of PM$_{2.5}$ content has raised increasing concern to the public in daily life. In many metropolises, daily plans, such as tours and exercise, are closely related to the PM$_{2.5}$ or the air quality index. The public was pleased to have access to precise PM$_{2.5}$ data on a website or smart phone. Appropriate methods for monitoring PM$_{2.5}$ seemed necessary. THz spectroscopy is an effective method for online monitoring and has been applied in some fields[21-24]. Figure 2 (a) shows the absorbance A spectra of ten selected PM$_{2.5}$. The maximum absorbance values occurred at ~6.5THz in all spectra. In the 1.5~6.5THz range, the absorbance increased monotonically; however, in the 6.5~9THz range, a monotonically decreasing trend was observed. The absorbance curves for $PM_{2.5}$ remained smooth over the entire range, proving a high and steady signal-to-noise ratio for the measurement. Based on the A spectra, we extracted all the $PM_{2.5}$ data and showed the actual $PM_{2.5}$ content from government websites versus THz absorbance at 6.5THz in Figure 2 (b). Strong agreement could be clearly observed from September 2014 to April 2015. During these period, the most serious pollution occurred in October 2014, and the second most serious pollution occurred in January 2015. Such phenomena were also observed via the THz response. Similarly, in the days with good weather, the corresponding absorbance was nearly zero due to the low absorption in the THz range.

A strong relationship between THz absorbance and $PM_{2.5}$ was clearly observed. For comparison, we plotted the time-dependent $PM_{2.5}$ from the government (top) and THz absorbance of $PM_{2.5}$ (bottom) in October and November 2014 in Figure 2 (c). Except for the absence of absorbance data due to the weather, the $PM_{2.5}$ data from the government website were covered by the THz absorbance in both October and November 2014. From 9th to 11th October, Beijing experienced dangerous levels of haze, with the largest $PM_{2.5}$ content exceeding 350 μg/m^3. During this period, China hosted APEC. To make a more comfortable condition for the overseas guests, the central and local governments made great efforts to reduce the pollutant emissions. As a by-product, the public also enjoyed the pleasant weather and hoped that such conditions would persist. The THz absorbance was low and changed only slightly during the APEC meeting. After APEC, some factories and beltlines were reopened, and the emergency restrictions on car-use were abandoned. As a result, Beijing returned to suffering from serious haze pollution. A sharp increasing trend was observed for the THz absorbance after 17th October, and absorbance reached a maximum (~0.22) on 20th October; this response was consistent with the actual $PM_{2.5}$ phenomenon. THz spectroscopy can be employed to monitor air pollution levels, especially during certain special conditions or events.

3.2 $PM_{2.5}$-THz grading

The urban air quality grade refers to the pollution index ratings and the corresponding limits on the pollutant concentrations according to the urban air environmental quality standards, the eco-environmental effects of pollutants and their effects on human health. The grade is typically confirmed by the Air Pollution Index (API) or the $PM_{2.5}$ concentration. Currently, the urban air quality pollution index grading standards include: (ⅰ) good or excellent (0~50μg/m^3) for daily mean values of national ambient air quality standards that satisfy the air quality requirements of nature reserves, scenic spots and other special protection area; (ⅱ) moderate (50~100μg/m^3) for the national secondary standard of the daily mean value of air quality, which satisfies the requirements in residential commercial cultural and rural areas; (ⅲ) light pollution, for the third level of standards, which satisfies the air quality requirements in specific industrial areas; long-term exposure to this level of pollution will make healthy people susceptible to symptoms of mild irritation; (ⅳ) moderate pollution (150~200μg/m³), the fourth level of standards; long-term exposure to air of this level will worsen symptoms in patients with heart and lung disease; (ⅴ) heavy pollution (200~300μg/m³), the fifth level of standards; under this level, tolerance is clearly reduced in healthy volunteers, and there are clear symptoms of certain diseases; and (ⅵ) severe pollution, the highest level of air pollution; in such an environment, people can be poisoned, or some other symptoms can occur after a short period of exposure. The classification criterion based on $PM_{2.5}$ concentration is the standard of urban air quality forecasts and the main basis for air quality assessments and urban environmental function division. A precise approach to air quality grading is necessary.

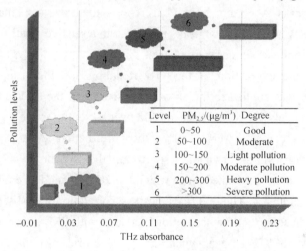

Figure 3 Hierarchic evaluation of $PM_{2.5}$ pollution in the atmosphere based on THz spectroscopy. According to the normal standard based on the $PM_{2.5}$ content in atmosphere, air quality can be divided into six levels by five datasets: 50μg/m³, 100μg/m³, 150μg/m³, 200μg/m³ and 300μg/m³. Based on the close relationship between THz absorbance and $PM_{2.5}$, six levels of pollution in the atmosphere can be classified by the five absorbance data. According to the hierarchical results, environmental departments and the public can directly evaluate the air quality and develop a plan to process the atmospheric pollution

Our analysis above suggests that air pollution in China can be graded by $PM_{2.5}$ parameters. Figure 3 shows a hierarchic evaluation of $PM_{2.5}$ pollution in the atmosphere based on THz absorbance. The inset describes the detailed grading of air pollution regarding the $PM_{2.5}$ content. The degree of air pollution can be graded into six levels, and the critical points included 0, 50μg/m³, 100μg/m³, 150μg/m³, 200μg/m³ and 300μg/m³, as discussed above. Due to the monotonically increasing trend between THz absorbance and $PM_{2.5}$ content, the THz parameter is suitable for the evaluation of air pollution level, which is of great significance for the pollution evaluation. Based on the forms of standards, the environmental authorities can evaluate pollution levels, air quality forecasts, and identify pollution accidents; in addition, the public can evaluate the impact of air quality on personal health, arrange times for outdoor exercise, and make appropriate travel plans. The THz parameter appears to be a simple approach to satisfy the requirements of air pollution grading. According to Figure 3 the THz absorbance was closely related to the different pollution levels, which were graded by the $PM_{2.5}$ content. More serious air pollution resulted in larger maxima of THz absorbance in the histogram.

3.3 $PM_{2.5}$ composite analysis based on THz spectra

Figure 2 (a) shows a broad absorption band whose central frequency is 6.5THz. According to the previous study, the THz absorption effect is related to the samples' components and chemical structure. The THz response depends on the molecular vibration modes and intensities[20]. Herein, the THz absorbance spectra had a broad absorption band in addition to one or more single peaks. According to previous reports, the broad band may be caused by the overlapping effect. Several absorption peaks may exist in a relative small frequency range due to the molecular vibration of similar components.

To investigate the interaction of the THz radiation and $PM_{2.5}$ data collected in the atmosphere, 2DCOS was applied to the absorbance spectra of the selected $PM_{2.5}$ in Figure 2 (a). In the 2DCOS models, the absorbance spectra over the entire range were employed as the input, and the $PM_{2.5}$ contents were used as the perturbation. After the two-dimensional correlation calculation of the dynamic spectrum, the synchronous and asynchronous correlation intensities were calculated, after which the 2DCOS models could be obtained. The synchronous case is plotted in Figure 4 (a); a positive correlation was observed over the entire frequency range, indicating that the absorption increases with $PM_{2.5}$ over the entire frequency range from 1.5THz to 9THz. The synchronous 2D correlation displayed obvious clear auto-peak located at (6.5, 6.5). The large correlation (6.73×10^{-3}) for this auto-peak represented a high degree of $PM_{2.5}$-induced dynamic fluctuations of the absorbance spectra. Thus, the auto-peak showed good agreement with Figure 2 (a) in that the absorption response was highly sensitive to the $PM_{2.5}$ level at selected frequencies. In addition, frequencies closer to 6.5THz yielded higher correlation intensities. In the asynchronous 2DCOS figure in Figure 4 (b), several strong cross peaks were observed in the absorption band. Overall, four cross peaks were observed in the asynchronous system, located at (7.5, 6.3), (7.5, 8.5), (7.8, 6.3) and (7.5, 4.2) with intensities of 6.3×10^{-5}, 5.5×10^{-5}, 5.2×10^{-5} and 4.0×10^{-5}, respectively. Combined with the broad band in Figure 2 (a), the absorption band was overlapped by five single absorption peaks, whose central frequencies were 7.5THz, 6.3THz, 8.5THz, 7.8THz and 4.2THz. The vibration modes at 7.5THz and 6.3THz reflected the primary information of the principal components, as they had the largest correlation intensities. The positive signs of the correlation illustrated the sequencing of intensity changes at the two frequencies. The peak at 7.5THz changed before that at 6.3THz, 8.5THz and 4.2THz, and the peak at 6.3THz changed after the peak at 7.8THz. Due to the small deviations from a frequency to the coherence, the five single peaks overlap and form a broad band in the absorbance spectra.

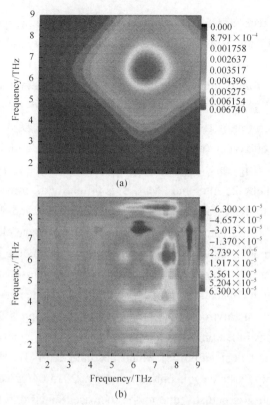

Figure 4 2DCOS analysis of the absorption effect of $PM_{2.5}$ collected in the atmosphere from 1.5THz to 9THz. (a) Synchronous 2D correlation plot of the selected $PM_{2.5}$ over the frequency range from 2.5THz to 7.5THz. There is a strong positive correlation at position (6.5, 6.5), validating the absorption feature at 6.5THz in Figure 1(a). (b) Asynchronous 2D correlation plot of $PM_{2.5}$ in the atmosphere over the frequency range from 1.5THz to 9THz. There are several new correlation peaks at approximately 6.5THz, proving that the absorbance spectra are overlapped by some absorption peaks

Table 1 Elemental analysis of $PM_{2.5}$ with different contents (unit: mol%)

Elements	$PM_{2.5}$		
	25μg/m³	80μg/m³	230μg/m³
O	52.76	52.91	52.90
Si	45.01	44.84	42.49
Na	0.68	0.63	0.69
Al	0.41	0.17	0.47
Ca	0.36	0.29	0.54
S	0.35	0.81	2.21
Mg	0.21	0.13	0.27
K	0.11	0.10	0.25
Cl	0.04	0.05	0.08
Fe	0.03	0.04	0.05
P	0.02	0.01	0.03
Ti	0.01	0.01	0.01
Zn	0.01	0.01	0.01

The absorption effect at 6.5THz was determined by the composites in $PM_{2.5}$. We measured the elemental compositions of three selected samples with different $PM_{2.5}$ contents using X-ray fluorescence (XRF). Filter together with $PM_{2.5}$ were measured, therefore the elements were obtained from both the filter and $PM_{2.5}$. However, the filter was only composed of SiO_2. The three $PM_{2.5}$ contents were $25\mu g/m^3$, $80\mu g/m^3$, and $230\mu g/m^3$. The respective elements and mass concentration are listed Table 1. The primary constituents of the filter with $PM_{2.5}$ had the elements of oxygen (O), silicon (Si), sodium (Na), aluminium (Al), calcium (Ca), sulphur (S), magnesium (Mg), and potassium (K). Within the three $PM_{2.5}$ contents, O was the highest, obtained from both the filter and $PM_{2.5}$. Si is primarily determined from the filter. The mass ratios of O/Si for the three samples were 1.172, 1.180 and 1.245, larger than the theoretical ratio of 1.143 in a blank filter. According to the primary elements of the selected $PM_{2.5}$, the concentration of elements, which can form cations, including Na, Al, Ca, Mg and K, maintained a steady relatively large content. In terms of the elements that could form anions, O and S were present in high concentrations of $PM_{2.5}$ in the atmosphere. According to an official report from the Ministry of Environmental protection of the People's Republic of China, the primary pollution sources in Beijing were motor vehicle exhaust emissions, coal combustion, industrial production and dust emission, which accounted for 31.1%, 22.4%, 18.1% and 14.3%, respectively (http://article.cyol.com/news/content/2015-04/02/content_11308646.htm). The increasing pollution was actually due to the increased emissions from vehicles and fire coal in factories. With an increasing number of motor vehicles driving in the city and large scale of combustion in the winter, increasing amounts of SO_2 will be discharged into atmosphere. Such SO_2 emissions remain in the atmosphere for a long time; as a result, they will form new pollutants under the physical, chemical, or biological action and re-generate new hazards to the environment. Such a process, which is called secondary pollution, generates various forms of sulphate. Herein, Na_2SO_4, $Al_2(SO_4)_3$, $CaSO_4$, $MgSO_4$ and K_2SO_4 were the primary pollutants according to the XRF results, which were in agreement with the previous reports. SO_4^{2-} was the main anion with a $52.5\mu g/m^3$ in the total $139.0\mu g/m^3$ of $PM_{2.5}$, and the Na^+, NH_4^+, K^+, Mg^{2+}, Ca^{2+} were the primary cations during haze days in Beijing[25]. Besides, in Miyun which is another district at the northeast of Beijing, anions such as SO_4^{2-} and cations such as Na^+, Cu^{2+}, K^+ were also the primary components of $PM_{2.5}$ in September, 2013[26]. Here, we combined the content of these components and the cross-peak intensities in the asynchronous plots. With higher $PM_{2.5}$ content, the concentrations of Ca and Mg increased from 0.36% to 0.54% and from 0.21% to 0.27%, respectively. The increasing tendency was agreement with that of the asynchronous intensities. Therefore, $CaSO_4$ and $MgSO_4$ contributed to the strong vibrations at 7.5THz and 6.3THz, respectively, in agreement with the 2DCOS results of $PM_{2.5}$ at a construction site. In the mentioned research, Ca^{2+} and Mg^{2+} corresponded to the 7.43THz and 6.36THz peaks according to the relation between THz features and elemental types[27]. Meanwhile, the percentage of Na and Al were relatively fixed, and Na_2SO_4

and $Al_2(SO_4)_3$ can be assigned to the vibration modes at 7.8THz and 8.5THz due to the similar asynchronous correlations at (7.8, 6.3) and (7.5, 8.5), respectively. Compare with other elements, K had a relatively small concentration in $PM_{2.5}$; thus, K_2SO_4 can be inferred to be related to the vibration at 4.2THz.

In the raw spectra, it was difficult to distinguish the absorbance of different components in the entire frequency range. However, the auto-peak at 6.5THz and the positive correlation in the entire synchronous range proved the steady composites in $PM_{2.5}$, which was also validated by XRF measurement. Furthermore, the cross peak in the asynchronous plot indicated that the absorption band centred at approximately 6.5THz was composed of overlapping peaks, which were located at 7.5THz, 6.3THz, 7.8THz, 8.5THz and 4.2THz. The peaks at these frequencies were the main component because of the agreement between the components' concentration and the correlation intensities when the $PM_{2.5}$ content was set as the external perturbation. According to the results herein, the features of THz absorption spectra were related to the $PM_{2.5}$ components and the pollution sources. THz spectroscopy can be a practical choice in the process of $PM_{2.5}$ control.

3.4 Quantification of $PM_{2.5}$ on a filter

Figure 2 illustrates that the parameters calculated from the THz technique can be used to evaluate the $PM_{2.5}$ content or pollution level in the atmosphere. The THz signals carried the molecular vibration information of $PM_{2.5}$, which was actually composed of various forms of chemical compounds. Different components and contents caused various vibration modes and intensities, and the trends can be reflected in the THz spectra. As shown in Figure 5 (a), we extracted the absorbance data of 128 $PM_{2.5}$ datasets at 6.5THz and list the corresponding contents. $PM_{2.5}$ content was set as the abscissa, and THz absorbance was confirmed as ordinates. The absorbance increased gradually with increasing $PM_{2.5}$. According to the observed trends in the plots, a line can be fitted so that a linear model can be built, which had the correlation coefficient R of 0.9548. All the points were located on both sides of the line and were well fitted by the line. In future predictions, THz absorbance can be used as a standard. We calculated the THz absorbance-dependent residuals of actual and predicted values in the linear model, shown in Figure 5 (b). The average errors were $19.6\mu g/m^3$. The first-degree function between THz absorption and $PM_{2.5}$ in the atmosphere demonstrated that the molecular vibration modes were constant but the vibration intensities changed linearly under different pollution levels [28-31].

$PM_{2.5}$ can be quantified using the linear model between THz absorbance at a selected frequency and the $PM_{2.5}$ content in the atmosphere. However, the average error of $19.6\mu g/m^3$ is still a change for the $PM_{2.5}$ content monitoring, especially in the condition of pollution with $PM_{2.5}$ less than $50\mu g/m^3$. Generally, linear method was mostly used according to the Lambert-Beer's law based on one frequency, and thus more fragile to environmental and instrumental noises[32]. To build a more robust and precise model to quantitatively predict the $PM_{2.5}$ content, we considered the nonlinear algorithms based on the absorption spectra over the entire range other than a single frequency. The general noise effect in the spectral data can be reduced or eliminated in the statistical analysis based on full-range-spectra calculation. Herein, two chemometric techniques, namely, BPANN and SVM, were employed to build the statistical models by using the entire range of absorption spectra as the input.

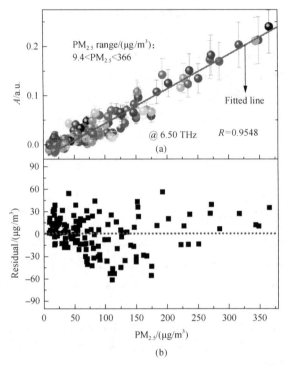

Figure 5 Quantitative monitoring of $PM_{2.5}$ on the filters. (a) A linear model between THz absorbance at 6.5 THz and $PM_{2.5}$. Error bars represent 20% of the absorbance data. (b) THz absorbance-dependent residuals of actual and predicted values plotted using a linear model

In the BPANN calculation, the THz absorbance spectra data of $PM_{2.5}$ were introduced into the input layer and calculated in the hidden layer, and then, the results were obtained in the output layer. Within all 128 samples, 32 $PM_{2.5}$ samples were selected randomly as the prediction set, and the remaining 96 $PM_{2.5}$ samples were used as the training set. The samples used for prediction were not used for training. Figure 6 (a) shows the predicted $PM_{2.5}$ values plotted against the actual values for both the training and prediction sets. All points in the two sets were located close to the reference line, which represented zero residuals between the actual and predicted values. In this BPANN model, for the training set, $R = 0.99996$ and RMSE = $0.6464\mu g/m^3$; for the prediction set, $R = 0.999471$ and RMSE = $8.2103\mu g/m^3$. We calculated the residuals of the actual and predicted $PM_{2.5}$ values in the BPANN model, which was shown in Figure 6 (b). The residuals were less than $21\mu g/m^3$, and the average was less than $10\mu g/m^3$. This result indicated that the predicted data were close to the actual values and that the BPANN model was valid, thereby improving the precision of the $PM_{2.5}$ prediction model in Figure 5.

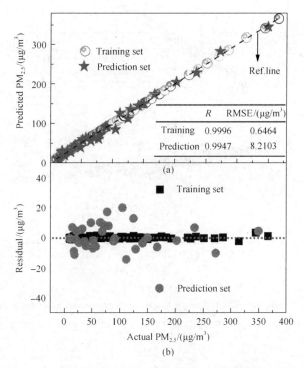

Figure 6　Quantitative monitoring of $PM_{2.5}$ on the filters. (a) Predicted $PM_{2.5}$ versus actual $PM_{2.5}$ in the BPANN quantitative model over the entire range. (b) Error analysis of the BPANN model over the entire range. The figure plots the absolute errors of both the training and prediction sets

In addition, SVM regression was employed here to quantitatively predict $PM_{2.5}$ based on THz absorbance over the entire frequency range. Similar to Figure 6 (a), we show the predicted $PM_{2.5}$ versus the actual $PM_{2.5}$ from the SVM quantitative model in Figure 7 (a), and the inset provided the absolute errors of both the training and prediction sets. The R value reached 0.9999 and the RMSE value reached $0.006\mu g/m^3$ in both the training and prediction sets. All the residuals were less than $0.02\mu g/m^3$ in Figure 7 (b). Figures 6 and 7 illustrate that the BPANN and SVM were able to precisely predict $PM_{2.5}$ levels. Relative to the BPANN, the SVM is most advantageous in $PM_{2.5}$ monitoring and is thus more useful in actual applications. Therefore, by comparing the linear model in Figure 5 and the statistical models in Figures 6 and 7, $PM_{2.5}$ seemed linear with THz absorption at 6.5THz which was the central frequency of the absorption band, but the relation was non-linear between $PM_{2.5}$ and the whole absorption effect over the entire frequency range.

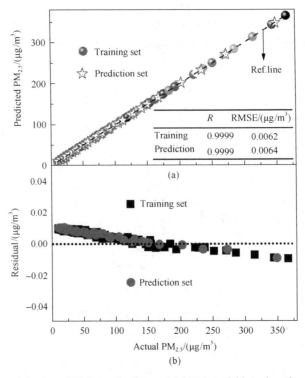

Figure 7   Quantitative monitoring of $PM_{2.5}$ on the filters. (a) SVM model based on the predicted versus actual $PM_{2.5}$ plots. (b) Error analysis of the BPANN model over the entire range

3.5 Discussion

This study suggests that in addition to previous common techniques (such as weighing), the methods of monitoring the $PM_{2.5}$ content and grading air pollution levels should also consider utilizing the THz technique, which can supply a non-contact and rapid means of monitoring the $PM_{2.5}$ content. The THz response depends on molecular vibration modes and intensities. The THz absorption of $PM_{2.5}$ in the same concentration from different regional source may be different because of the different components of $PM_{2.5}$ from different regional sources, thus the model established of quantification and $PM_{2.5}$-THz grading could be for the PM pollution of Beijing. Due to its unique advantages, the THz method has attracted considerable attention in a number of countries. With the development of THz science and technology, the application of THz technology in physics, chemical engineering, electronic information, material science, astronomy, communications, and national security will continue to emerge due to the realization of the potential and advantages of THz technology. Environmental protection is now one of the issues drawing the most public concern, and this field requires more support from various techniques. The THz response depends on the molecular vibration modes of the analysed samples. The THz absorption is appropriate to characterize the particles with complex components, such as $PM_{2.5}$. An advantage of this method is its multiple functionality in the process of $PM_{2.5}$ monitoring. According to the results above, THz absorbance and its statistical models can precisely predict the $PM_{2.5}$ content in air; the absorption features are closely related to the principal components in $PM_{2.5}$, which is important for environmental departments to determine the sources of pollution; more importantly, THz absorbance can be used as a standard for pollution grading, and based on the grading results, we can easily determine the pollution level in the atmosphere and make outdoor plans; this capability is useful for all individuals living in the city. Thus, THz waves should be used for $PM_{2.5}$ monitoring and grading.

4 Conclusions

In summary, the relationship between the THz response and $PM_{2.5}$ pollution was studied and validated. The tendency of the THz parameter in different periods was in agreement with that of $PM_{2.5}$ from the government for September 2014 to April 2015. A distinguishing feature of THz spectroscopy is the grading of the atmospheric pollution level shown in this study. Precise grading is significant to the general public for the health of the elderly and children. Based on the 2DCOS analysis, the broad absorption band was found to be overlapped by five single absorption peaks, which was caused by various forms of sulphates in the air. The principal components in $PM_{2.5}$ can be rapidly characterized by THz features. The absorbance was linearly monotonously increasing with changes in $PM_{2.5}$, with a correlation coefficient of 0.96. Thus, THz absorbance can be used as a parameter to predict $PM_{2.5}$. The average residual of prediction was ~20μg/m^3, and the average relative error was close to 12% according to the THz absorbance. From October 2014 to January 2015, the THz absorbance was high, indicating a large number of days with air pollution during this period, in agreement with the actual phenomena. Analytical methods were employed to analyse the THz absorbance. Mathematical algorithms, such as the BPANN and SVM, can process the THz absorbance data and improve the prediction precision. In the BPANN model, the RMSE can reach ~8μg/m^3, which is considerably smaller than that in the absorbance linear model. The relative error was close to 4%. The SVM model provided a higher precision due to the high R and small RMSE values, showing that this combination can monitor $PM_{2.5}$ precisely. Compared with the existing methods, THz spectroscopy can not only reveal the component information for pollution source determination, but quantitatively monitor the $PM_{2.5}$ content for pollution level evaluation, therefore, THz can be considered as a normal tool, and the database should be built. The collection of $PM_{2.5}$ values and measurement will be combined, and the other setups will be controlled by the program and computer. Thus, THz spectroscopy monitoring system represent a new way to determine the $PM_{2.5}$ levels in the atmosphere.

Acknowledgements

This work was supported by the National Nature Science Foundation of China (Grant No. 11574401), the Science Foundation of China University of Petroleum, Beijing (Grant Nos. 2462017YJRC029 and yjs2017019), and the Beijing Natural Science Foundation (Grant No. 1184016).

References

[1] Huang R J, Zhang Y, Bozzetti C, et al. High secondary aerosol contribution to particulate pollution during haze events in China. Nature, 2014, 514(7521): 218-222.

[2] Zhang Y L, Cao F. Fine particulate matter ($PM_{2.5}$) in China at a city level. Scientific Reports, 2015, 5: 14884.

[3] Wang Y, Zhang R, Saravanan R. Asian pollution climatically modulates mid-latitude cyclones following hierarchical modelling and observational analysis. Nature Communications, 2014, 5: 3098.

[4] Zhang Y L, Huang R J, El Haddad I, et al. Fossil vs. non-fossil sources of fine carbonaceous aerosols in four Chinese cities during the extreme winter haze episode of 2013. Atmospheric Chemistry and Physics, 2015, 15(3): 1299-1312.

[5] Wang Q Y, Huang R J, Cao J J, et al. Black carbon aerosol in winter northeastern Qinghai-Tibetan Plateau, China: the source, mixing state and optical property. Atmospheric Chemistry and Physics, 2015, 15(22): 13059-13069.

[6] Wuebbles D J, Lei H, Lin J. Intercontinental transport of aerosols and photochemical oxidants from Asia and its consequences. Environmental Pollution, 2007, 150(1): 0-84.

[7] Chinese State Council. Atmospheric pollution prevention and control action plan (in Chinese) (http://www.gov.cn/zwgk/2013-09/12/content_2486773.htm), accessed on September 12, 2013.

[8] Huang K, Zhang X, Lin Y. The "APEC Blue" phenomenon: regional emission control effects observed from space. Atmospheric Research, 2015, 164-165: 65-75.

[9] People's Daily Online. Beijing Wants to keep "APEC blue" (http://en.people.cn/n/2014/1114/c90882-8808691.html), published online on November 14, 2014.

[10] People's Daily Online. How the sky turned blue in Beijing? (http://en.people.cn/n/2015/0907/c98649-8946581.html), Published online on September 7, 2015.

[11] Austen K. Environmental science: Pollution patrol. Nature, 2015, 517(7533): 136-138.

[12] Akimoto H. Global air quality and pollution. Science, 2003, 302(5651): 1716-1719.

[13] Meng Z, Dabdub D, Seinfeld J H, et al. Chemical coupling between atmospheric ozone and particulate matter. Science, 1997, 277(5322): 116-119.

[14] Pui D Y H, Chen S C, Zuo Z. $PM_{2.5}$ in China: Measurements, sources, visibility and health effects, and mitigation. Particuology, 2014, 13: 1-26.

[15] Miao X Y, Sun S N, Li Y Z, et al. Real-time monitoring the formation and decomposition processes of methane hydrate with THz spectroscopy. Science China: Physics, Mechanics & Astronomy, 2017(01): 78-79.

[16] Guan L M, Zhan H L, Miao X Y, et al. Terahertz-dependent evaluation of water content in high-water-cut crude oil using additive-manufactured samplers. Science China: Physics, Mechanics & Astronomy, 2017(04): 46-50.

[17] Hoshina H, Ishii S, Morisawa Y, et al. Isothermal crystallization of poly(3-hydroxybutyrate) studied by terahertz two-dimensional correlation spectroscopy. Applied Physics Letters, 2012, 100(1): 011907.

[18] Cai C B, Yang H W, Wang B, et al. Using near-infrared process analysis to study gas-solid adsorption process as well as its data treatment based on artificial neural network and partial least squares. Vibrational Spectroscopy, 2011, 56(2): 202-209.

[19] Moré J J. The Levenberg-Marquardt algorithm: Implementation and theory. Lecture Notes in Mathematics, 1978, 630: 105-116.

[20] Zhan H L, Zhao K, Zhao H, et al. The spectral analysis of fuel oils using terahertz radiation and chemometric methods. Journal of Physics D Applied Physics, 2016, 49(39): 395101.

[21] Popovic Z, Grossman E N. THz metrology and instrumentation. IEEE Transactions on Terahertz Science & Technology, 2011, 1(1): 133-144.

[22] Qian L I, Zhao K, Zhang L W, et al. Probing $PM_{2.5}$ with terahertz wave. Science China: Physics Mechanics & Astronomy, 2014, 57(12): 2354-2356.

[23] Zhan H L, Wu S X, Bao R M, et al. Qualitative identification of crude oils from different oil fields using terahertz time-domain spectroscopy. Fuel, 2015, 143: 189-193.

[24] Burnett A D, Fan W, Upadhya P C, et al. Broadband terahertz time-domain spectroscopy of drugs-of-abuse and the use of principal component analysis. Analyst, 2009, 134(8): 1658-1668.

[25] Zhang Y, Huang W, Cai T, et al. Concentrations and chemical compositions of fine particles ($PM_{2.5}$) during haze and non-haze days in Beijing. Atmospheric Research, 2016, 174: 62-69.

[26] Gao Y, Guo X, Li C, et al. Characteristics of $PM_{2.5}$ in Miyun, the northeastern suburb of Beijing: Chemical composition and evaluation of health risk. Environmental Science & Pollution Research International, 2015, 22(21): 16688-16699.

[27] Zhan H L, Qian L, Zhao K, et al. Evaluating $PM_{2.5}$ at a construction site using terahertz radiation. IEEE Transactions on Terahertz Science & Technology, 2015, 5(6): 1028-1034.

[28] Zhan H L, Zhao K, Bao R M, et al. Monitoring $PM_{2.5}$ in the atmosphere by using terahertz time-domain spectroscopy. Journal of Infrared Millimeter & Terahertz Waves, 2016, 37(9):1-10.

[29] Zhan H L, Zhao K, Xiao L Z. Non-contacting characterization of $PM_{2.5}$ in dusty environment with THz-TDS. Science China: Physics, Mechanics & Astronomy, 2016, 59(4): 644201.

[30] Zhan H. L, Li N, Zhao K, et al. Terahertz assessment of the atmospheric pollution during the first-ever red alert period in Beijing. Science China: Physics, Mechanics & Astronomy, 2017, 60(4): 044221.

[31] Miao X Y, Zhan H L, Zhao K. Application of THz technology in oil and gas optics. Science China: Physics, Mechanics & Astronomy, 2017, 60(2): 024231.

[32] Hua Y, Zhang H, Zhou H. Quantitative determination of cyfluthrin in *n*-hexane by terahertz time-domain spectroscopy with chemometrics methods. IEEE Transactions on Instrumentation & Measurement, 2010, 59(5): 1414-1423.

Fuel properties determination of biodiesel-diesel blends by terahertz spectrum

Hui Zhao[1]   Kun Zhao[1,2]   Rima Bao[1]

(1.Laboratory of Optic Sensing and Detecting Technology, College of Science, China University of Petroleum, Beijing 102249, China; 2.State Key Laboratory of Heavy Oil Processing, China University of Petroleum, Beijing 102249, China)

Abstract: The frequency-dependent absorption characteristics of biodiesel and its blends with conventional diesel fuel have been researched in the spectral range of 0.2~1.5THz by the terahertz time-domain spectroscopy (THz-TDS). The absorption coefficient presented a regular increasing with biodiesel content. A nonlinear multivariate model that correlating cetane number and solidifying point of biodiesel blends with absorption coefficient has been established, making the quantitative analysis of fuel properties simple. The results made the cetane number and solidifying point prediction possible by THz-TDS technology and indicated a bright future in practical application.

Keywords: THz time-domain spectroscopy; cetane number; solidifying point; biodiesel blends

1 Introduction

In recent years, there is much attention on biodiesel, the alternative source of petroleum-based fuel. Biodiesel is produced by a relatively simple reaction known as transesterification, where natural oils and fats make up mainly of triglycerides react with a simple alcohol, such as methanol or ethanol, in the presence of a catalyst to produce mono alkyl esters of fatty acids, which is biodiesel[1-4]. It presents some advantages, such as non-toxic, renewable energy, biodegradable, high flash point and less polluting gas emissions, etc. The physical chemical characteristics of biodiesel were similar to those of common diesel, so that biodiesel and its blends with conventional diesel fuel always were used in diesel engines without significant modification on the engine[5-7].

Cetane number (CN) and solidifying point (SP) are prime indicators of fuel quality in diesel engines, which have been described as measures of its ignition delay time and low temperature flow property. American society of testing materials (ASTM) defines the CN of a diesel fuel as the percentage by volume of normal cetane ($C_{16}H_{34}$), in a blend with 2,2,4,4,6,8,8-heptamethyl nonane (sometimes called HMN or *iso*-cetane), which matches the ignition quality of the diesel fuel being rated under the specified test conditions, measured in a specially designed test engine (ASTM D613) or in a constant volume combustion apparatus (ASTM D6890)[8-10]. The SP is the temperature in which diesel has no fluid properties and always be measured by GB 510—1983 standard method[11]. However, these tests are awkward and expensive. For this reason there have been many attempts to develop methods to estimate the indexes economically, rapidly and effectively.

In this letter, terahertz spectroscopy has been used to research the CN and SP of biodiesel-diesel blends. Terahertz spectroscopy contains rich physical, chemical, and structural information of the materials. Most low-frequency vibrational and rotational spectra of organics and related compounds lie in this frequency range. Recently, there has been a remarkable effort in employing terahertz time-domain spectroscopy (THz-TDS) for investigating the petroleum products and organics, including fuel oil, fatty acid, polymers, oil-water complexes, and gases[12-17]. This paper illustrated the feasibility of CN and SP prediction of biodiesel-diesel blends by THz-TDS.

2 Experiments

A repetition rate of 80 MHz, diode-pump mode-locked Ti-sapphire laser (MaiTai, Spectra Physics) provided the femtosecond pulses with duration of 100 fs and center wavelength of 810nm[18,19]. A p-type InAs wafer with <100> orientation was used as the THz emitter and a 2.8mm thickness <110> ZnTe was employed as the sensor. A standard lock-in technology was used in this system. A femtosecond laser pulse was split into two beams. The pump beam was used to generate THz radiation and the probe beam acted as a gated detector to monitor the temporal waveform of THz field. A silicon lens and parabolic mirrors were used to collimate and focus the THz beam through free space onto the detector. A balanced photodiode detector detected the probe beam, and the signal was amplified by a lock-in amplifier and sent to the computer for processing. The THz beam path was purged with dry nitrogen to minimize the absorption of water vapor and enhance the signal to noise ratio (SNR). The humidity was kept less than 1% and the temperature was kept at 298 K. Here, the focus diameter of the THz beam is about 1mm. The samples are located in the focus of the two Si lens and are held in a 3mm thickness polyethylene cells, which are transparent for visible light and have a low refractive index and THz absorption. Both the time-domain sample and reference spectra were obtained by testing the polyethylene cell holding the sample and empty cell, respectively. After applying fast Fourier transform, we will get the frequency-domain sample and reference spectra and calculate the absorption characters of samples[20].

Table 1 The proportion of bio-diesel mixture and measured CN and SP

	Biodiesel/%	Diesel/%	Measured CN	Measured SP
B0	0	100	45.6	−23.4
B5	5	95	46.1	−22.7

	Biodiesel/%	Diesel/%	Measured CN	Continued Measured SP
B10	10	90	47.7	−20.2
B20	20	80	50	−19.9
B40	40	60	54.2	−15.8
B60	60	40	58.2	−9.3
B80	80	20	61	−7.1
B100	100	0	66.1	−2.6

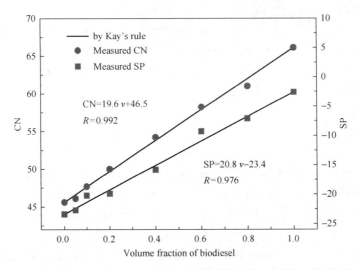

Figure 1　The diagram of measured CN and SP of blends complies with Kay's rule

The drainage oil biodiesel and conventional diesel used in this work were blended each other at room temperature and stirred for 30 min. The biodiesel content varied from 5% to 80% by volume, as shown in Table 1. Biodiesel and neat fuel are designated as B100 and B0, respectively. The CNs and SPs of blends have been listed in Table 1, measured by ASTM D613 and GB 510—1983. The variation of CN and SP complied with Kay's mixing rule, which have been widely used in the hydrocarbon mixtures industry in the literature to predict the density, viscosity and CN of binary biodiesel-diesel fuel mixtures. Kay's law has been expressed as

$$X = v_1 x_1 + v_2 x_2 \tag{1}$$

where x is the CN or SP of the mixture, v_1, v_2 ($v_1+v_2=1$ in binary system) and x_1, x_2 are volume fractions and CN or SP of pure biodiesel and diesel respectively. x_1 and x_2 will be confirmed by Table 1 and an empirical equation was obtained by Equation (1) as expressed in Figure 1, where v represents the volume fraction of biodiesel. Figure 1 presented the accordance of CN and SP obtained by measured and Kay's rule, indicating the stability and accuracy of samples.

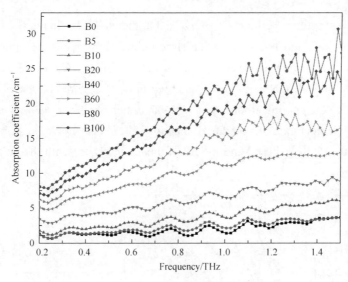

Figure 2　Terahertz spectra of biodiesel blends

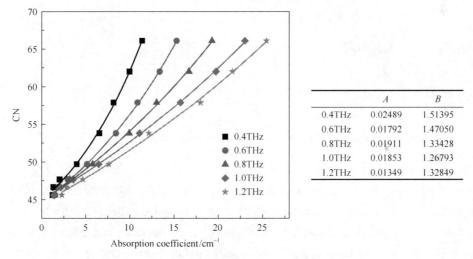

	A	B
0.4THz	0.02489	1.51395
0.6THz	0.01792	1.47050
0.8THz	0.01911	1.33428
1.0THz	0.01853	1.26793
1.2THz	0.01349	1.32849

Figure 3　The absorption coefficient dependence of CN

3　Results and discussion

In Table 1, biodiesel showed a higher CN than diesel which was determinated by its compositions and volume fraction. Diesel consists of paraffins, olefins, branched i-paraffins and aromatic hydrocarbon, and each pure-component's contribution toward the fuel cetane number is distinguishing, reported in References [21] and [22]. Paraffins have the highest CNs of average 75.6, followed closely by olefins of 63.1, and aromatics and naphthenes typically have lower CNs of average 31.3 and 14.7. Then the CN of diesel is always 45 near by[8,21]. Biodiesel is the mixture of long-chain fatty acids alkyl esters, e.g. myristate, palmitate, stearate and oleate, with a substituent groups (—$COOCH_3$) at the end of C-chain. The ester group will rupture from the chain and break down into free radical, improving the ignition of fuel. So myristate, palmitate, stearate and oleate have a relatively higher CN of 66.2, 74.3, 75.6 and 64.8 respectively, and the CN of biodiesel is about 66[22]. Furthermore, esters showed higher kinematic viscosity than hydrocarbons, leading to an easier paraffin crystallization and worse flow property at low temperature of biodiesel than diesel[23]. As shown in Table 1, biodiesel presented a higher SP than diesel, and the SP of blends increased with the volume fraction of biodiesel.

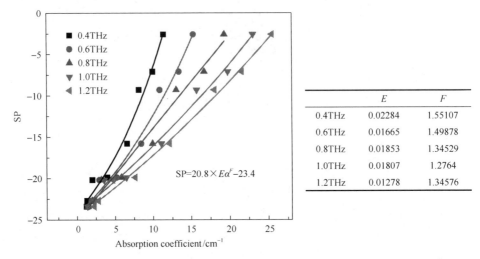

Figure 4　The absorption coefficient dependence of CN

Figure 2 is the absorption spectra of mixture in 0.2~1.5THz, showing the same trend of slow upward, comprehended to be the scattering of samples or wide and structureless absorption. In this region, there were not remarkable peaks because of the complex components of sample. The absorption coefficients (α) of biodiesel was higher than that of diesel. It also presents regularly increasing of α with biodiesel concentration, as well as the CN and SP, indicating the absorption activity of blends in THz region and the probability of the CN and SP prediction based on the absorption effect of the samples. Terahertz wave is sensitive to intermolecular forces and collective vibration. Comparing to diesel, biodiesel has higher viscosity, meaning the much greater tension that exists in the molecules and stronger absorption by terahertz wave. The molecule vibration modes have been observed by visualization software GaussView, manifesting that esters show a higher vibration tension than hydrocarbons molecule due to the vibration coupling between C-chain and ester group at the end of it. For these reason, biodiesel had a higher terahertz absorption than conventional diesel fuel and α increased with the volume fraction of biodiesel, which was similar to CN and SP[24,25].

To further reveal the correlation of CN and α, the α with various concentrations at 0.4THz, 0.6THz, 0.8THz, 1.0THz and 1.2THz THz have been collected as independent variable, and CN as dependent variable. As shown in Figure 3, α and CN exhibited the nonlinear features at different frequencies, and a fitted regression curved equation has been presented as below:

$$CN = 19.6A\alpha^B + 46.5 \qquad (2)$$

where α is the absorption coefficient, and the parameters A, B for different frequency are listed in Figure 3.

Similarly, the α-dependent SP equation has been obtained as shown in Figure 4.

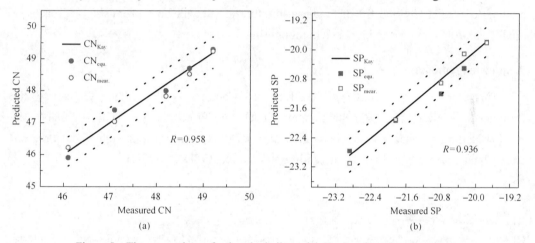

Figure 5 The comparison of values by built model, Kay's rule and measurement

To verify the performance of fitted model, the THz absorption coefficients of B2.5, B7.5, B12.5, B15 and B17.5, as testing set, have been measured in 0.2~1.2THz. The CNs and SPs of these samples also have been measured by ASTM D613 and GB 510—1983. Put α at 0.4 THz, 0.6 THz, 0.8 THz, 1.0 THz and 1.2 THz into empirical equation, the CN and SPs will be calculated. Table 2 is the result of calculation. Averaging the values at different frequencies, the calculated CN ($CN_{equ.}$) and SP ($SP_{equ.}$) were obtained, which were close to $CN_{mear.}$ and $SP_{mear.}$ (measured CN by ASTM D613 and SP by GB 510—1983) and CN_{Kay} and SP_{Kay} (calculated by Kay's rule). Figure 5(a) showed that both calculated and measured results are close to the values by Kay's rule, with the error less than ±1% (the two dotted lines), demonstrating

Table 2 The calculated CN comparing with the CN obtained by ASTM D613 and Kay's rule

	Average $CN_{equ.}$	CN_{Kay}	$CN_{mear.}$	Average $SP_{equ.}$	SP_{Kay}	$SP_{mear.}$
B2.5	46.21827	46.1	45.9	−22.76077	−22.88	−23.1
B7.5	47.03961	47.1	47.4	−21.92771	−21.84	−21.9
B12.5	47.83141	48.2	48.0	−21.19986	−20.8	−20.9
B15	48.52917	48.7	48.7	−20.50069	−20.28	−20.1
B17.5	49.24561	49.2	49.3	−19.78093	−19.76	−19.8

a good agreement and the effectiveness of the built model. The same fitting about SP model will be found in Figure 5(b).

4 Conclusions

The biodiesel blends have been researched using THz-TDS technology in this study. The THz absorption coefficient and CN and SP of biodiesel blends increase regularly with the biodiesel content, by which the CN and SP can be calculated easily by the built absorption-CN and absorption-SP models. The investigations had illustrated that THz-TDS was an effective method to predict CN and SP of biodiesel mixture and it will be a promising approach for quality control no matter on line or field monitoring.

Acknowledgements

This work was supported by THz Research Centre of China Jiliang University, the Program for New Century Excellent Talents in University (Grant No. NCET-08-0841), the Beijing Natural Science Foundation (Grant No. 4122064), and the Foresight Fund Program from China University of Petroleum (Beijing) (Grant No. QZDX-2010-01).

References

[1] Chisti Y. Biodiesel from microalgae. Biotechnolo. Adv., 2007, 25: 294.
[2] Kusdiana D, Saka S. Kinetics of transesterification in rapeseed oil to biodiesel fuel as treated in supercritical methanol. Fuel.2001, 80: 693.
[3] Meher L C, Sagar D V, Naik S N. Technical aspects of biodiesel production by transesterification:A review. Renew. Sust. Enery. Rev. , 2006, 10: 248.
[4] Zhang Y, Dube M A, McLean D D, et al. Biodiesel production from waste cooking oil: 1. Process design and technological assessment. Bioresource Technol.,2003, 89: 1.
[5] Nabi M N, Akhter M S, Zaglul Shahadat M M. Improvement of engine emissions with conventional diesel fuel and diesel-biodiesel blends. Bioresource Technol, 2006, 97: 372.
[6] Boehman A L, Song J H, Alam M. Impact of biodiesel blending on diesel soot and the regeneration of particulate filters. Energ Fuel ,2005, 19: 1857.
[7] Benjumeaa P, Agudelob J, Agudelob A. Basic properties of palm oil biodiesel-diesel blends. Fuel, 2008, 87: 2069.
[8] Ghosh P. Predicting the effect of cetane improvers on diesel fuels. Energy & Fuels, 2008, 22: 1073.
[9] Knothe G, Matheaus A C, Ryan T W. Cetane numbers of branched and straight-chain fatty esters determined in an ignition quality tester. Fuel, 2003, 82: 971.
[10] Knothe G. Dependence of biodiesel fuel properties on the structure of fatty acid alkyl esters. Fuel Process. Technol., 2005, 86: 1059.
[11] Han S, Song Y P, Ren T H. Impact of alkyl methacrylate-maleic anhydride copolymers as pour point depressant on crystallization behavior of simulated diesel fuel. Energy & Fuels, 2009, 23: 2576.
[12] Kim G J, Jeon S G, Kim J, et al. Terahertz time domain spectroscopy of petroleum products and organic solvents. International Conference on Infrared, Millimeter and Terahertz Waves. Pasadena, 2008.
[13] Al-Douseri F M. Applications of T-ray spectroscopy in the petroleum field, New York: Rensselaer Polytechnic Institute, 2005.

[14] Jiang F L, Lkeda L, Ogawa Y, et al. Terahertz absorption spectra of fatty acids and their analogues. Oleo Sci.,2011, 60: 339.
[15] Jin Y S, Kim G J, Jeon S G, et al. Terahertz dielectric properties of polymers. J Kor. Phys. Soc., 2006, 49: 13.
[16] Gorenflo S, Tauer U, Hinkov I, et al. Dielectric properties of oil-water complexes using terahertz transmission spectroscopy. Phys. Lett., 2006, 421: 494.
[17] Jacobsen R H, Mittleman D M, Nuss M C. Chemical recognition of gases and gas mixtures with terahertz waves. Opt. Lett., 1996, 21: 2011.
[18] Li J S, Li X J.Determination principal component content of seed oils by THz-TDS.Chem. Phys. Lett., 2009, 47: 92.
[19] Li X J, Hong Z, He J L, et al. Precisely optical material parameter determination by time domain waveform rebuilding with THz time-domain spectroscopy. Opt. Comm.,2010, 283: 4701.
[20] Tian L, Zhou Q L, Jin B, et al. Optical property and spectroscopy studies on the selected lubricating oil in the terahertz range. Sci. China Ser. G, 2009, 52: 1938.
[21] Ghosh P, Jaffe S B, Ind. Detailed composition-based model for predicting the cetane number of diesel fuels. Eng. Chem. Res., 2006, 45: 346.
[22] Tong D M, Hu C W, Jiang K H, et al. Cetane number prediction of biodiesel from the composition of the fatty acid methyl esters. Oil Chem. Soc., 2001, 88: 415 .
[23] Franco Z, Nguyen Q D. Rheological properties of vegetable oil-diesel fuel blends. Fuel, 2011, 90: 38.
[24] Nita I, Geacai S, Iulian O. Measurements and correlations of physico-chemical properties to composition of pseudo-binary mixtures with biodiesel. Renewable Energ., 2011, 36: 3417.
[25] Benjumea P, Agudelo J, Agudelo A. Basic properties of palm oil biodiesel-diesel blends. Fuel, 2008, 87: 2069.